Praise for *Enlight...* P9-ELX-559

"An excellent book, lucidly written, timely, rich in data, and eloquent in its championing of a rational humanism that is—it turns out—really quite cool."　　　　　　　　　　—Sarah Bakewell, *New York Times Book Review*

"For years, I've been saying Steven Pinker's *The Better Angels of Our Nature* was the best book I'd read in a decade. . . . Pinker has managed to top himself. . . . I was already familiar with a lot of the information he shares—especially about health and energy—but he understands each subject so deeply that he's able to articulate his case in a way that feels fresh and new. . . . The world *is* getting better, even if it doesn't always feel that way. I'm glad we have brilliant thinkers like Steven Pinker to help us see the big picture. *Enlightenment Now* is not only the best book Pinker's ever written. It's my new favorite book of all time."　　　　　　　　—Bill Gates

"A terrific book . . . [Pinker] recounts the progress across a broad array of metrics, from health to wars, the environment to happiness, equal rights to quality of life."　　　　　　　　—Nicholas Kristof, *New York Times*

"Steven Pinker's book is full of vigor and vim, and it sets out to inspire a similar energy in its readers. An exhaustive, compelling, and uplifting account of human progress. . . . Beautifully written."
　　　　　　　—Charles Kenny, *Democracy: A Journal of Ideas*

"Magnificent, uplifting, and makes you want to rush to your laptop and close your Twitter account."　　　　　　　　—*The Economist*

"[A] magisterial new book . . . *Enlightenment Now* is the most uplifting work of science I've ever read."　　　　　　　—Michael Shermer, *Science*

"Awesome. The confidence with which Pinker tears through the issues that cause such deep anxiety today is compelling."
　　　　　　　　—William Davies, *The Guardian* (UK)

"Extols the amazing achievements of modernity, and demonstrates that humankind has never been so peaceful, healthy, and prosperous."
　　　　　　　—Yuval Noah Harari, *The Guardian* (UK)

"Pinker is ahead of his critics . . . To accuse him of smugly sipping cocktails at the End of History café is simply to ignore his repeated calls to work for the better future that is there for the taking, but also for the losing."
　　　　　　　　—Julian Baggini, *The Literary Review*

"It's about the many ways in which the world is improving . . . and why we don't believe it. . . . This is the biggest story of our time."
—Fraser Nelson, *The Spectator* (UK)

"[Pinker] is right. Not just a bit right, but completely, utterly, incontrovertibly right."
—Dominic Sandbrook, *Daily Mail* (UK)

"A characteristically fluent, decisive, and data-rich demonstration of why, given the chance to live at any point in human history, only a stone-cold idiot would choose any time other than the present."
—Sam Leith, *The Spectator* (UK)

"Compelling . . . At a moment when [liberal Enlightenment] values are under attack, from the right and the left, this is a very important contribution."
—*The Atlantic*

"Remarkable, heart-warming, and long overdue."
—Rayyan Al-Shawaf, *The Christian Science Monitor*

"With *Enlightenment Now,* Steven Pinker has written the best book on progress that is currently in print. Or ebook. Or audiobook."
—Joshua Kim, *Inside Higher Ed*

"Gripping, provocative. . . . An important and timely book."
—David Wootton, *Times Literary Supplement*

"Pinker is a paragon of exactly the kind of intellectual honesty and courage we need to restore conversation and community, and the students are right to revere him."
—David Brooks, *New York Times*

"Steven Pinker's mind bristles with pure, crystalline intelligence, deep knowledge, and human sympathy. And he writes as he thinks, with a sinewy mot-justery of language which I find irresistible."
—Richard Dawkins

"A passionate and persuasive defense of reason and science . . . [and] an urgently needed reminder that progress is, to no small extent, a result of values that have served us—and can serve us—extraordinarily well."
—Glenn C. Altschuler, *The Philadelphia Inquirer*

ABOUT THE AUTHOR

Steven Pinker is Johnstone Family Professor of Psychology at Harvard, where he conducts research on cognition, language, and social relations. His prizewinning books include *The Language Instinct, How the Mind Works, The Blank Slate, The Stuff of Thought, The Better Angels of Our Nature,* and *The Sense of Style.* He has been elected to the National Academy of Sciences, has won many prizes for his teaching and research, and is listed as one of the world's most influential thinkers by *Time, Foreign Policy,* and other magazines. He also chairs the usage panel of *The American Heritage Dictionary.*

ENLIGHTENMENT
NOW

THE CASE FOR
REASON,
SCIENCE,
HUMANISM,
AND PROGRESS

STEVEN PINKER

PENGUIN BOOKS

PENGUIN BOOKS
An imprint of Penguin Random House LLC
375 Hudson Street
New York, New York 10014
penguinrandomhouse.com

First published in the United States of America by Viking Penguin,
an imprint of Penguin Random House LLC, 2018
Published in Penguin Books 2019

Charts rendered by Ilavenil Subbiah

ISBN 9780525427575 (hardcover)
ISBN 9780143111382 (paperback)
ISBN 9780698177888 (ebook)

Printed in the United States of America
10 9 8 7 6 5 4

Set in Palatino LT Pro
Designed by Nancy Resnick

TO

Harry Pinker (1928–2015)
optimist

Solomon Lopez (2017–)
and the 22nd century

Those who are governed by reason desire nothing for themselves which they do not also desire for the rest of humankind.

—Baruch Spinoza

Everything that is not forbidden by laws of nature is achievable, given the right knowledge.

—David Deutsch

CONTENTS

LIST OF FIGURES

PREFACE

The second half of the second decade of the third millennium would not seem to be an auspicious time to publish a book on the historical sweep of progress and its causes. At the time of this writing, my country is led by people with a dark vision of the current moment: "mothers and children trapped in poverty . . . an education system which leaves our young and beautiful students deprived of all knowledge . . . and the crime, and the gangs, and the drugs that have stolen too many lives." We are in an "outright war" that is "expanding and metastasizing." The blame for this nightmare may be placed on a "global power structure" that has eroded "the underlying spiritual and moral foundations of Christianity."[1]

In the pages that follow, I will show that this bleak assessment of the state of the world is wrong. And not just a little wrong—wrong wrong, flat-earth wrong, couldn't-*be*-more-wrong. But this book is not about the forty-fifth president of the United States and his advisors. It was conceived some years before Donald Trump announced his candidacy, and I hope it will outlast his administration by many more. The ideas that prepared the ground for his election are in fact widely shared among intellectuals and laypeople, on both the left and the right. They include pessimism about the way the world is heading, cynicism about the institutions of modernity, and an inability to conceive of a higher purpose in anything other than religion. I will present a different understanding of the world, grounded in fact and inspired by the ideals of the Enlightenment: reason, science, humanism, and progress. Enlightenment ideals, I hope to show, are timeless, but they have never been more relevant than they are right now.

~

The sociologist Robert Merton identified Communalism as a cardinal scientific virtue, together with Universalism, Disinterestedness, and Organized Skepticism: CUDOS.[2] Kudos indeed goes to the many scientists who shared their data in a communal spirit and responded to my queries thoroughly and swiftly. First among these is Max Roser, proprietor of the mind-expanding *Our World in Data* Web site, whose insight and generosity were indispensable to many discussions in part II, the section on progress. I am grateful as well to Marian Tupy of *HumanProgress* and to Ola Rosling and Hans Rosling of *Gapminder*, two other invaluable resources for understanding the state of humanity. Hans was an inspiration, and his death in 2017 a tragedy for those who are committed to reason, science, humanism, and progress.

My gratitude goes as well to the other data scientists I pestered and to the institutions that collect and maintain their data: Karlyn Bowman, Daniel Cox (PRRI), Tamar Epner (Social Progress Index), Christopher Fariss, Chelsea Follett (*HumanProgress*), Andrew Gelman, Yair Ghitza, April Ingram (Science Heroes), Jill Janocha (Bureau of Labor Statistics), Gayle Kelch (US Fire Administration/FEMA), Alaina Kolosh (National Safety Council), Kalev Leetaru (Global Database of Events, Language, and Tone), Monty Marshall (Polity Project), Bruce Meyer, Branko Milanović (World Bank), Robert Muggah (Homicide Monitor), Pippa Norris (World Values Survey), Thomas Olshanski (US Fire Administration/ FEMA), Amy Pearce (Science Heroes), Mark Perry, Therese Pettersson (Uppsala Conflict Data Program), Leandro Prados de la Escosura, Steven Radelet, Auke Rijpma (OECD Clio Infra), Hannah Ritchie (*Our World in Data*), Seth Stephens-Davidowitz (Google Trends), James X. Sullivan, Sam Taub (Uppsala Conflict Data Program), Kyla Thomas, Jennifer Truman (Bureau of Justice Statistics), Jean Twenge, Bas van Leeuwen (OECD Clio Infra), Carlos Vilalta, Christian Welzel (World Values Survey), Justin Wolfers, and Billy Woodward (Science Heroes).

David Deutsch, Rebecca Newberger Goldstein, Kevin Kelly, John Mueller, Roslyn Pinker, Max Roser, and Bruce Schneier read a draft of the entire manuscript and offered invaluable advice. I also profited from comments by experts who read chapters or excerpts, including Scott Aaronson, Leda Cosmides, Jeremy England, Paul Ewald, Joshua Goldstein, A. C. Grayling, Joshua Greene, Cesar Hidalgo, Jodie Jackson, Lawrence Krauss, Branko Milanović, Robert Muggah, Jason Nemirow, Matthew Nock, Ted Nordhaus, Anthony Pagden, Robert Pinker, Susan Pinker, Steven Radelet, Peter Scoblic, Martin Seligman, Michael Shellenberger, and Christian Welzel.

Other friends and colleagues answered questions or made important suggestions, including Charleen Adams, Rosalind Arden, Andrew Balmford, Nicolas Baumard, Brian Boutwell, Stewart Brand, David Byrne, Richard Dawkins, Daniel Dennett, Gregg Easterbrook, Emily-

Rose Eastop, Nils Petter Gleditsch, Jennifer Jacquet, Barry Latzer, Mark Lilla, Karen Long, Andrew Mack, Michael McCullough, Heiner Rindermann, Jim Rossi, Scott Sagan, Sally Satel, and Michael Shermer. Special thanks go to my Harvard colleagues Mahzarin Banaji, Mercè Crosas, James Engell, Daniel Gilbert, Richard McNally, Kathryn Sikkink, and Lawrence Summers.

I thank Rhea Howard and Luz Lopez for their heroic efforts in obtaining, analyzing, and plotting data, and Keehup Yong for several regression analyses. I thank as well Ilavenil Subbiah for designing the elegant graphs and for her suggestions on form and substance.

I am deeply grateful to my editors, Wendy Wolf and Thomas Penn, and to my literary agent, John Brockman, for their guidance and encouragement throughout the project. Katya Rice has now copyedited eight of my books, and I have learned and profited from her handiwork every time.

Special thanks go to my family: Roslyn, Susan, Martin, Eva, Carl, Eric, Robert, Kris, Jack, David, Yael, Solomon, Danielle, and most of all Rebecca, my teacher and partner in appreciating the ideals of the Enlightenment.

PART I
ENLIGHTENMENT

The common sense of the eighteenth century, its grasp of the obvious facts of human suffering, and of the obvious demands of human nature, acted on the world like a bath of moral cleansing.

—Alfred North Whitehead

In the course of several decades giving public lectures on language, mind, and human nature, I have been asked some mighty strange questions. Which is the best language? Are clams and oysters conscious? When will I be able to upload my mind to the Internet? Is obesity a form of violence?

But the most arresting question I have ever fielded followed a talk in which I explained the commonplace among scientists that mental life consists of patterns of activity in the tissues of the brain. A student in the audience raised her hand and asked me:

"Why should I live?"

The student's ingenuous tone made it clear that she was neither suicidal nor sarcastic but genuinely curious about how to find meaning and purpose if traditional religious beliefs about an immortal soul are undermined by our best science. My policy is that there is no such thing as a stupid question, and to the surprise of the student, the audience, and most of all myself, I mustered a reasonably creditable answer. What I recall saying—embellished, to be sure, by the distortions of memory and *l'esprit de l'escalier*, the wit of the staircase—went something like this:

In the very act of asking that question, you are seeking *reasons* for your convictions, and so you are committed to reason as the means to discover and justify what is important to you. And there are so many reasons to live!

As a sentient being, you have the potential to *flourish*. You can refine your faculty of reason itself by learning and debating. You can seek explanations of the natural world through science, and insight into the human condition through the arts and humanities. You can make the most of your capacity for pleasure and satisfaction, which allowed your ancestors to thrive and thereby allowed you to exist. You can appreciate the beauty and richness of the natural and cultural world. As the heir to billions of years of life perpetuating itself, you can perpetuate life in turn. You have been endowed with a sense of *sympathy*—

the ability to like, love, respect, help, and show kindness—and you can enjoy the gift of mutual benevolence with friends, family, and colleagues.

And because reason tells you that none of this is particular to *you*, you have the responsibility to provide to others what you expect for yourself. You can foster the welfare of other sentient beings by enhancing life, health, knowledge, freedom, abundance, safety, beauty, and peace. History shows that when we sympathize with others and apply our ingenuity to improving the human condition, we can make progress in doing so, and you can help to continue that progress.

Explaining the meaning of life is not in the usual job description of a professor of cognitive science, and I would not have had the gall to take up her question if the answer depended on my arcane technical knowledge or my dubious personal wisdom. But I knew I was channeling a body of beliefs and values that had taken shape more than two centuries before me and that are now more relevant than ever: the ideals of the Enlightenment.

The Enlightenment principle that we can apply reason and sympathy to enhance human flourishing may seem obvious, trite, old-fashioned. I wrote this book because I have come to realize that it is not. More than ever, the ideals of reason, science, humanism, and progress need a wholehearted defense. We take its gifts for granted: newborns who will live more than eight decades, markets overflowing with food, clean water that appears with a flick of a finger and waste that disappears with another, pills that erase a painful infection, sons who are not sent off to war, daughters who can walk the streets in safety, critics of the powerful who are not jailed or shot, the world's knowledge and culture available in a shirt pocket. But these are human accomplishments, not cosmic birthrights. In the memories of many readers of this book—and in the experience of those in less fortunate parts of the world—war, scarcity, disease, ignorance, and lethal menace are a natural part of existence. We know that countries can slide back into these primitive conditions, and so we ignore the achievements of the Enlightenment at our peril.

In the years since I took the young woman's question, I have often been reminded of the need to restate the ideals of the Enlightenment (also called humanism, the open society, and cosmopolitan or classical liberalism). It's not just that questions like hers regularly appear in my inbox. ("Dear Professor Pinker, What advice do you have for someone who has taken ideas in your books and science to heart, and sees himself as a collection of atoms? A machine with a limited scope of intelligence, sprung out of selfish genes, inhabiting spacetime?") It's also that an obliviousness to the scope of human progress can lead to symptoms that are worse than existential angst. It can make people cynical about the Enlightenment-inspired institutions that are securing this progress, such as liberal democracy and organizations of international cooperation, and turn them toward atavistic alternatives.

The ideals of the Enlightenment are products of human reason, but they always struggle with other strands of human nature: loyalty to tribe, deference to authority, magical thinking, the blaming of misfortune on evildoers. The second decade of the 21st century has seen the rise of political movements that depict their countries as being pulled into a hellish dystopia by malign factions that can be resisted only by a strong leader who wrenches the country backward to make it "great again." These movements have been abetted by a narrative shared by many of their fiercest opponents, in which the institutions of modernity have failed and every aspect of life is in deepening crisis—the two sides in macabre agreement that wrecking those institutions will make the world a better place. Harder to find is a positive vision that sees the world's problems against a background of progress that it seeks to build upon by solving those problems in their turn.

If you still are unsure whether the ideals of Enlightenment humanism need a vigorous defense, consider the diagnosis of Shiraz Maher, an analyst of radical Islamist movements. "The West is shy of its values—it doesn't speak up for classical liberalism," he says. "We are unsure of them. They make us feel uneasy." Contrast that with the Islamic State, which "knows exactly what it stands for," a certainty that is "incredibly seductive"—and he should know, having once been a regional director of the jihadist group Hizb ut-Tahrir.[1]

Reflecting on liberal ideals in 1960, not long after they had withstood their greatest trial, the economist Friedrich Hayek observed, "If old truths are to retain their hold on men's minds, they must be restated in the language and concepts of successive generations" (inadvertently proving his point with the expression *men's minds*). "What at one time are their most effective expressions gradually become so worn with use that they cease to carry a definite meaning. The underlying ideas may be as valid as ever, but the words, even when they refer to problems that are still with us, no longer convey the same conviction."[2]

This book is my attempt to restate the ideals of the Enlightenment in the language and concepts of the 21st century. I will first lay out a framework for understanding the human condition informed by modern science—who we are, where we came from, what our challenges are, and how we can meet them. The bulk of the book is devoted to defending those ideals in a distinctively 21st-century way: with data. This evidence-based take on the Enlightenment project reveals that it was not a naïve hope. The Enlightenment has *worked*—perhaps the greatest story seldom told. And because this triumph is so unsung, the underlying ideals of reason, science, and humanism are unappreciated as well. Far from being an insipid consensus, these ideals are treated by today's intellectuals with indifference, skepticism, and sometimes contempt. When properly appreciated, I will suggest, the ideals of the Enlightenment are in fact stirring, inspiring, noble—a reason to live.

CHAPTER 1

DARE TO UNDERSTAND!

What is enlightenment? In a 1784 essay with that question as its title, Immanuel Kant answered that it consists of "humankind's emergence from its self-incurred immaturity," its "lazy and cowardly" submission to the "dogmas and formulas" of religious or political authority.[1] Enlightenment's motto, he proclaimed, is "Dare to understand!" and its foundational demand is freedom of thought and speech. "One age cannot conclude a pact that would prevent succeeding ages from extending their insights, increasing their knowledge, and purging their errors. That would be a crime against human nature, whose proper destiny lies precisely in such progress."[2]

A 21st-century statement of the same idea may be found in the physicist David Deutsch's defense of enlightenment, *The Beginning of Infinity*. Deutsch argues that if we dare to understand, progress is possible in all fields, scientific, political, and moral:

> Optimism (in the sense that I have advocated) is the theory that all failures—all evils—are due to insufficient knowledge. . . . Problems are inevitable, because our knowledge will always be infinitely far from complete. Some problems are hard, but it is a mistake to confuse hard problems with problems unlikely to be solved. Problems are soluble, and each particular evil is a problem that can be solved. An optimistic civilization is open and not afraid to innovate, and is based on traditions of criticism. Its institutions keep improving, and the most important knowledge that they embody is knowledge of how to detect and eliminate errors.[3]

What is *the* Enlightenment?[4] There is no official answer, because the era named by Kant's essay was never demarcated by opening and closing ceremonies like the Olympics, nor are its tenets stipulated in an oath or creed. The Enlightenment is conventionally placed in the last two-thirds of the 18th century, though it flowed out of the Scientific Revolution and the Age of Reason in the 17th century and spilled into the heyday of classical liberalism of the first half of the 19th. Provoked by challenges to conventional wisdom from science and exploration, mindful of the bloodshed of recent wars of religion, and abetted by the easy movement of ideas and people, the thinkers of the Enlightenment sought a new understanding of the human condition. The era was a cornucopia of ideas, some of them contradictory, but four themes tie them together: reason, science, humanism, and progress.

Foremost is reason. Reason is nonnegotiable. As soon as you show up to discuss the question of what we should live for (or any other question), as long as you insist that your answers, whatever they are, are reasonable or justified or true and that therefore other people ought to believe them too, then you have committed yourself to reason, and to holding your beliefs accountable to objective standards.[5] If there's anything the Enlightenment thinkers had in common, it was an insistence that we energetically apply the standard of reason to understanding our world, and not fall back on generators of delusion like faith, dogma, revelation, authority, charisma, mysticism, divination, visions, gut feelings, or the hermeneutic parsing of sacred texts.

It was reason that led most of the Enlightenment thinkers to repudiate a belief in an anthropomorphic God who took an interest in human affairs.[6] The application of reason revealed that reports of miracles were dubious, that the authors of holy books were all too human, that natural events unfolded with no regard to human welfare, and that different cultures believed in mutually incompatible deities, none of them less likely than the others to be products of the imagination. (As Montesquieu wrote, "If triangles had a god they would give him three sides.") For all that, not all of the Enlightenment thinkers were atheists. Some were deists (as opposed to theists): they thought that God set the universe in motion and then stepped back, allowing it to unfold according to the laws of nature. Others were pantheists, who used "God" as a *synonym* for the laws of nature. But few appealed to the law-giving, miracle-conjuring, son-begetting God of scripture.

Many writers today confuse the Enlightenment endorsement of reason with the implausible claim that humans are perfectly rational agents. Nothing could be further from historical reality. Thinkers such as Kant, Baruch Spinoza, Thomas Hobbes, David Hume, and Adam Smith were inquisitive psychologists and all too aware of our irrational passions and foibles. They insisted that it was only by calling out the common sources of folly that we could hope to overcome them. The deliberate application of reason was necessary precisely because our common habits of thought are not particularly reasonable.

That leads to the second ideal, science, the refining of reason to understand the world. The Scientific Revolution was revolutionary in a way that is hard to appreciate today, now that its discoveries have become second nature to most of us. The historian David Wootton reminds us of the understanding of an educated Englishman on the eve of the Revolution in 1600:

> He believes witches can summon up storms that sink ships at sea. . . .
> He believes in werewolves, although there happen not to be any in England—he knows they are to be found in Belgium. . . . He believes Circe really did turn Odysseus's crew into pigs. He believes mice are spontaneously generated in piles of straw. He believes in contemporary magicians. . . . He has seen a unicorn's horn, but not a unicorn.
>
> He believes that a murdered body will bleed in the presence of the murderer. He believes that there is an ointment which, if rubbed on a dagger which has caused a wound, will cure the wound. He believes that the shape, colour and texture of a plant can be a clue to how it will work as a medicine because God designed nature to be interpreted by mankind. He believes that it is possible to turn base metal into gold, although he doubts that anyone knows how to do it. He believes that nature abhors a vacuum. He believes the rainbow is a sign from God and that comets portend evil. He believes that dreams predict the future, if we know how to interpret them. He believes, of course, that the earth stands still and the sun and stars turn around the earth once every twenty-four hours.[7]

A century and a third later, an educated descendant of this Englishman would believe none of these things. It was an escape not just from ignorance but from terror. The sociologist Robert Scott notes that in the Middle Ages "the belief that an external force controlled daily life contributed to a kind of collective paranoia":

> Rainstorms, thunder, lightning, wind gusts, solar or lunar eclipses, cold snaps, heat waves, dry spells, and earthquakes alike were considered signs and signals of God's displeasure. As a result, the "hobgob-

lins of fear" inhabited every realm of life. The sea became a satanic realm, and forests were populated with beasts of prey, ogres, witches, demons, and very real thieves and cutthroats. . . . After dark, too, the world was filled with omens portending dangers of every sort: comets, meteors, shooting stars, lunar eclipses, the howls of wild animals.[8]

To the Enlightenment thinkers the escape from ignorance and superstition showed how mistaken our conventional wisdom could be, and how the methods of science—skepticism, fallibilism, open debate, and empirical testing—are a paradigm of how to achieve reliable knowledge.

That knowledge includes an understanding of ourselves. The need for a "science of man" was a theme that tied together Enlightenment thinkers who disagreed about much else, including Montesquieu, Hume, Smith, Kant, Nicolas de Condorcet, Denis Diderot, Jean-Baptiste d'Alembert, Jean-Jacques Rousseau, and Giambattista Vico. Their belief that there was such a thing as universal human nature, and that it could be studied scientifically, made them precocious practitioners of sciences that would be named only centuries later.[9] They were cognitive neuroscientists, who tried to explain thought, emotion, and psychopathology in terms of physical mechanisms of the brain. They were evolutionary psychologists, who sought to characterize life in a state of nature and to identify the animal instincts that are "infused into our bosoms." They were social psychologists, who wrote of the moral sentiments that draw us together, the selfish passions that divide us, and the foibles of shortsightedness that confound our best-laid plans. And they were cultural anthropologists, who mined the accounts of travelers and explorers for data both on human universals and on the diversity of customs and mores across the world's cultures.

The idea of a universal human nature brings us to a third theme, humanism. The thinkers of the Age of Reason and the Enlightenment saw an urgent need for a secular foundation for morality, because they were haunted by a historical memory of centuries of religious carnage: the Crusades, the Inquisition, witch hunts, the European wars of religion. They laid that foundation in what we now call humanism, which privileges the well-being of individual men, women, and children over the glory of the tribe, race, nation, or religion. It is individuals, not groups, who are *sentient*—who feel pleasure and pain, fulfillment and anguish. Whether it is framed as the goal of providing the greatest happiness for the greatest number or as a categorical imperative to treat people as ends rather than means, it was the universal capacity of a person to suffer and flourish, they said, that called on our moral concern.

Fortunately, human nature prepares us to answer that call. That is because we are endowed with the sentiment of *sympathy*, which they also called benevolence, pity, and commiseration. Given that we are equipped with the capacity to sympathize with others, nothing can prevent the circle of sympathy from expanding from the family and tribe to embrace all of humankind, particularly as reason goads us into realizing that there can be nothing uniquely deserving about ourselves or any of the groups to which we belong.[10] We are forced into cosmopolitanism: accepting our citizenship in the world.[11]

A humanistic sensibility impelled the Enlightenment thinkers to condemn not just religious violence but also the secular cruelties of their age, including slavery, despotism, executions for frivolous offenses such as shoplifting and poaching, and sadistic punishments such as flogging, amputation, impalement, disembowelment, breaking on the wheel, and burning at the stake. The Enlightenment is sometimes called the Humanitarian Revolution, because it led to the abolition of barbaric practices that had been commonplace across civilizations for millennia.[12]

If the abolition of slavery and cruel punishment is not progress, nothing is, which brings us to the fourth Enlightenment ideal. With our understanding of the world advanced by science and our circle of sympathy expanded through reason and cosmopolitanism, humanity could make intellectual and moral progress. It need not resign itself to the miseries and irrationalities of the present, nor try to turn back the clock to a lost golden age.

The Enlightenment belief in progress should not be confused with the 19th-century Romantic belief in mystical forces, laws, dialectics, struggles, unfoldings, destinies, ages of man, and evolutionary forces that propel mankind ever upward toward utopia.[13] As Kant's remark about "increasing knowledge and purging errors" indicates, it was more prosaic, a combination of reason and humanism. If we keep track of how our laws and manners are doing, think up ways to improve them, try them out, and keep the ones that make people better off, we can gradually make the world a better place. Science itself creeps forward through this cycle of theory and experiment, and its ceaseless headway, superimposed on local setbacks and reversals, shows how progress is possible.

The ideal of progress also should not be confused with the 20th-century movement to re-engineer society for the convenience of techno-

crats and planners, which the political scientist James Scott calls
Authoritarian High Modernism.[14] The movement denied the existence of
human nature, with its messy needs for beauty, nature, tradition, and
social intimacy.[15] Starting from a "clean tablecloth," the modernists designed urban renewal projects that replaced vibrant neighborhoods with
freeways, high-rises, windswept plazas, and brutalist architecture.
"Mankind will be reborn," they theorized, and "live in an ordered relation to the whole."[16] Though these developments were sometimes linked
to the word *progress,* the usage was ironic: "progress" unguided by humanism is not progress.

Rather than trying to shape human nature, the Enlightenment hope
for progress was concentrated on human institutions. Human-made systems like governments, laws, schools, markets, and international bodies
are a natural target for the application of reason to human betterment.

In this way of thinking, government is not a divine fiat to reign, a
synonym for "society," or an avatar of the national, religious, or racial
soul. It is a human invention, tacitly agreed to in a social contract, designed to enhance the welfare of citizens by coordinating their behavior
and discouraging selfish acts that may be tempting to every individual
but leave everyone worse off. As the most famous product of the Enlightenment, the Declaration of Independence, put it, in order to secure the
right to life, liberty, and the pursuit of happiness, governments are insti-
tuted among people, deriving their just powers from the consent of the
governed.

Among the powers of government is meting out punishment, and
writers such as Montesquieu, Cesare Beccaria, and the American founders thought afresh about the government's license to harm its citizens.[17]
Criminal punishment, they argued, is not a mandate to implement cosmic justice but part of an incentive structure that discourages antisocial
acts without causing more suffering than it deters. The reason the punishment should fit the crime, for example, is not to balance some mystical
scale of justice but to ensure that a wrongdoer stops at a minor crime
rather than escalating to a more harmful one. Cruel punishments,
whether or not they are in some sense "deserved," are no more effective
at deterring harm than moderate but surer punishments, and they de-
sensitize spectators and brutalize the society that implements them.

The Enlightenment also saw the first rational analysis of prosperity.
Its starting point was not how wealth is distributed but the prior ques-
tion of how wealth comes to exist in the first place.[18] Smith, building on
French, Dutch, and Scottish influences, noted that an abundance of use-

ful stuff cannot be conjured into existence by a farmer or craftsman working in isolation. It depends on a network of specialists, each of whom learns how to make something as efficiently as possible, and who combine and exchange the fruits of their ingenuity, skill, and labor. In a famous example, Smith calculated that a pin-maker working alone could make at most one pin a day, whereas in a workshop in which "one man draws out the wire, another straights it, a third cuts it, a fourth points it, a fifth grinds it at the top for receiving the head," each could make almost five thousand.

Specialization works only in a market that allows the specialists to exchange their goods and services, and Smith explained that economic activity was a form of mutually beneficial cooperation (a positive-sum game, in today's lingo): each gets back something that is more valuable to him than what he gives up. Through voluntary exchange, people benefit others by benefiting themselves; as he wrote, "It is not from the benevolence of the butcher, the brewer, or the baker that we expect our dinner, but from their regard to their own interest. We address ourselves, not to their humanity but to their self-love." Smith was not saying that people are ruthlessly selfish, or that they ought to be; he was one of history's keenest commentators on human sympathy. He only said that in a market, whatever tendency people have to care for their families and themselves can work to the good of all.

Exchange can make an entire society not just richer but nicer, because in an effective market it is cheaper to buy things than to steal them, and other people are more valuable to you alive than dead. (As the economist Ludwig von Mises put it centuries later, "If the tailor goes to war against the baker, he must henceforth bake his own bread.") Many Enlightenment thinkers, including Montesquieu, Kant, Voltaire, Diderot, and the Abbé de Saint-Pierre, endorsed the ideal of *doux commerce*, gentle commerce.[19] The American founders—George Washington, James Madison, and especially Alexander Hamilton—designed the institutions of the young nation to nurture it.

This brings us to another Enlightenment ideal, peace. War was so common in history that it was natural to see it as a permanent part of the human condition and to think peace could come only in a messianic age. But now war was no longer thought of as a divine punishment to be endured and deplored, or a glorious contest to be won and celebrated, but a practical problem to be mitigated and someday solved. In "Perpetual Peace," Kant laid out measures that would discourage leaders from dragging their countries into war.[20] Together with international commerce, he recommended representative republics (what we would call democracies), mutual transparency, norms against conquest and internal interference, freedom of travel and immigration, and a federation of states that would adjudicate disputes between them.

For all the prescience of the founders, framers, and *philosophes*, this is not a book of Enlightenolatry. The Enlightenment thinkers were men and women of their age, the 18th century. Some were racists, sexists, anti-Semites, slaveholders, or duelists. Some of the questions they worried about are almost incomprehensible to us, and they came up with plenty of daffy ideas together with the brilliant ones. More to the point, they were born too soon to appreciate some of the keystones of our modern understanding of reality.

They of all people would have been the first to concede this. If you extol reason, then what matters is the integrity of the thoughts, not the personalities of the thinkers. And if you're committed to progress, you can't very well claim to have it all figured out. It takes nothing away from the Enlightenment thinkers to identify some critical ideas about the human condition and the nature of progress that we know and they didn't. Those ideas, I suggest, are entropy, evolution, and information.

CHAPTER 2

ENTRO, EVO, INFO

The first keystone in understanding the human condition is the concept of entropy or disorder, which emerged from 19th-century physics and was defined in its current form by the physicist Ludwig Boltzmann.[1] The Second Law of Thermodynamics states that in an isolated system (one that is not interacting with its environment), entropy never decreases. (The First Law is that energy is conserved; the Third, that a temperature of absolute zero is unreachable.) Closed systems inexorably become less structured, less organized, less able to accomplish interesting and useful outcomes, until they slide into an equilibrium of gray, tepid, homogeneous monotony and stay there.

In its original formulation the Second Law referred to the process in which usable energy in the form of a difference in temperature between two bodies is inevitably dissipated as heat flows from the warmer to the cooler body. (As the musical team Flanders & Swann explained, "You can't pass heat from the cooler to the hotter; Try it if you like but you far better notter.") A cup of coffee, unless it is placed on a plugged-in hot plate, will cool down. When the coal feeding a steam engine is used up, the cooled-off steam on one side of the piston can no longer budge it because the warmed-up steam and air on the other side are pushing back just as hard.

Once it was appreciated that heat is not an invisible fluid but the energy in moving molecules, and that a difference in temperature between two bodies consists of a difference in the average speeds of those molecules, a more general, statistical version of the concept of entropy and the Second Law took shape. Now order could be characterized in terms of the set of all microscopically distinct states of a system (in the original example involving heat, the possible speeds and positions of all the molecules in the two bodies). Of all these states, the ones that we find useful from a bird's-eye view (such as one body being hotter than the other, which translates into the average speed of the molecules in one body being higher than the average speed in the other) make up a tiny fraction of the possibilities, while all the disorderly or useless states (the ones without a temperature difference, in which the average speeds in the two bodies are the same) make up the vast majority. It follows that any perturbation of the system, whether it is a random jiggling of its parts or a whack from the outside, will, by the laws of probability, nudge the system toward disorder or uselessness—not because nature strives for disorder, but because there are so many more ways of being disorderly than of being orderly. If you walk away from a sandcastle, it won't be there tomorrow, because as the wind, waves, seagulls, and small children push the grains of sand around, they're more likely to arrange them into one of the vast number of configurations that don't look like a castle than into the tiny few that do. I'll often refer to the statistical version of the Second Law, which does not apply specifically to temperature differences evening out but to order dissipating, as the Law of Entropy.

How is entropy relevant to human affairs? Life and happiness depend on an infinitesimal sliver of orderly arrangements of matter amid the astronomical number of possibilities. Our bodies are improbable assemblies of molecules, and they maintain that order with the help of other improbabilities: the few substances that can nourish us, the few materials in the few shapes that can clothe us, shelter us, and move things around to our liking. Far more of the arrangements of matter found on Earth are of no worldly use to us, so when things change without a human agent directing the change, they are likely to change for the worse. The Law of Entropy is widely acknowledged in everyday life in sayings such as "Things fall apart," "Rust never sleeps," "Shit happens," "Whatever can go wrong will go wrong," and (from the Texas lawmaker Sam Rayburn) "Any jackass can kick down a barn, but it takes a carpenter to build one."

Scientists appreciate that the Second Law is far more than an explanation of everyday nuisances. It is a foundation of our understanding of the universe and our place in it. In 1928 the physicist Arthur Eddington wrote:

The law that entropy always increases . . . holds, I think, the supreme position among the laws of Nature. If someone points out to you that your pet theory of the universe is in disagreement with Maxwell's equations—then so much the worse for Maxwell's equations. If it is found to be contradicted by observation—well, these experimentalists do bungle things sometimes. But if your theory is found to be against the second law of thermodynamics I can give you no hope; there is nothing for it but to collapse in deepest humiliation.[2]

In his famous 1959 Rede lectures, published as *The Two Cultures and the Scientific Revolution*, the scientist and novelist C. P. Snow commented on the disdain for science among educated Britons in his day:

A good many times I have been present at gatherings of people who, by the standards of the traditional culture, are thought highly educated and who have with considerable gusto been expressing their incredulity at the illiteracy of scientists. Once or twice I have been provoked and have asked the company how many of them could describe the Second Law of Thermodynamics. The response was cold: it was also negative. Yet I was asking something which is about the scientific equivalent of: *Have you read a work of Shakespeare's?*[3]

The chemist Peter Atkins alludes to the Second Law in the title of his book *Four Laws That Drive the Universe*. And closer to home, the evolutionary psychologists John Tooby, Leda Cosmides, and Clark Barrett entitled a recent paper on the foundations of the science of mind "The Second Law of Thermodynamics Is the First Law of Psychology."[4]

Why the awe for the Second Law? From an Olympian vantage point, it defines the fate of the universe and the ultimate purpose of life, mind, and human striving: to deploy energy and knowledge to fight back the tide of entropy and carve out refuges of beneficial order. From a terrestrial vantage point we can get more specific, but before we get to familiar ground I need to lay out the other two foundational ideas.

～

At first glance the Law of Entropy would seem to allow for only a discouraging history and a depressing future. The universe began in a state of low entropy, the Big Bang, with its unfathomably dense concentration of energy. From there everything went downhill, with the universe dispersing—as it will continue to do—into a thin gruel of particles evenly and sparsely distributed through space. In reality, of course, the universe as we find it is not a featureless gruel. It is enlivened with galaxies, planets, mountains, clouds, snowflakes, and an efflorescence of flora and fauna, including us.

One reason the cosmos is filled with so much interesting stuff is a set of processes called self-organization, which allow circumscribed zones of order to emerge.⁵ When energy is poured into a system, and the system dissipates that energy in its slide toward entropy, it can become poised in an orderly, indeed beautiful, configuration—a sphere, spiral, starburst, whirlpool, ripple, crystal, or fractal. The fact that we find these configurations beautiful, incidentally, suggests that beauty may not just be in the eye of the beholder. The brain's aesthetic response may be a receptiveness to the counter-entropic patterns that can spring forth from nature.

But there is another kind of orderliness in nature that also must be explained: not the elegant symmetries and rhythms in the physical world, but the functional design in the living world. Living things are made of organs that have heterogeneous parts which are uncannily shaped and arranged to do things that keep the organism alive (that is, continuing to absorb energy to resist entropy).⁶

The customary illustration of biological design is the eye, but I will make the point with my second-favorite sense organ. The human ear contains an elastic drumhead that vibrates in response to the slightest puff of air, a bony lever that multiplies the vibration's force, a piston that impresses the vibration into the fluid in a long tunnel (conveniently coiled to fit inside the wall of the skull), a tapering membrane that runs down the length of the tunnel and physically separates the waveform into its harmonics, and an array of cells with tiny hairs that are flexed back and forth by the vibrating membrane, sending a train of electrical impulses to the brain. It is impossible to explain why these membranes and bones and fluids and hairs are arranged in that improbable way without noting that this configuration allows the brain to register patterned sound. Even the fleshy outer ear—asymmetrical top to bottom and front to back, and crinkled with ridges and valleys—is shaped in a way that sculpts the incoming sound to inform the brain whether the soundmaker is above or below, in front or behind.

Organisms are replete with improbable configurations of flesh like eyes, ears, hearts, and stomachs which cry out for an explanation. Before Charles Darwin and Alfred Russel Wallace provided one in 1859, it was reasonable to think they were the handiwork of a divine designer—one of the reasons, I suspect, that so many Enlightenment thinkers were deists rather than outright atheists. Darwin and Wallace made the designer unnecessary. Once self-organizing processes of physics and chemistry gave rise to a configuration of matter that could replicate itself, the copies would make copies, which would make copies of the copies, and so on, in an exponential explosion. The replicating systems would compete for the material to make their copies and the energy to power the replication. Since no copying process is perfect—the Law of Entropy sees to that—errors will crop up, and though most of these mutations will degrade the replicator (entropy again), occasionally dumb luck will throw one up that's more effective at replicating, and its descendants will swamp the competition. As copying errors that enhance stability and replication accumulate over the generations, the replicating system—we call it an organism—will appear to have been engineered for survival and reproduction in the future, though it only preserved the copying errors that led to survival and reproduction in the past.

Creationists commonly doctor the Second Law of Thermodynamics to claim that biological evolution, an increase in order over time, is physically impossible. The part of the law they omit is "in a closed system." Organisms are open systems: they capture energy from the sun, food, or ocean vents to carve out temporary pockets of order in their bodies and nests while they dump heat and waste into the environment, increasing disorder in the world as a whole. Organisms' use of energy to maintain their integrity against the press of entropy is a modern explanation of the principle of *conatus* (effort or striving), which Spinoza defined as "the endeavor to persist and flourish in one's own being," and which was a foundation of several Enlightenment-era theories of life and mind.[7]

The ironclad requirement to suck energy out of the environment leads to one of the tragedies of living things. While plants bask in solar energy, and a few creatures of the briny deep soak up the chemical broth spewing from cracks in the ocean floor, animals are born exploiters: they live off the hard-won energy stored in the bodies of plants and other animals by eating them. So do the viruses, bacteria, and other pathogens and parasites that gnaw at bodies from the inside. With the exception of fruit, everything we call "food" is the body part or energy store of some other organism, which would just as soon keep that treasure for itself. Nature is a war, and much of what captures our attention in the natural world is an arms race. Prey animals protect themselves with shells, spines, claws, horns, venom, camouflage, flight, or self-defense; plants have thorns, rinds, bark, and irritants and poisons saturating their tissues. Animals evolve weapons to penetrate these defenses: carnivores have speed, talons, and eagle-eyed vision, while herbivores have grinding teeth and livers that detoxify natural poisons.

~

And now we come to the third keystone, information.[8] Information may be thought of as a reduction in entropy—as the ingredient that distinguishes an orderly, structured system from the vast set of random, useless ones.[9] Imagine pages of random characters tapped out by a monkey at a typewriter, or a stretch of white noise from a radio tuned between channels, or a screenful of confetti from a corrupted computer file. Each of these objects can take trillions of different forms, each as boring as the next. But now suppose that the devices are controlled by a signal that arranges the characters or sound waves or pixels into a pattern that correlates with something in the world: the Declaration of Independence, the opening bars of "Hey Jude," a cat wearing sunglasses. We say that the signal transmits *information* about the Declaration or the song or the cat.[10]

The information contained in a pattern depends on how coarsely or finely grained our view of the world is. If we cared about the *exact* sequence of characters in the monkey's output, or the precise difference between one burst of noise and another, or the particular pattern of pixels in just one of the haphazard displays, then we would have to say that each of the items contains the same amount of information as the others. Indeed, the interesting ones would contain *less* information, because when you look at one part (like the letter q) you can guess others (such as the following letter, u) without needing the signal. But more commonly we lump together the immense majority of random-looking configurations as equivalently boring, and distinguish them all from the tiny few that correlate with something else. From that vantage point the cat photo contains more information than the confetti of pixels, because it takes a garrulous message to pinpoint a rare orderly configuration out of the vast number of equivalently disorderly ones. To say that the universe is orderly rather than random is to say that it contains information in this sense. Some physicists enshrine information as one of the basic constituents of the universe, together with matter and energy.[11]

Information is what gets accumulated in a genome in the course of evolution. The sequence of bases in a DNA molecule correlates with the sequence of amino acids in the proteins that make up the organism's body, and they got that sequence by structuring the organism's ancestors— reducing their entropy—into the improbable configurations that allowed them to capture energy and grow and reproduce.

Information is also collected by an animal's nervous system as it lives its life. When the ear transduces sound into neural firings, the two physical processes—vibrating air and diffusing ions—could not be more different. But thanks to the correlation between them, the pattern of neural activity in the animal's brain carries information about the sound in the world. From there the information can switch from electrical to chemical and back as it crosses the synapses connecting one neuron to the next; through all these physical transformations, the information is preserved.

A momentous discovery of 20th-century theoretical neuroscience is that networks of neurons not only can preserve information but can transform it in ways that allow us to explain how brains can be *intelligent*. Two input neurons can be connected to an output neuron in such a way that their firing patterns correspond to logical relations such as AND, OR, and NOT, or to a statistical decision that depends on the weight of the incoming evidence. That gives neural networks the power to engage in information processing or computation. Given a large enough network built out of these logical and statistical circuits (and with billions of neurons, the brain has room for plenty), a brain can compute complex functions, the prerequisite for intelligence. It can transform the information about the world that it receives from the sense organs in a way that mirrors the laws governing that world, which in turn allows it to make useful inferences and predictions.[12] Internal representations that reliably correlate with states of the world, and that participate in inferences that tend to derive true implications from true premises, may be called knowledge.[13] We say that someone knows what a robin is if she thinks the thought "robin" whenever she sees one, and if she can infer that it is a kind of bird which appears in the spring and pulls worms out of the ground.

Getting back to evolution, a brain wired by information in the genome to perform computations on information coming in from the senses could organize the animal's behavior in a way that allowed it to capture energy and resist entropy. It could, for example, implement the rule "If it squeaks, chase it; if it barks, flee from it."

Chasing and fleeing, though, are not just sequences of muscle contractions—they are *goal-directed*. Chasing may consist of running or climbing or leaping or ambushing, depending on the circumstances, as long as it increases the chances of snagging the prey; fleeing may include hiding or freezing or zigzagging. And that brings up another momentous 20th-century idea, sometimes called cybernetics, feedback, or control. The idea explains how a physical system can appear to be teleological, that is, directed by purposes or goals. All it needs are a way of sensing the state of itself and its environment, a representation of a goal state (what it "wants," what it's "trying for"), an ability to compute the difference between the current state and the goal state, and a repertoire of actions that are tagged with their typical effects. If the system is wired so that it triggers actions that typically reduce the difference between the current state and the goal state, it can be said to pursue goals (and when the world is sufficiently predictable, it will attain them). The principle was discovered by natural selection in the form of homeostasis, as when our bodies regulate their temperature by shivering and sweating. When it was discovered by humans, it was engineered into analog systems like thermostats and cruise control and then into digital systems like chess-playing programs and autonomous robots.

The principles of information, computation, and control bridge the chasm between the physical world of cause and effect and the mental world of knowledge, intelligence, and purpose. It's not just a rhetorical aspiration to say that ideas can change the world; it's a fact about the physical makeup of brains. The Enlightenment thinkers had an inkling that thought could consist of patterns in matter—they likened ideas to impressions in wax, vibrations in a string, or waves from a boat. And some, like Hobbes, proposed that "reasoning is but reckoning," in the original sense of *reckoning* as calculation. But before the concepts of information and computation were elucidated, it was reasonable for someone to be a mind-body dualist and attribute mental life to an immaterial soul (just as before the concept of evolution was elucidated, it was reasonable to be a creationist and attribute design in nature to a cosmic designer). That's another reason, I suspect, that so many Enlightenment thinkers were deists.

Of course it's natural to think twice about whether your cell phone truly "knows" a favorite number, your GPS is really "figuring out" the best route home, and your Roomba is genuinely "trying" to clean the floor. But as information-processing systems become more sophisticated— as their representations of the world become richer, their goals are arranged into hierarchies of subgoals within subgoals, and their actions for attaining the goals become more diverse and less predictable—it starts to look like hominid chauvinism to insist that they don't. (Whether information and computation explain *consciousness*, in addition to knowledge, intelligence, and purpose, is a question I'll turn to in the final chapter.)

Human intelligence remains the benchmark for the artificial kind, and what makes *Homo sapiens* an unusual species is that our ancestors invested in bigger brains that collected more information about the world, reasoned about it in more sophisticated ways, and deployed a greater variety of actions to achieve their goals. They specialized in the cognitive niche, also called the cultural niche and the hunter-gatherer niche.[14] This embraced a suite of new adaptations, including the ability to manipulate mental models of the world and predict what would happen if one tried out new things; the ability to cooperate with others, which allowed teams of people to accomplish what a single person could not; and language, which allowed them to coordinate their actions and to pool the fruits of their experience into the collections of skills and norms we call cultures.[15] These investments allowed early hominids to defeat the defenses of a wide range of plants and animals and reap the bounty in energy, which stoked their expanding brains, giving them still more know-how and access to still more energy. A well-studied contemporary hunter-gatherer tribe, the Hadza of Tanzania, who live in the ecosystem where modern humans first evolved and probably preserve much of their lifestyle, extract 3,000 calories daily per person from more than 880 species.[16] They create this menu through ingenious and uniquely human ways of foraging, such as felling large animals with poison-tipped arrows, smoking bees out of their hives to steal their honey, and enhancing the nutritional value of meat and tubers by cooking them.

Energy channeled by knowledge is the elixir with which we stave off entropy, and advances in energy capture are advances in human destiny. The invention of farming around ten thousand years ago multiplied the availability of calories from cultivated plants and domesticated animals, freed a portion of the population from the demands of hunting and gathering, and eventually gave them the luxury of writing, thinking, and accumulating their ideas. Around 500 BCE, in what the philosopher Karl Jaspers called the Axial Age, several widely separated cultures pivoted from systems of ritual and sacrifice that merely warded off misfortune to systems of philosophical and religious belief that promoted selflessness and promised spiritual transcendence.[17] Taoism and Confucianism in China, Hinduism, Buddhism, and Jainism in India, Zoroastrianism in Persia, Second Temple Judaism in Judea, and classical Greek philosophy and drama emerged within a few centuries of one another. (Confucius, Buddha, Pythagoras, Aeschylus, and the last of the Hebrew prophets walked the earth at the same time.) Recently an interdisciplinary team of scholars identified a common cause.[18] It was not an aura of spirituality that descended on the planet but something more prosaic: energy capture. The Axial Age was when agricultural and economic advances provided a burst of energy: upwards of 20,000 calories per person per day in food, fodder, fuel, and raw materials. This surge allowed the civilizations to afford larger cities, a scholarly and priestly class, and a reorientation of their priorities from short-term survival to long-term harmony. As Bertolt Brecht put it millennia later: Grub first, then ethics.[19]

When the Industrial Revolution released a gusher of usable energy from coal, oil, and falling water, it launched a Great Escape from poverty, disease, hunger, illiteracy, and premature death, first in the West and increasingly in the rest of the world (as we shall see in chapters 5–8). And the next leap in human welfare—the end of extreme poverty and spread of abundance, with all its moral benefits—will depend on technological advances that provide energy at an acceptable economic and environmental cost to the entire world (chapter 10).

~

Entro, evo, info. These concepts define the narrative of human progress: the tragedy we were born into, and our means for eking out a better existence.

The first piece of wisdom they offer is that *misfortune may be no one's fault.* A major breakthrough of the Scientific Revolution—perhaps its biggest breakthrough—was to refute the intuition that the universe is saturated with purpose. In this primitive but ubiquitous understanding, everything happens for a reason, so when bad things happen—accidents, disease, famine, poverty—some agent must have *wanted* them to happen. If a person can be fingered for the misfortune, he can be punished or squeezed for damages. If no individual can be singled out, one might blame the nearest ethnic or religious minority, who can be lynched or massacred in a pogrom. If no mortal can plausibly be indicted, one might cast about for witches, who may be burned or drowned. Failing that, one points to sadistic gods, who cannot be punished but can be placated with prayers and sacrifices. And then there are disembodied forces like karma, fate, spiritual messages, cosmic justice, and other guarantors of the intuition that "everything happens for a reason."

Galileo, Newton, and Laplace replaced this cosmic morality play with a clockwork universe in which events are caused by conditions in the present, not goals for the future.[20] *People* have goals, of course, but projecting goals onto the workings of nature is an illusion. Things can happen without anyone taking into account their effects on human happiness.

This insight of the Scientific Revolution and the Enlightenment was deepened by the discovery of entropy. Not only does the universe not care about our desires, but in the natural course of events it will appear to thwart them, because there are so many more ways for things to go wrong than for them to go right. Houses burn down, ships sink, battles are lost for want of a horseshoe nail.

Awareness of the indifference of the universe was deepened still fur-

ther by an understanding of evolution. Predators, parasites, and pathogens are constantly trying to eat us, and pests and spoilage organisms try
to eat our stuff. It may make us miserable, but that's not their problem.

Poverty, too, needs no explanation. In a world governed by entropy
and evolution, it is the default state of humankind. Matter does not arrange itself into shelter or clothing, and living things do everything they
can to avoid becoming our food. As Adam Smith pointed out, what
needs to be explained is wealth. Yet even today, when few people believe
that accidents or diseases have perpetrators, discussions of poverty consist mostly of arguments about whom to blame for it.

None of this is to say that the natural world is free of malevolence. On
the contrary, evolution guarantees there will be plenty of it. Natural selection consists of competition among genes to be represented in the next
generation, and the organisms we see today are descendants of those
that edged out their rivals in contests for mates, food, and dominance.
This does not mean that all creatures are always rapacious; modern evolutionary theory explains how selfish genes can give rise to unselfish
organisms. But the generosity is measured. Unlike the cells in a body or
the individuals in a colonial organism, humans are genetically unique,
each having accumulated and recombined a different set of mutations
that arose over generations of entropy-prone replication in their lineage.
Genetic individuality gives us our different tastes and needs, and it also
sets the stage for strife. Families, couples, friends, allies, and societies
seethe with partial conflicts of interest, which are played out in tension,
arguments, and sometimes violence. Another implication of the Law of
Entropy is that a complex system like an organism can easily be disabled,
because its functioning depends on so many improbable conditions being satisfied at once. A rock against the head, a hand around the neck, a
well-aimed poisoned arrow, and the competition is neutralized. More
tempting still to a language-using organism, a *threat* of violence may be
used to coerce a rival, opening the door to oppression and exploitation.

Evolution left us with another burden: our cognitive, emotional, and
moral faculties are adapted to individual survival and reproduction in
an archaic environment, not to universal thriving in a modern one. To
appreciate this burden, one doesn't have to believe that we are cavemen
out of time, only that evolution, with its speed limit measured in generations, could not possibly have adapted our brains to modern technology
and institutions. Humans today rely on cognitive faculties that worked
well enough in traditional societies, but which we now see are infested
with bugs.

People are by nature illiterate and innumerate, quantifying the world by "one, two, many" and by rough guesstimates.[21] They understand physical things as having hidden essences that obey the laws of sympathetic magic or voodoo rather than physics and biology: objects can reach across time and space to affect things that resemble them or that had been in contact with them in the past (remember the beliefs of pre–Scientific Revolution Englishmen).[22] They think that words and thoughts can impinge on the physical world in prayers and curses. They underestimate the prevalence of coincidence.[23] They generalize from paltry samples, namely their own experience, and they reason by stereotype, projecting the typical traits of a group onto any individual that belongs to it. They infer causation from correlation. They think holistically, in black and white, and physically, treating abstract networks as concrete stuff. They are not so much intuitive scientists as intuitive lawyers and politicians, marshaling evidence that confirms their convictions while dismissing evidence that contradicts them.[24] They overestimate their own knowledge, understanding, rectitude, competence, and luck.[25]

The human moral sense can also work at cross-purposes to our well-being.[26] People demonize those they disagree with, attributing differences of opinion to stupidity and dishonesty. For every misfortune they seek a scapegoat. They see morality as a source of grounds for condemning rivals and mobilizing indignation against them.[27] The grounds for condemnation may consist in the defendants' having harmed others, but they also may consist in their having flouted custom, questioned authority, undermined tribal solidarity, or engaged in unclean sexual or dietary practices. People see violence as moral, not immoral: across the world and throughout history, more people have been murdered to mete out justice than to satisfy greed.[28]

~

But we're not all bad. Human cognition comes with two features that give it the means to transcend its limitations.[29] The first is abstraction. People can co-opt their concept of an object at a place and use it to conceptualize an entity in a circumstance, as when we take the pattern of a thought like *The deer ran from the pond to the hill* and apply it to *The child went from sick to well*. They can co-opt the concept of an agent exerting physical force and use it to conceptualize other kinds of causation, as when we extend the image in *She forced the door to open* to *She forced Lisa to join her* or *She forced herself to be polite*. These formulas give people the means to think about a variable with a value and about a cause and its effect—just the conceptual machinery one needs to frame theories and laws. They can do this not just with the elements of thought but with more complex assemblies, allowing them to think in metaphors and analogies: heat is a fluid, a message is a container, a society is a family, obligations are bonds.

The second stepladder of cognition is its combinatorial, recursive power. The mind can entertain an explosive variety of ideas by assembling basic concepts like thing, place, path, actor, cause, and goal into propositions. And it can entertain not only propositions, but propositions about the propositions, and propositions about the propositions about the propositions. Bodies contain humors; illness is an imbalance in the humors that bodies contain; I no longer believe the theory that illness is an imbalance in the humors that bodies contain.

Thanks to language, ideas are not just abstracted and combined inside the head of a single thinker but can be pooled across a community of thinkers. Thomas Jefferson explained the power of language with the help of an analogy: "He who receives an idea from me, receives instruction himself without lessening mine; as he who lights his taper at mine, receives light without darkening me."[30] The potency of language as the original sharing app was multiplied by the invention of writing (and again in later epochs by the printing press, the spread of literacy, and electronic media). The networks of communicating thinkers expanded over time as populations grew, mixed, and became concentrated in cities. And the availability of energy beyond the minimum needed for survival gave more of them the luxury to think and talk.

When large and connected communities take shape, they can come up with ways of organizing their affairs that work to their members' mutual advantage. Though everyone wants to be right, as soon as people start to air their incompatible views it becomes clear that not everyone can be right about everything. Also, the desire to be right can collide with a second desire, to know the truth, which is uppermost in the minds of bystanders to an argument who are not invested in which side wins. Communities can thereby come up with rules that allow true beliefs to emerge from the rough-and-tumble of argument, such as that you have to provide reasons for your beliefs, you're allowed to point out flaws in the beliefs of others, and you're not allowed to forcibly shut people up who disagree with you. Add in the rule that you should allow the world to show you whether your beliefs are true or false, and we can call the rules science. With the right rules, a community of less than fully rational thinkers can cultivate rational thoughts.[31]

The wisdom of crowds can also elevate our moral sentiments. When a wide enough circle of people confer on how best to treat each other, the conversation is bound to go in certain directions. If my starting offer is "I get to rob, beat, enslave, and kill you and your kind, but you don't get to rob, beat, enslave, or kill me or my kind," I can't expect you to agree to the deal or third parties to ratify it, because there's no good reason that I should get privileges just because I'm me and you're not.[32] Nor are we likely to agree to the deal "I get to rob, beat, enslave, and kill you and your kind, and you get to rob, beat, enslave, and kill me and my kind," despite its symmetry, because the advantages either of us might get in harming the other are massively outweighed by the disadvantages we would suffer in being harmed (yet another implication of the Law of Entropy: harms are easier to inflict and have larger effects than benefits). We'd be wiser to negotiate a social contract that puts us in a positive-sum game: neither gets to harm the other, and both are encouraged to help the other.

So for all the flaws in human nature, it contains the seeds of its own improvement, as long as it comes up with norms and institutions that channel parochial interests into universal benefits. Among those norms are free speech, nonviolence, cooperation, cosmopolitanism, human rights, and an acknowledgment of human fallibility, and among the institutions are science, education, media, democratic government, international organizations, and markets. Not coincidentally, these were the major brainchildren of the Enlightenment.

CHAPTER 3

COUNTER-ENLIGHTENMENTS

W ho could be against reason, science, humanism, or progress? The words seem saccharine, the ideals unexceptionable. They define the missions of all the institutions of modernity— schools, hospitals, charities, news agencies, democratic governments, international organizations. Do these ideals really need a defense?

They absolutely do. Since the 1960s, trust in the institutions of modernity has sunk, and the second decade of the 21st century saw the rise of populist movements that blatantly repudiate the ideals of the Enlightenment.[1] They are tribalist rather than cosmopolitan, authoritarian rather than democratic, contemptuous of experts rather than respectful of knowledge, and nostalgic for an idyllic past rather than hopeful for a better future. But these reactions are by no means confined to 21st-century political populism (a movement we will examine in chapters 20 and 23). Far from sprouting from the grass roots or channeling the anger of know-nothings, the disdain for reason, science, humanism, and progress has a long pedigree in elite intellectual and artistic culture.

Indeed, a common criticism of the Enlightenment project—that it is a Western invention, unsuited to the world in all its diversity—is doubly wrongheaded. For one thing, all ideas have to come from somewhere, and their birthplace has no bearing on their merit. Though many Enlightenment ideas were articulated in their clearest and most influential form in 18th-century Europe and America, they are rooted in reason and human nature, so any reasoning human can engage with them. That's why Enlightenment ideals have been articulated in non-Western civilizations at many times in history.[2]

But my main reaction to the claim that the Enlightenment is the guiding ideal of the West is: If only! The Enlightenment was swiftly followed by a counter-Enlightenment, and the West has been divided ever since.[3]

No sooner did people step into the light than they were advised that
darkness wasn't so bad after all, that they should stop daring to understand so much, that dogmas and formulas deserved another chance, and
that human nature's destiny was not progress but decline.

The Romantic movement pushed back particularly hard against Enlightenment ideals. Rousseau, Johann Herder, Friedrich Schelling, and
others denied that reason could be separated from emotion, that individuals could be considered apart from their culture, that people should
provide reasons for their acts, that values applied across times and
places, and that peace and prosperity were desirable ends. A human is a
part of an organic whole—a culture, race, nation, religion, spirit, or historical force—and people should creatively channel the transcendent
unity of which they are a part. Heroic struggle, not the solving of problems, is the greatest good, and violence is inherent to nature and cannot
be stifled without draining life of its vitality. "There are but three groups
worthy of respect," wrote Charles Baudelaire, "the priest, the warrior,
and the poet. To know, to kill, and to create."

It sounds mad, but in the 21st century those counter-Enlightenment
ideals continue to be found across a surprising range of elite cultural and
intellectual movements. The notion that we should apply our collective
reason to enhance flourishing and reduce suffering is considered crass,
naïve, wimpy, square. Let me introduce some of the popular alternatives
to reason, science, humanism, and progress; they will reappear in other
chapters, and in part III of the book I will confront them head on.

The most obvious is religious faith. To take something on faith means
to believe it without good reason, so by definition a faith in the existence
of supernatural entities clashes with reason. Religions also commonly
clash with humanism whenever they elevate some moral good above the
well-being of humans, such as accepting a divine savior, ratifying a sacred narrative, enforcing rituals and taboos, proselytizing other people
to do the same, and punishing or demonizing those who don't. Religions
can also clash with humanism by valuing *souls* above *lives*, which is not
as uplifting as it sounds. Belief in an afterlife implies that health and
happiness are not such a big deal, because life on earth is an infinitesimal
portion of one's existence; that coercing people into accepting salvation
is doing them a favor; and that martyrdom may be the best thing that can
ever happen to you. As for incompatibilities with science, these are the
stuff of legend and current events, from Galileo and the Scopes Monkey
Trial to stem-cell research and climate change.

A second counter-Enlightenment idea is that people are the expendable cells of a superorganism—a clan, tribe, ethnic group, religion, race, class, or nation—and that the supreme good is the glory of this collectivity rather than the well-being of the people who make it up. An obvious example is nationalism, in which the superorganism is the nation-state, namely an ethnic group with a government. We see the clash between nationalism and humanism in morbid patriotic slogans like "Dulce et decorum est pro patria mori" (Sweet and right it is to die for your country) and "Happy those who with a glowing faith in one embrace clasped death and victory."[4] Even John F. Kennedy's less gruesome "Ask not what your country can do for you; ask what you can do for your country" makes the tension clear.

Nationalism should not be confused with civic values, public spirit, social responsibility, or cultural pride. Humans are a social species, and the well-being of every individual depends on patterns of cooperation and harmony that span a community. When a "nation" is conceived as a tacit social contract among people sharing a territory, like a condominium association, it is an essential means for advancing its members' flourishing. And of course it is genuinely admirable for one individual to sacrifice his or her interests for those of many individuals. It's quite another thing when a person is forced to make the supreme sacrifice for the benefit of a charismatic leader, a square of cloth, or colors on a map. Nor is it sweet and right to clasp death in order to prevent a province from seceding, expand a sphere of influence, or carry out an irredentist crusade.

Religion and nationalism are signature causes of political conservatism, and continue to affect the fate of billions of people in the countries under their influence. Many left-wing colleagues who learned that I was writing a book on reason and humanism egged me on, relishing the prospect of an arsenal of talking points against the right. But not so long ago the left was sympathetic to nationalism when it was fused with Marxist liberation movements. And many on the left encourage identity politicians and social justice warriors who downplay individual rights in favor of equalizing the standing of races, classes, and genders, which they see as being pitted in zero-sum competition.

Religion, too, has defenders on both halves of the political spectrum. Even writers who are unwilling to defend the literal content of religious beliefs may be fiercely defensive of religion and hostile to the idea that science and reason have anything to say about morality (most of them show little awareness that humanism even exists).[5] Defenders of the faith insist that religion has the exclusive franchise for questions about what matters. Or that even if we sophisticated people don't need religion to be moral, the teeming masses do. Or that even if everyone would be better off without religious faith, it's pointless to talk about the place of religion in the world because religion is a part of human nature, which is why, mocking Enlightenment hopes, it is more tenacious than ever. In chapter 23 I will examine all these claims.

The left tends to be sympathetic to yet another movement that subordinates human interests to a transcendent entity, the ecosystem. The romantic Green movement sees the human capture of energy not as a way of resisting entropy and enhancing human flourishing but as a heinous crime against nature, which will exact a dreadful justice in the form of resource wars, poisoned air and water, and civilization-ending climate change. Our only salvation is to repent, repudiate technology and economic growth, and revert to a simpler and more natural way of life. Of course, no informed person can deny that damage to natural systems from human activity has been harmful and that if we do nothing about it the damage could become catastrophic. The question is whether a complex, technologically advanced society *is* condemned to do nothing about it. In chapter 10 we will explore a humanistic environmentalism, more Enlightened than Romantic, sometimes called ecomodernism or ecopragmatism.[6]

Left-wing and right-wing political ideologies have themselves become secular religions, providing people with a community of like-minded brethren, a catechism of sacred beliefs, a well-populated demonology, and a beatific confidence in the righteousness of their cause. In chapter 21 we will see how political ideology undermines reason and science.[7] It scrambles people's judgment, inflames a primitive tribal mindset, and distracts them from a sounder understanding of how to improve the world. Our greatest enemies are ultimately not our political adversaries but entropy, evolution (in the form of pestilence and the flaws in human nature), and most of all ignorance—a shortfall of knowledge of how best to solve our problems.

The last two counter-Enlightenment movements cut across the left–right divide. For almost two centuries, a diverse array of writers has proclaimed that modern civilization, far from enjoying progress, is in steady decline and on the verge of collapse. In *The Idea of Decline in Western History,* the historian Arthur Herman recounts two centuries of doomsayers who have sounded the alarm of racial, cultural, political, or ecological degeneration. Apparently the world has been coming to an end for a long time indeed.[8]

One form of declinism bemoans our Promethean dabbling with tech-nology.[9] By wresting fire from the gods, we have only given our species the means to end its own existence, if not by poisoning our environment then by loosing nuclear weapons, nanotechnology, cyberterror, bioterror, artificial intelligence, and other existential threats upon the world (chap-ter 19). And even if our technological civilization manages to escape out-right annihilation, it is spiraling into a dystopia of violence and injustice: a brave new world of terrorism, drones, sweatshops, gangs, trafficking, refugees, inequality, cyberbullying, sexual assault, and hate crimes.

Another variety of declinism agonizes about the opposite problem—not that modernity has made life too harsh and dangerous, but that it has made it too pleasant and safe. According to these critics, health, peace, and prosperity are bourgeois diversions from what truly matters in life. In serving up these philistine pleasures, technological capitalism has only damned people to an atomized, conformist, consumerist, material-ist, other-directed, rootless, routinized, soul-deadening wilderness. In this absurd existence, people suffer from alienation, angst, anomie, apa-thy, bad faith, ennui, malaise, and nausea; they are "hollow men eating their naked lunches in the wasteland while waiting for Godot."[10] (I will examine these claims in chapters 17 and 18.) In the twilight of a decadent, degenerate civilization, true liberation is to be found not in sterile ratio-nality or effete humanism but in an authentic, heroic, holistic, organic, sacred, vital being-in-itself and will to power. In case you are wondering what this sacred heroism consists of, Friedrich Nietzsche, who coined the term *will to power*, recommends the aristocratic violence of the "blond Teuton beasts" and the samurai, Vikings, and Homeric heroes: "hard, cold, terrible, without feelings and without conscience, crushing every-thing, and bespattering everything with blood."[11] (We'll take a closer look at this morality in the final chapter.)

Herman notes that the intellectuals and artists who foresee the col-lapse of civilization react to their prophecy in either of two ways. The historical pessimists dread the downfall but lament that we are power-less to stop it. The cultural pessimists welcome it with a "ghoulish schadenfreude." Modernity is so bankrupt, they say, that it cannot be improved, only transcended. Out of the rubble of its collapse, a new or-der will emerge that can only be superior.

A final alternative to Enlightenment humanism condemns its em-brace of science. Following C. P. Snow, we can call it the Second Culture, the worldview of many literary intellectuals and cultural critics, as dis-tinguished from the First Culture of science.[12] Snow decried the iron curtain between the two cultures and called for a greater integration of science into intellectual life. It was not just that science was, "in its intellectual depth, complexity, and articulation, the most beautiful and wonderful collective work of the mind of man."[13] Knowledge of science, he argued, was a moral imperative, because it could alleviate suffering on a global scale by curing disease, feeding the hungry, saving the lives of infants and mothers, and allowing women to control their fertility.

Though Snow's argument seems prescient today, a famous 1962 rebuttal from the literary critic F. R. Leavis was so vituperative that *The Spectator* had to ask Snow to promise not to sue for libel before they would publish it.[14] After noting Snow's "utter lack of intellectual distinction and . . . embarrassing vulgarity of style," Leavis scoffed at a value system in which " 'standard of living' is the ultimate criterion, its raising an ultimate aim."[15] As an alternative, he suggested that "in coming to terms with great literature we discover what at bottom we really believe. What for—what ultimately for? What do men live by?—the questions work and tell at what I can only call a religious depth of thought and feeling." (Anyone whose "depth of thought and feeling" extends to a woman in a poor country who has lived to see her newborn because her standard of living has risen, and then multiplied that sympathy by a few hundred million, might wonder why "coming to terms with great literature" is morally superior to "raising the standard of living" as a criterion for "what at bottom we really believe"—or why the two should be seen as alternatives in the first place.)

As we shall see in chapter 22, Leavis's outlook may be found in a wide swath of the Second Culture today. Many intellectuals and critics express a disdain for science as anything but a fix for mundane problems. They write as if the consumption of elite art is the ultimate moral good. Their methodology for seeking the truth consists not in framing hypotheses and citing evidence but in issuing pronouncements that draw on their breadth of erudition and lifetime habits of reading. Intellectual magazines regularly denounce "scientism," the intrusion of science into the territory of the humanities such as politics and the arts. In many colleges and universities, science is presented not as the pursuit of true explanations but as just another narrative or myth. Science is commonly blamed for racism, imperialism, world wars, and the Holocaust. And it is accused of robbing life of its enchantment and stripping humans of freedom and dignity.

Enlightenment humanism, then, is far from being a crowd-pleaser. The idea that the ultimate good is to use knowledge to enhance human

COUNTER-ENLIGHTENMENTS 35

welfare leaves people cold. Deep explanations of the universe, the planet, life, the brain? Unless they use magic, we don't want to believe them! Saving the lives of billions, eradicating disease, feeding the hungry? *Bo*-ring. People extending their compassion to all of humankind? Not good enough—we want *the laws of physics* to care about us! Longevity, health, understanding, beauty, freedom, love? There's got to be more to life than that!

But it's the idea of progress that sticks most firmly in the craw. Even people who think it is a fine idea in theory to use knowledge to improve well-being insist it will never work in practice. And the daily news offers plenty of support for their cynicism: the world is depicted as a vale of tears, a tale of woe, a slough of despond. Since any defense of reason, science, and humanism would count for nothing if, two hundred and fifty years after the Enlightenment, we're no better off than our ancestors in the Dark Ages, an appraisal of human progress is where the case must begin.

PART II
PROGRESS

If you had to choose a moment in history to be born, and you did not know ahead of time who you would be—you didn't know whether you were going to be born into a wealthy family or a poor family, what country you'd be born in, whether you were going to be a man or a woman—if you had to choose blindly what moment you'd want to be born, you'd choose now.

—Barack Obama, 2016

CHAPTER 4

PROGRESSOPHOBIA

Intellectuals hate progress. Intellectuals who call themselves "progressive" *really* hate progress. It's not that they hate the *fruits* of progress, mind you: most pundits, critics, and their *bien-pensant* readers use computers rather than quills and inkwells, and they prefer to have their surgery with anesthesia rather than without it. It's the *idea* of progress that rankles the chattering class—the Enlightenment belief that by understanding the world we can improve the human condition.

An entire lexicon of abuse has grown up to express their scorn. If you think knowledge can help solve problems, then you have a "blind faith" and a "quasi-religious belief" in the "outmoded superstition" and "false promise" of the "myth" of the "onward march" of "inevitable progress." You are a "cheerleader" for "vulgar American can-doism" with the "rah-rah" spirit of "boardroom ideology," "Silicon Valley," and the "Chamber of Commerce." You are a practitioner of "Whig history," a "naïve optimist," a "Pollyanna," and of course a "Pangloss," a modern-day version of the philosopher in Voltaire's *Candide* who asserts that "all is for the best in the best of all possible worlds."

Professor Pangloss, as it happens, is what we would now call a pessimist. A modern optimist believes that the world can be *much, much* better than it is today. Voltaire was satirizing not the Enlightenment hope for progress but its opposite, the religious rationalization for suffering called theodicy, according to which God had no choice but to allow epidemics and massacres because a world without them is metaphysically impossible.

Epithets aside, the idea that the world is better than it was and can get better still fell out of fashion among the clerisy long ago. In *The Idea of Decline in Western History*, Arthur Herman shows that prophets of doom are the all-stars of the liberal arts curriculum, including Nietzsche, Arthur Schopenhauer, Martin Heidegger, Theodor Adorno, Walter Benjamin, Herbert Marcuse, Jean-Paul Sartre, Frantz Fanon, Michel Foucault, Edward Said, Cornel West, and a chorus of eco-pessimists.[1] Surveying the intellectual landscape at the end of the 20th century, Herman lamented a "grand recessional" of "the luminous exponents" of Enlightenment humanism, the ones who believed that "since people generate conflicts and problems in society, they can also resolve them." In *History of the Idea of Progress*, the sociologist Robert Nisbet agreed: "The skepticism regarding Western progress that was once confined to a very small number of intellectuals in the nineteenth century has grown and spread to not merely the large majority of intellectuals in this final quarter of the century, but to many millions of other people in the West."[2]

Yes, it's not just those who intellectualize for a living who think the world is going to hell in a handcart. It's ordinary people when they switch into intellectualizing mode. Psychologists have long known that people tend to see their own lives through rose-colored glasses: they think they're less likely than the average person to become the victim of a divorce, layoff, accident, illness, or crime. But change the question from the people's *lives* to their *society*, and they transform from Pollyanna to Eeyore.

Public opinion researchers call it the Optimism Gap.[3] For more than two decades, through good times and bad, when Europeans were asked by pollsters whether their *own* economic situation would get better or worse in the coming year, more of them said it would get better, but when they were asked about their *country's* economic situation, more of them said it would get worse.[4] A large majority of Britons think that immigration, teen pregnancy, litter, unemployment, crime, vandalism, and drugs are a problem in the United Kingdom as a whole, while few think they are problems in their area.[5] Environmental quality, too, is judged in most nations to be worse in the nation than in the community, and worse in the world than in the nation.[6] In almost every year from 1992 through 2015, an era in which the rate of violent crime plummeted, a majority of Americans told pollsters that crime was rising.[7] In late 2015, large majorities in eleven developed countries said that "the world is getting worse," and in most of the last forty years a solid majority of Americans have said that the country is "heading in the wrong direction."[8]

Are they right? Is pessimism correct? Could the state of the world, like the stripes on a barbershop pole, keep sinking lower and lower? It's easy to see why people feel that way: every day the news is filled with stories about war, terrorism, crime, pollution, inequality, drug abuse, and oppression. And it's not just the headlines we're talking about; it's the op-eds and long-form stories as well. Magazine covers warn us of coming anarchies, plagues, epidemics, collapses, and so many "crises" (farm, health, retirement, welfare, energy, deficit) that copywriters have had to escalate to the redundant "serious crisis."

Whether or not the world really is getting worse, the nature of news will interact with the nature of cognition to make us think that it is. News is about things that happen, not things that don't happen. We never see a journalist saying to the camera, "I'm reporting live from a country where a war has not broken out"—or a city that has not been bombed, or a school that has not been shot up. As long as bad things have not vanished from the face of the earth, there will always be enough incidents to fill the news, especially when billions of smartphones turn most of the world's population into crime reporters and war correspondents.

And among the things that do happen, the positive and negative ones unfold on different time lines. The news, far from being a "first draft of history," is closer to play-by-play sports commentary. It focuses on discrete events, generally those that took place since the last edition (in earlier times, the day before; now, seconds before).[9] Bad things can happen quickly, but good things aren't built in a day, and as they unfold, they will be out of sync with the news cycle. The peace researcher Johan Galtung pointed out that if a newspaper came out once every fifty years, it would not report half a century of celebrity gossip and political scandals. It would report momentous global changes such as the increase in life expectancy.[10]

The nature of news is likely to distort people's view of the world because of a mental bug that the psychologists Amos Tversky and Daniel Kahneman called the Availability heuristic: people estimate the probability of an event or the frequency of a kind of thing by the ease with which instances come to mind.[11] In many walks of life this is a serviceable rule of thumb. Frequent events leave stronger memory traces, so stronger memories generally indicate more-frequent events: you really are on solid ground in guessing that pigeons are more common in cities than orioles, even though you're drawing on your memory of encountering them rather than on a bird census. But whenever a memory turns up high in the result list of the mind's search engine for reasons other than frequency—because it is recent, vivid, gory, distinctive, or upsetting—people will overestimate how likely it is in the world. Which are more numerous in the English language, words that begin with *k* or words with *k* in the third position? Most people say the former. In fact, there are three times as many words with *k* in the third position (*ankle, ask, awkward, bake, cake, make, take* . . .), but we retrieve words by their initial sounds, so *keep, kind, kill, kid,* and *king* are likelier to pop into mind on demand.

Availability errors are a common source of folly in human reasoning. First-year medical students interpret every rash as a symptom of an exotic disease, and vacationers stay out of the water after they have read about a shark attack or if they have just seen *Jaws*.[12] Plane crashes always make the news, but car crashes, which kill far more people, almost never do. Not surprisingly, many people have a fear of flying, but almost no one has a fear of driving. People rank tornadoes (which kill about fifty Americans a year) as a more common cause of death than asthma (which kills more than four thousand Americans a year), presumably because tornadoes make for better television.

It's easy to see how the Availability heuristic, stoked by the news policy "If it bleeds, it leads," could induce a sense of gloom about the state of the world. Media scholars who tally news stories of different kinds, or present editors with a menu of possible stories and see which they pick and how they display them, have confirmed that the gatekeepers prefer negative to positive coverage, holding the events constant.[13] That in turn provides an easy formula for pessimists on the editorial page: make a list of all the worst things that are happening anywhere on the planet that week, and you have an impressive-sounding case that civilization has never faced greater peril.

The consequences of negative news are themselves negative. Far from being better informed, heavy newswatchers can become miscalibrated. They worry more about crime, even when rates are falling, and sometimes they part company with reality altogether: a 2016 poll found that a large majority of Americans follow news about ISIS closely, and 77 percent agreed that "Islamic militants operating in Syria and Iraq pose a serious threat to the existence or survival of the United States," a belief that is nothing short of delusional.[14] Consumers of negative news, not surprisingly, become glum: a recent literature review cited "misperception of risk, anxiety, lower mood levels, learned helplessness, contempt and hostility towards others, desensitization, and in some cases, . . . complete avoidance of the news."[15] And they become fatalistic, saying things like "Why should I vote? It's not gonna help," or "I could donate money, but there's just gonna be another kid who's starving next week."[16]

Seeing how journalistic habits and cognitive biases bring out the worst in each other, how can we soundly appraise the state of the world?

The answer is to *count*. How many people are victims of violence as a proportion of the number of people alive? How many are sick, how many starving, how many poor, how many oppressed, how many illiterate, how many unhappy? And are those numbers going up or down? A quantitative mindset, despite its nerdy aura, is in fact the morally enlightened one, because it treats every human life as having equal value rather than privileging the people who are closest to us or most photogenic. And it holds out the hope that we might identify the causes of suffering and thereby know which measures are most likely to reduce it.

That was the goal of my 2011 book *The Better Angels of Our Nature*, which presented a hundred graphs and maps showing how violence and the conditions that foster it have declined over the course of history. To emphasize that the declines took place at different times and had different causes, I gave them names. The Pacification Process was a fivefold reduction in the rate of death from tribal raiding and feuding, the consequence of effective states exerting control over a territory. The Civilizing Process was a fortyfold reduction in homicide and other violent crimes which followed upon the entrenchment of the rule of law and norms of self-control in early modern Europe. The Humanitarian Revolution is another name for the Enlightenment-era abolition of slavery, religious persecution, and cruel punishments. The Long Peace is the historians' term for the decline of great-power and interstate war after World War II. Following the end of the Cold War, the world has enjoyed a New Peace with fewer civil wars, genocides, and autocracies. And since the 1950s the world has been swept by a cascade of Rights Revolutions: civil rights, women's rights, gay rights, children's rights, and animal rights.

Few of these declines are contested among experts who are familiar with the numbers. Historical criminologists, for example, agree that homicide plummeted after the Middle Ages, and it's a commonplace among international-relations scholars that major wars tapered off after 1945. But they come as a surprise to most people in the wider world.[17]

I had thought that a parade of graphs with time on the horizontal axis, body counts or other measures of violence on the vertical, and a line that meandered from the top left to the bottom right would cure audiences of the Availability bias and persuade them that at least in this sphere of well-being the world has made progress. But I learned from their questions and objections that resistance to the idea of progress runs deeper than statistical fallacies. Of course, any dataset is an imperfect reflection of reality, so it is legitimate to question how accurate and representative the numbers truly are. But the objections revealed not just a skepticism about the data but also an unpreparedness for the *possibility* that the human condition has improved. Many people lack the conceptual tools to ascertain whether progress has taken place or not; the very idea that things can get better just doesn't compute. Here are stylized versions of dialogues I have often had with questioners.

So violence has declined linearly since the beginning of history! Awesome!

No, not "linearly"—it would be astonishing if any measure of human behavior with all its vicissitudes ticked downward by a constant amount per unit of time, decade after decade and century after century. And not monotonically, either (which is probably what the questioners have in mind)—that would mean that it always decreased or stayed the same, never increased. Real historical curves have wiggles, upticks, spikes, and sometimes sickening lurches. Examples include the two world wars, a boom in crime in Western countries from the mid-1960s to the early 1990s, and a bulge of civil wars in the developing world following decolonization in the 1960s and 1970s. Progress consists of trends in violence on which these fluctuations are superimposed—a downward swoop or drift, a return from a temporary swelling to a low baseline. Progress cannot always be monotonic because solutions to problems create new problems.[18] But progress can resume when the new problems are solved in their turn.

By the way, the nonmonotonicity of social data provides an easy formula for news outlets to accentuate the negative. If you ignore all the years in which an indicator of some problem declines, and report every uptick (since, after all, it's "news"), readers will come away with the impression that life is getting worse and worse even as it gets better and better. In the first six months of 2016 the *New York Times* pulled this trick three times, with figures for suicide, longevity, and automobile fatalities.

Well, if levels of violence don't always go down, that means they're cyclical, so even if they're low right now it's only a matter of time before they go back up.

No, changes over time may be *statistical*, with unpredictable fluctuations, without being *cyclical*, namely oscillating like a pendulum between two extremes. That is, even if a reversal is possible at any time, that does not mean it becomes more likely as time passes. (Many investors have lost their shirts betting on a misnamed "business cycle" that in fact consists of unpredictable swings.) Progress can take place when the reversals in a positive trend become less frequent, become less severe, or, in some cases, cease altogether.

How can you say that violence has decreased? Didn't you read about the school shooting (or terrorist bombing, or artillery shelling, or soccer riot, or barroom stabbing) in the news this morning?

A decline is not the same thing as a disappearance. (The statement "$x > y$" is different from the statement "$y = 0$.") Something can decrease a lot without vanishing altogether. That means that the level of violence today is *completely irrelevant* to the question of whether violence has declined over the course of history. The only way to answer that question is to compare the level of violence now with the level of violence in the past. And whenever you look at the level of violence in the past, you find a lot of it, even if it isn't as fresh in memory as the morning's headlines.

All your fancy statistics about violence going down don't mean anything if you're one of the victims.

True, but they do mean that you're less likely to *be* a victim. For that reason they mean the world to the millions of people who are not victims but would have been if rates of violence had stayed the same.

So you're saying that we can all sit back and relax, that violence will just take care of itself.

Illogical, Captain. If you see that a pile of laundry has gone down, it does not mean the clothes washed themselves; it means someone washed the clothes. If a type of violence has gone down, then some change in the social, cultural, or material milieu has caused it to go down. If the conditions persist, violence could remain low or decline even further; if they don't, it won't. That makes it important to find out what the causes are, so we can try to intensify them and apply them more widely to ensure that the decline of violence continues.

To say that violence has gone down is to be naïve, sentimental, idealistic, romantic, starry-eyed, Whiggish, utopian, a Pollyanna, a Pangloss.

No, to look at data showing that violence has gone down and say "Violence has gone down" is to describe a fact. To look at data showing that violence has gone down and say "Violence has gone up" is to be delusional. To ignore data on violence and say "Violence has gone up" is to be a know-nothing.

As for accusations of romanticism, I can reply with some confidence. I am also the author of the staunchly unromantic, anti-utopian *The Blank Slate: The Modern Denial of Human Nature,* in which I argued that human beings are fitted by evolution with a number of destructive motives such as greed, lust, dominance, vengeance, and self-deception. But I believe that people are also fitted with a sense of sympathy, an ability to reflect on their predicament, and faculties to think up and share new ideas—the better angels of our nature, in the words of Abraham Lincoln. Only by looking at the facts can we tell to what extent our better angels have prevailed over our inner demons at a given time and place.

How can you predict that violence will keep going down? Your theory could be refuted by a war breaking out tomorrow.

A statement that some measure of violence has gone down is not a "theory" but an observation of a fact. And yes, the fact that a measure has changed over time is not the same as a prediction that it will continue to change in that way at all times forever. As the investment ads are required to say, past performance is no guarantee of future results.

In that case, what good are all those graphs and analyses? Isn't a scientific theory supposed to make testable predictions?

A scientific theory makes predictions in *experiments* in which the causal influences are controlled. No theory can make a prediction about the world at large, with its seven billion people spreading viral ideas in global networks and interacting with chaotic cycles of weather and resources. To declare what the future holds in an uncontrollable world, and without an explanation of why events unfold as they do, is not prediction but *prophecy*, and as David Deutsch observes, "The most important of all limitations on knowledge-creation is that we cannot prophesy: we cannot predict the content of ideas yet to be created, or their effects. This limitation is not only consistent with the unlimited growth of knowledge, it is entailed by it."[19]

Our inability to prophesy is not, of course, a license to ignore the facts. An improvement in some measure of human well-being suggests that, overall, more things have pushed in the right direction than in the wrong direction. Whether we should expect progress to continue depends on whether we know what those forces are and how long they will remain in place. That will vary from trend to trend. Some may turn out to be like Moore's Law (the number of transistors per computer chip doubles every two years) and give grounds for confidence (though not certainty) that the fruits of human ingenuity will accumulate and progress will continue. Some may be like the stock market and foretell short-term fluctuations but long-term gains. Some of these may reel in a statistical distribution with a "thick tail," in which extreme events, even if less likely, cannot be ruled out.[20] Still others may be cyclical or chaotic. In chapters 19 and 21 we will examine rational forecasting in an uncertain world. For now we should keep in mind that a positive trend suggests (but does not prove) that we have been doing something right, and that we should seek to identify what it is and do more of it.

When all these objections are exhausted, I often see people racking their brains to find *some* way in which the news cannot be as good as the data suggest. In desperation, they turn to semantics.

Isn't Internet trolling a form of violence? Isn't strip-mining a form of violence? Isn't inequality a form of violence? Isn't pollution a form of violence? Isn't poverty a form of violence? Isn't consumerism a form of violence? Isn't divorce a form of violence? Isn't advertising a form of violence? Isn't keeping statistics on violence a form of violence?

As wonderful as metaphor is as a rhetorical device, it is a poor way to assess the state of humanity. Moral reasoning requires proportionality. It may be upsetting when someone says mean things on Twitter, but it is not the same as the slave trade or the Holocaust. It also requires distinguishing rhetoric from reality. Marching into a rape crisis center and demanding to know what they have done about the rape of the environment does nothing for rape victims and nothing for the environment. Finally, improving the world requires an understanding of cause and effect. Though primitive moral intuitions tend to lump bad things together and find a villain to blame them on, there is no coherent phenomenon of "bad things" that we can seek to understand and eliminate. (Entropy and evolution will generate them in profusion.) War, crime, pollution, poverty, disease, and incivility are evils that may have little in common, and if we want to reduce them, we can't play word games that make it impossible even to discuss them individually.

~

I have run through these objections to prepare the way for my presentation of other measures of human progress. The incredulous reaction to *Better Angels* convinced me that it isn't just the Availability heuristic that makes people fatalistic about progress. Nor can the media's fondness for bad news be blamed entirely on a cynical chase for eyeballs and clicks. No, the psychological roots of progressophobia run deeper.

The deepest is a bias that has been summarized in the slogan "Bad is stronger than good."[21] The idea can be captured in a set of thought experiments suggested by Tversky.[22] How much better can you imagine yourself feeling than you are feeling right now? How much *worse* can you imagine yourself feeling? In answering the first hypothetical, most of us can imagine a bit more of a spring in our step or a twinkle in our eye, but the answer to the second one is: it's bottomless. This asymmetry in mood can be explained by an asymmetry in life (a corollary of the Law of Entropy). How many things could happen to you today that would leave you much better off? How many things could happen that would leave you much *worse* off? Once again, to answer the first question, we can all come up with the odd windfall or stroke of good luck, but the answer to the second one is: it's endless. But we needn't rely on our imaginations. The psychological literature confirms that people dread losses more than they look forward to gains, that they dwell on setbacks more than they savor good fortune, and that they are more stung by criticism than they are heartened by praise. (As a psycholinguist I am compelled to add that the English language has far more words for negative emotions than for positive ones.)[23]

One exception to the Negativity bias is found in autobiographical memory. Though we tend to remember bad events as well as we remember good ones, the negative coloring of the misfortunes fades with time, particularly the ones that happened to us.[24] We are wired for nostalgia: in human memory, time heals most wounds. Two other illusions mislead us into thinking that things ain't what they used to be: we mistake the growing burdens of maturity and parenthood for a less innocent world, and we mistake a decline in our own faculties for a decline in the times.[25] As the columnist Franklin Pierce Adams pointed out, "Nothing is more responsible for the good old days than a bad memory."

Intellectual culture should strive to counteract our cognitive biases, but all too often it reinforces them. The cure for the Availability bias is quantitative thinking, but the literary scholar Steven Connor has noted that "there is in the arts and humanities an exceptionless consensus about the encroaching horror of the domain of number."[26] This "ideological rather than accidental innumeracy" leads writers to notice, for example, that wars take place today and wars took place in the past and to conclude that "nothing has changed"—failing to acknowledge the difference between an era with a handful of wars that collectively kill in the thousands and an era with dozens of wars that collectively killed in the millions. And it leaves them unappreciative of systemic processes that eke out incremental improvements over the long term.

Nor is intellectual culture equipped to treat the Negativity bias. Indeed, our vigilance for bad things around us opens up a market for professional curmudgeons who call our attention to bad things we may have missed. Experiments have shown that a critic who pans a book is perceived as more competent than a critic who praises it, and the same may be true of critics of society.[27] "Always predict the worst, and you'll be hailed as a prophet," the musical humorist Tom Lehrer once advised. At least since the time of the Hebrew prophets, who blended their social criticism with forewarnings of disaster, pessimism has been equated with moral seriousness. Journalists believe that by accentuating the negative they are discharging their duty as watchdogs, muckrakers, whistleblowers, and afflicters of the comfortable. And intellectuals know they can attain instant gravitas by pointing to an unsolved problem and theorizing that it is a symptom of a sick society.

The converse is true as well. The financial writer Morgan Housel has observed that while pessimists sound like they're trying to help you, optimists sound like they're trying to sell you something.[28] Whenever someone offers a solution to a problem, critics will be quick to point out that it is not a panacea, a silver bullet, a magic bullet, or a one-size-fits-all solution; it's just a Band-Aid or a quick technological fix that fails to get at the root causes and will blow back with side effects and unintended consequences. Of course, since nothing is a panacea and everything has side effects (you can't do just one thing), these common tropes are little more than a refusal to entertain the possibility that anything can ever be improved.[29]

Pessimism among the intelligentsia can also be a form of one-upmanship. A modern society is a league of political, industrial, financial, technological, military, and intellectual elites, all competing for prestige and influence, and with differing responsibilities for making the society run. Complaining about modern society can be a backhanded way of putting down one's rivals—for academics to feel superior to businesspeople, businesspeople to feel superior to politicians, and so on. As Thomas Hobbes noted in 1651, "Competition of praise inclineth to a reverence of antiquity. For men contend with the living, not with the dead."

Pessimism, to be sure, has a bright side. The expanding circle of sympathy makes us concerned about harms that would have passed unnoticed in more callous times. Today we recognize the Syrian civil war as a humanitarian tragedy. The wars of earlier decades, such as the Chinese Civil War, the partition of India, and the Korean War, are seldom remembered that way, though they killed and displaced more people. When I grew up, bullying was considered a natural part of boyhood. It would have strained belief to think that someday the president of the United States would deliver a speech about its evils, as Barack Obama did in 2011. As we care about more of humanity, we're apt to mistake the harms around us for signs of how low the world has sunk rather than how high our standards have risen.

But relentless negativity can itself have unintended consequences, and recently a few journalists have begun to point them out. In the wake of the 2016 American election, the *New York Times* writers David Bornstein and Tina Rosenberg reflected on the media's role in its shocking outcome:

> Trump was the beneficiary of a belief—near universal in American journalism—that "serious news" can essentially be defined as "what's going wrong." . . . For decades, journalism's steady focus on problems and seemingly incurable pathologies was preparing the soil that allowed Trump's seeds of discontent and despair to take root. . . . One consequence is that many Americans today have difficulty imagining, valuing or even believing in the promise of incremental system change, which leads to a greater appetite for revolutionary, smash-the-machine change.[30]

Bornstein and Rosenberg don't blame the usual culprits (cable TV, social media, late-night comedians) but instead trace it to the shift during the Vietnam and Watergate eras from glorifying leaders to checking their power—with an overshoot toward indiscriminate cynicism, in which everything about America's civic actors invites an aggressive takedown.

If the roots of progressophobia lie in human nature, is my suggestion that it is on the rise itself an illusion of the Availability bias? Anticipating the methods I will use in the rest of the book, let's look at an objective measure. The data scientist Kalev Leetaru applied a technique called sentiment mining to every article published in the *New York Times* between 1945 and 2005, and to an archive of translated articles and broadcasts from 130 countries between 1979 and 2010. Sentiment mining assesses the emotional tone of a text by tallying the number and contexts of words with positive and negative connotations, like *good, nice, terrible,* and *horrific.* Figure 4-1 shows the results. Putting aside the wiggles and waves that reflect the crises of the day, we see that the impression that the news has become more negative over time is real. The *New York Times* got steadily more morose from the early 1960s to the early 1970s, lightened up a bit (but just a bit) in the 1980s and 1990s, and then sank into a progressively worse mood in the first decade of the new century. News outlets in the rest of the world, too, became gloomier and gloomier from the late 1970s to the present day.

So has the world really gone steadily downhill during these decades?

Keep figure 4-1 in mind as we examine the state of humanity in the chapters to come.

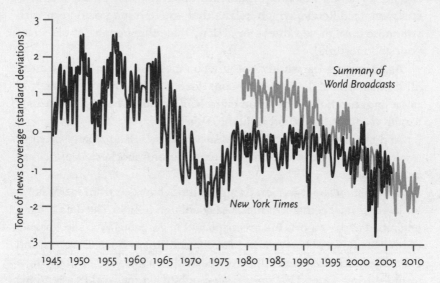

Figure 4-1: Tone of the news, 1945–2010
Source: Leetaru 2011. Plotted by month, beginning in January.

~

What is progress? You might think that the question is so subjective and culturally relative as to be forever unanswerable. In fact, it's one of the easier questions to answer.

Most people agree that life is better than death. Health is better than sickness. Sustenance is better than hunger. Abundance is better than poverty. Peace is better than war. Safety is better than danger. Freedom is better than tyranny. Equal rights are better than bigotry and discrimination. Literacy is better than illiteracy. Knowledge is better than ignorance. Intelligence is better than dull-wittedness. Happiness is better than misery. Opportunities to enjoy family, friends, culture, and nature are better than drudgery and monotony.

All these things can be measured. If they have increased over time, that is progress.

Granted, not everyone would agree on the exact list. The values are avowedly humanistic, and leave out religious, romantic, and aristocratic virtues like salvation, grace, sacredness, heroism, honor, glory, and authenticity. But most would agree that it's a necessary start. It's easy to extoll transcendent values in the abstract, but most people prioritize life,

health, safety, literacy, sustenance, and stimulation for the obvious reason that these goods are a prerequisite to everything else. If you're reading this, you are not dead, starving, destitute, moribund, terrified,
enslaved, or illiterate, which means that you're in no position to turn
your nose up at these values—or to deny that other people should share
your good fortune.

As it happens, the world does agree on these values. In the year 2000,
all 189 members of the United Nations, together with two dozen international organizations, agreed on eight Millennium Development Goals
for the year 2015 that blend right into this list.[31]

And here is a shocker: *The world has made spectacular progress in every
single measure of human well-being.* Here is a second shocker: *Almost no one
knows about it.*

Information about human progress, though absent from major news
outlets and intellectual forums, is easy enough to find. The data are not
entombed in dry reports but are displayed in gorgeous Web sites, particularly Max Roser's *Our World in Data,* Marian Tupy's *HumanProgress,* and
Hans Rosling's *Gapminder.* (Rosling learned that not even swallowing a
sword during a 2007 TED talk was enough to get the world's attention.)
The case has been made in beautifully written books, some by Nobel
laureates, which flaunt the news in their titles—*Progress, The Progress
Paradox, Infinite Progress, The Infinite Resource, The Rational Optimist, The
Case for Rational Optimism, Utopia for Realists, Mass Flourishing, Abundance,
The Improving State of the World, Getting Better, The End of Doom, The Moral
Arc, The Big Ratchet, The Great Escape, The Great Surge, The Great Conver-
gence.*[32] (None was recognized with a major prize, but over the period in
which they appeared, Pulitzers in nonfiction were given to four books
on genocide, three on terrorism, two on cancer, two on racism, and one
on extinction.) And for those whose reading habits tend toward listicles,
recent years have offered "Five Amazing Pieces of Good News Nobody
Is Reporting," "Five Reasons Why 2013 Was the Best Year in Human
History," "Seven Reasons the World Looks Worse Than It Really Is," "26
Charts and Maps That Show the World Is Getting Much, Much Better,"
"40 Ways the World Is Getting Better," and my favorite, "50 Reasons
We're Living Through the Greatest Period in World History." Let's look
at some of those reasons.

CHAPTER 5

LIFE

The struggle to stay alive is the primal urge of animate beings, and humans deploy their ingenuity and conscious resolve to stave off death as long as possible. "Choose life, so that you and your children may live," commanded the God of the Hebrew Bible; "Rage, rage against the dying of the light," adjured Dylan Thomas. A long life is the ultimate blessing.

How long do you think an average person in the world can be expected to live today? Bear in mind that the global average is dragged down by the premature deaths from hunger and disease in the populous countries in the developing world, particularly by the deaths of infants, who mix a lot of zeroes into the average.

The answer for 2015 is 71.4 years.[1] How close is that to your guess? In a recent survey Hans Rosling found that less than one in four Swedes guessed that it was that high, a finding consistent with the results of other multinational surveys of opinions on longevity, literacy, and poverty in what Rosling dubbed the Ignorance Project. The logo of the project is a chimpanzee, because, as Rosling explained, "If for each question I wrote the alternatives on bananas, and asked chimpanzees in the zoo to pick the right answers, they'd have done better than the respondents." The respondents, including students and professors of global health, were not so much ignorant as fallaciously pessimistic.[2]

Figure 5-1, a plot from Max Roser of life expectancy over the centuries, displays a general pattern in world history. At the time when the lines begin, in the mid-18th century, life expectancy in Europe and the Americas was around 35, where it had been parked for the 225 previous years for which we have data.[3] Life expectancy for the world as a whole was 29. These numbers are in the range of expected life spans for most of human history. The life expectancy of hunter-gatherers is around 32.5, and it probably decreased among the peoples who first took up farming because of their starchy diet and the diseases they caught from their livestock and each other. It returned to the low 30s by the Bronze Age, where it stayed put for thousands of years, with small fluctuations across centuries and regions.[4] This period in human history may be called the Malthusian Era, when any advance in agriculture or health was quickly canceled by the resulting bulge in population, though "era" is an odd term for 99.9 percent of our species' existence.

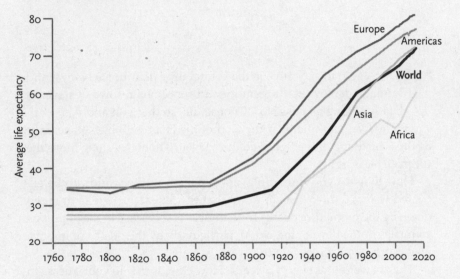

Figure 5-1: Life expectancy, 1771–2015

Sources: *Our World in Data*, Roser 2016n, based on data from Riley 2005 for the years before 2000 and from the World Health Organization and the World Bank for the subsequent years. Updated with data provided by Max Roser.

But starting in the 19th century, the world embarked on the Great Escape, the economist Angus Deaton's term for humanity's release from its patrimony of poverty, disease, and early death. Life expectancy began to rise, picked up speed in the 20th century, and shows no signs of slowing down. As the economic historian Johan Norberg points out, we tend to think that "we approach death by one year for every year we age, but during the twentieth century, the average person approached death by just seven months for every year they aged." Thrillingly, the gift of longevity is spreading to all of humankind, including the world's poorest countries, and at a much faster pace than it did in the rich ones. "Life expectancy in Kenya increased by almost ten years between 2003 and 2013," Norberg writes. "After having lived, loved and struggled for a whole decade, the average person in Kenya had not lost a single year of their remaining lifetime. Everyone got ten years older, yet death had not come a step closer."[5]

As a result, inequality in life expectancy, which opened up during the Great Escape when a few fortunate countries broke away from the pack, is shrinking as the rest catch up. In 1800, no country in the world had a life expectancy above 40. By 1950, it had grown to around 60 in Europe and the Americas, leaving Africa and Asia far behind. But since then Asia has shot up at twice the European rate, and Africa at one and a half times the rate. An African born today can expect to live as long as a person born in the Americas in 1950 or in Europe in the 1930s. The average would have been longer still were it not for the calamity of AIDS, which caused the terrible trough in the 1990s before antiretroviral drugs started to bring it under control.

The African AIDS dip is a reminder that progress is not an escalator that inexorably raises the well-being of every human everywhere all the time. That would be magic, and progress is an outcome not of magic but of problem-solving. Problems are inevitable, and at times particular sectors of humanity have suffered terrible setbacks. In addition to the African AIDS epidemic, longevity went into reverse for young adults worldwide during the Spanish flu pandemic of 1918–19 and for middle-aged, non-college-educated, non-Hispanic white Americans in the early 21st century.[6] But problems are solvable, and the fact that longevity continues to increase in every other Western demographic means that solutions to the problems facing this one exist as well.

Average life spans are stretched the most by decreases in infant and child mortality, both because children are fragile and because the death of a child brings down the average more than the death of a 60-year-old. Figure 5-2 shows what has happened to child mortality since the Age of Enlightenment in five countries that are more or less representative of their continents.

Look at the numbers on the vertical axis: they refer to the percentage of children who die before reaching the age of 5. Yes, well into the 19th century, in Sweden, one of the world's wealthiest countries, between *a quarter and a third* of all children died before their fifth birthday, and in some years the death toll was close to half. This appears to be typical in human history: a fifth of hunter-gatherer children die in their first year, and almost half before they reach adulthood.[7] The spikiness in the curve before the 20th century reflects not just noise in the data but the parlous nature of life: an epidemic, war, or famine could bring death to one's

door at any time. Even the well-to-do could be struck by tragedy: Charles
Darwin lost two children in infancy and his beloved daughter Annie at
the age of 10.

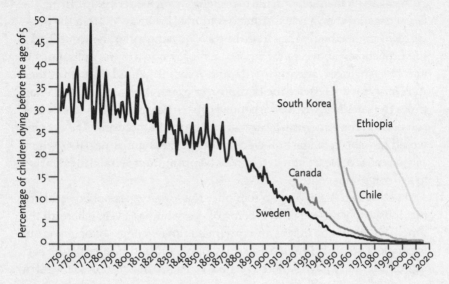

Figure 5-2: Child mortality, 1751–2013

Sources: *Our World in Data,* Roser 2016a, based on data from the UN Child Mortality estimates, http://
www.childmortality.org/, and the *Human Mortality Database,* http://www.mortality.org/.

Then a remarkable thing happened. The rate of child mortality
plunged a hundredfold, to a fraction of a percentage point in developed
countries, and the plunge went global. As Deaton observed in 2013,
"There is not a single country in the world where infant or child mortality today is not lower than it was in 1950."[8] In sub-Saharan Africa, the
child mortality rate has fallen from around one in four in the 1960s to
less than one in ten in 2015, and the global rate has fallen from 18 to
4 percent—still too high, but sure to come down if the current thrust to
improve global health continues.

Remember two facts behind the numbers. One is demographic: when
fewer children die, parents have fewer children, since they no longer
have to hedge their bets against losing their entire families. So contrary
to the worry that saving children's lives would only set off a "population
bomb" (a major eco-panic of the 1960s and 1970s, which led to calls for
reducing health care in the developing world), the decline in child mortality has defused it.[9]

The other is personal. The loss of a child is among the most devastating experiences. Imagine the tragedy; then try to imagine it another million times. That's a quarter of the number of children who did not die *last year alone* who would have died had they been born fifteen years earlier. Now repeat, two hundred times or so, for the years since the decline in child mortality began. Graphs like figure 5-2 display a triumph of human well-being whose magnitude the mind cannot begin to comprehend.

Just as difficult to appreciate is humanity's impending triumph over another of nature's cruelties, the death of a mother in childbirth. The God of the Hebrew Bible, ever merciful, told the first woman, "I will multiply your pain in childbearing; in pain you shall bring forth children." Until recently about one percent of mothers died in the process; for an American woman, being pregnant a century ago was almost as dangerous as having breast cancer today.[10] Figure 5-3 shows the trajectory of maternal mortality since 1751 in four countries that are representative of their regions.

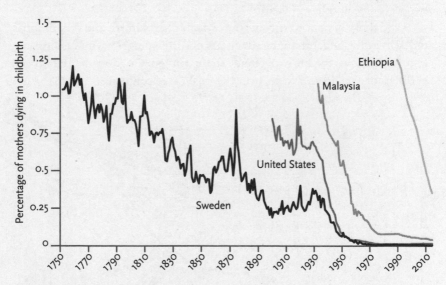

Figure 5-3: Maternal mortality, 1751–2013

Source: *Our World in Data,* Roser 2016p, based partly on data from Claudia Hanson of *Gapminder,* https://www.gapminder.org/data/documentation/gd010/.

Starting in the late 18th century in Europe, the mortality rate plummeted three hundredfold, from 1.2 to 0.004 percent. The declines have spread to the rest of the world, including the poorest countries, where the death rate has fallen even faster, though for a shorter time because of their later start. The rate for the entire world, after dropping almost in half in just twenty-five years, is now about 0.2 percent, around where Sweden was in 1941.[11]

You may be wondering whether the drops in child mortality explain all the gains in longevity shown in figure 5-1. Are we really living longer, or are we just surviving infancy in greater numbers? After all, the fact that people before the 19th century had an average life expectancy at birth of around 30 years doesn't mean that everyone dropped dead on their thirtieth birthday. The many children who died pulled the average down, canceling the boost of the people who died of old age, and these seniors can be found in every society. In the time of the Bible, the days of our years were said to be threescore and ten, and that's the age at which Socrates's life was cut short in 399 BCE, not by natural causes but by a cup of hemlock. Most hunter-gatherer tribes have plenty of people in their seventies and even some in their eighties. Though a Hadza woman's life expectancy at birth is 32.5 years, if she makes it to 45 she can expect to live another 21 years.[12]

So do those of us who survive the ordeals of childbirth and childhood today live any longer than the survivors of earlier eras? Yes, much longer. Figure 5-4 shows the life expectancy in the United Kingdom at birth, and at different ages from 1 to 70, over the past three centuries.

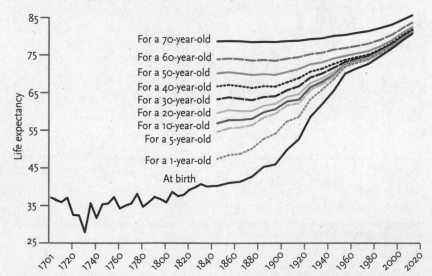

Figure 5-4: Life expectancy, UK, 1701–2013

Sources: *Our World in Data*, Roser 2016n. Data before 1845 are for England and Wales and come from OECD Clio Infra, van Zanden et al. 2014. Data from 1845 on are for mid-decade years only, and come from the *Human Mortality Database*, http://www.mortality.org/.

No matter how old you are, you have more years ahead of you than people of your age did in earlier decades and centuries. A British baby who had survived the hazardous first year of life would have lived to 47 in 1845, 57 in 1905, 72 in 1955, and 81 in 2011. A 30-year-old could look forward to another thirty-three years of life in 1845, another thirty-six in 1905, another forty-three in 1955, and another fifty-two in 2011. If Socrates had been acquitted in 1905, he could have expected to live another nine years; in 1955, another ten; in 2011, another sixteen. An 80-year-old in 1845 had five more years of life; an 80-year-old in 2011, nine years.

Similar trends, though with lower numbers (so far), have occurred in every part of the world. For example, a 10-year-old Ethiopian in 1950 could expect to live to 44; a 10-year-old Ethiopian today can expect to live to 61. The economist Steven Radelet has pointed out that "the improvements in health among the global poor in the last few decades are so large and widespread that they rank among the greatest achievements in human history. Rarely has the basic well-being of so many people around the world improved so substantially, so quickly. Yet few people are even aware that it is happening."[13]

And no, the extra years of life will not be spent senile in a rocking chair. Of course the longer you live, the more of those years you'll live as an older person, with its inevitable aches and pains. But bodies that are better at resisting a mortal blow are also better at resisting the lesser assaults of disease, injury, and wear. As the life span is stretched, our run of vigor is stretched out as well, even if not by the same number of years. A heroic project called the Global Burden of Disease has tried to measure this improvement by tallying not just the number of people who drop dead of each of 291 diseases and disabilities, but how many years of healthy life they lose, weighted by the degree to which each condition compromises the quality of their lives. For the world in 1990, the project estimated that 56.8 of the 64.5 years of life that an average person could be expected to live were years of *healthy* life. And at least in developed countries, where estimates are available for 2010 as well, we know that out of the 4.7 years of additional expected life we gained in those two decades, 3.8 were healthy years.[14] Numbers like these show that people today live far more years in the pink of health than their ancestors lived altogether, healthy and infirm years combined. For many people the greatest fear raised by the prospect of a longer life is dementia, but another pleasant surprise has come to light: between 2000 and 2012, the rate among Americans over 65 fell by a quarter, and the average age at diagnosis rose from 80.7 to 82.4 years.[15]

There is still more good news. The curves in figure 5-4 are not tapestries of your life that have been drawn out and measured by two of the Fates and will someday be cut by the third. Rather, they are projections from today's vital statistics, based on the assumption that medical knowledge will be frozen at its current state. It's not that anyone believes that assumption, but in the absence of clairvoyance about future medical advances we have no other choice. That means you will almost certainly live longer—perhaps much longer—than the numbers you read off the vertical axis.

People will complain about anything, and in 2001 George W. Bush appointed a President's Council on Bioethics to deal with the looming threat of biomedical advances that promise longer and healthier lives.[16] Its chairman, the physician and public intellectual Leon Kass, decreed that "the desire to prolong youthfulness is an expression of a childish and narcissistic wish incompatible with a devotion to posterity," and that the years that would be added to other people's lives were not worth living ("Would professional tennis players really enjoy playing 25 percent more games of tennis?" he asks). Most people would rather decide that for themselves, and even if he is right that "mortality makes life matter," longevity is not the same as immortality.[17] But the fact that experts' assertions about maximum possible life expectancy have repeatedly been shattered (on average five years after they were published) raises the question of whether longevity will increase indefinitely and someday slip the surly bonds of mortality entirely.[18] Should we worry about a world of stodgy multicentenarians who will resist the innovations of ninety-something upstarts and perhaps ban the begetting of pesky children altogether?

A number of Silicon Valley visionaries are trying to bring that world closer.[19] They have funded research institutes which aim not to chip away at mortality one disease at a time but to reverse-engineer the aging process itself and upgrade our cellular hardware to a version without that bug. The result, they hope, will be an increase in the human life span of fifty, a hundred, even a thousand years. In his 2005 bestseller *The Singularity Is Near*, the inventor Ray Kurzweil forecasts that those of us who make it to 2045 will live forever, thanks to advances in genetics, nanotechnology (such as nanobots that will course through our bloodstream and repair our bodies from the inside), and artificial intelligence, which will not just figure out how to do all this but recursively improve its own intelligence without limit.

To readers of medical newsletters and other hypochondriacs, the prospects for immortality look rather different. We certainly find incremental improvements to celebrate, such as a decline in the death rate from cancer over the past twenty-five years of around a percentage point a year, saving a million lives in the United States alone.[20] But we also are regularly disappointed by miracle drugs that work no better than the placebo, treatments with side effects worse than the disease, and trumpeted benefits that wash out in the meta-analysis. Medical progress today is more Sisyphus than Singularity.

Lacking the gift of prophecy, no one can say whether scientists will ever find a cure for mortality. But evolution and entropy make it unlikely. Senescence is baked into our genome at every level of organization, because natural selection favors genes that make us vigorous when we are young over those that make us live as long as possible. That bias is built in because of the asymmetry of time: there is a nonzero probability at any moment that we will be felled by an unpreventable accident like a lightning strike or landslide, making the advantage of any costly longevity gene moot. Biologists would have to reprogram thousands of genes or molecular pathways, each with a small and uncertain effect on longevity, to launch the leap to immortality.[21]

And even if we were fitted with perfectly tuned biological hardware, the march of entropy would degrade it. As the physicist Peter Hoffman points out, "Life pits biology against physics in mortal combat." Violently thrashing molecules constantly collide with the machinery of our cells, including the very machinery that staves off entropy by correcting errors and repairing damage. As damage to the various damage-control systems accumulates, the risk of collapse increases exponentially, sooner or later swamping whatever protections biomedical science has given us against constant risks like cancer and organ failure.[22]

In my view the best projection of the outcome of our multicentury war on death is Stein's Law—"Things that can't go on forever don't"—as amended by Davies's Corollary—"Things that can't go on forever can go on much longer than you think."

CHAPTER 6

HEALTH

How do we explain the gift of life that has been granted to more and more of our species since the end of the 18th century? The timing offers a clue. In *The Great Escape*, Deaton writes, "Ever since people rebelled against authority in the Enlightenment, and set about using the force of reason to make their lives better, they have found a way to do so, and there is little doubt that they will continue to win victories against the forces of death."[1] The gains in longevity celebrated in the previous chapter are the spoils of victory against several of those forces—disease, starvation, war, homicide, accidents—and in this chapter and subsequent ones I will tell the story of each.

For most of human history, the strongest force of death was infectious disease, the nasty feature of evolution in which small, rapidly reproducing organisms make their living at our expense and hitch a ride from body to body in bugs, worms, and bodily effluvia. Epidemics killed by the millions, wiping out entire civilizations, and visited sudden misery on local populations. To take just one example, yellow fever, a viral disease transmitted by mosquitoes, was so named because its victims turned that color before dying in agony. According to an account of an 1878 Memphis epidemic, the sick had "crawled into holes twisted out of shape, their bodies discovered later only by the stench of their decaying flesh. . . . [A mother was found dead] with her body sprawled across the bed . . . black vomit like coffee grounds spattered all over . . . the children rolling on the floor, groaning."[2]

The rich were not spared: in 1836, the wealthiest man in the world, Nathan Mayer Rothschild, died of an infected abscess. Nor the powerful: various British monarchs were cut down by dysentery, smallpox, pneumonia, typhoid, tuberculosis, and malaria. American presidents, too, were vulnerable: William Henry Harrison fell ill shortly after his inauguration in 1841 and died of septic shock thirty-one days later, and James Polk succumbed to cholera three months after leaving office in 1849. As recently as 1924, the sixteen-year-old son of a sitting president, Calvin Coolidge Jr., died of an infected blister he got while playing tennis.

Ever-creative *Homo sapiens* had long fought back against disease with quackery such as prayer, sacrifice, bloodletting, cupping, toxic metals, homeopathy, and squeezing a hen to death against an infected body part. But starting in the late 18th century with the invention of vaccination, and accelerating in the 19th with acceptance of the germ theory of disease, the tide of battle began to turn. Handwashing, midwifery, mosquito control, and especially the protection of drinking water by public sewerage and chlorinated tap water would come to save billions of lives. Before the 20th century, cities were piled high in excrement, their rivers and lakes viscous with waste, and their residents drinking and washing their clothes in putrid brown liquid.[3] Epidemics were blamed on miasmas—foul-smelling air—until John Snow (1813–1858), the first epidemiologist, determined that cholera-stricken Londoners got their water from an intake pipe that was downstream from an outflow of sewage. Doctors themselves used to be a major health hazard as they went from autopsy to examining room in black coats encrusted with dried blood and pus, probed their patients' wounds with unwashed hands, and sewed them up with sutures they kept in their buttonholes, until Ignaz Semmelweis (1818–1865) and Joseph Lister (1827–1912) got them to sterilize their hands and equipment. Antisepsis, anesthesia, and blood transfusions allowed surgery to cure rather than torture and mutilate, and antibiotics, antitoxins, and countless other medical advances further beat back the assault of pestilence.

The sin of ingratitude may not have made the Top Seven, but according to Dante it consigns the sinners to the ninth circle of Hell, and that's where post-1960s intellectual culture may find itself because of its amnesia for the conquerors of disease. It wasn't always that way. When I was a boy, a popular literary genre for children was the heroic biography of a medical pioneer such as Edward Jenner, Louis Pasteur, Joseph Lister, Frederick Banting, Charles Best, William Osler, or Alexander Fleming. On April 12, 1955, a team of scientists announced that Jonas Salk's vaccine against polio—the disease that had killed thousands a year, paralyzed Franklin Roosevelt, and sent many children into iron lungs—was proven safe. According to Richard Carter's history of the discovery, on that day "people observed moments of silence, rang bells, honked horns, blew factory whistles, fired salutes, . . . took the rest of the day off, closed their schools or convoked fervid assemblies therein, drank toasts, hugged children, attended church, smiled at strangers, and forgave enemies."[4] The city of New York offered to honor Salk with a ticker-tape parade, which he politely declined.

And how much thought have you given lately to Karl Landsteiner? Karl who? He only saved *a billion lives* by his discovery of blood groups. Or how about these other heroes?

Scientist	Discovery	Lives Saved
Abel Wolman (1892–1989) and Linn Enslow (1891–1957)	chlorination of water	177 million
William Foege (1936–)	smallpox eradication strategy	131 million
Maurice Hilleman (1919–2005)	eight vaccines	129 million
John Enders (1897–1985)	measles vaccine	120 million
Howard Florey (1898–1968)	penicillin	82 million
Gaston Ramon (1886–1963)	diphtheria and tetanus vaccines	60 million
David Nalin (1941–)	oral rehydration therapy	54 million
Paul Ehrlich (1854–1915)	diphtheria and tetanus antitoxins	42 million
Andreas Grüntzig (1939–1985)	angioplasty	15 million
Grace Eldering (1900–1988) and Pearl Kendrick (1890–1980)	whooping cough vaccine	14 million
Gertrude Elion (1918–1999)	rational drug design	5 million

The researchers who assembled these conservative estimates calculate that more than *five billion* lives have been saved (so far) by the hundred or so scientists they selected.[5] Of course hero stories don't do justice to the way science is really done. Scientists stand on the shoulders of giants, collaborate in teams, toil in obscurity, and aggregate ideas across worldwide webs. But whether it's the scientists or the science that is ignored, the neglect of the discoveries that transformed life for the better is an indictment of our appreciation of the modern human condition.

As a psycholinguist who once wrote an entire book on the past tense, I can single out my favorite example in the history of the English language.[6] It comes from the first sentence of a *Wikipedia* entry:

Smallpox was an infectious disease caused by either of two virus variants, *Variola major* and *Variola minor*.

Yes, "smallpox *was*." The disease that got its name from the painful pustules that cover the victim's skin, mouth, and eyes and that killed more than 300 million people in the 20th century has ceased to exist. (The last case was diagnosed in Somalia in 1977.) For this astounding moral triumph we can thank, among others, Edward Jenner, who discovered vaccination in 1796, the World Health Organization, which in 1959 set the audacious goal of eradicating the disease, and William Foege, who figured out that vaccinating small but strategically chosen portions of the vulnerable populations would do the job. In *Getting Better*, the economist Charles Kenny comments:

> The total cost of the program over those ten years . . . was in the region of $312 million—perhaps 32 cents per person in infected countries. The eradication program cost about the same as producing five recent Hollywood blockbusters, or the wing of a B-2 bomber, or a little under one-tenth the cost of Boston's recent road-improvement project nicknamed the Big Dig. However much one admires the improved views of the Boston waterfront, the lines of the stealth bomber, or the acting skills of Keira Knightley in *Pirates of the Caribbean*, or indeed of the gorilla in *King Kong*, this still seems like a very good deal.[7]

Even as a resident of the Boston waterfront, I'd have to agree. But this stupendous achievement was only the beginning. *Wikipedia*'s definition of rinderpest (cattle plague), which starved millions of farmers and herders throughout history by wiping out their livestock, is also in the past tense. And four other sources of misery in the developing world are slated for eradication. Jonas Salk did not live to see the Global Polio Eradication Initiative approach its goal: by 2016 the disease had been beaten back to just thirty-seven cases in three countries (Afghanistan, Pakistan, and Nigeria), the lowest in history, with an even lower rate thus far in 2017.[8] Guinea worm is a three-foot-long parasite that worms its way into the victim's lower limbs and diabolically forms a painful blister. When the sufferer soaks his or her foot for relief, the blister bursts, releasing thousands of larvae into the water, which other people drink, continuing the cycle. The only treatment consists of pulling the worm out over several days or weeks. But thanks to a three-decade campaign of education and water treatment by the Carter Center, the number of cases fell from 3.5 million in twenty-one countries in 1986 to just twenty-five cases in three countries in 2016 (and just three in one country in the first quarter of 2017).[9] Elephantiasis, river blindness, and blinding trachoma, whose symptoms are as bad as they sound, may also be defined in the past tense by 2030, and measles, rubella, yaws, sleeping sickness, and hookworm are in epidemiologists' sights as well.[10] (Will any of these triumphs be heralded with moments of silence, ringing bells, honking horns, people smiling at strangers and forgiving their enemies?)

Even diseases that are not obliterated are being decimated. Between 2000 and 2015, the number of deaths from malaria (which in the past killed half the people who had ever lived) fell by 60 percent. The World Health Organization has adopted a plan to reduce the rate by another 90 percent by 2030, and to eliminate it from thirty-five of the ninety-seven countries in which it is endemic today (just as it was eliminated from the United States, where it had been endemic until 1951).[11] The Bill & Melinda Gates Foundation has adopted the goal of eradicating it altogether.[12] As we saw in chapter 5, in the 1990s HIV/AIDS in Africa was a setback for humanity's progress in lengthening life spans. But the tide turned in the next decade, and the global death rate for children was cut in half, emboldening the UN to agree in 2016 to a plan to end the AIDS epidemic (though not necessarily to eradicate the virus) by 2030.[13] Figure 6-1 shows that between 2000 and 2013 the world also saw massive reductions in the number of children dying from the five most lethal infectious diseases. In all, the control of infectious disease since 1990 has saved the lives of more than a hundred million children.[14]

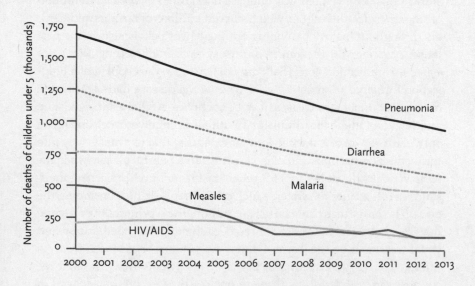

Figure 6-1: Childhood deaths from infectious disease, 2000–2013

Source: Child Health Epidemiology Reference Group of the World Health Organization, Liu et al. 2014, supplementary appendix.

And in the most ambitious plan of all, a team of global health experts led by the economists Dean Jamison and Lawrence Summers have laid out a roadmap for "a grand convergence in global health" by 2035, when infectious, maternal, and child deaths everywhere in the world could be reduced to the levels found in the healthiest middle-income countries today.[15]

As impressive as the conquest of infectious disease in Europe and America was, the ongoing progress among the global poor is even more astonishing. Part of the explanation lies in economic development (chapter 8), because a richer world is a healthier world. Part lies in the expanding circle of sympathy, which inspired global leaders such as Bill Gates, Jimmy Carter, and Bill Clinton to make their legacy the health of the poor in distant continents rather than glittering buildings close to home. George W. Bush, for his part, has been praised by even his harshest critics for his policy on African AIDS relief, which saved millions of lives.

But the most powerful contributor was science. "It is knowledge that is the key," Deaton argues. "Income—although important both in and of itself and as a component of wellbeing . . .—is not the ultimate cause of wellbeing."[16] The fruits of science are not just high-tech pharmaceuticals such as vaccines, antibiotics, antiretrovirals, and deworming pills. They also comprise *ideas*—ideas that may be cheap to implement and obvious in retrospect, but which save millions of lives. Examples include boiling, filtering, or adding bleach to water; washing hands; giving iodine supplements to pregnant women; breast-feeding and cuddling infants; defecating in latrines rather than in fields, streets, and waterways; protecting sleeping children with insecticide-impregnated bed nets; and treating diarrhea with a solution of salt and sugar in clean water. Conversely, progress can be reversed by bad ideas, such as the conspiracy theory spread by the Taliban and Boko Haram that vaccines sterilize Muslim girls, or the one spread by affluent American activists that vaccines cause autism. Deaton notes that even the idea that lies at the core of the Enlightenment—knowledge can make us better off—may come as a revelation in the parts of the world where people are resigned to their poor health, never dreaming that changes to their institutions and norms could improve it.[17]

CHAPTER 7

SUSTENANCE

Together with senescence, childbirth, and pathogens, another mean trick has been played on us by evolution and entropy: our ceaseless need for energy. Famine has long been part of the human condition. The Hebrew Bible tells of seven lean years in Egypt; the Christian Bible has Famine as one of the four horsemen of the apocalypse. Well into the 19th century a crop failure could bring sudden misery even to privileged parts of the world. Johan Norberg quotes the childhood reminiscence of a contemporary of one of his ancestors in Sweden in the winter of 1868:

> We often saw mother weeping to herself, and it was hard on a mother, not having any food to put on the table for her hungry children. Emaciated, starving children were often seen going from farm to farm, begging for a few crumbs of bread. One day three children came to us, crying and begging for something to still the pangs of hunger. Sadly, her eyes brimming with tears, our mother was forced to tell them that we had nothing but a few crumbs of bread which we ourselves needed. When we children saw the anguish in the unknown children's supplicatory eyes, we burst into tears and begged mother to share with them what crumbs we had. Hesitantly she acceded to our request, and the unknown children wolfed down the food before going on to the next farm, which was a good way off from our home. The following day all three were found dead between our farm and the next.[1]

The historian Fernand Braudel has documented that premodern Europe suffered from famines every few decades.[2] Desperate peasants would harvest grain before it was ripe, eat grass or human flesh, and pour into cities to beg. Even in good times, many would get the bulk of their calories from bread or gruel, and not many at that: in *The Escape from Hunger and Premature Death, 1700–2100,* the economist Robert Fogel noted that "the energy value of the typical diet in France at the start of the eighteenth century was as low as that of Rwanda in 1965, the most malnourished nation for that year."[3] Many of those who were not starving were too weak to work, which locked them into poverty. Hungry Europeans titillated themselves with food pornography, such as tales of Cockaigne, a country where pancakes grew on trees, the streets were paved with pastry, roasted pigs wandered around with knives in their backs for easy carving, and cooked fish jumped out of the water and landed at one's feet.

Today we live in Cockaigne, and our problem is not too few calories but too many. As the comedian Chris Rock observed, "This is the first society in history where the poor people are fat." With the usual first-world ingratitude, modern social critics rail against the obesity epidemic with a level of outrage that might be appropriate for a famine (that is, when they are not railing at fat-shaming, slender fashion models, or eating disorders). Though obesity surely is a public health problem, by the standards of history it's a good problem to have.

What about the rest of the world? The hunger that many Westerners associate with Africa and Asia is by no means a modern phenomenon. India and China have always been vulnerable to famine, because millions of people subsisted on rice that was watered by erratic monsoons or fragile irrigation systems and had to be transported across great distances. Braudel recounts the testimony of a Dutch merchant who was in India during a famine in 1630–31:

"Men abandoned towns and villages and wandered helplessly. It was easy to recognize their condition: eyes sunk deep in the head, lips pale and covered with slime, the skin hard, with the bones showing through, the belly nothing but a pouch hanging down empty. . . . One would cry and howl for hunger, while another lay stretched on the ground dying in misery." The familiar human dramas followed: wives and children abandoned, children sold by parents, who either abandoned them or sold themselves in order to survive, collective suicides. . . . Then came the stage when the starving split open the stomachs of the dead or dying and "drew at the entrails to fill their own bellies." "Many hundred thousands of men died of hunger, so that the whole country was covered with corpses lying unburied, which caused such a stench that the whole air was filled and infected with it. . . . In the village of Susuntra . . . human flesh was sold in open market."[4]

But in recent times the world has been blessed with another remark-able and little-noticed advance: in spite of burgeoning numbers, the developing world is feeding itself. This is most obvious in China, whose 1.3 billion people now have access to an average of 3,100 calories per person per day, which, according to US government guidelines, is the number needed by a highly active young man.[5] India's billion peo-ple get an average of 2,400 calories a day, the number recommended for a highly active young woman or an active middle-aged man. The figure for the continent of Africa comes in between the two at 2,600.[6] Fig-ure 7-1, which plots available calories for a representative sample of developed and developing nations and for the world as a whole, shows a pattern familiar from earlier graphs: hardship everywhere before the 19th century, rapid improvement in Europe and the United States over the next two centuries, and, in recent decades, the developing world catching up.

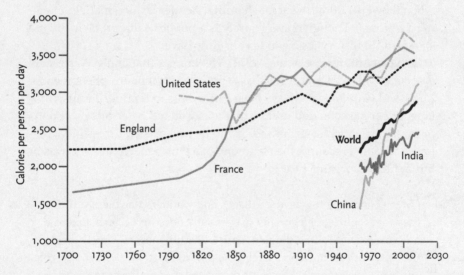

Figure 7-1: Calories, 1700–2013

Sources: United States, England, and France: *Our World in Data*, Roser 2016d, based on data from Fogel 2004. **China, India, and the World:** Food and Agriculture Organization of the United Nations, http://www.fao.org/faostat/en/#data.

The numbers plotted in figure 7-1 are averages, and they would be a misleading index of well-being if they were just lifted by rich people scarfing down more calories (if no one was getting fat except Mama Cass). Fortunately, the numbers reflect an increase in the availability of calories throughout the range, including the bottom. When children are underfed, their growth is stunted, and throughout their lives they have a higher risk of getting sick and dying. Figure 7-2 shows the proportion of children who are stunted in a representative sample of countries which have data for the longest spans of time. Though the proportion of stunted children in poor countries like Kenya and Bangladesh is deplorable, we see that in just two decades the rate of stunting has been cut in half. Countries like Colombia and China also had high rates of stunting not long ago and have managed to bring them even lower.

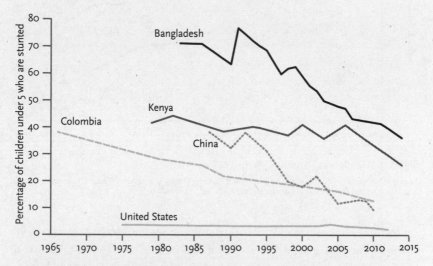

Figure 7-2: Childhood stunting, 1966–2014

Source: *Our World in Data,* Roser 2016j, based on data from the World Health Organization's *Nutrition Landscape Information System,* http://www.who.int/nutrition/nlis/en/.

Figure 7-3 offers another look at how the world has been feeding the hungry. It shows the rate of undernourishment (a year or more of insufficient food) for developing countries in five regions and for the world as a whole. In developed countries, which are not included in the estimates, the rate of undernourishment was less than 5 percent during the entire period, statistically indistinguishable from zero. Though 13 percent of people in the developing world being undernourished is far too much, it's better than 35 percent, which was the level forty-five years earlier, or for that matter 50 percent, an estimate for the entire world in 1947 (not shown on the graph).[7] Remember that these figures are proportions. The world added almost *five billion* people in those seventy years, which means that as the world was reducing the rate of hunger it was also feeding billions of additional mouths.

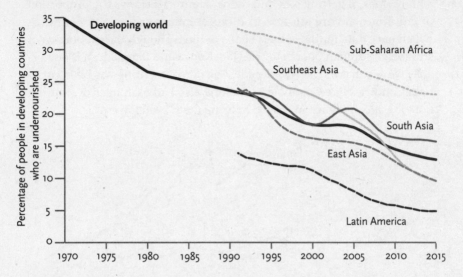

Figure 7-3: Undernourishment, 1970–2015

Source: *Our World in Data*, Roser 2016j, based on data from the Food and Agriculture Organization 2014, also reported in http://www.fao.org/economic/ess/ess-fs/ess-fadata/en/.

Not only has chronic undernourishment been in decline, but so have catastrophic famines—the crises that kill people in large numbers and cause widespread wasting (the condition of being two standard deviations below one's expected weight) and kwashiorkor (the protein deficiency which causes the swollen bellies of the children in photographs that have become icons of famine).[8] Figure 7-4 shows the number of deaths in major famines in each decade for the past 150 years, scaled by world population at the time.

Writing in 2000, the economist Stephen Devereux summarized the world's progress in the 20th century:

Vulnerability to famine appears to have been virtually eradicated from all regions outside Africa. . . . Famine as an endemic problem in Asia and Europe seems to have been consigned to history. The grim label "land of famine" has left China, Russia, India and Bangladesh, and since the 1970s has resided only in Ethiopia and Sudan.

[In addition,] the link from crop failure to famine has been broken.

Most recent drought- or flood-triggered food crises have been ade-
quately met by a combination of local and international humanitarian
response. . . .

If this trend continues, the 20th century should go down as the last
during which tens of millions of people died for lack of access to food.[9]

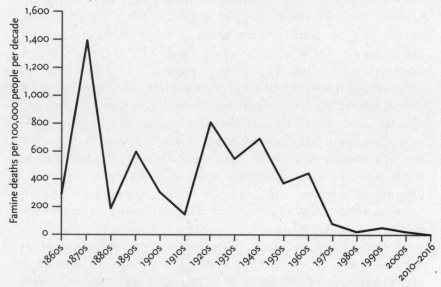

Figure 7-4: Famine deaths, 1860–2016

Sources: *Our World in Data*, Hasell & Roser 2017, based on data from Devereux 2000; Ó Gráda 2009; White
2011, and EM-DAT, *The International Disaster Database*, http://www.emdat.be/; and other sources. "Famine" is defined as in Ó Gráda 2009.

So far, the trend *has* continued. There is still hunger (including among
the poor in developed countries), and there were famines in East Africa
in 2011, the Sahel in 2012, and South Sudan in 2016, together with near-
famines in Somalia, Nigeria, and Yemen. But they did not kill on the scale
of the catastrophes that were regular occurrences in earlier centuries.

None of this was supposed to happen. In 1798 Thomas Malthus explained that the frequent famines of his era were unavoidable and would
only get worse, because "population, when unchecked, increases in a
geometrical ratio. Subsistence increases only in an arithmetic ratio. A
slight acquaintance with numbers will show the immensity of the first
power in comparison with the second." The implication was that efforts
to feed the hungry would only lead to more misery, because they would
breed more children who were doomed to hunger in their turn.

Not long ago, Malthusian thinking was revived with a vengeance. In 1967 William and Paul Paddock wrote *Famine 1975!*, and in 1968 the biologist Paul R. Ehrlich wrote *The Population Bomb*, in which he proclaimed that "the battle to feed all of humanity is over" and predicted that by the 1980s sixty-five million Americans and four billion other people would starve to death. *New York Times Magazine* readers were introduced to the battlefield term *triage* (the emergency practice of separating wounded soldiers into the savable and the doomed) and to philosophy-seminar arguments about whether it is morally permissible to throw someone overboard from a crowded lifeboat to prevent it from capsizing and drowning everyone.[10] Ehrlich and other environmentalists argued for cutting off food aid to countries they deemed basket cases.[11] Robert Mc-Namara, president of the World Bank from 1968 to 1981, discouraged financing of health care "unless it was very strictly related to population control, because usually health facilities contributed to the decline of the death rate, and thereby to the population explosion." Population-control programs in India and China (especially under China's one-child policy) coerced women into sterilizations, abortions, and being implanted with painful and septic IUDs.[12]

Where did Malthus's math go wrong? Looking at the first of his curves, we already saw that population growth needn't increase in a geometric ratio indefinitely, because when people get richer and more of their babies survive, they have fewer babies (see also figure 10-1). Conversely, famines don't reduce population growth for long. They disproportionately kill children and the elderly, and when conditions improve, the survivors quickly replenish the population.[13] As Hans Rosling put it, "You can't stop population growth by letting poor children die."[14]

Looking at the second curve, we discover that the food supply *can* grow geometrically when *knowledge* is applied to increase the amount of food that can be coaxed out of a patch of land. Since the birth of agriculture ten thousand years ago, humans have been genetically engineering plants and animals by selectively breeding the ones that had the most calories and fewest toxins and that were the easiest to plant and harvest. The wild ancestor of corn was a grass with a few tough seeds; the ancestor of carrots looked and tasted like a dandelion root; the ancestors of many wild fruits were bitter, astringent, and more stone than flesh. Clever farmers also tinkered with irrigation, plows, and organic fertilizers, but Malthus always had the last word.

It was only at the time of the Enlightenment and the Industrial Revolution that people figured out how to bend the curve upward.[15] In Jonathan Swift's 1726 novel, the moral imperative was explained to Gulliver by the King of Brobdingnag: "Whoever makes two ears of corn, or two blades of grass to grow where only one grew before, deserves better of humanity, and does more essential service to his country than the whole race of politicians put together." Soon after that, as figure 7-1 shows, more ears of corn were indeed made to grow, in what has been called the British Agricultural Revolution.[16] Crop rotation and improvements to plows and seed drills were followed by mechanization, with fossil fuels replacing human and animal muscle. In the mid-19th century it took twenty-five men a full day to harvest and thresh a ton of grain; today one person operating a combine harvester can do it in six minutes.[17]

Machines also solve an inherent problem with food. As any zucchini gardener in August knows, a lot becomes available all at once, and then it quickly rots or gets eaten by vermin. Railroads, canals, trucks, granaries, and refrigeration evened out the peaks and troughs in the supply and matched it with demand, coordinated by the information carried in prices. But the truly gargantuan boost would come from chemistry. The N in SPONCH, the acronym taught to schoolchildren for the chemical elements that make up the bulk of our bodies, stands for nitrogen, a major ingredient of protein, DNA, chlorophyll, and the energy carrier ATP. Nitrogen atoms are plentiful in the air but bound in pairs (hence the chemical formula N_2), which are hard to split apart so that plants can use them. In 1909 Carl Bosch perfected a process invented by Fritz Haber which used methane and steam to pull nitrogen out of the air and turn it into fertilizer on an industrial scale, replacing the massive quantities of bird poop that had previously been needed to return nitrogen to depleted soils. Those two chemists top the list of the 20th-century scientists who saved the greatest number of lives in history, with 2.7 billion.[18]

So forget arithmetic ratios: over the past century, grain yields per hectare have swooped upward while real prices have plunged. The savings are mind-boggling. If the food grown today had to be grown with pre-nitrogen-farming techniques, an area the size of Russia would go under the plow.[19] In the United States in 1901, an hour's wages could buy around three quarts of milk; a century later, the same wages would buy *sixteen* quarts. The amount of every other foodstuff that can be bought with an hour of labor has multiplied as well: from a pound of butter to five pounds, a dozen eggs to twelve dozen, two pounds of pork chops to five pounds, and nine pounds of flour to forty-nine pounds.[20]

In the 1950s and '60s, another giga-lifesaver, Norman Borlaug, outsmarted evolution to foment the Green Revolution in the developing world.[21] Plants in nature invest a lot of energy and nutrients in woody stalks that raise their leaves and blossoms above the shade of neighboring weeds and of each other. Like fans at a rock concert, everyone stands up, but no one gets a better view. That's the way evolution works: it myopically selects for individual advantage, not the greater good of the species, let alone the good of some other species. From a farmer's perspective, not only do tall wheat plants waste energy in inedible stalks, but when they are enriched with fertilizer they collapse under the weight of the heavy seedhead. Borlaug took evolution into his own hands, crossing thousands of strains of wheat and then selecting the offspring with dwarfed stalks, high yields, resistance to rust, and an insensitivity to day length. After several years of this "mind-warpingly tedious work," Borlaug evolved strains of wheat (and then corn and rice) with many times the yield of their ancestors. By combining these strains with modern techniques of irrigation, fertilization, and crop management, Borlaug turned Mexico and then India, Pakistan, and other famine-prone countries into grain exporters almost overnight. The Green Revolution continues—it has been called "Africa's best-kept secret"—driven by improvements in sorghum, millet, cassava, and tubers.[22]

Thanks to the Green Revolution, the world needs less than a third of the land it used to need to produce a given amount of food.[23] Another way of stating the bounty is that between 1961 and 2009 the amount of land used to grow food increased by 12 percent, but the amount of food that was grown increased by 300 percent.[24] In addition to beating back hunger, the ability to grow more food from less land has been, on the whole, good for the planet. Despite their bucolic charm, farms are biological deserts which sprawl over the landscape at the expense of forests and grasslands. Now that farms have receded in some parts of the world, temperate forests have been bouncing back, a phenomenon we will return to in chapter 10.[25] If agricultural efficiency had remained the same over the past fifty years while the world grew the same amount of food, an area the size of the United States, Canada, and China combined would have had to be cleared and plowed.[26] The environmental scientist Jesse Ausubel has estimated that the world has reached Peak Farmland: we may never again need as much as we use today.[27]

Like all advances, the Green Revolution came under attack as soon as it began. High-tech agriculture, the critics said, consumes fossil fuels and groundwater, uses herbicides and pesticides, disrupts traditional subsistence agriculture, is biologically unnatural, and generates profits for corporations. Given that it saved a billion lives and helped consign major famines to the dustbin of history, this seems to me like a reasonable price to pay. More important, the price need not be with us forever. The beauty of scientific progress is that it never locks us into a technology but can develop new ones with fewer problems than the old ones (a dynamic we will return to in chapter 10, p. 127).

Genetic engineering can now accomplish in days what traditional farmers accomplished in millennia and Borlaug accomplished in his years of "mind-warping tedium." Transgenic crops are being developed with high yields, lifesaving vitamins, tolerance of drought and salinity, resistance to disease, pests, and spoilage, and reduced need for land, fertilizer, and plowing. Hundreds of studies, every major health and science organization, and more than a hundred Nobel laureates have testified to their safety (unsurprisingly, since there is no such thing as a genetically unmodified crop).[28] Yet traditional environmentalist groups, with what the ecology writer Stewart Brand has called their "customary indifference to starvation," have prosecuted a fanatical crusade to keep transgenic crops from people—not just from whole-food gourmets in rich countries but from poor farmers in developing ones.[29] Their opposition begins with a commitment to the sacred yet meaningless value of "naturalness," which leads them to decry "genetic pollution" and "playing with nature" and to promote "real food" based on "ecological agriculture." From there they capitalize on primitive intuitions of essentialism and contamination among the scientifically illiterate public. Depressing studies have shown that about half of the populace believes that ordinary tomatoes don't have genes but genetically modified ones do, that a gene inserted into a food might migrate into the genomes of people who eat it, and that a spinach gene inserted into an orange would make it taste like spinach. Eighty percent favored a law that would mandate labels on all foods "containing DNA."[30] As Brand put it, "I daresay the environmental movement has done more harm with its opposition to genetic engineering than with any other thing we've been wrong about. We've starved people, hindered science, hurt the natural environment, and denied our own practitioners a crucial tool."[31]

One reason for Brand's harsh judgment is that opposition to transgenic crops has been perniciously effective in the part of the world that could most benefit from it. Sub-Saharan Africa has been cursed by nature with thin soil, capricious rainfall, and a paucity of harbors and navigable rivers, and it never developed an extensive network of roads, rails, or canals.[32] Like all farmed land, its soils have been depleted, but unlike those in the rest of the world, Africa's have not been replenished with synthetic fertilizer. Adoption of transgenic crops, both those already in use and ones customized for Africa, grown with other modern practices such as no-till farming and drip irrigation, could allow Africa to leapfrog the more invasive practices of the first Green Revolution and eliminate its remaining undernourishment.

For all the importance of agronomy, food security is not just about farming. Famines are caused not only when food is scarce but when people can't afford it, when armies prevent them from getting it, or when their governments don't care how much of it they have.[33] The pinnacles and valleys in figure 7-4 show that the conquest of famine was not a story of steady gains in agricultural efficiency. In the 19th century, famines were triggered by the usual droughts and blights, but they were exacerbated in colonial India and Africa by the callousness, bungling, and sometimes deliberate policies of administrators who had no benevolent interest in their subjects' welfare.[34] By the early 20th century, colonial policies had become more responsive to food crises, and advances in agriculture had taken a bite out of hunger.[35] But then a horror show of political catastrophes triggered sporadic famines for the rest of the century.

Of the seventy million people who died in major 20th-century famines, 80 percent were victims of Communist regimes' forced collectivization, punitive confiscation, and totalitarian central planning.[36] These included famines in the Soviet Union in the aftermaths of the Russian Revolution, the Russian Civil War, and World War II; Stalin's Holodomor (terror-famine) in Ukraine in 1932–33; Mao's Great Leap Forward in 1958–61; Pol Pot's Year Zero in 1975–79; and Kim Jong-il's Arduous March in North Korea as recently as the late 1990s. The first governments in postcolonial Africa and Asia often implemented ideologically fashionable but economically disastrous policies such as the mass collectivization of farming, import restrictions to promote "self-sufficiency," and artificially low food prices which benefited politically influential city-dwellers at the expense of farmers.[37] When the countries fell into civil war, as they so often did, not only was food distribution disrupted, but both sides could use hunger as a weapon, sometimes with the complicity of their Cold War patrons.

Fortunately, since the 1990s the prerequisites to plenty have been falling into place in more of the world. Once the secrets to growing food in abundance are unlocked and the infrastructure to move it around is in place, the decline of famine depends on the decline of poverty, war, and autocracy. Let's turn to the progress that has been made against each of these scourges.

CHAPTER 8

WEALTH

P overty has no causes," wrote the economist Peter Bauer. "Wealth has causes." In a world governed by entropy and evolution, the streets are not paved with pastry, and cooked fish do not land at our feet. But it's easy to forget this truism and think that wealth has always been with us. History is written not so much by the victors as by the affluent, the sliver of humanity with the leisure and education to write about it. As the economist Nathan Rosenberg and the legal scholar L. E. Birdzell Jr. point out, "We are led to forget the dominating misery of other times in part by the grace of literature, poetry, romance, and legend, which celebrate those who lived well and forget those who lived in the silence of poverty. The eras of misery have been mythologized and may even be remembered as golden ages of pastoral simplicity. They were not."[1]

Norberg, drawing on Braudel, offers vignettes of this era of misery, when the definition of poverty was simple: "if you could afford to buy bread to survive another day, you were not poor."

> In wealthy Genoa, poor people sold themselves as galley slaves every winter. In Paris the very poor were chained together in pairs and forced to do the hard work of cleaning the drains. In England, the poor had to work in workhouses to get relief, where they worked long hours for almost no pay. Some were instructed to crush dog, horse and cattle bones for use as fertilizer, until an inspection of a workhouse in 1845 showed that hungry paupers were fighting over the rotting bones to suck out the marrow.[2]

Another historian, Carlo Cipolla, noted:

In preindustrial Europe, the purchase of a garment or of the cloth for a garment remained a luxury the common people could only afford a few times in their lives. One of the main preoccupations of hospital administration was to ensure that the clothes of the deceased should not be usurped but should be given to lawful inheritors. During epidemics of plague, the town authorities had to struggle to confiscate the clothes of the dead and to burn them: people waited for others to die so as to take over their clothes—which generally had the effect of spreading the epidemic.[3]

The need to explain the creation of wealth is obscured yet again by political debates within modern societies on how wealth ought to be distributed, which presuppose that wealth worth distributing exists in the first place. Economists speak of a "lump fallacy" or "physical fallacy" in which a finite amount of wealth has existed since the beginning of time, like a lode of gold, and people have been fighting over how to divide it up ever since.[4] Among the brainchildren of the Enlightenment is the realization that *wealth is created*.[5] It is created primarily by knowledge and cooperation: networks of people arrange matter into improbable but useful configurations and combine the fruits of their ingenuity and labor. The corollary, just as radical, is that we can figure out how to make more of it.

The endurance of poverty and the transition to modern affluence can be shown in a simple but stunning graph. It plots, for the past two thousand years, a standard measure of wealth creation, the Gross World Product, measured in 2011 international dollars. (An international dollar is a hypothetical unit of currency equivalent to a US dollar in a particular reference year, adjusted for inflation and for purchasing-power parity. The latter compensates for differences in the prices of comparable goods and services in different places—the fact that a haircut, for example, is cheaper in Dhaka than in London.)

The story of the growth of prosperity in human history depicted in figure 8-1 is close to: nothing . . . nothing . . . nothing . . . (repeat for a few thousand years) . . . *boom!* A millennium after the year 1 CE, the world was barely richer than it was at the time of Jesus. It took another half-millennium for income to double. Some regions enjoyed spurts now and again, but they did not lead to sustained, cumulative growth. Starting in the 19th century, the increments turned into leaps and bounds. Between 1820 and 1900, the world's income tripled. It tripled again in a bit more than fifty years. It took only twenty-five years for it to triple again, and another thirty-three years to triple yet another time.

The Gross World Product today has grown almost a hundredfold since the Industrial Revolution was in place in 1820, and almost two hundred-fold from the start of the Enlightenment in the 18th century. Debates on economic distribution and growth often contrast dividing a pie with baking a larger one (or as George W. Bush mangled it, "making the pie higher"). If the pie we were dividing in 1700 was baked in a standard nine-inch pan, then the one we have today would be more than ten feet in diameter. If we were to surgically carve out the teensiest slice imaginable—say, one that was two inches at its widest point—it would be the size of the entire pie in 1700.

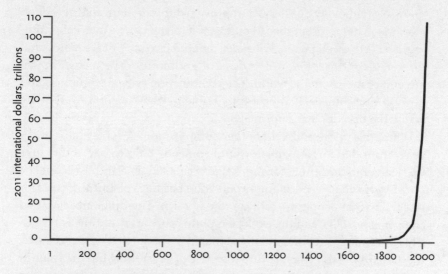

Figure 8-1: Gross World Product, 1–2015

Source: *Our World in Data*, Roser 2016c, based on data from the World Bank and from Angus Maddison and Maddison Project 2014.

Indeed, the Gross World Product is a gross *underestimate* of the expansion of prosperity.[6] How does one count units of currency, like pounds or dollars, across the centuries, so they can be plotted in a single line? Is one hundred dollars in the year 2000 more or less than one dollar in 1800? They're just pieces of paper with numbers on them; their value depends on what people can buy with them at the time, which changes with inflation and revaluations. The only way to compare a dollar in 1800 with a dollar in 2000 is to look up how many one would have to fork over to buy a standard market basket of goods: a fixed amount of food, clothing, health care, fuel, and so on. That's how the numbers in figure 8-1, and in other graphs denominated in dollars or pounds, are converted into a single scale such as "2011 international dollars."

The problem is that the advance of technology confounds the very idea of an unchanging market basket. To start with, the quality of the goods in the basket improves over time. An item of "clothing" in 1800 might be a rain cape made of stiff, heavy, and leaky oilcloth; in 2000 it would be a zippered raincoat made of a light, breathable synthetic. "Dental care" in 1800 meant pliers and wooden dentures; in 2000 it meant Novocain and implants. It's misleading, then, to say that the $300 it would take to buy a certain amount of clothing and medical care in 2000 can be equated with the $10 it would take to buy "the same amount" in 1800.

Also, technology doesn't just improve old things; it invents new ones. How much did it cost in 1800 to purchase a refrigerator, a musical recording, a bicycle, a cell phone, *Wikipedia*, a photo of your child, a laptop and printer, a contraceptive pill, a dose of antibiotics? The answer is: no amount of money in the world. The combination of better products and new products makes it almost impossible to track material well-being across the decades and centuries.

Plunging prices add yet another complication. A refrigerator today costs around $500. How much would someone have to pay you to give up refrigeration? Surely far more than $500! Adam Smith called it the paradox of value: when an important good becomes plentiful, it costs far less than what people are willing to pay for it. The difference is called consumer surplus, and the explosion of this surplus over time is impossible to tabulate. Economists are the first to point out that their measures, like Oscar Wilde's cynic, capture the price of everything but the value of nothing.[7]

This doesn't mean that comparisons of wealth across times and places in currency adjusted for inflation and purchasing power are meaningless—they are better than ignorance, or guesstimates—but it does mean that they shortchange our accounting of progress. A person whose wallet contains the cash equivalent of a hundred 2011 international dollars today is fantastically richer than her ancestor with the equivalent wallet's worth two hundred years ago. As we'll see, this also affects our assessment of prosperity in the developing world (this chapter), of income inequality in the developed world (next chapter), and of the future of economic growth (chapter 20).

~

What launched the Great Escape? The most obvious cause was the application of science to the improvement of material life, leading to what the economic historian Joel Mokyr calls "the enlightened economy."[8] The machines and factories of the Industrial Revolution, the productive farms of the Agricultural Revolution, and the water pipes of the Public Health Revolution could deliver more clothes, tools, vehicles, books, furniture, calories, clean water, and other things that people want than the craftsmen and farmers of a century before. Many early innovations, such as in steam engines, looms, spinning frames, foundries, and mills, came out of the workshops and backyards of atheoretical tinkerers.[9] But trial and error is a profusely branching tree of possibilities, most of which lead nowhere, and the tree can be pruned by the application of science, accelerating the rate of discovery. As Mokyr notes, "After 1750 the epistemic base of technology slowly began to expand. Not only did new products and techniques emerge; it became better understood why and how the old ones worked, and thus they could be refined, debugged, improved, combined with others in novel ways and adapted to new uses."[10] The invention of the barometer in 1643, which proved the existence of atmospheric pressure, eventually led to the invention of steam engines, known at the time as "atmospheric engines." Other two-way streets between science and technology included the application of chemistry, facilitated by the invention of the battery, to synthesize fertilizer, and the application of the germ theory of disease, made possible by the microscope, to keep pathogens out of drinking water and off doctors' hands and instruments.

The applied scientists would not have been motivated to apply their ingenuity to ease the pains of everyday life, and their gadgets would have remained in their labs and garages, were it not for two other innovations.

One was the development of *institutions* that lubricated the exchange of goods, services, and ideas—the dynamic singled out by Adam Smith as the generator of wealth. The economists Douglass North, John Wallis, and Barry Weingast argue that the most natural way for states to function, both in history and in many parts of the world today, is for elites to agree not to plunder and kill each other, in exchange for which they are awarded a fief, franchise, charter, monopoly, turf, or patronage network that allows them to control some sector of the economy and live off the rents (in the economist's sense of income extracted from exclusive access to a resource).[11] In 18th-century England this cronyism gave way to *open* economies in which anyone could sell anything to anyone, and their transactions were protected by the rule of law, property rights, enforceable contracts, and institutions like banks, corporations, and government agencies that run by fiduciary duties rather than personal connections. Now an enterprising person could introduce a new kind of product to the market, or undersell other merchants if he could provide a product at lower cost, or accept money now for something he would not deliver until later, or invest in equipment or land that might not return a profit for years. Today I take it for granted that if I want some milk, I can walk into a convenience store and a quart will be on the shelves, the milk won't be diluted or tainted, it will be for sale at a price I can afford, and the owner will let me walk out with it after a swipe of a card, even though we have never met, may never see each other again, and have no friends in common who can testify to our bona fides. A few doors down and I could do the same with a pair of jeans, a power drill, a computer, or a car. A lot of institutions have to be in place for these and the millions of other anonymous transactions that make up a modern economy to be consummated so easily.

The third innovation, after science and institutions, was a change in values: an endorsement of what the economic historian Deirdre Mc-Closkey calls bourgeois virtue.[12] Aristocratic, religious, and martial cultures have always looked down on commerce as tawdry and venal. But in 18th-century England and the Netherlands, commerce came to be seen as moral and uplifting. Voltaire and other Enlightenment *philosophes* valorized the spirit of commerce for its ability to dissolve sectarian hatreds:

> Take a view of the *Royal Exchange in London*, a place more venerable than many courts of justice, where the representatives of all nations meet for the benefit of mankind. There the Jew, the Mahometan, and the Christian transact together as tho' they all profess'd the same religion, and give the name of Infidel to none but bankrupts. There the Presbyterian confides in the Anabaptist, and the Churchman depends on the Quaker's word. And all are satisfied.[13]

Commenting on this passage, the historian Roy Porter noted that "by depicting men content, and content to be content—differing, but agreeing to differ—the *philosophe* pointed towards a rethinking of the *summum bonum*, a shift from God-fearingness to a selfhood more psychologically oriented. The Enlightenment thus translated the ultimate question 'How can I be saved?' into the pragmatic 'How can I be happy?'—thereby heralding a new praxis of personal and social adjustment."[14] This praxis included norms of propriety, thrift, and self-restraint, an orientation toward the future rather than the past, and a conferral of dignity and prestige upon merchants and inventors rather than just on soldiers, priests, and courtiers. Napoleon, that exponent of martial glory, sniffed at England as "a nation of shopkeepers." But at the time Britons earned 83 percent more than Frenchmen and enjoyed a third more calories, and we all know what happened at Waterloo.[15]

The Great Escape in Britain and the Netherlands was quickly followed by escapes in the Germanic states, the Nordic countries, and Britain's colonial offshoots in Australia, New Zealand, Canada, and the United States. In 1905 the sociologist Max Weber proposed that capitalism depended on a "Protestant ethic" (a hypothesis with the intriguing prediction that Jews should fare poorly in capitalist societies, particularly in business and finance). In any case the Catholic countries of Europe soon zoomed out of poverty too, and a succession of other escapes shown in figure 8-2 have put the lie to various theories explaining why Buddhism, Confucianism, Hinduism, or generic "Asian" or "Latin" values were incompatible with dynamic market economies.

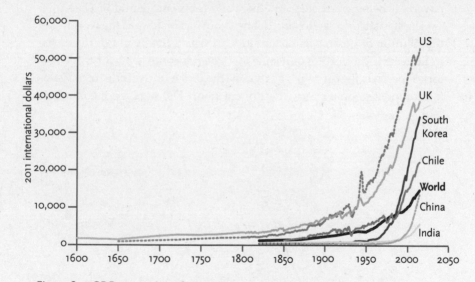

Figure 8-2: GDP per capita, 1600–2015

Source: *Our World in Data*, Roser 2016c, based on data from the World Bank and from Maddison Project 2014.

The non-British curves in figure 8-2 tell of a second astonishing chapter in the story of prosperity: starting in the late 20th century, poor countries have been escaping from poverty in their turn. The Great Escape is becoming the Great Convergence.[16] Countries that until recently were miserably poor have become comfortably rich, such as South Korea, Taiwan, and Singapore. (My Singaporean former mother-in-law recalls a childhood dinner at which her family split an egg four ways.) Since 1995, 30 of the world's 109 developing countries, including countries as diverse as Bangladesh, El Salvador, Ethiopia, Georgia, Mongolia, Mozambique, Panama, Rwanda, Uzbekistan, and Vietnam, have enjoyed economic growth rates that amount to a doubling of income every eighteen years. Another 40 countries have had rates that would double income every thirty-five years, which is comparable to the historical growth rate of the United States.[17] It's remarkable enough to see that by 2008 China and India had the same per capita income that Sweden had in 1950 and 1920, respectively, but more remarkable still when we remember how many capitas this income was per: 1.3 and 1.2 billion people. By 2008 the world's population, all 6.7 billion of them, had an average income equivalent to that of Western Europe in 1964. And no, it's not just because the rich are getting even richer (though of course they are, a topic we will examine in the next chapter). Extreme poverty is being eradicated, and the world is becoming middle class.[18]

The statistician Ola Rosling (Hans's son) has displayed the worldwide distribution of income as histograms, in which the height of the curve indicates the proportion of people at a given income level, for three historical periods (figure 8-3).[19] In 1800, at the dawn of the Industrial Revolution, most people everywhere were poor. The average income was

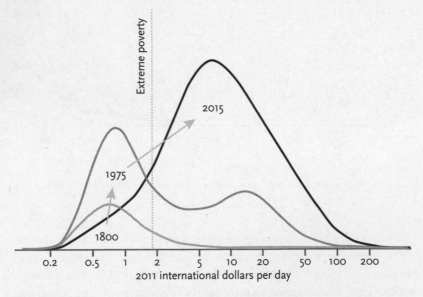

Figure 8-3: World income distribution, 1800, 1975, and 2015

Source: *Gapminder*, via Ola Rosling, http://www.gapminder.org/tools/mountain. The scale is in 2011 international dollars.

equivalent to that in the poorest countries in Africa today (about $500 a year in international dollars), and almost 95 percent of the world lived in what counts today as "extreme poverty" (less than $1.90 a day). By 1975, Europe and its offshoots had completed the Great Escape, leaving the rest of the world behind, with one-tenth their income, in the lower hump of a camel-shaped curve.[20] In the 21st century the camel has become a dromedary, with a single hump shifted to the right and a much lower tail on the left: the world had become richer and more equal.[21]

The slices to the left of the dotted line deserve their own picture. Figure 8-4 shows the percentage of the world's population that lives in "extreme poverty." Admittedly, any cutoff for that condition must be arbitrary, but the United Nations and the World Bank do their best by combining the national poverty lines from a sample of developing countries, which are in turn based on the income of a typical family that manages to feed itself. In 1996 it was the alliterative "a dollar a day" per person; currently it's set at $1.90 a day in 2011 international dollars.[22] (Curves with more generous cutoffs are higher and shallower but also skitter downward.)[23] Notice not just the shape of the curve but how low it has sunk—to 10 percent. In two hundred years the rate of extreme poverty in the world has tanked from 90 percent to 10, with almost half that decline occurring in the last thirty-five years.

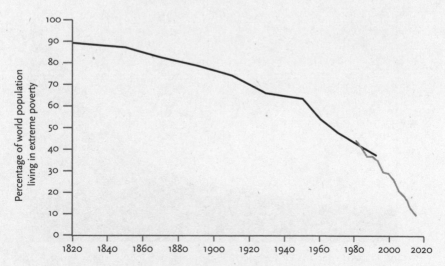

Figure 8-4: Extreme poverty (proportion), 1820–2015

Sources: *Our World in Data,* Roser & Ortiz-Ospina 2017, based on data from Bourguignon & Morrisson 2002 (1820–1992), averaging their "Extreme poverty" and "Poverty" percentages for commensurability with data on "Extreme poverty" for 1981–2015 from the World Bank 2016g.

The world's progress can be appreciated in two ways. By one reckoning, the proportions and per capita rates I have been plotting are the morally relevant measure of progress, because they fit with John Rawls's thought experiment for defining a just society: specify a world in which you would agree to be incarnated as a random citizen from behind a veil of ignorance as to that citizen's circumstances.[24] A world with a higher percentage of long-lived, healthy, well-fed, well-off people is a world in which one would prefer to play the lottery of birth. But by another reckoning, absolute numbers matter, too. Every additional long-lived, healthy, well-fed, well-off person is a sentient being capable of happiness, and the world is a better place for having more of them. Also, an increase in the number of people who can withstand the grind of entropy and the struggle of evolution is a testimonial to the sheer magnitude of the benevolent powers of science, markets, good government, and other modern institutions. In the stacked layer graph in figure 8-5, the thickness of the bottom slab represents the number of people living in extreme poverty, the thickness of the top slab represents the number not living in poverty, and the height of the stack represents the population of the world. It shows that the number of poor people declined just as the number of all people exploded, from 3.7 billion in 1970 to 7.3 billion in 2015. (Max Roser points out that if news outlets truly reported the changing state of the world,

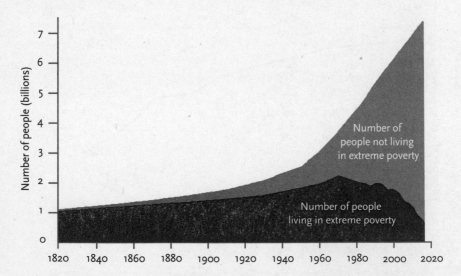

Figure 8-5: Extreme poverty (number), 1820–2015

Sources: *Our World in Data,* Roser & Ortiz-Ospina 2017, based on data from Bourguignon & Morrisson 2002 (1820–1992) and the World Bank 2016g (1981–2015).

they could have run the headline NUMBER OF PEOPLE IN EXTREME POVERTY FELL BY 137,000 SINCE YESTERDAY every day for the last twenty-five years.) We live in a world not just with a smaller proportion of extremely poor people but with a smaller number of them, and with 6.6 billion people who are not extremely poor.

Most surprises in history are unpleasant surprises, but this news came as a pleasant shock even to the optimists. In 2000 the United Nations laid out eight Millennium Development Goals, their starting lines backdated to 1990.[25] At the time, cynical observers of that underperforming organization dismissed the targets as aspirational boilerplate. Cut the global poverty rate in half, lifting a billion people out of poverty, in twenty-five years? Yeah, yeah. But the world reached the goal *five years ahead of schedule.* Development experts are still rubbing their eyes. Deaton writes, "This is perhaps the most important fact about wellbeing in the world since World War II."[26] The economist Robert Lucas (like Deaton, a Nobel laureate) said, "The consequences for human welfare involved [in understanding rapid economic development] are simply staggering: once one starts to think about them, it is hard to think about anything else."[27]

Let's not stop thinking about tomorrow. Though it's always dangerous to extrapolate a historical curve, what happens when we try? If we align a ruler with the World Bank data in figure 8-4, we find that it crosses the x-axis (indicating a poverty rate of 0) in 2026. The UN gave itself a cushion in its 2015 Sustainable Development Goals (the successor to its Millennium Development Goals) and set a target of "ending extreme poverty for all people everywhere" by 2030.[28] Ending extreme poverty for all people everywhere! May I live to see the day. (Not even Jesus was that optimistic: he told a supplicant, "The poor you will always have with you.")

Of course that day is a ways off. Hundreds of millions of people remain in extreme poverty, and getting to zero will require a greater effort than just extrapolating along a ruler. Though the numbers are dwindling in countries like India and Indonesia, they are increasing in the poorest of the poor countries, like Congo, Haiti, and Sudan, and the last pockets of poverty will be the hardest to eliminate.[29] Also, as we approach the goal we should move the goalposts, since not-so-extreme poverty is still poverty. In introducing the concept of progress I warned against confusing hard-won headway with a process that magically takes place by itself. The point of calling attention to progress is not self-congratulation but identifying the causes so we can do more of what works. And since we know that something has worked, it's unnecessary to keep depicting the developing world as a basket case to shake people out of their apathy—with the danger that they will think that additional support would just be throwing money down a rat hole.[30]

So what *is* the world doing right? As with most forms of progress, a lot of good things happen at once and reinforce one another, so it's hard to identify a first domino. Cynical explanations, such as that the enrichment is a one-time dividend of a surge in the price of oil and other commodities, or that the statistics are inflated by the rise of populous China, have been examined and dismissed. Radelet and other development experts point to five causes.[31]

"In 1976," Radelet writes, "Mao single-handedly and dramatically changed the direction of global poverty with one simple act: he died."[32] Though China's rise is not exclusively responsible for the Great Convergence, the country's sheer bulk is bound to move the totals around, and the explanations for its progress apply elsewhere. The death of Mao Zedong is emblematic of three of the major causes of the Great Convergence.

The first is the decline of communism (together with intrusive socialism). For reasons we have seen, market economies can generate wealth prodigiously while totalitarian planned economies impose scarcity, stagnation, and often famine. Market economies, in addition to reaping the benefits of specialization and providing incentives for people to produce things that other people want, solve the problem of coordinating the efforts of hundreds of millions of people by using prices to propagate information about need and availability far and wide, a computational problem that no planner is brilliant enough to solve from a central bureau.[33] A shift from collectivization, centralized control, government monopolies, and suffocating permit bureaucracies (what in India was called "the license raj") to open economies took place on a number of fronts beginning in the 1980s. They included Deng Xiaoping's embrace of capitalism in China, the collapse of the Soviet Union and its domination of Eastern Europe, and the liberalization of the economies of India, Brazil, Vietnam, and other countries.

Though intellectuals are apt to do a spit take when they read a defense of capitalism, its economic benefits are so obvious that they don't need to be shown with numbers. They can literally be seen from space. A satellite photograph of Korea showing the capitalist South aglow in light and the Communist North a pit of darkness vividly illustrates the contrast in the wealth-generating capability between the two economic systems, holding geography, history, and culture constant. Other matched pairs with an experimental group and a control group lead to the same conclusion: West and East Germany when they were divided by the Iron Curtain; Botswana versus Zimbabwe under Robert Mugabe; Chile versus Venezuela under Hugo Chávez and Nicolás Maduro—the latter a once-wealthy, oil-rich country now suffering from widespread hunger and a critical shortage of medical care.[34] It's important to add that the market economies which blossomed in the more fortunate parts of the developing world were not the laissez-faire anarchies of right-wing fantasies and left-wing nightmares. To varying degrees, their governments invested in education, public health, infrastructure, and agricultural and job training, together with social insurance and poverty-reduction programs.[35]

Radelet's second explanation of the Great Convergence is leadership. Mao imposed more than communism on China. He was a mercurial megalomaniac who foisted crackbrained schemes on the country, such as the Great Leap Forward (with its gargantuan communes, useless backyard smelters, and screwball agronomic practices) and the Cultural Revolution (which turned the younger generation into gangs of thugs who terrorized teachers, managers, and descendants of "rich peasants").[36] During the decades of stagnation from the 1970s to the early 1990s, many other developing countries were commandeered by psychopathic strongmen with ideological, religious, tribal, paranoid, or self-aggrandizing agendas rather than a mandate to enhance the well-being of their citizens. Depending on their sympathy or antipathy for communism, they were propped up by the Soviet Union or the United States under the principle "He may be a son of a bitch, but he's *our* son of a bitch."[37] The 1990s and 2000s saw a spread of democracy (chapter 14) and the rise of levelheaded, humanistic leaders—not just national statesmen like Nelson Mandela, Corazon Aquino, and Ellen Johnson Sirleaf but local religious and civil-society leaders acting to improve the lives of their compatriots.[38]

A third cause was the end of the Cold War. It not only pulled the rug out from under a number of tinpot dictators but snuffed out many of the civil wars that had racked developing countries since they attained independence in the 1960s. Civil war is both a humanitarian disaster and an economic one, as facilities are destroyed, resources are diverted, children are kept out of school, and managers and workers are pulled away from work or killed. The economist Paul Collier, who calls war "development in reverse," has estimated that a typical civil war costs a country $50 billion.[39]

A fourth cause is globalization, in particular the explosion in trade made possible by container ships and jet airplanes and by the liberalization of tariffs and other barriers to investment and trade. Classical economics and common sense agree that a larger trading network should make everyone, on average, better off. As countries specialize in different goods and services, they can produce them more efficiently, and it doesn't cost them much more to offer their wares to billions of people than to thousands. At the same time buyers, shopping for the best price in a global bazaar, can get more of what they want. (Common sense is less likely to appreciate a corollary called comparative advantage, which predicts that, on average, everyone is better off when each country sells the goods and services that it can produce most efficiently *even if* the buyers could produce them still more efficiently themselves.) Notwithstanding the horror that the word elicits in many parts of the political spectrum, globalization, development analysts agree, has been a bonanza for the poor. Deaton notes, "Some argue that globalization is a neoliberal conspiracy designed to enrich a very few at the expense of many. If so, that conspiracy was a disastrous failure—or at least, it helped more than a billion people as an unintended consequence. If only unintended consequences always worked so favorably."[40]

To be sure, the industrialization of the developing world, like the Industrial Revolution two centuries before it, has produced working conditions that are harsh by the standards of modern rich countries and have elicited bitter condemnation. The Romantic movement in the 19th century was partly a reaction to the "dark satanic mills" (as William Blake called them), and since that time a loathing of industry has been a sacred value of C. P. Snow's Second Culture of literary intellectuals.[41] Nothing in Snow's essay enraged his assailant F. R. Leavis as much as this passage:

> It is all very well for us, sitting pretty, to think that material standards of living don't matter all that much. It is all very well for one, as a personal choice, to reject industrialisation—do a modern Walden if you like, and if you go without much food, see most of your children die in infancy, despise the comforts of literacy, accept twenty years off your own life, then I respect you for the strength of your aesthetic revulsion. But I don't respect you in the slightest if, even passively, you try to impose the same choice on others who are not free to choose. In fact, we know what their choice would be. For, with singular unanimity, in any country where they have had the chance, the

poor have walked off the land into the factories as fast as the factories could take them.[42]

As we have seen, Snow was accurate in his claims about advances in life and health, and he was also right that the appropriate standard in considering the plight of the poor in industrializing countries is the set of alternatives available to them where and when they live. Snow's argument is being echoed fifty years later by development experts such as Radelet, who observes that "while working on the factory floor is often referred to as sweatshop labor, it is often better than the granddaddy of all sweatshops: working in the fields as an agricultural day laborer."

When I lived in Indonesia in the early 1990s, I arrived with a somewhat romanticized view of the beauty of people working in rice paddies, together with reservations about the rapidly growing factory jobs. The longer I was there, the more I recognized how incredibly difficult it is to work in the rice fields. It's a backbreaking grind, with people eking out the barest of livings by bending over for hours in the hot sun to terrace the fields, plant the seeds, pull the weeds, transplant the seedlings, chase the pests, and harvest the grain. Standing in the pools of water brings leeches and the constant risk of malaria, encephalitis, and other diseases. And, of course, it is hot, all the time. So, it was not too much of a surprise that when factory jobs opened offering wages of $2 a day, hundreds of people lined up just to get a shot at applying.[43]

The benefits of industrial employment can go beyond material living standards. For the women who get these jobs, it can be a liberation. In her article "The Feminist Side of Sweatshops," Chelsea Follett (the managing editor of *HumanProgress*) recounts that factory work in the 19th century offered women an escape from the traditional gender roles of farm and village life, and so was held by some men at the time "sufficient to damn to infamy the most worthy and virtuous girl." The girls themselves did not always see it that way. A textile mill worker in Lowell, Massachusetts, wrote in 1840:

We are collected . . . to get money, as much of it and as fast as we can. . . . Strange would it be, if in money-loving New England, one of the most lucrative female employments should be rejected because it is toilsome, or because some people are prejudiced against it. Yankee girls have too much *independence* for *that*.[44]

Here again, experiences during the Industrial Revolution prefigure those in the developing world today. Kavita Ramdas, the head of the Global Fund for Women, said in 2001 that in an Indian village "all there is for a woman is to obey her husband and relatives, pound millet, and sing. If she moves to town, she can get a job, start a business, and get education for her children."[45] An analysis in Bangladesh confirmed that the women who worked in the garment industry (as my grandparents did in 1930s Canada) enjoyed rising wages, later marriage, and fewer and better-educated children.[46] Over the course of a generation, slums, barrios, and favelas can morph into suburbs, and the working class can become middle class.[47]

To appreciate the long-term benefits of industrialization one does not have to accept its cruelties. One can imagine an alternative history of the Industrial Revolution in which modern sensibilities applied earlier and the factories operated without children and with better working conditions for the adults. Today there are doubtless factories in the developing world that could offer as many jobs and still turn a profit while treating their workers more humanely. Pressure from trade negotiators and consumer protests has measurably improved working conditions in many places, and it is a natural progression as countries get richer and more integrated into the global community (as we will see in chapters 12 and 17 when we look at the history of working conditions in our own society).[48] Progress consists not in accepting every change as part of an indivisible package—as if we had to make a yes-or-no decision on whether the Industrial Revolution, or globalization, is a good thing or bad thing, exactly as each has unfolded in every detail. Progress consists of unbundling the features of a social process as much as we can to maximize the human benefits while minimizing the harms.

The last, and in many analyses the most important, contributor to the Great Convergence is science and technology.[49] Life is getting cheaper, in a good way. Thanks to advances in know-how, an hour of labor can buy more food, health, education, clothing, building materials, and small necessities and luxuries than it used to. Not only can people eat cheaper food and take cheaper medicines, but children can wear cheap plastic sandals instead of going barefoot, and adults can hang out together getting their hair done or watching a soccer game using cheap solar panels and appliances. As for good advice on health, farming, and business: it's better than cheap; it's free.

Today about half the adults in the world own a smartphone, and there are as many subscriptions as people. In parts of the world without roads, landlines, postal service, newspapers, or banks, mobile phones are more than a way to share gossip and cat photos; they are a major generator of wealth. They allow people to transfer money, order supplies, track the weather and markets, find day labor, get advice on health and farming practices, even obtain a primary education.[50] An analysis by the economist Robert Jensen subtitled "The Micro and Mackerel Economics of Information" showed how South Indian small fishermen increased their income and lowered the local price of fish by using their mobile phones at sea to find the market which offered the best price that day, sparing them from having to unload their perishable catch on fish-glutted towns while other towns went fishless.[51] In this way mobile phones are allowing hundreds of millions of small farmers and fishers to become the omniscient rational actors in the ideal frictionless markets of economics textbooks. According to one estimate, every cell phone adds $3,000 to the annual GDP of a developing country.[52]

The beneficent power of knowledge has rewritten the rules of global development. Development experts differ on the wisdom of foreign aid. Some argue that it does more harm than good by enriching corrupt governments and competing with local commerce.[53] Others cite recent numbers which suggest that intelligently allocated aid has in fact done tremendous good.[54] But while they disagree on the effects of donated food and dollars, all agree that donated technology—medicines, electronics, crop varieties, and best practices in agriculture, business, and public health—has been an unalloyed boon. (As Jefferson noted, he who receives an idea from me receives instruction without lessening mine.) And for all the emphasis I've placed on GDP per capita, the value of knowledge has made that measure less relevant to what we really care about, quality of life. If I had squeezed a line for Africa into the lower right corner of figure 8-2, it would look unimpressive: the line would curve upward, to be sure, but without the exponential blastoff of the lines for Europe and Asia. Charles Kenny emphasizes that the actual progress of Africa belies the shallow slope, because health, longevity, and education are so much more affordable than they used to be. Though in general people in richer countries live longer (a relationship called the Preston curve, after the economist who discovered it), the whole curve is being pushed upward, as everyone is living longer regardless of income.[55] In the richest country two centuries ago (the Netherlands), life expectancy was just forty, and in no country was it above forty-five. Today, life expectancy in the *poorest* country in the world (the Central African Republic) is fifty-four, and in no country is it *below* forty-five.[56]

Though it's easy to sneer at national income as a shallow and materialistic measure, it correlates with every indicator of human flourishing, as we will repeatedly see in the chapters to come. Most obviously, GDP per capita correlates with longevity, health, and nutrition.[57] Less obviously, it correlates with higher ethical values like peace, freedom, human rights, and tolerance.[58] Richer countries, on average, fight fewer wars with each other (chapter 11), are less likely to be riven by civil wars (chapter 11), are more likely to become and stay democratic (chapter 14), and have greater respect for human rights (chapter 14—on average, that is; Arab oil states are rich but repressive). The citizens of richer countries have greater respect for "emancipative" or liberal values such as women's equality, free speech, gay rights, participatory democracy, and protection of the environment (chapters 10 and 15). Not surprisingly, as countries get richer they get happier (chapter 18); more surprisingly, as countries get richer they get smarter (chapter 16).[59]

In explaining this Somalia-to-Sweden continuum, with poor violent repressive unhappy countries at one end and rich peaceful liberal happy ones at the other, correlation is not causation, and other factors like education, geography, history, and culture may play roles.[60] But when the quants try to tease them apart, they find that economic development does seem to be a major mover of human welfare.[61] In an old academic joke, a dean is presiding over a faculty meeting when a genie appears and offers him one of three wishes—money, fame, or wisdom. The dean replies, "That's easy. I'm a scholar. I've devoted my life to understanding. Of course I'll take wisdom." The genie waves his hand and vanishes in a puff of smoke. The smoke clears to reveal the dean with his head in his hands, lost in thought. A minute elapses. Ten minutes. Fifteen. Finally a professor calls out, "Well? Well?" The dean mutters, "I should have taken the money."

CHAPTER 9

INEQUALITY

"B ut is it all going to the rich?" That's a natural question to ask in developed countries in the second decade of the 21st century, when economic inequality has become an obsession. Pope Francis called it "the root of social evil"; Barack Obama, "the defining challenge of our time." Between 2009 and 2016, the proportion of articles in the *New York Times* containing the word *inequality* soared tenfold, reaching 1 in 73.[1] The new conventional wisdom is that the richest one percent have skimmed off all the economic growth of recent decades, and everyone else is treading water or slowly sinking. If so, the explosion of wealth documented in the previous chapter would no longer be worth celebrating, since it would have ceased contributing to overall human welfare.

Economic inequality has long been a signature issue of the left, and it rose in prominence after the Great Recession began in 2007. It ignited the Occupy Wall Street movement in 2011 and the presidential candidacy of the self-described socialist Bernie Sanders in 2016, who proclaimed that "a nation will not survive morally or economically when so few have so much, while so many have so little."[2] But in that year the revolution devoured its children and propelled the candidacy of Donald Trump, who claimed that the United States had become "a third-world country" and blamed the declining fortunes of the working class not on Wall Street and the one percent but on immigration and foreign trade. The left and right ends of the political spectrum, incensed by economic inequality for their different reasons, curled around to meet each other, and their shared cynicism about the modern economy helped elect the most radical American president in recent times.

Has rising inequality really immiserated the majority of citizens? Economic inequality undoubtedly has increased in most Western countries since its low point around 1980, particularly in the United States and other English-speaking countries, and especially in the contrast between the very richest and everyone else.[3] Economic inequality is usually measured by the Gini coefficient, a number that can vary between 0, when everyone has the same as everyone else, and 1, when one person has everything and everyone else has nothing. (Gini values generally range from .25 for the most egalitarian income distributions, such as in Scandinavia after taxes and benefits, to .7 for a highly unequal distribution such as the one in South Africa.) In the United States, the Gini index for market income (before taxes and benefits) rose from .44 in 1984 to .51 in 2012. Inequality can also be measured by the proportion of total income that is earned by a given fraction (quantile) of the population. In the United States, the share of income going to the richest one percent grew from 8 percent in 1980 to 18 percent in 2015, while the share going to the richest *tenth* of one percent grew from 2 percent to 8 percent.[4]

There's no question that some of the phenomena falling under the inequality rubric (there are many) are serious and must be addressed, if only to defuse the destructive agendas they have incited, such as abandoning market economies, technological progress, and foreign trade. Inequality is devilishly complicated to analyze (in a population of one million, there are 999,999 ways in which they can be unequal), and the subject has filled many books. I need a chapter on the topic because so many people have been swept up in the dystopian rhetoric and see inequality as a sign that modernity has failed to improve the human condition. As we will see, this is wrong, and for many reasons.

~

The starting point for understanding inequality in the context of human progress is to recognize that income inequality is not a fundamental component of well-being. It is not like health, prosperity, knowledge, safety, peace, and the other areas of progress I examine in these chapters. The reason is captured in an old joke from the Soviet Union. Igor and Boris are dirt-poor peasants, barely scratching enough crops from their small plots of land to feed their families. The only difference between them is that Boris owns a scrawny goat. One day a fairy appears to Igor and grants him a wish. Igor says, "I wish that Boris's goat should die."

The point of the joke, of course, is that the two peasants have become more equal but that neither is better off, aside from Igor's indulging his spiteful envy. The point is made with greater nuance by the philosopher Harry Frankfurt in his 2015 book *On Inequality*.[5] Frankfurt argues that inequality itself is not morally objectionable; what is objectionable is *poverty.* If a person lives a long, healthy, pleasurable, and stimulating life, then how much money the Joneses earn, how big their house is, and how many cars they drive are morally irrelevant. Frankfurt writes, "From the point of view of morality, it is not important everyone should have *the same.* What is morally important is that each should have *enough.*"[6] Indeed, a narrow focus on economic inequality can be destructive if it distracts us into killing Boris's goat instead of figuring out how Igor can get one.

The confusion of inequality with poverty comes straight out of the lump fallacy—the mindset in which wealth is a finite resource, like an antelope carcass, which has to be divvied up in zero-sum fashion, so that if some people end up with more, others must have less. As we just saw, wealth is not like that: since the Industrial Revolution, it has expanded exponentially.[7] That means that when the rich get richer, the poor can get richer, too. Even experts repeat the lump fallacy, presumably out of rhetorical zeal rather than conceptual confusion. Thomas Piketty, whose 2014 bestseller *Capital in the Twenty-First Century* became a talisman in the uproar over inequality, wrote, "The poorer half of the population are as poor today as they were in the past, with barely 5 percent of total wealth in 2010, just as in 1910."[8] But total wealth today is vastly greater than it was in 1910, so if the poorer half own the same proportion, they are far richer, not "as poor."

A more damaging consequence of the lump fallacy is the belief that if some people get richer, they must have stolen more than their share from everyone else. A famous illustration by the philosopher Robert Nozick, updated for the 21st century, shows why this is wrong.[9] Among the world's billionaires is J. K. Rowling, author of the *Harry Potter* novels, which have sold more than 400 million copies and have been adapted into a series of films seen by a similar number of people.[10] Suppose that a billion people have handed over $10 each for the pleasure of a *Harry Potter* paperback or movie ticket, with a tenth of the proceeds going to Rowling. She has become a billionaire, increasing inequality, but she has made people better off, not worse off (which is not to say that every rich person has made people better off). This doesn't mean that Rowling's wealth is just deserts for her effort or skill, or a reward for the literacy and happiness she added to the world; no committee ever judged that she deserved to be that rich. Her wealth arose as a by-product of the voluntary decisions of billions of book buyers and moviegoers.

To be sure, there may be reasons to worry about inequality itself, not just poverty. Perhaps most people are like Igor and their happiness is determined by how they compare with their fellow citizens rather than how well-off they are in absolute terms. When the rich get too rich, everyone else feels poor, so inequality lowers well-being even if everyone gets richer. This is an old idea in social psychology, variously called the theory of social comparison, reference groups, status anxiety, or relative deprivation.[11] But the idea must be kept in perspective. Imagine Seema, an illiterate woman in a poor country who is village-bound, has lost half her children to disease, and will die at fifty, as do most of the people she knows. Now imagine Sally, an educated person in a rich country who has visited several cities and national parks, has seen her children grow up, and will live to eighty, but is stuck in the lower middle class. It's conceivable that Sally, demoralized by the conspicuous wealth she will never attain, is not particularly happy, and she might even be unhappier than Seema, who is grateful for small mercies. Yet it would be mad to suppose that Sally is not better off, and positively depraved to conclude that one may as well not try to improve Seema's life because it might improve her neighbors' lives even more and leave her no happier.[12]

In any case, the thought experiment is moot, because in real life Sally almost certainly *is* happier. Contrary to an earlier belief that people are so mindful of their richer compatriots that they keep resetting their internal happiness meter to the baseline no matter how well they are doing, we will see in chapter 18 that richer people and people in richer countries are (on average) happier than poorer people and people in poorer countries.[13]

But even if people are happier when they and their countries get richer, might they become more miserable if others around them are still richer than they are—that is, as economic inequality increases? In their well-known book *The Spirit Level*, the epidemiologists Richard Wilkinson and Kate Pickett claim that countries with greater income inequality also have higher rates of homicide, imprisonment, teen pregnancy, infant mortality, physical and mental illness, social distrust, obesity, and substance abuse.[14] The economic inequality *causes* the ills, they argue: unequal societies make people feel that they are pitted in a winner-take-all competition for dominance, and the stress makes them sick and self-destructive.

The Spirit Level theory has been called "the left's new theory of everything," and it is as problematic as any other theory that leaps from a tangle of correlations to a single-cause explanation. For one thing, it's not obvious that people are whipped into competitive anxiety by the existence of J. K. Rowling and Sergey Brin as opposed to their own, local rivals for professional, romantic, and social success. Worse, economically egalitarian countries like Sweden and France differ from lopsided countries like Brazil and South Africa in many ways other than their income distribution. The egalitarian countries are, among other things, richer, better educated, better governed, and more culturally homogeneous, so a raw correlation between inequality and happiness (or any other social good) may show only that there are many reasons why it's better to live in Denmark than in Uganda. Wilkinson and Pickett's sample was restricted to developed countries, but even within that sample the correlations are evanescent, coming and going with choices about which countries to include.[15] Wealthy but unequal countries, such as Singapore and Hong Kong, are often socially healthier than poorer but more equal countries, such as those of ex-Communist Eastern Europe.

Most damagingly, the sociologists Jonathan Kelley and Mariah Evans have snipped the causal link joining inequality to happiness in a study of two hundred thousand people in sixty-eight societies over three decades.[16] (We will examine how happiness and life satisfaction are measured in chapter 18.) Kelley and Evans held constant the major factors that are known to affect happiness, including GDP per capita, age, sex, education, marital status, and religious attendance, and found that the theory that inequality causes unhappiness "comes to shipwreck on the rock of the facts." In developing countries, inequality is not dispiriting but heartening: people in the more unequal societies are *happier*. The authors suggest that whatever envy, status anxiety, or relative deprivation people may feel in poor, unequal countries is swamped by *hope*. Inequality is seen as a harbinger of opportunity, a sign that education and other routes to upward mobility might pay off for them and their children. Among developed countries (other than formerly Communist ones), inequality made no difference one way or another. (In formerly Communist countries, the effects were also equivocal: inequality hurt the aging generation that grew up under communism, but helped or made no difference to the younger generations.)

The fickle effects of inequality on well-being bring up another common confusion in these discussions: the conflation of inequality with *unfairness*. Many studies in psychology have shown that people, including young children, prefer windfalls to be split evenly among participants, even if everyone ends up with less overall. That led some psychologists to posit a syndrome called inequity aversion: an apparent desire to spread the wealth. But in their recent article "Why People Prefer Unequal Societies," the psychologists Christina Starmans, Mark Sheskin, and Paul Bloom took another look at the studies and found that people prefer *un-*

equal distributions, both among fellow participants in the lab and among citizens in their country, as long as they sense that the allocation is *fair:* that the bonuses go to harder workers, more generous helpers, or even the lucky winners of an impartial lottery.[17] "There is no evidence so far," the authors conclude, "that children or adults possess any general aversion to inequality." People are content with economic inequality as long as they feel that the country is meritocratic, and they get angry when they feel it isn't. Narratives about the *causes* of inequality loom larger in people's minds than the *existence* of inequality. That creates an opening for politicians to rouse the rabble by singling out cheaters who take more than their fair share: welfare queens, immigrants, foreign countries, bankers, or the rich, sometimes identified with ethnic minorities.[18]

In addition to effects on individual psychology, inequality has been linked to several kinds of society-wide dysfunction, including economic stagnation, financial instability, intergenerational immobility, and political influence-peddling. These harms must be taken seriously, but here too the leap from correlation to causation has been contested.[19] Either way, I suspect that it's less effective to aim at the Gini index as a deeply buried root cause of many social ills than to zero in on solutions to each problem: investment in research and infrastructure to escape economic stagnation, regulation of the finance sector to reduce instability, broader access to education and job training to facilitate economic mobility, electoral transparency and finance reform to eliminate illicit influence, and so on. The influence of money on politics is particularly pernicious because it can distort every government policy, but it's not the same issue as income inequality. After all, in the absence of electoral reform the richest donors can get the ear of politicians whether they earn 2 percent of national income or 8 percent of it.[20]

Economic inequality, then, is not itself a dimension of human wellbeing, and it should not be confused with unfairness or with poverty. Let's now turn from the moral significance of inequality to the question of why it has changed over time.

～

The simplest narrative of the history of inequality is that it comes with modernity. We must have begun in a state of original equality, because when there is no wealth, everyone has equal shares of nothing, and then, when wealth is created, some can have more of it than others. Inequality, in this story, started at zero, and as wealth increased over time, inequality grew with it. But the story is not quite right.

Hunter-gatherers are by all appearances highly egalitarian, a fact that inspired Marx and Engels's theory of "primitive communism." But ethnographers point out that the image of forager egalitarianism is misleading. For one thing, the hunter-gatherer bands that are still around for us to study are not representative of an ancestral way of life, because they have been pushed into marginal lands and lead nomadic lives that make the accumulation of wealth impossible, if for no other reason than that it would be a nuisance to carry around. But sedentary hunter-gatherers, such as the natives of the Pacific Northwest, which is flush with salmon, berries, and fur-bearing animals, were florid inegalitarians, and developed a hereditary nobility who kept slaves, hoarded luxuries, and flaunted their wealth in gaudy potlatches. Also, while nomadic hunter-gatherers share meat, since hunting is largely a matter of luck and sharing a windfall insures everyone against days in which they come home empty-handed, they are less likely to share plant foods, since gathering is a matter of effort, and indiscriminate sharing would allow free-riding.[21] Some degree of inequality is universal across societies, as is an awareness of inequality.[22] A recent survey of inequality in the forms of wealth that are possible for hunter-gatherers (houses, boats, and hunting and foraging returns) found that they were "far from a state of 'primitive communism'": the Ginis averaged .33, close to the value for disposable income in the United States in 2012.[23]

What happens when a society starts to generate substantial wealth? An increase in *absolute* inequality (the difference between the richest and poorest) is almost a mathematical necessity. In the absence of an Income Distribution Authority that parcels out identical shares, some people are bound to take greater advantage of the new opportunities than others, whether by luck, skill, or effort, and they will reap disproportionate rewards.

An increase in *relative* inequality (measured by the Gini or income shares) is not mathematically necessary, but it is highly likely. According to a famous conjecture by the economist Simon Kuznets, as countries get richer they should get less equal, because some people leave farming for higher-paying lines of work while the rest stay in rural squalor. But eventually a rising tide lifts all the boats. As more of the population gets swept into the modern economy, inequality should decline, tracing out an inverted U. This hypothetical arc of inequality over time is called the Kuznets curve.[24]

In the preceding chapter we saw hints of a Kuznets curve for inequality between countries. As the Industrial Revolution gathered steam, European countries made a Great Escape from universal poverty, leaving the other countries behind. As Deaton observes, "A better world makes

for a world of differences; escapes make for inequality."[25] Then, as globalization proceeded and wealth-generating know-how spread, poor
countries started catching up in a Great Convergence. We saw hints of a
drop in global inequality in the blastoff of GDP in Asian countries (figure
8-2), in the morphing of the world income distribution from snail to two-
humped camel to one-humped dromedary (figure 8-3), and in the plunging proportion (figure 8-4) and number (figure 8-5) of people living in
extreme poverty.

To confirm that these gains really constitute a decline in inequality—
that poor countries are getting richer faster than the rich countries are
getting richer—we need a single measure that combines them, an international Gini, which treats each country like a person. Figure 9-1 shows
that the international Gini rose from a low of .16 in 1820, when all countries were poor, to a high of .56 in 1970, when some were rich, and then,
as Kuznets predicted, it plateaued and began to droop in the 1980s.[26]
But an international Gini is a bit misleading, because it counts an improvement in the living standards of a billion Chinese as equivalent to
an improvement in the standards of, say, four million Panamanians.
Figure 9-1 also shows an international Gini calculated by the economist
Branko Milanović in which every country counts in proportion to its

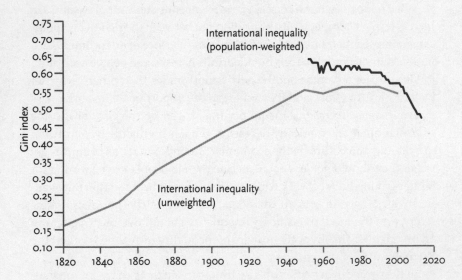

Figure 9-1: International inequality, 1820–2013

Sources: International inequality: OECD Clio Infra Project, Moatsos et al. 2014; data are for market
household income across countries. Population-weighted international inequality: Milanović 2012;
data for 2012 and 2013 provided by Branko Milanović, personal communication.

population, making the human impact of the drop in inequality more apparent.

Still, an international Gini treats all the Chinese as if they earned the same amount, all the Americans as if they earned the American average, and so on, and as a result it underestimates inequality across the human race. A global Gini, in which every *person* counts the same, regardless of country, is harder to calculate, because it requires mixing the incomes from disparate countries into a single bowl, but two estimates are shown in figure 9-2. The lines float at different heights because they were calibrated in dollars adjusted for purchasing parity in different years, but their slopes trace out a kind of Kuznets curve: after the Industrial Revolution, global inequality rose steadily until around 1980, then started to fall. The international and global Gini curves show that despite the anxiety about rising inequality within Western countries, *inequality in the world is declining*. That's a circuitous way to state the progress, though: what's significant about the decline in inequality is that it's a decline in poverty.

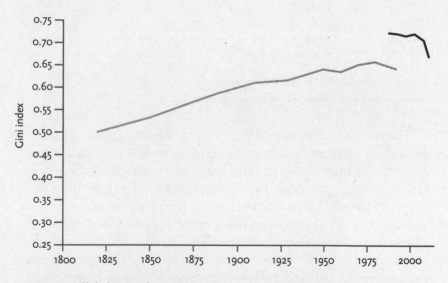

Figure 9-2: Global inequality, 1820–2011

Source: Milanović 2016, fig. 3.1. The left-hand curve shows 1990 international dollars of disposable income per capita; the right-hand curve shows 2005 international dollars, and combines household surveys of per capita disposable income and consumption.

The version of inequality that has generated the recent alarm is the inequality within developed countries like the United States and the

United Kingdom. The long view of these countries is shown in figure 9-3. Until recently, both countries traveled a Kuznets arc. Inequality rose during the Industrial Revolution and then began to fall, first gradually in the late 19th century, then steeply in the middle decades of the 20th. But then, starting around 1980, inequality bounced into a decidedly un-Kuznetsian rise. Let's examine each segment in turn.

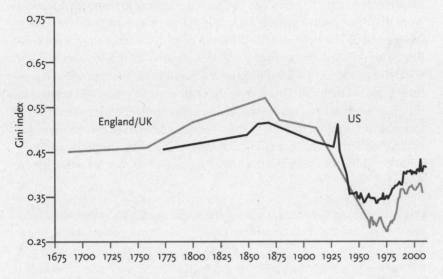

Figure 9-3: Inequality, UK and US, 1688–2013

Source: Milanović 2016, fig. 2.1, disposable income per capita.

The rise and fall in inequality in the 19th century reflects Kuznets's expanding economy, which gradually pulls more people into urban, skilled, and thus higher-paying occupations. But the 20th-century plunge—which has been called the Great Leveling or the Great Compression—had more sudden causes. The plunge overlaps the two world wars, and that is no coincidence: major wars often level the income distribution.[27] Wars destroy wealth-generating capital, inflate away the assets of creditors, and induce the rich to put up with higher taxes, which the government redistributes into the paychecks of soldiers and munition workers, in turn increasing the demand for labor in the rest of the economy.

Wars are just one kind of catastrophe that can generate equality by the logic of Igor and Boris. The historian Walter Scheidel identifies "Four Horsemen of Leveling": mass-mobilization warfare, transformative revolution, state collapse, and lethal pandemics. In addition to obliterating wealth (and, in the communist revolutions, the people who owned it), the four horsemen reduce inequality by killing large numbers of workers, driving up the wages of those who survive. Scheidel concludes, "All of us who prize greater economic equality would do well to remember that with the rarest of exceptions it was only ever brought forth in sorrow. Be careful what you wish for."[28]

Scheidel's warning applies to the long run of history. But modernity has brought a more benign way to reduce inequality. As we have seen, a market economy is the best poverty-reduction program we know of for an entire country. It is ill-equipped, however, to provide for individuals within that country who have nothing to exchange: the young, the old, the sick, the unlucky, and others whose skills and labor are not valuable enough to others for them to earn a decent living in return. (Another way of putting it is that a market economy maximizes the average, but we also care about the variance and the range.) As the circle of sympathy in a country expands to encompass the poor (and as people want to insure themselves should they ever become poor), they increasingly allocate a portion of their pooled resources—that is, government funds—to alleviating that poverty. Those resources have to come from somewhere. They may come from a corporate or sales tax, or a sovereign wealth fund, but in most countries they largely come from a graduated income tax, in which richer citizens pay at a higher rate because they don't feel the loss as sharply. The net result is "redistribution," but that is something of a misnomer, because the goal is to raise the bottom, not lower the top, even if in practice the top is lowered.

Those who condemn modern capitalist societies for callousness toward the poor are probably unaware of how little the pre-capitalist societies of the past spent on poor relief. It's not just that they had less to spend in absolute terms; they spent a smaller proportion of their wealth. A *much* smaller proportion: from the Renaissance through the early 20th century, European countries spent an average of 1.5 percent of their GDP on poor relief, education, and other social transfers. In many countries and periods, they spent nothing at all.[29]

In another example of progress, sometimes called the Egalitarian Revolution, modern societies now devote a substantial chunk of their wealth to health, education, pensions, and income support.[30] Figure 9-4 shows that social spending took off in the middle decades of the 20th century (in the United States, with the New Deal in the 1930s; in other

developed countries, with the rise of the welfare state after World War II).
Social spending now takes up a median of 22 percent of their GDP.[31]

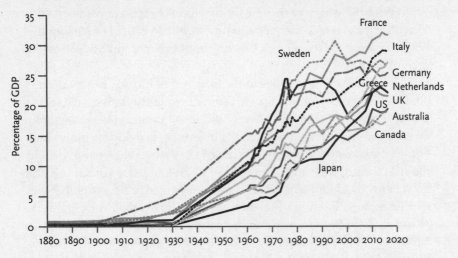

Figure 9-4: Social spending, OECD countries, 1880–2016

Source: *Our World in Data,* Ortiz-Ospina & Roser 2016b, based on data from Lindert 2004 and OECD 1985, 2014, 2017. The Organisation for Economic Co-operation and Development includes thirty-five democratic states with market economies.

The explosion in social spending has redefined the mission of government: from warring and policing to also nurturing.[32] Governments underwent this transformation for several reasons. Social spending inoculates citizens against the appeal of communism and fascism. Some of the benefits, like universal education and public health, are public goods that accrue to everyone, not just the direct beneficiaries. Many of the programs indemnify citizens against misfortunes for which they can't or won't insure themselves (hence the euphemism "social safety net"). And assistance to the needy assuages the modern conscience, which cannot bear the thought of the Little Match Girl freezing to death, Jean Valjean imprisoned for stealing bread to save his starving sister, or the Joads burying Grampa by the side of Route 66.

Since there's no point in everyone sending money to the government and getting it right back (minus the bureaucracy's cut), social spending is designed to help people who have less money, with the bill footed by people who have more money. This is the principle known as redistribution, the welfare state, social democracy, or socialism (misleadingly, because free-market capitalism is compatible with any amount of social spending). Whether or not the social spending is *designed* to reduce inequality, that is one of its effects, and the rise in social expenditures from the 1930s through the 1970s explains part of the decline in the Gini.

Social spending demonstrates an uncanny aspect of progress that we'll encounter again in subsequent chapters.[33] Though I am skittish about any notion of historical inevitability, cosmic forces, or mystical arcs of justice, some kinds of social change really do seem to be carried along by an inexorable tectonic force. As they proceed, certain factions oppose them hammer and tongs, but resistance turns out to be futile. Social spending is an example. The United States is famously resistant to anything smacking of redistribution. Yet it allocates 19 percent of its GDP to social services, and despite the best efforts of conservatives and libertarians the spending has continued to grow. The most recent expansions are a prescription drug benefit introduced by George W. Bush and the eponymous health insurance plan known as Obamacare introduced by his successor.

Indeed, social spending in the United States is even higher than it appears, because many Americans are forced to pay for health, retirement, and disability benefits through their employers rather than the government. When this privately administered social spending is added to the public portion, the United States vaults from twenty-fourth into second place among the thirty-five OECD countries, just behind France.[34]

For all their protestations against big government and high taxes, people *like* social spending. Social Security has been called the third rail of American politics, because if politicians touch it they die. According to legend, an irate constituent at a town-hall meeting warned his representative, "Keep your government hands off my Medicare" (referring to the government health insurance program for seniors).[35] No sooner did Obamacare pass than the Republican Party made it a sacred cause to repeal it, but each of their assaults on it after gaining control of the presidency in 2017 was beaten back by angry citizens at town-hall meetings and legislators afraid of their ire. In Canada the top two national pastimes (after hockey) are complaining about their health care system and boasting about their health care system.

Developing countries today, like developed countries a century ago, stint on social spending. Indonesia, for example, spends 2 percent of its GDP, India 2.5 percent, and China 7 percent. But as they get richer they become more munificent, a phenomenon called Wagner's Law.[36] Between 1985 and 2012 Mexico quintupled its proportion of social spending, and Brazil's now stands at 16 percent.[37] Wagner's Law appears to be not a cautionary tale about overweening government and bureaucratic bloat but a manifestation of progress. The economist Leandro Prados de la Escosura found a strong correlation between the percentage of GDP that an OECD country allocated to social transfers as it developed between 1880 and 2000 and its score on a composite measure of prosperity, health, and education.[38] And tellingly, the number of libertarian paradises in the world—developed countries without substantial social spending—is zero.[39]

The correlation between social spending and social well-being holds only up to a point: the curve levels off starting at around 25 percent and may even drop off at higher proportions. Social spending, like everything, has downsides. As with all insurance, it can create a "moral hazard" in which the insured slack off or take foolish risks, counting on the insurer to bail them out if they fail. And since the premiums have to cover the payouts, if the actuaries get the numbers wrong or the numbers change so that more money is taken out than put in, the system can collapse. In reality social spending is never exactly like insurance but is a combination of insurance, investment, and charity. Its success thus depends on the degree to which the citizens of a country sense they are part of one community, and that fellow feeling can be strained when the beneficiaries are disproportionately immigrants or ethnic minorities.[40] These tensions are inherent to social spending and will always be politically contentious. Though there is no "correct amount," all developed states have decided that the benefits of social transfers outweigh the costs and have settled on moderately large amounts, cushioned by their massive wealth.

∼

Let's complete our tour of the history of inequality by turning to the final segment in figure 9-3, the rise of inequality in wealthy nations that began around 1980. This is the development that inspired the claim that life has gotten worse for everyone but the richest. The rebound defies the Kuznets curve, in which inequality was supposed to have settled into a low equilibrium. Many explanations have been proffered for this surprise.[41] Wartime restrictions on economic competition may have been sticky, outlasting World War II, but they finally dissipated, freeing the rich to get richer from their investment income and opening up an arena of dynamic economic competition with winner-take-all payoffs. The ideological shift associated with Ronald Reagan and Margaret Thatcher slowed the movement toward greater social spending financed by taxes on the rich while eroding social norms against extravagant salaries and conspicuous wealth. As more people stayed single or got divorced, and at the same time more power couples pooled two fat paychecks, the variance in income from household to household was bound to increase, even if the paychecks had stayed the same. A "second industrial revolution" driven by electronic technologies replayed the Kuznets rise by creating a demand for highly skilled professionals, who pulled away from the less educated at the same time that the jobs requiring less education were eliminated by automation. Globalization allowed workers in China, India, and elsewhere to underbid their American competitors in a worldwide labor market, and the domestic companies that failed to take advantage of these offshoring opportunities were outcompeted on price. At the same time, the intellectual output of the most successful analysts, entrepreneurs, investors, and creators was increasingly available to a gargantuan worldwide market. The Pontiac worker is laid off, while J. K. Rowling becomes a billionaire.

Milanović has combined the two inequality trends of the past thirty years—declining inequality worldwide, increasing inequality within rich countries—into a single graph which pleasingly takes the shape of an elephant (figure 9-5). This "growth incidence curve" sorts the world's population into twenty numerical bins or quantiles, from poorest to richest, and plots how much each bin gained or lost in real income per capita between 1988 (just before the fall of the Berlin Wall) and 2008 (just before the Great Recession).

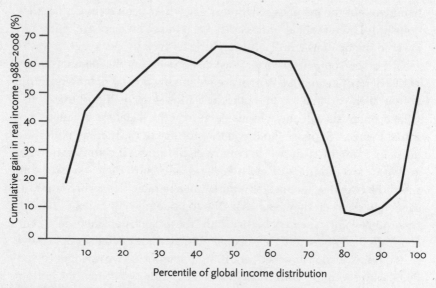

Figure 9-5: Income gains, 1988–2008

Source: Milanović 2016, fig. 1.3.

The cliché about globalization is that it creates winners and losers, and the elephant curve displays them as peaks and valleys. It reveals . that the winners include most of humanity. The elephant's bulk (its body and head), which includes about seven-tenths of the world's population, consists of the "emerging global middle class," mainly in Asia. Over this period they saw cumulative gains of 40 to 60 percent in their real incomes. The nostrils at the tip of the trunk consist of the world's richest one percent, who also saw their incomes soar. The rest of the trunk tip, which includes the next 4 percent down, didn't do badly either. Where the bend of the trunk hovers over the floor around the 85th percentile we see globalization's "losers": the lower middle classes of the rich world, who gained less than 10 percent. These are the focus of the new concern about inequality: the "hollowed-out middle class," the Trump supporters, the people globalization left behind.

I couldn't resist plotting the most recognizable elephant in Milanović's herd, because it serves as a vivid mnemonic for the effects of globalization (and it rounds out a nice menagerie with the camel and dromedary in figure 8-3). But the curve makes the world look more unequal than it really is, for two reasons. One is that the financial crisis of 2008, which postdated the graph, had a strangely equalizing effect on the world. The Great Recession, Milanović points out, was really a recession in North Atlantic countries. The incomes of the world's richest one percent were trimmed, but the incomes of workers elsewhere soared (in China, they doubled). Three years after the crisis we still see an elephant, but it has lowered the tip of its trunk while arching its back twice as high.[42]

The other elephant-distorter is a conceptual point that bedevils many discussions of inequality. Whom are we talking about when we say "the bottom fifth" or "the top one percent"? Most income distributions use what economists call anonymous data: they track statistical ranges, not actual people.[43] Suppose I told you that the age of the median American declined from thirty in 1950 to twenty-eight in 1970. If your first thought is "Wow, how did that guy get two years younger?" then you have confused the two: the "median" is a rank, not a person. Readers commit the same fallacy when they read that "the top one percent in 2008" had incomes that were 50 percent higher than "the top one percent in 1988" and conclude that a bunch of rich people got half again richer. People move in and out of income brackets, shuffling the order, so we're not necessarily talking about the same individuals. The same is true for "the bottom fifth" and every other statistical bin.

Nonanonymous or longitudinal data, which track people over time, are unavailable in most countries, so Milanović did the next best thing and tracked individual quantiles in particular countries, so that, say, poor Indians in 1988 were no longer being compared with poor Ghanaians in 2008.[44] He still got an elephantoid, but with a much higher tail and haunches, because the poorer classes of so many countries rose out of extreme poverty. The pattern remains—globalization helped the lower and middle classes of poor countries, and the upper class of rich countries, much more than it helped the lower middle class of rich countries— but the differences are less extreme.

~

Now that we have run through the history of inequality and seen the forces that push it around, we can evaluate the claim that the growing inequality of the past three decades means that the world is getting worse—that only the rich have prospered, while everyone else is stagnating or suffering. The rich certainly have prospered more than anyone else, perhaps more than they should have, but the claim about everyone else is not accurate, for a number of reasons.

Most obviously, it's false for the world as a whole: the majority of the human race has become much better off. The two-humped camel has become a one-humped dromedary; the elephant has a body the size of, well, an elephant; extreme poverty has plummeted and may disappear; and both international and global inequality coefficients are in decline. Now, it's true that the world's poor have gotten richer in part at the expense of the American lower middle class, and if I were an American politician I would not publicly say that the tradeoff was worth it. But as citizens of the world considering humanity as a whole, we have to say that the tradeoff is worth it.

But even in the lower and lower middle classes of rich countries, moderate income gains are not the same as a decline in living standards. Today's discussions of inequality often compare the present era unfavorably with a golden age of well-paying, dignified, blue-collar jobs that have been made obsolete by automation and globalization. This idyllic image is belied by contemporary depictions of the harshness of working-class life in that era, both in journalistic exposés (such as Michael Harrington's 1962 *The Other America*) and in realistic films (such as *On the Waterfront, Blue Collar, Coal Miner's Daughter,* and *Norma Rae*). The historian Stephanie Coontz, a debunker of 1950s nostalgia, puts some numbers to the depictions:

A full 25 percent of Americans, 40 to 50 million people, were poor in the mid-1950s, and in the absence of food stamps and housing programs, this poverty was searing. Even at the end of the 1950s, a third of American children were poor. Sixty percent of Americans over sixty-five had incomes below $1,000 in 1958, considerably below the $3,000 to $10,000 level considered to represent middle-class status. A majority of elders also lacked medical insurance. Only half the population had savings in 1959; one-quarter of the population had no liquid assets at all. Even when we consider only native-born, white families, one-third could not get by on the income of the household head.[45]

How do we reconcile the obvious improvements in living standards in recent decades with the conventional wisdom of economic stagnation? Economists point to four ways in which inequality statistics can paint a misleading picture of the way people live their lives, each depending on a distinction we have examined.

The first is the difference between relative and absolute prosperity. Just as not all children can be above average, it's not a sign of stagnation if the proportion of income earned by the bottom fifth does not increase over time. What's relevant to well-being is how much people earn, not how high they rank. A recent study by the economist Stephen Rose divided the American population into classes using fixed milestones rather than quantiles. "Poor" was defined as an income of $0–$30,000 (in 2014 dollars) for a family of three, "lower middle class" as $30,000–$50,000, and so on.[46] The study found that in absolute terms, Americans have been moving on up. Between 1979 and 2014, the percentage of poor Americans dropped from 24 to 20, the percentage in the lower middle class dropped from 24 to 17, and the percentage in the middle class shrank from 32 to 30. Where did they go? Many ended up in the upper middle class ($100,000–$350,000), which grew from 13 to 30 percent of the population, and in the upper class, which grew from 0.1 percent to 2 percent. The middle class is being hollowed out in part because so many Americans are becoming affluent. Inequality undoubtedly increased— the rich got richer faster than the poor and middle class got richer—but everyone (on average) got richer.

The second confusion is the one between anonymous and longitudinal data. If (say) the bottom fifth of the American population gained no ground in twenty years, it does not mean that Joe the Plumber got the same paycheck in 1988 that he did in 2008 (or one that's a bit higher, owing to cost-of-living increases). People earn more as they get older and gain experience, or switch from a lower-paying job to a higher-paying one, so Joe may have moved from the bottom fifth into, say, the middle fifth, while a younger man or woman or an immigrant took his place at the bottom. The turnover is by no means small. A recent study using longitudinal data showed that half of Americans will find themselves among the top tenth of income earners for at least one year of their working lives, and that one in nine will find themselves in the top one percent (though most don't stay there for long).[47] This may be one of the reasons that economic opinions are subject to the Optimism Gap (the "I'm OK, They're Not" bias): a majority of Americans believe that the standard of living of the middle class has declined in recent years but that their own standard of living has improved.[48]

A third reason that rising inequality has not made the lower classes worse off is that low incomes have been mitigated by social transfers. For all its individualist ideology, the United States has a lot of redistribution. The income tax is still graduated, and low incomes are buffered by a "hidden welfare state" that includes unemployment insurance, Social Security, Medicare, Medicaid, Temporary Assistance for Needy Families, food stamps, and the Earned Income Tax Credit, a kind of negative income tax in which the government boosts the income of low earners. Put them together and America becomes far less unequal. In 2013 the Gini index for American market income (before taxes and transfers) was a high .53; for disposable income (after taxes and transfers) it was a moderate .38.[49] The United States has not gone as far as countries like Germany and Finland, which start off with a similar market income distribution but level it more aggressively, pushing their Ginis down into the high .2s and sidestepping most of the post-1980s inequality rise. Whether or not the generous European welfare state is sustainable over the long run and transplantable to the United States, some kind of welfare state may be found in all developed countries, and it reduces inequality even when it is hidden.[50]

These transfers have not just reduced income inequality (in itself a dubious accomplishment) but boosted the incomes of the nonrich (a real one). An analysis by the economist Gary Burtless has shown that between 1979 and 2010 the disposable incomes of the lowest four income quintiles grew by 49, 37, 36, and 45 percent, respectively.[51] And that was before the long-delayed recovery from the Great Recession: between 2014 and 2016, median wages leapt to an all-time high.[52]

Even more significant is what has happened at the bottom of the scale. Both the left and the right have long expressed cynicism about antipoverty programs, as in Ronald Reagan's famous quip, "Some years ago, the federal government declared war on poverty, and poverty won." In reality, poverty is losing. The sociologist Christopher Jencks has calculated that when the benefits from the hidden welfare state are added up, and the cost of living is estimated in a way that takes into account the improving quality and falling price of consumer goods, the poverty rate has fallen in the past fifty years by more than three-quarters, and in 2013 stood at 4.8 percent.[53] Three other analyses have come to the same conclusion; data from one of them, by the economists Bruce Meyer and James Sullivan, are shown in the upper line in figure 9-6. The progress stagnated around the time of the Great Recession, but it picked up in 2015 and 2016 (not shown in the graph), when middle-class income reached a record high and the poverty rate showed its largest drop since 1999.[54] And in yet another unsung accomplishment, the poorest of the poor—the unsheltered homeless—fell in number between 2007 and 2015 by almost a third, despite the Great Recession.[55]

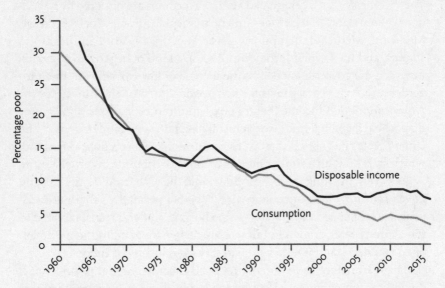

Figure 9-6: Poverty, US, 1960–2016

Sources: Meyer & Sullivan 2017a, b. "Disposable income" refers to their "After-tax money income," including credits, adjusted for inflation using the bias-corrected CPI-U-RS, and representing a family with two adults and two children. "Consumption" refers to data from the BLS Consumer Expenditure Survey on food, housing, vehicles, appliances, furnishings, clothing, jewelry, insurance, and other expenses. "Poverty" corresponds to the US Census definition for 1980, adjusted for inflation; anchoring the poverty line in other years would result in different absolute numbers but the same trends. See Meyer & Sullivan 2011, 2012, and 2017a, b for details.

The lower line in figure 9-6 highlights the fourth way in which inequality measures understate the progress of the lower and middle classes in rich countries.[56] Income is just a means to an end: a way of paying for things that people need, want, and like, or as economists gracelessly call it, consumption. When poverty is defined in terms of what people consume rather than what they earn, we find that the American poverty rate has declined by *ninety percent* since 1960, from 30 percent of the population to just 3 percent. The two forces that have famously increased inequality in income have at the same time decreased inequality in what matters. The first, globalization, may produce winners and losers in income, but in consumption it makes almost everyone a winner. Asian factories, container ships, and efficient retailing bring goods to the masses that were formerly luxuries for the rich. (In 2005 the economist Jason Furman estimated that Walmart saved the typical American family $2,300 a year.)[57] The second force, technology, continually revolutionizes the meaning of income (as we saw in the discussion of the paradox of value in chapter 8). A dollar today, no matter how heroically adjusted for inflation, buys far more betterment of life than a dollar yesterday. It buys things that didn't exist, like refrigeration, electricity, toilets, vaccinations, telephones, contraception, and air travel, and it transforms things that do exist, such as a party line patched by a switchboard operator to a smartphone with unlimited talk time.

Together, technology and globalization have transformed what it means to be a poor person, at least in developed countries. The old stereotype of poverty was an emaciated pauper in rags. Today, the poor are likely to be as overweight as their employers, and dressed in the same fleece, sneakers, and jeans. The poor used to be called the have-nots. In 2011, more than 95 percent of American households below the poverty line had electricity, running water, flush toilets, a refrigerator, a stove, and a color TV.[58] (A century and a half before, the Rothschilds, Astors, and Vanderbilts had none of these things.) Almost half of the households below the poverty line had a dishwasher, 60 percent had a computer, around two-thirds had a washing machine and a clothes dryer, and more than 80 percent had an air conditioner, a video recorder, and a cell phone. In the golden age of economic equality in which I grew up, middle-class "haves" had few or none of these things. As a result, the most precious resources of all—time, freedom, and worthy experiences—are rising across the board, a topic we will explore in chapter 17.

The rich have gotten richer, but their lives haven't gotten *that* much better. Warren Buffett may have more air conditioners than most people, or better ones, but by historical standards the fact that a majority of poor Americans even *have* an air conditioner is astonishing. When the Gini index is calculated over consumption rather than income, it has remained shallow or flat.[59] Inequality in self-reported happiness in the American population has actually declined.[60] And though I find it distasteful, even grotesque, to celebrate declining Ginis for life, health, and education (as if killing off the healthiest and keeping the smartest out of school would be good for humanity), they have in fact declined for the right reasons: the lives of the poor are improving more rapidly than the lives of the rich.[61]

~

To acknowledge that the lives of the lower and middle classes of developed countries have improved in recent decades is not to deny the formidable problems facing 21st-century economies. Though disposable income has increased, the pace of the increase is slow, and the resulting lack of consumer demand may be dragging down the economy as a whole.[62] The hardships faced by one sector of the population—middle-aged, less-educated, non-urban white Americans—are real and tragic, manifested in higher rates of drug overdose (chapter 12) and suicide (chapter 18). Advances in robotics threaten to make millions of additional jobs obsolete. Truck drivers, for example, make up the most common occupation in a majority of states, and self-driving vehicles may send them the way of scriveners, wheelwrights, and switchboard operators. Education, a major driver of economic mobility, is not keeping up with the demands of modern economies: tertiary education has soared in cost (defying the inexpensification of almost every other good), and in poor American neighborhoods, primary and secondary education are unconscionably substandard. Many parts of the American tax system are regressive, and money buys too much political influence. Perhaps most damaging, the impression that the modern economy has left most people behind encourages Luddite and beggar-thy-neighbor policies that would make everyone worse off.

Still, a narrow focus on income inequality and a nostalgia for the mid-20th-century Great Compression are misplaced. The modern world can continue to improve even if the Gini index or top income shares stay high, as they may well do, because the forces that lifted them are not going away. Americans cannot be forced to buy Pontiacs instead of Priuses. The *Harry Potter* books will not be kept out of the hands of the world's children just because they turn J. K. Rowling into a billionaire. It makes little sense to make tens of millions of poor Americans pay more for clothing to save tens of thousands of jobs in the apparel industry.[63] Nor does it make sense, in the long term, to have people do boring and dangerous jobs that could be carried out more effectively by machines just to give them remunerable work.[64]

Rather than tilting at inequality per se it may be more constructive to target the specific problems lumped with it.[65] An obvious priority is to boost the rate of economic growth, since it would increase everyone's slice of the pie and provide more pie to redistribute.[66] The trends of the past century, and a survey of the world's countries, point to governments playing an increasing role in both. They are uniquely suited to invest in education, basic research, and infrastructure, to underwrite health and retirement benefits (relieving American corporations of their enervating mandate to provide social services), and to supplement incomes to a level above their market price, which for millions of people may decline even as overall wealth rises.[67]

The next step in the historic trend toward greater social spending may be a universal basic income (or its close relative, a negative income tax). The idea has been bruited for decades, and its day may be coming.[68] Despite its socialist aroma, the idea has been championed by economists (such as Milton Friedman), politicians (such as Richard Nixon), and states (such as Alaska) that are associated with the political right, and today analysts across the political spectrum are toying with it. Though implementing a universal basic income is far from easy (the numbers have to add up, and incentives for education, work, and risk-taking have to be maintained), its promise cannot be ignored. It could rationalize the kludgy patchwork of the hidden welfare state, and it could turn the slow-motion disaster of robots replacing workers into a horn of plenty. Many of the jobs that robots will take over are jobs that people don't particularly enjoy, and the dividend in productivity, safety, and leisure could be a boon to humanity as long as it is widely shared. The specter of anomie and meaninglessness is probably exaggerated (according to studies of regions that have experimented with a guaranteed income), and it could be met with public jobs that markets won't support and robots can't do, or with new opportunities in meaningful volunteering and other forms of effective altruism.[69] The net effect might be to reduce inequality, but that would be a side effect of raising everyone's standard of living, particularly that of the economically vulnerable.

∼

Income inequality, in sum, is not a counterexample to human progress, and we are not living in a dystopia of falling incomes that has reversed the centuries-long rise in prosperity. Nor does it call for smashing the robots, raising the drawbridge, switching to socialism, or bringing back the 50s. Let me sum up my complicated story on a complicated topic.

Inequality is not the same as poverty, and it is not a fundamental dimension of human flourishing. In comparisons of well-being across countries, it pales in importance next to overall wealth. An increase in inequality is not necessarily bad: as societies escape from universal poverty, they are bound to become more unequal, and the uneven surge may be repeated when a society discovers new sources of wealth. Nor is a decrease in inequality always good: the most effective levelers of economic disparities are epidemics, massive wars, violent revolutions, and state collapse.

For all that, the long-term trend in history since the Enlightenment is for everyone's fortunes to rise. In addition to generating massive amounts of wealth, modern societies have devoted an increasing proportion of that wealth to benefiting the less well-off.

As globalization and technology have lifted billions out of poverty and created a global middle class, international and global inequality have decreased, at the same time that they enrich elites whose analytical, creative, or financial impact has global reach. The fortunes of the lower classes in developed countries have not improved nearly as much, but they have improved, often because their members rise into the upper classes. The improvements are enhanced by social spending, and by the falling cost and rising quality of the things people want. In some ways the world has become less equal, but in more ways the world's people have become better off.

CHAPTER 10

THE ENVIRONMENT

B ut is progress sustainable? A common response to the good news about our health, wealth, and sustenance is that it cannot continue. As we infest the world with our teeming numbers, guzzle the earth's bounty heedless of its finitude, and foul our nests with pollution and waste, we are hastening an environmental day of reckoning. If over-population, resource depletion, and pollution don't finish us off, then climate change will.

As in the chapter on inequality, I won't pretend that all the trends are positive or that the problems facing us are minor. But I will present a way of thinking about these problems that differs from the lugubrious conventional wisdom and offers a constructive alternative to the radicalism or fatalism it encourages. The key idea is that environmental problems, like other problems, are solvable, given the right knowledge.

To be sure, the very idea that there *are* environmental problems cannot be taken for granted. From the vantage point of an individual, the Earth seems infinite, and our effects on it inconsequential. From the vantage points of science, the view is more troubling. The microscopic vantage point reveals pollutants that insidiously poison us and the species we admire and depend on; the macroscopic one reveals effects on ecosystems that may be imperceptible one action at a time but add up to tragic despoliation. Beginning in the 1960s, the environmental movement grew out of scientific knowledge (from ecology, public health, and earth and atmospheric sciences) and a Romantic reverence for nature. The movement made the health of the planet a permanent priority on humanity's agenda, and as we shall see, it deserves credit for substantial achievements—another form of human progress.

Ironically, many voices in the traditional environmental movement refuse to acknowledge that progress, or even that human progress is a worthy aspiration. In this chapter I will present a newer conception of environmentalism which shares the goal of protecting the air and water, species, and ecosystems but is grounded in Enlightenment optimism rather than Romantic declinism.

~

Starting in the 1970s, the mainstream environmental movement latched onto a quasi-religious ideology, greenism, which can be found in the manifestoes of activists as diverse as Al Gore, the Unabomber, and Pope Francis.[1] Green ideology begins with an image of the Earth as a pristine ingénue which has been defiled by human rapacity. As Francis put it in his 2015 encyclical *Laudato Si'* (Praise be to you), "Our common home is like a sister with whom we share our life . . . [who] now cries out to us because of the harm we have inflicted on her." The harm, according to this narrative, has been inexorably worsening: "The earth, our home, is beginning to look more and more like an immense pile of filth." The root cause is the Enlightenment commitment to reason, science, and progress: "Scientific and technological progress cannot be equated with the progress of humanity and history," wrote Francis. "The way to a better future lies elsewhere," namely in an appreciation of "the mysterious network of relations between things" and (of course) "the treasure of Christian spiritual experience." Unless we repent our sins by degrowth, deindustrialization, and a rejection of the false gods of science, technology, and progress, humanity will face a ghastly reckoning in an environmental Judgment Day.

As with many apocalyptic movements, greenism is laced with misanthropy, including an indifference to starvation, an indulgence in ghoulish fantasies of a depopulated planet, and Nazi-like comparisons of human beings to vermin, pathogens, and cancer. For example, Paul Watson of the Sea Shepherd Conservation Society wrote, "We need to radically and intelligently reduce human populations to fewer than one billion. . . . Curing a body of cancer requires radical and invasive therapy, and therefore, curing the biosphere of the human virus will also require a radical and invasive approach."[2]

Recently an alternative approach to environmental protection has been championed by John Asafu-Adjaye, Jesse Ausubel, Andrew Balmford, Stewart Brand, Ruth DeFries, Nancy Knowlton, Ted Nordhaus, Michael Shellenberger, and others. It has been called Ecomodernism, Ecopragmatism, Earth Optimism, and the Blue-Green or Turquoise movement, though we can also think of it as Enlightenment Environmentalism or Humanistic Environmentalism.[3]

Ecomodernism begins with the realization that some degree of pollution is an inescapable consequence of the Second Law of Thermodynamics. When people use energy to create a zone of structure in their bodies and homes, they must increase entropy elsewhere in the environment in the form of waste, pollution, and other forms of disorder. The human species has always been ingenious at doing this—that's what differentiates us from other mammals—and it has never lived in harmony with the environment. When native peoples first set foot in an ecosystem, they typically hunted large animals to extinction, and often burned and cleared vast swaths of forest.[4] A dirty secret of the conservation movement is that wilderness preserves are set up only after indigenous peoples have been decimated or forcibly removed from them, including the national parks in the United States and the Serengeti in East Africa.[5] As the environmental historian William Cronon writes, "wilderness" is not a pristine sanctuary; it is itself a product of civilization.

When humans took up farming, they became more disruptive still. According to the paleoclimatologist William Ruddiman, the adoption of wet rice cultivation in Asia some five thousand years ago may have released so much methane into the atmosphere from rotting vegetation as to have changed the climate. "A good case can be made," he suggests, that "the people in the Iron Age and even the late Stone Age had a much greater per-capita impact on the earth's landscape than the average modern-day person."[6] And as Brand has pointed out (chapter 7), "natural farming" is a contradiction in terms. Whenever he hears the words *natural food*, he is tempted to rail:

No product of agriculture is the slightest bit natural to an ecologist! You take a nice complex ecosystem, chop it into rectangles, clear it to the ground, and hammer it into perpetual early succession! You bust its sod, flatten it flat, and drench it with vast quantities of constant water! Then you populate it with uniform monocrops of profoundly damaged plants incapable of living on their own! Every food plant is a pathetic narrow specialist in one skill, inbred for thousands of years to a state of genetic idiocy! Those plants are so fragile, they had to domesticate humans just to take endless care of them![7]

A second realization of the ecomodernist movement is that industrialization has been good for humanity.[8] It has fed billions, doubled life spans, slashed extreme poverty, and, by replacing muscle with machinery, made it easier to end slavery, emancipate women, and educate children (chapters 7, 15, and 17). It has allowed people to read at night, live where they want, stay warm in winter, see the world, and multiply human contact. Any costs in pollution and habitat loss have to be weighed against these gifts. As the economist Robert Frank has put it, there is an optimal amount of pollution in the environment, just as there is an optimal amount of dirt in your house. Cleaner is better, but not at the expense of everything else in life.

The third premise is that the tradeoff that pits human well-being against environmental damage can be renegotiated by technology. How to enjoy more calories, lumens, BTUs, bits, and miles with less pollution and land is itself a technological problem, and one that the world is increasingly solving. Economists speak of the environmental Kuznets curve, a counterpart to the U-shaped arc for inequality as a function of economic growth. As countries first develop, they prioritize growth over environmental purity. But as they get richer, their thoughts turn to the environment.[9] If people can afford electricity only at the cost of some smog, they'll live with the smog, but when they can afford both electricity *and* clean air, they'll spring for the clean air. This can happen all the faster as technology makes cars and factories and power plants cleaner and thus makes clean air more affordable.

Economic growth bends the environmental Kuznets curve by advances not just in technology but in values. Some environmental concerns are entirely practical: people complain about smog in their city, or green space getting paved over. But other concerns are more spiritual. The fate of the black rhinoceros and the well-being of our descendants in the year 2525 are significant moral concerns, but worrying about them now is something of a luxury. As societies get richer and people no longer think about putting food on the table or a roof over their heads, their values climb a hierarchy of needs, and the scope of their concern expands in space and time. Ronald Inglehart and Christian Welzel, using data from the World Values Survey, have found that people with stronger emancipative values—tolerance, equality, freedom of thought and speech—which tend to go with affluence and education, are also more likely to recycle and to pressure governments and businesses into protecting the environment.[10]

～

Ecopessimists commonly dismiss this entire way of thinking as the "faith that technology will save us." In fact it is a skepticism that the status quo will doom us—that knowledge will be frozen in its current state and people will robotically persist in their current behavior regardless of circumstances. Indeed, a naïve faith in stasis has repeatedly led to prophecies of environmental doomsdays that never happened.

The first is the "population bomb," which (as we saw in chapter 7) defused itself. When countries get richer and better educated, they pass through what demographers call the demographic transition.[11] First, death rates decline as nutrition and health improve. This does swell the population, but that is hardly something to bewail: as Johan Norberg notes, it happens not because people in poor countries start breeding like rabbits but because they stop dying like flies. In any case, the increase is temporary: birth rates peak and then decline, for at least two reasons. Parents no longer breed large broods as insurance against some of their children dying, and women, when they become better educated, marry later and delay having children. Figure 10-1 shows that the world population growth rate peaked at 2.1 percent a year in 1962, fell to 1.2 percent by 2010, and will probably fall to less than 0.5 percent by 2050 and be close to zero around 2070, when the population is projected to level off and then decline. Fertility rates have fallen most noticeably in developed regions like Europe and Japan, but they can

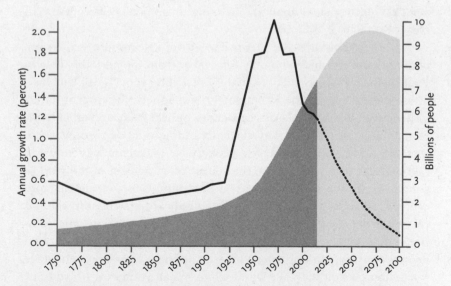

Figure 10-1: Population and population growth, 1750–2015 and projected to 2100

Sources: *Our World in Data*, Ortiz-Ospina & Roser 2016d. **1750–2015:** United Nations Population Division and *History Database of the Global Environment* (HYDE), PBL Netherlands Environmental Assessment Agency (undated). **Post-2015 projections:** Annual growth rate, same as for 1750–2015. Billions of people, International Institute for Applied Systems Analysis, Medium Projection (aggregate of country-specific estimates, taking education into account), Lutz, Butz, & Samir 2014.

suddenly collapse, often to demographers' surprise, in other parts of the world. Despite the widespread belief that Muslim societies are resistant to the social changes that have transformed the West and will be indefinitely rocked by youthquakes, Muslim countries have seen a 40 percent decline in fertility over the past three decades, including a 70 percent drop in Iran and 60 percent drops in Bangladesh and in seven Arab countries.[12]

The other scare from the 1960s was that the world would run out of resources. But resources just refuse to run out. The 1980s came and went without the famines that were supposed to starve tens of millions of Americans and billions of people worldwide. Then the year 1992 passed and, contrary to projections from the 1972 bestseller *The Limits to Growth* and similar philippics, the world did not exhaust its aluminum, copper, chromium, gold, nickel, tin, tungsten, or zinc. (In 1980 Paul Ehrlich famously bet the economist Julian Simon that five of these metals would become scarcer and hence more expensive by the end of the decade; he lost all five bets. Indeed, most metals and minerals are cheaper today than they were in 1960.)[13] From the 1970s to the early 2000s newsmagazines periodically illustrated cover stories on the world's oil supply with a gas gauge pointing to Empty. In 2013 *The Atlantic* ran a cover story about the fracking revolution entitled "We Will Never Run Out of Oil."

And then there are rare earths like yttrium, scandium, europium, and lanthanum, which you may remember from the periodic table in your chemistry classroom or from the Tom Lehrer song "The Elements." These metals are a critical component of magnets, fluorescent lights, video screens, catalysts, lasers, capacitors, optical glass, and other high-tech applications. When they started running out, we were warned, there would be critical shortages, a collapse of the technology industry, and perhaps war with China, the source of 95 percent of the world's supply. That's what led to the Great Europium Crisis of the late 20th century, when the world ran out of the critical ingredient in the red phosphor dots in the cathode-ray tubes in color televisions and computer monitors and society was divided between the haves, who hoarded the last working color TVs, and the angry have-nots, who were forced to make do with black-and-white. What, you never heard of it? Among the reasons there was no such crisis was that cathode-ray tubes were superseded by liquid crystal displays made of common elements.[14] And the Rare Earths War? In reality, when China squeezed its exports in 2010 (not because of shortages but as a geopolitical and mercantilist weapon), other countries started extracting rare earths from their own mines, recycling them from industrial waste, and re-engineering products so they no longer needed them.[15]

When predictions of apocalyptic resource shortages repeatedly fail to come true, one has to conclude either that humanity has miraculously escaped from certain death again and again like a Hollywood action hero or that there is a flaw in the thinking that predicts apocalyptic resource shortages. The flaw has been pointed out many times.[16] Humanity does not suck resources from the earth like a straw in a milkshake until a gurgle tells it that the container is empty. Instead, as the most easily extracted supply of a resource becomes scarcer, its price rises, encouraging people to conserve it, get at the less accessible deposits, or find cheaper and more plentiful substitutes.

Indeed, it's a fallacy to think that people "need resources" in the first place.[17] They need ways of growing food, moving around, lighting their homes, displaying information, and other sources of well-being. They satisfy these needs with *ideas:* with recipes, formulas, techniques, blueprints, and algorithms for manipulating the physical world to give them what they want. The human mind, with its recursive combinatorial power, can explore an infinite space of ideas, and is not limited by the quantity of any particular kind of stuff in the ground. When one idea no longer works, another can take its place. This doesn't defy the laws of probability but obeys them. Why should the laws of nature have allowed *exactly one* physically possible way of satisfying a human desire, no more and no less?[18]

Admittedly, this way of thinking does not sit well with the ethic of "sustainability." In figure 10-2, the cartoonist Randall Munroe illustrates what's wrong with this vogue word and sacred value. The doctrine of sustainability assumes that the current rate of use of a resource may be extrapolated into the future until it rams into a ceiling. The implication is that we must switch to a renewable resource that can be replenished at the rate we use it, indefinitely. In reality, societies have always abandoned a resource for a better one long before the old one was exhausted. It's often said that the Stone Age did not end because the world ran out of stones, and that has been true of energy as well. "Plenty of wood and hay remained to be exploited when the world shifted to coal," Ausubel notes. "Coal abounded when oil rose. Oil abounds now as methane [natural gas] rises."[19] As we will see, gas in turn may be replaced by energy sources still lower in carbon well before the last cubic foot goes up in a blue flame.

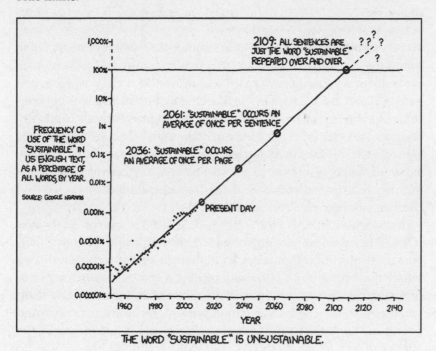

THE WORD "SUSTAINABLE" IS UNSUSTAINABLE.

Figure 10-2: Sustainability, 1955–2109

Source: Randall Munroe, *XKCD*, http://xkcd.com/1007/. Credit: Randall Munroe, xkcd.com.

The supply of food, too, has grown exponentially (as we saw in chapter 7), even though no single method of growing it has ever been sustainable. In *The Big Ratchet: How Humanity Thrives in the Face of Natural Crisis*, the geographer Ruth DeFries describes the sequence as "ratchet-hatchet-pivot." People discover a way of growing more food, and the population ratchets upward. The method fails to keep up with the demand or develops unpleasant side effects, and the hatchet falls. People then pivot to a new method. At various times, farmers have pivoted to slash-and-burn horticulture, night soil (a euphemism for human feces), crop rotation, guano, saltpeter, ground-up bison bones, chemical fertilizer, hybrid crops, pesticides, and the Green Revolution.[20] Future pivots may include genetically modified organisms, hydroponics, aeroponics, urban vertical farms, robotic harvesting, meat cultured in vitro, artificial intelligence algorithms fed by GPS and biosensors, the recovery of energy and fertilizer from sewage, aquaculture with fish that eat tofu instead of other fish, and who knows what else—as long as people are allowed to indulge their ingenuity.[21] Though water is one resource that people will never pivot away from, farmers could save massive amounts if they switched to Israeli-style precision farming. And if the world develops abundant carbon-free energy sources (a topic we will explore later), it could get what it needs by desalinating seawater.[22]

~

Not only have the disasters prophesied by 1970s greenism failed to take place, but improvements that it deemed impossible *have* taken place. As the world has gotten richer and crested the environmental curve, nature has begun to rebound.[23] Pope Francis's "immense pile of filth" is the vision of someone who has woken up thinking it's 1965, the era of belching smokestacks, waterfalls of sewage, rivers catching fire, and jokes about New Yorkers not liking to breathe air they can't see. Figure 10-3 shows that since 1970, when the Environmental Protection Agency was established, the United States has slashed its emissions of five air pollutants by almost two-thirds. Over the same period, the population grew by more than 40 percent, and those people drove twice as many miles and became two and a half times richer. Energy use has leveled off, and even

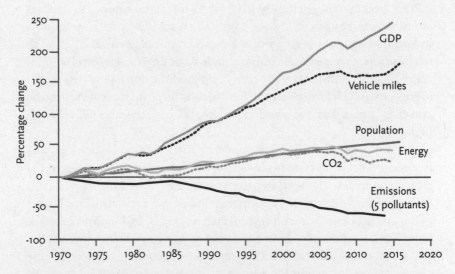

Figure 10-3: Pollution, energy, and growth, US, 1970–2015

Sources: US Environmental Protection Agency 2016, based on the following sources. GDP: Bureau of Economic Analysis. Vehicle miles traveled: Federal Highway Administration. Population: US Census Bureau. Energy Consumption: US Department of Energy. CO₂: US Greenhouse Gas Inventory Report. Emissions (carbon monoxide, oxides of nitrogen, particulate matter smaller than 10 micrometers, sulfur dioxide, and volatile organic compounds): EPA, https://www.epa.gov/air-emissions-inventories/air-pollutant-emissions-trends-data.

carbon dioxide emissions have turned a corner, a point to which we will return. The declines don't just reflect an offshoring of heavy industry to the developing world, because the bulk of energy use and emissions comes from transportation, heating, and electricity generation, which cannot be outsourced. Rather, they mainly reflect gains in efficiency and emission control. These diverging curves refute both the orthodox Green claim that only degrowth can curb pollution and the orthodox right-wing claim that environmental protection must sabotage economic growth and people's standard of living.

Many of the improvements can be seen with the naked eye. Cities are less often shrouded in purple-brown haze, and London no longer has the fog—actually coal smoke—that was immortalized in Impressionist paintings, gothic novels, the Gershwin song, and the brand of raincoats. Urban waterways that had been left for dead—including Puget Sound, Chesapeake Bay, Boston Harbor, Lake Erie, and the Hudson, Potomac, Chicago, Charles, Seine, Rhine, and Thames rivers (the last described by Disraeli as "a Stygian pool reeking with ineffable and intolerable horrors")—have been recolonized by fish, birds, marine mammals, and sometimes swimmers. Suburbanites are seeing wolves, foxes, bears, bobcats, badgers, deer, ospreys, wild turkeys, and bald eagles. As agriculture becomes more efficient (chapter 7), farmland returns to temperate forest, as any hiker knows who has stumbled upon a stone wall incongruously running through a New England woodland. Though tropical forests are still, alarmingly, being cut down, between the middle of the 20th century and the turn of the 21st the rate fell by two-thirds (figure 10-4).[24] Deforestation of the world's largest tropical forest, the Amazon, peaked in 1995, and from 2004 to 2013 the rate fell by four-fifths.[25]

The time-lagged decline of deforestation in the tropics is one sign that environmental protection is spreading from developed countries to the rest of the world. The world's progress can be tracked in a report card called the Environmental Performance Index, a composite of indicators of the quality of air, water, forests, fisheries, farms, and natural habitats. Out of 180 countries that have been tracked for a decade or more, all but two show an improvement.[26] The wealthier the country, on average, the cleaner its environment: the Nordic countries were cleanest; Afghanistan, Bangladesh, and several sub-Saharan African countries, the most compromised. Two of the deadliest forms of pollution—contaminated drinking water and indoor cooking smoke—are afflictions of poor countries.[27] But as poor countries have gotten richer in recent decades, they are escaping these blights: the proportion of the world's population that drinks tainted water has fallen by five-eighths, the proportion breathing cooking smoke by a third.[28] As Indira Gandhi said, "Poverty is the greatest polluter."[29]

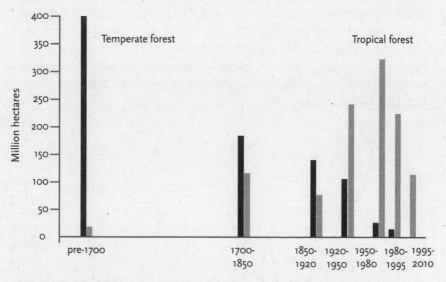

Figure 10-4: Deforestation, 1700–2010

Source: United Nations Food and Agriculture Organization 2012, p. 9. The bars represent totals over intervals of different durations, not annual rates, and thus are not directly commensurable.

The epitome of environmental insults is the oil spill from tanker ships, which coats pristine beaches with toxic black sludge and fouls the plumage of seabirds and the fur of otters and seals. The most notorious accidents, such as the breakup of the *Torrey Canyon* in 1967 and the *Exxon Valdez* in 1989, linger in our collective memory, and few people are aware that seaborne oil transport has become vastly safer. Figure 10-5 shows that the annual number of oil spills has fallen from more than a hundred in 1973 to just five in 2016 (and the number of *major* spills fell from thirty-two in 1978 to one in 2016). The graph also shows that even as less oil was spilled, more oil was shipped; the crossing curves provide additional evidence that environmental protection is compatible with economic growth. It's no mystery that oil companies should *want* to reduce tanker accidents, because their interests and those of the environment coincide: oil spills are a public-relations disaster (especially when the name of the company is emblazoned on a cracked-up ship), bring on huge fines, and of course waste valuable oil. More interesting is the fact that the companies have largely succeeded. Technologies follow a learning curve and become less hazardous over time as the boffins design out the most dangerous vulnerabilities (a point we'll return to in chapter 12). But people remember the accidents and are unaware of the incremental improvements. The improvements in different technologies unfold on different timetables: in 2010, when seaborne oil spills had fallen to an all-time low, the third-worst spill from stationary rigs took place. The *Deepwater Horizon* accident in the Gulf of Mexico led in turn to new regulations for blowout preventers, well design, monitoring, and containment.[30]

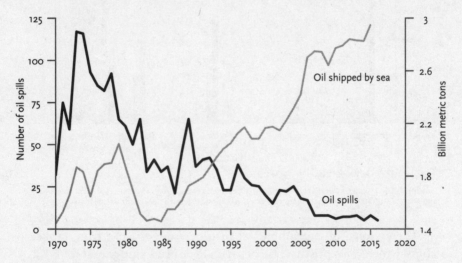

Figure 10-5: Oil spills, 1970–2016

Source: *Our World in Data,* Roser 2016r, based on data (updated) from the International Tanker Owners Pollution Federation, http://www.itopf.com/knowledge-resources/data-statistics/statistics/. Oil spills include all those that result in the loss of at least 7 metric tons of oil. Oil shipped consists of "total crude oil, petroleum product, and gas loaded."

In another advance, entire swaths of land and ocean have been protected from human use altogether. Conservation experts are unanimous in their assessment that the protected areas are still inadequate, but the momentum is impressive. Figure 10-6 shows that the proportion of the Earth's land set aside as national parks, wildlife reserves, and other protected areas has grown from 8.2 percent in 1990 to 14.8 percent in 2014—an area double the size of the United States. Marine conservation areas have grown as well, more than doubling during this period and now protecting more than 12 percent of the world's oceans.

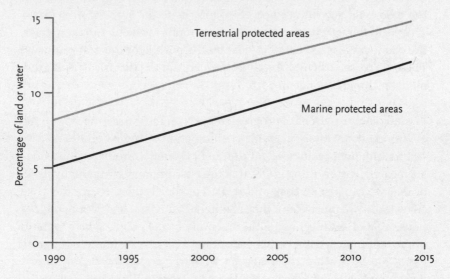

Figure 10-6: Protected areas, 1990–2014

Source: World Bank 2016h and 2017, based on data from the United Nations Environment Programme and the World Conservation Monitoring Centre, compiled by the World Resources Institute.

Thanks to habitat protection and targeted conservation efforts, many beloved species have been pulled from the brink of extinction, including albatrosses, condors, manatees, oryxes, pandas, rhinoceroses, Tasmanian devils, and tigers; according to the ecologist Stuart Pimm, the rate of bird extinctions has been reduced by 75 percent.[31] Though many species remain in precarious straits, a number of ecologists and paleontologists believe that the claim that humans are causing a mass extinction like the Permian and Cretaceous is hyperbolic. As Brand notes, "No end of specific wildlife problems remain to be solved, but describing them too often as extinction crises has led to a general panic that nature is extremely fragile or already hopelessly broken. That is not remotely the case. Nature as a whole is exactly as robust as it ever was—maybe more so. . . . Working with that robustness is how conservation's goals get reached."[32]

Other improvements are global in scope. The 1963 treaty banning atmospheric nuclear testing eliminated the most terrifying form of pollution of all, radioactive fallout, and proved that the world's nations could agree on measures to protect the planet even in the absence of a world government. Global cooperation has dealt with several other challenges since. International treaties on the reduction of sulfur emissions and other forms of "long-range transboundary air pollution" signed in the 1980s and 1990s have helped to eliminate the scare of acid rain.[33] Thanks to the 1987 ban on chlorofluorocarbons ratified by 197 countries, the ozone layer is expected to heal by the middle of the 21st century.[34] These successes, as we will see, set the stage for the historic Paris Agreement on climate change in 2015.

～

Like all demonstrations of progress, reports on the improving state of the environment are often met with a combination of anger and illogic. The fact that many measures of environmental quality are improving does *not* mean that everything is OK, that the environment got better by itself, or that we can just sit back and relax. For the cleaner environment we enjoy today we must thank the arguments, activism, legislation, regulations, treaties, and technological ingenuity of the people who sought to improve it in the past.[35] We'll need more of each to sustain the progress we've made, prevent reversals (particularly under the Trump presidency), and extend it to the wicked problems that still face us, such as the health of the oceans and, as we shall see, atmospheric greenhouse gases.

But for many reasons, it's time to retire the morality play in which modern humans are a vile race of despoilers and plunderers who will hasten the apocalypse unless they undo the Industrial Revolution, renounce technology, and return to an ascetic harmony with nature. Instead, we can treat environmental protection as a problem to be solved: how can people live safe, comfortable, and stimulating lives with the least possible pollution and loss of natural habitats? Far from licensing complacency, our progress so far at solving this problem emboldens us to strive for more. It also points to the forces that pushed this progress along.

One key is to decouple productivity from resources: to get more human benefit from less matter and energy. This puts a premium on *density*.[36] As agriculture becomes more intensive by growing crops that are bred or engineered to produce more protein, calories, and fiber with less land, water, and fertilizer, farmland is spared, and it can morph back to natural habitats. (Ecomodernists point out that organic farming, which needs far more land to produce a kilogram of food, is neither green nor sustainable.) As people move to cities, they not only free up land in the countryside but need fewer resources for commuting, building, and heating, because one man's ceiling is another man's floor. As trees are harvested from dense plantations, which have five to ten times the yield of natural forests, forest land is spared, together with its feathered, furry, and scaly inhabitants.

All these processes are helped along by another friend of the Earth, *dematerialization*. Progress in technology allows us to do more with less. An aluminum soda can used to weigh three ounces; today it weighs less than half an ounce. Mobile phones don't need miles of telephone poles and wires. The digital revolution, by replacing atoms with bits, is dematerializing the world in front of our eyes. The cubic yards of vinyl that used to be my music collection gave way to cubic inches of compact discs and then to the nothingness of MP3s. The river of newsprint flowing through my apartment has been stanched by an iPad. With a terabyte of storage on my laptop I no longer buy paper by the ten-ream box. And just think of all the plastic, metal, and paper that no longer go into the forty-odd consumer products that can be replaced by a single smartphone, including a telephone, answering machine, phone book, camera, camcorder, tape recorder, radio, alarm clock, calculator, dictionary, Rolodex, calendar, street maps, flashlight, fax, and compass—even a metronome, outdoor thermometer, and spirit level.

Digital technology is also dematerializing the world by enabling the sharing economy, so that cars, tools, and bedrooms needn't be made in huge numbers that sit around unused most of the time. The advertising analyst Rory Sutherland has noted that dematerialization is also being helped along by changes in the criteria of social status.[37] The most expensive London real estate today would have seemed impossibly cramped to wealthy Victorians, but the city center is now more fashionable than the suburbs. Social media have encouraged younger people to show off their experiences rather than their cars and wardrobes, and hipsterization leads them to distinguish themselves by their tastes in beer, coffee, and music. The era of the Beach Boys and *American Graffiti* is over: half of American eighteen-year-olds do not have a driver's license.[38]

The expression "Peak Oil," which became popular after the energy crises of the 1970s, refers to the year that the world would reach its maximum extraction of petroleum. Ausubel notes that because of the demographic transition, densification, and dematerialization, we may have reached Peak Children, Peak Farmland, Peak Timber, Peak Paper, and Peak Car. Indeed, we may be reaching Peak Stuff: of a hundred commodities Ausubel plotted, thirty-six have peaked in absolute use in the United States, and another fifty-three may be poised to drop (including water, nitrogen, and electricity), leaving only eleven that are still growing. Britons, too, have reached Peak Stuff, having reduced their annual use of material from 15.1 metric tons per person in 2001 to 10.3 metric tons in 2013.[39]

These remarkable trends required no coercion, legislation, or moralization; they spontaneously unfolded as people made choices about how to live their lives. The trends certainly don't show that environmental legislation is dispensable—by all accounts, environmental protection agencies, mandated energy standards, endangered species protection, and national and international clean air and water acts have had enormously beneficial effects.[40] But they suggest that the tide of modernity does not sweep humanity headlong toward ever more unsustainable use of resources. Something in the nature of technology, particularly information technology, works to decouple human flourishing from the exploitation of physical stuff.

~

Just as we must not accept the narrative that humanity inexorably despoils every part of the environment, we must not accept the narrative that every part of the environment will rebound under our current practices. An enlightened environmentalism must face the facts, hopeful or alarming, and one set of facts is unquestionably alarming: the effect of greenhouse gases on the earth's climate.[41]

Whenever we burn wood, coal, oil, or gas, the carbon in the fuel is oxidized to form carbon dioxide (CO_2), which wafts into the atmosphere. Though some of the CO_2 dissolves in the ocean, chemically combines with rocks, or is taken up by photosynthesizing plants, these natural sinks cannot keep up with the 38 billion tons we dump into the atmosphere each year. As gigatons of carbon laid down during the Carboniferous Period have gone up in smoke, the concentration of CO_2 in the atmosphere has risen from about 270 parts per million before the Industrial Revolution to more than 400 parts today. Since CO_2 traps heat radiating from the Earth's surface, the global average temperature has risen as well, by about .8° Celsius (1.4° Fahrenheit), and 2016 was the hottest year on record, with 2015 coming in second and 2014 coming in third. The atmosphere has also been warmed by the clearing of carbon-eating forests and by the release of methane (an even more potent greenhouse gas) from leaky gas wells, melting permafrost, and the orifices at both ends of cattle. It could become warmer still in a runaway feedback loop if white, heat-reflecting snow and ice are replaced by dark, heat-absorbing land and water, if the melting of permafrost accelerates, and if more water vapor (yet another greenhouse gas) is sent into the air.

If the emission of greenhouse gases continues, the Earth's average temperature will rise to at least 1.5°C (2.7°F) above the preindustrial level by the end of the 21st century, and perhaps to 4°C (7.2°F) above that level or more. That will cause more frequent and more severe heat waves, more floods in wet regions, more droughts in dry regions, heavier storms, more severe hurricanes, lower crop yields in warm regions, the extinction of more species, the loss of coral reefs (because the oceans will be both warmer and more acidic), and an average rise in sea level of between 0.7 and 1.2 meters (2 and 4 feet) from both the melting of land ice and the expansion of seawater. (Sea level has already risen almost eight inches since 1870, and the rate of the rise appears to be accelerating.) Low-lying areas would be flooded, island nations would disappear beneath the waves, large stretches of farmland would no longer be arable, and millions of people would be displaced. The effects could get still worse in the 22nd century and beyond, and in theory could trigger upheavals such as a diversion of the Gulf Stream (which would turn Europe into Siberia) or a collapse of Antarctic ice sheets. A rise of 2°C is considered the most that the world could reasonably adapt to, and a rise of 4°C, in the words of a 2012 World Bank report, "simply must not be allowed to occur."[42]

To keep the rise to 2°C or less, the world would, at a minimum, have to reduce its greenhouse gas emissions by half or more by the middle of the 21st century and eliminate them altogether before the turn of the 22nd.[43] The challenge is daunting. Fossil fuels provide 86 percent of the world's energy, powering almost every car, truck, train, plane, ship, tractor, furnace, and factory on the planet, together with most of its electricity plants.[44] Humanity has never faced a problem like it.

One response to the prospect of climate change is to deny that it is occurring or that human activity is the cause. It's completely appropriate, of course, to challenge the hypothesis of anthropogenic climate change on scientific grounds, particularly given the extreme measures it calls for if it is true. The great virtue of science is that a true hypothesis will, in the long run, withstand attempts to falsify it. Anthropogenic climate change is the most vigorously challenged scientific hypothesis in history. By now, all the major challenges—such as that global temperatures have stopped rising, that they only seem to be rising because they were measured in urban heat islands, or that they really are rising but only because the sun is getting hotter—have been refuted, and even many skeptics have been convinced.[45] A recent survey found that exactly *four* out of 69,406 authors of peer-reviewed articles in the scientific literature rejected the hypothesis of anthropogenic global warming, and that "the peer-reviewed literature contains no convincing evidence against [the hypothesis]."[46]

Nonetheless, a movement within the American political right, heavily underwritten by fossil fuel interests, has prosecuted a fanatical and mendacious campaign to deny that greenhouse gases are warming the planet.[47] In doing so they have advanced the conspiracy theory that the scientific community is fatally infected with political correctness and ideologically committed to a government takeover of the economy. As someone who considers himself something of a watchdog for politically correct dogma in academia, I can state that this is nonsense: physical scientists have no such agenda, and the evidence speaks for itself.[48] (And it's precisely because of challenges like this that scholars in all fields have a duty to secure the credibility of the academy by *not* enforcing political orthodoxies.)

To be sure, there are judicious climate change skeptics, sometimes called lukewarmers, who accept the mainstream science but accentuate the positive.[49] They favor the fringe of the envelope of possibilities with the slowest temperature rise, note that the worst-case scenarios with runaway feedback are hypothetical, point out that moderately higher temperatures and CO_2 have benefits in crop yields that should be traded off against their costs, and argue that if countries are allowed to get as rich as possible (without growth-sapping restrictions on fossil fuels) they will be better equipped to adapt to the climate change that does occur. But as the economist William Nordhaus points out, this is a rash gamble in what he calls the Climate Casino.[50] If the status quo presents, say, an even chance that the world will get significantly worse, and a 5 percent chance that it will pass a tipping point and face a catastrophe, it would be prudent to take preventive action even if the catastrophic outcome is not certain, just as we buy fire extinguishers and insurance for our houses and don't keep open cans of gasoline in our garages. Since dealing with climate change will be a multidecade effort, there's plenty of time to back off if temperature, sea level, and ocean acidity happily stop rising.

Another response to climate change, from the far left, seems designed to vindicate the conspiracy theories of the far right. According to the "climate justice" movement popularized by the journalist Naomi Klein in her 2014 bestseller *This Changes Everything: Capitalism vs. the Climate*, we should not treat the threat of climate change as a challenge to prevent climate change. No, we should treat it as an opportunity to abolish free markets, restructure the global economy, and remake our political system.[51] In one of the more surreal episodes in the history of environmental politics, Klein joined the infamous David and Charles Koch, the billionaire oil industrialists and bankrollers of climate change denial, in helping to defeat a 2016 Washington state ballot initiative that would have implemented the country's first carbon tax, the policy measure which almost every analyst endorses as a prerequisite to dealing with climate change.[52] Why? Because the measure was "right-wing friendly," and it did not "make the polluters pay, and put their immoral profits to work repairing the damage they have knowingly created." In a 2015 interview Klein even opposed analyzing climate change quantitatively:

> We're not going to win this as bean counters. We can't beat the bean counters at their own game. We're going to win this because this is an issue of values, human rights, right and wrong. We just have this brief period where we also have to have some nice stats that we can wield, but we shouldn't lose sight of the fact that what actually moves people's hearts are the arguments based on the value of life.[53]

Blowing off quantitative analysis as "bean-counting" is not just anti-intellectual but works *against* "values, human rights, right and wrong." Someone who values human life will favor the policies that have the greatest chance of saving people from being displaced or starved while furnishing them with the means to live healthy and fulfilled lives.[54] In a universe governed by the laws of nature rather than magic and deviltry, that requires "bean-counting." Even when it comes to the purely rhetorical challenge of "moving people's hearts," efficacy matters: people are likelier to accept the fact of global warming when they are told that the problem is solvable by innovations in policy and technology than when they are given dire warnings about how awful it will be.[55]

Another common sentiment about how to prevent climate change is expressed in this letter, of a kind I receive every now and again:

Dear Professor Pinker

We need to do something about global warming. Why don't the Nobel prize winning scientists sign a petition? Why don't they tell the blunt truth, that the politicians are pigs who don't care how many people get killed in floods and droughts?

Why don't you and some friends start a movement on the Internet to get people to sign a pledge that they will make real sacrifices to fight global warming. Because that's the problem. Nobody wants to make any sacrifices. People should pledge to never fly in airplanes except in dire emergencies, because airplanes burn so much fuel. People should pledge to eat no meat on at least three days per week, because meat production adds so much carbon to the atmosphere. People should pledge to buy no jewelry, ever, because refining gold and silver is so energy-intensive. We should abolish artistic pottery, because it burns so much carbon. The potters in university art departments are just going to have to accept the fact that we can't go on like this.

Forgive the bean-counting, but even if everyone gave up their jewelry, it would not make a scratch in the world's emission of greenhouse gases, which are dominated by heavy industry (29 percent), buildings (18 percent), transport (15 percent), land-use change (15 percent), and the energy needed to supply energy (13 percent). (Livestock is responsible for 5.5 percent, mostly methane rather than CO_2, and aviation for 1.5 percent.)[56] Of course my correspondent suggested forgoing jewelry and pottery not because of the *effect* but because of the *sacrifice*, and it's no surprise that she singled out jewelry, the quintessential luxury. I bring up her ingenuous suggestion to illustrate two psychological impediments we face in dealing with climate change.

The first is cognitive. People have trouble thinking in scale: they don't differentiate among actions that would reduce CO_2 emissions by thousands of tons, millions of tons, and billions of tons.[57] Nor do they distinguish among level, rate, acceleration, and higher-order derivatives—between actions that would affect the rate of *increase* in CO_2 emissions, affect the *rate* of CO_2 emissions, affect the *level* of CO_2 in the atmosphere, and affect global *temperatures* (which will rise even if the level of CO_2 remains constant). Only the last of these matters, but if one doesn't think in scale and in orders of change, one can be satisfied with policies that accomplish nothing.

The other impediment is moralistic. As I mentioned in chapter 2, the human moral sense is not particularly moral; it encourages dehumanization ("politicians are pigs") and punitive aggression ("make the polluters pay"). Also, by conflating profligacy with evil and asceticism with virtue, the moral sense can sanctify pointless displays of sacrifice.[58] In many cultures people flaunt their righteousness with vows of fasting, chastity, self-abnegation, bonfires of the vanities, and animal (or sometimes human) sacrifice. Even in modern societies—according to studies I've done with the psychologists Jason Nemirow, Max Krasnow, and Rhea Howard—people esteem others according to how much time or money they forfeit in their altruistic acts rather than by how much good they accomplish.[59]

Much of the public chatter about mitigating climate change involves voluntary sacrifices like recycling, reducing food miles, unplugging chargers, and so on. (I myself have posed for posters in several of these campaigns led by Harvard students.)[60] But however virtuous these displays may feel, they are a distraction from the gargantuan challenge facing us. The problem is that carbon emissions are a classic public goods game, also known as a Tragedy of the Commons. People benefit from everyone else's sacrifices and suffer from their own, so everyone has an incentive to be a free rider and let everyone else make the sacrifice, and everyone suffers. A standard remedy for public goods dilemmas is a coercive authority that can punish free riders. But any government with the totalitarian power to abolish artistic pottery is unlikely to restrict that power to maximizing the common good. One can, alternatively, daydream that moral suasion is potent enough to induce everyone to make the necessary sacrifices. But while humans do have public sentiments, it's unwise to let the fate of the planet hinge on the hope that billions of people will simultaneously volunteer to act against their interests. Most important, the sacrifice needed to bring carbon emissions down by half and then to zero is far greater than forgoing jewelry: it would require forgoing electricity, heating, cement, steel, paper, travel, and affordable food and clothing.

Climate justice warriors, indulging the fantasy that the developing world will do just that, advocate a regime of "sustainable development." As Shellenberger and Ted Nordhaus satirize it, that consists of "small co-ops in the Amazon forest where peasant farmers and Indians would pick nuts and berries to sell to Ben and Jerry's for their 'Rainforest Crunch' flavor."[61] They would be allowed solar panels that could light an LED or charge a cell phone, but nothing more. Needless to say, the people who actually live in those countries have a different idea. Escaping from poverty requires abundant energy. The proprietor of *HumanProgress*, Marian Tupy, points out that in 1962 Botswana and Burundi were equally destitute, with an annual per capita income of $70, and neither emitted much CO_2. By 2010, Botswanans earned $7,650 a year, 32 times as much as the still-poor Burundians, and they emitted 89 times as much CO_2.[62]

Faced with such facts, climate justice warriors reply that rather than enriching poor nations, we should impoverish rich ones, switching back, for example, to "labor-intensive agriculture" (to which an appropriate reply is: You first). Shellenberger and Nordhaus note how far progressive politics has moved from the days in which rural electrification and economic development were among its signature projects: "In the name of democracy it now offers the global poor not what they want—cheap electricity—but more of what they don't want, namely intermittent and expensive power."[63]

Economic progress is an imperative in rich and poor countries alike precisely because it will be needed to adapt to the climate change that does occur. Thanks in good part to prosperity, humanity has been getting healthier (chapters 5 and 6), better fed (chapter 7), more peaceful (chapter 11), and better protected from natural hazards and disasters (chapter 12). These advances have made humanity more resilient to natural and human-made threats: disease outbreaks don't become pandemics, crop failures in one region are alleviated by surpluses in another, local skirmishes are defused before they erupt into war, populations are better protected against storms, floods, and droughts. Part of our response to climate change must be to ensure that these gains in resilience continue to outpace the threats that a warming planet will throw at it. Every year that developing countries get richer, they will have more resources for building seawalls and reservoirs, improving public health services, and moving people away from rising seas. For that reason they must not be kept in energy poverty—but neither does it make sense for them to raise incomes with massive coal burning that will overwhelm everyone later with weather disasters.[64]

⁓

How, then, *should* we deal with climate change? Deal with it we must. I agree with Pope Francis and the climate justice warriors that preventing climate change is a moral issue because it has the potential to harm billions, particularly the world's poor. But morality is different from moralizing, and is often poorly served by it. (The Pope's encyclical backfired, *decreasing* concern about climate change among the conservative Catholics who were aware of it.)[65] It may be satisfying to demonize the fossil fuel corporations that sell us the energy we want, or to signal our virtue by making conspicuous sacrifices, but these indulgences won't prevent destructive climate change.

The enlightened response to climate change is to figure out how to get the most energy with the least emission of greenhouse gases. There is, to be sure, a tragic view of modernity in which this is impossible: industrial society, powered by flaming carbon, contains the fuel of its own destruction. But the tragic view is incorrect. Ausubel notes that the modern world has been progressively *de*carbonizing.

The hydrocarbons in the stuff we burn are composed of hydrogen and carbon, which release energy as they combine with oxygen to form

H_2O and CO_2. The oldest hydrocarbon fuel, dry wood, has a ratio of combustible carbon atoms to hydrogen atoms of about 10 to 1.[66] The coal which replaced it during the Industrial Revolution has an average carbon-to-hydrogen ratio of 2 to 1.[67] A petroleum fuel such as kerosene may have a ratio of 1 to 2. Natural gas is composed mainly of methane, whose chemical formula is CH_4, with a ratio of 1 to 4.[68] So as the industrial world climbed an energy ladder from wood to coal to oil to gas (the last transition accelerated in the 21st century by the abundance of shale gas from fracking), the ratio of carbon to hydrogen in its energy source steadily fell, and so did the amount of carbon that had to be burned to release a unit of energy (from 30 kg of carbon per gigajoule in 1850 to about 15 today).[69] Figure 10-7 shows that carbon emissions follow a Kuznets arc: when rich countries such as the United States and the United Kingdom first industrialized, they emitted more and more CO_2 to produce a dollar of GDP, but they turned a corner in the 1950s and since then have been emitting less and less. China and India are following suit, cresting in the late 1970s and mid-1990s, respectively. (China flew off the charts in the late 1950s because of Mao's boneheaded schemes like backyard iron smelters with copious emissions and zero economic output.) Carbon intensity for the world as a whole has been declining for half a century.[70]

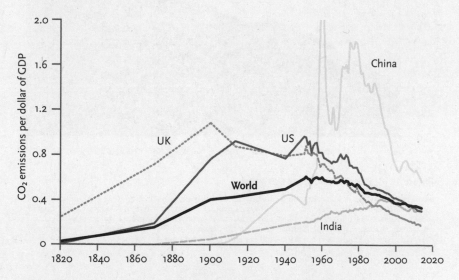

Figure 10-7: Carbon intensity (CO_2 emissions per dollar of GDP), 1820–2014

Source: Ritchie & Roser 2017, based on data from the Carbon Dioxide Information Analysis Center, http://cdiac.ornl.gov/trends/emis/tre_coun.html. GDP is in 2011 international dollars; for the years before 1990, GDP comes from Maddison Project 2014.

Decarbonization is a natural consequence of people's preferences. "Carbon blackens miners' lungs, endangers urban air, and threatens climate change," Ausubel explains. "Hydrogen is as innocent as an element can be, ending combustion as water."[71] People want their energy dense and clean, and as they move into cities, they accept only electricity and gas, delivered right to their bedside and stovetop. Remarkably, this natural development has brought the world to Peak Coal and maybe even Peak Carbon. As figure 10-8 shows, global emissions plateaued from 2014 to 2015 and declined among the top three emitters, namely China, the European Union, and the United States. (As we saw for the United States in figure 10-3, carbon emissions plateaued while prosperity rose: between 2014 and 2016, the Gross World Product grew by 3 percent annually.)[72] Some of the carbon was reduced by the growth of wind and solar power, but most of it, particularly in the United States, was reduced by the replacement of $C_{137}H_{97}O_9NS$ coal with CH_4 gas.

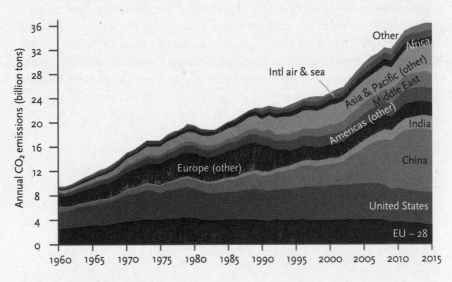

Figure 10-8: CO_2 emissions, 1960–2015

Sources: *Our World in Data*, Ritchie & Roser 2017 and https://ourworldindata.org/grapher/annual-co2-emissions-by-region, based on data from the Carbon Dioxide Information Analysis Center, http://cdiac.ornl.gov/CO2_Emission/, and Le Quéré et al. 2016. "International air & sea" refers to aviation and sea transport; it corresponds to "Bunker fuels" in the original sources. "Other" refers to the difference between estimated global CO2 emissions and the sum of the regional and national totals; it corresponds to the "Statistical difference" component.

The long sweep of decarbonization shows that economic growth is not synonymous with burning carbon. Some optimists believe that if the trend is allowed to evolve into its next phase—from low-carbon natural gas to zero-carbon nuclear energy, a process abbreviated as "N2N"—the climate will have a soft landing. But only the sunniest believe this will happen by itself. Annual CO_2 emissions may have leveled off for the time being at around 36 billion tons, but that's still a *lot* of CO_2 added to the atmosphere every year, and there is no sign of the precipitous plunge we would need to stave off the harmful outcomes. Instead, decarbonization needs to be helped along with pushes from policy and technology, an idea called deep decarbonization.[73]

It begins with carbon pricing: charging people and companies for the damage they do when they dump their carbon into the atmosphere, either as a tax on carbon or as a national cap with tradeable credits. Economists across the political spectrum endorse carbon pricing because it combines the unique advantages of governments and markets.[74] No one owns the atmosphere, so people (and companies) have no reason to stint on emissions that allow each of them to enjoy their energy while harming everyone else, a perverse outcome that economists call a negative externality (another name for the collective costs in a public goods game, or the damage to the commons in the Tragedy of the Commons). A carbon tax, which only governments can impose, "internalizes" the public costs, forcing people to factor the harm into every carbon-emitting decision they make. Having billions of people decide how best to conserve, given their values and the information conveyed by prices, is bound to be more efficient and humane than having government analysts try to divine the optimal mixture from their desks. The potters don't have to hide their kilns from the Carbon Police; they can do their part in saving the planet by taking shorter showers, forgoing Sunday drives, and switching from beef to eggplant. Parents don't have to calculate whether diaper services, with their trucks and laundries, emit more carbon than the makers of disposable diapers; the difference will be folded into the prices, and each company has an incentive to lower its emissions to compete with the other. Inventors and entrepreneurs can take risks on carbon-free energy sources that would compete against fossil fuels on a level playing field rather than the tilted one we have now, in which the fossils get to spew their waste into the atmosphere for free. Without carbon pricing, fossil fuels—which are uniquely abundant, portable, and energy-dense—have too great an advantage over the alternatives.

Carbon taxes, to be sure, hit the poor in a way that concerns the left, and they transfer money from the private to the public sector in a way that annoys the right. But these effects can be neutralized by adjusting sales, payroll, income, and other taxes and transfers. (As Al Gore put it:

Tax what you burn, not what you earn.) And if the tax starts low and increases steeply and predictably over time, people can factor the increase into their long-term purchases and investments, and by favoring low-carbon technologies as they evolve, escape most of the tax altogether.[75]

A second key to deep decarbonization brings up an inconvenient truth for the traditional Green movement: nuclear power is the world's most abundant and scalable carbon-free energy source.[76] Although renewable energy sources, particularly solar and wind, have become drastically cheaper, and their share of the world's energy has more than tripled in the past five years, that share is still a paltry 1.5 percent, and there are limits on how high it can go.[77] The wind is often becalmed, and the sun sets every night and may be clouded over. But people need energy around the clock, rain or shine. Batteries that could store and release large amounts of energy from renewables will help, but ones that could work on the scale of cities are years away. Also, wind and solar sprawl over vast acreage, defying the densification process that is friendliest to the environment. The energy analyst Robert Bryce estimates that simply keeping up with the world's increase in energy use would require turning an area the size of Germany into wind farms every year.[78] To satisfy the world's needs with renewables by 2050 would require tiling windmills and solar panels over an area the size of the United States (including Alaska), plus Mexico, Central America, and the inhabited portion of Canada.[79]

Nuclear energy, in contrast, represents the ultimate in density, because, in a nuclear reaction, $E = mc^2$: you get an immense amount of energy (proportional to the speed of light squared) from a small bit of mass. Mining the uranium for nuclear energy leaves a far smaller environmental scar than mining coal, oil, or gas, and the power plants themselves take up about one five-hundredth of the land needed by wind or solar.[80] Nuclear energy is available around the clock, and it can be plugged into power grids that provide concentrated energy where it is needed. It has a lower carbon footprint than solar, hydro, and biomass, and it's safer than them, too. The sixty years with nuclear power have seen thirty-one deaths in the 1986 Chernobyl disaster, the result of extraordinary Soviet-era bungling, together with a few thousand early deaths from cancer above the 100,000 natural cancer deaths in the exposed population.[81] The other two famous accidents, at Three Mile Island in 1979 and Fukushima in 2011, killed no one. Yet vast numbers of people are killed day in, day out by the pollution from burning combustibles and by accidents in mining and transporting them, none of which make headlines. Compared with nuclear power, natural gas kills 38 times as many people per kilowatt-hour of electricity generated, biomass 63 times as many, petroleum 243 times as many, and coal 387 times as many—perhaps a million deaths a year.[82]

Nordhaus and Shellenberger summarize the calculations of an increasing number of climate scientists: "There is no credible path to reducing global carbon emissions without an enormous expansion of nuclear power. It is the only low carbon technology we have today with the demonstrated capability to generate large quantities of centrally generated electric power."[83] The Deep Decarbonization Pathways Project, a consortium of research teams that have worked out roadmaps for countries to reduce their emissions enough to meet the 2°C target, estimates that the United States will have to get between 30 and 60 percent of its electricity from nuclear power by 2050 (1.5 to 3 times the current fraction), at the same time that it generates far more of that electricity to take over from fossil fuels in heating homes, powering vehicles, and producing steel, cement, and fertilizer.[84] In one scenario, this would require quadrupling its nuclear capacity. Similar expansions would be necessary in China, Russia, and other countries.[85]

Unfortunately, the use of nuclear power has been shrinking just when it should be growing. In the United States, eleven nuclear reactors have recently been closed or are threatened with closure, which would cancel the entire carbon savings from the expanded use of solar and wind. Germany, which has relied on nuclear energy for much of its electricity, is shutting down its plants as well, increasing its carbon emissions from the coal-fired plants that replace them, and France and Japan may follow its lead.

Why are Western countries going the wrong way? Nuclear power presses a number of psychological buttons—fear of poisoning, ease of imagining catastrophes, distrust of the unfamiliar and the man-made—and the dread has been amplified by the traditional Green movement and its dubiously "progressive" supporters.[86] One commentator blames global warming on the Doobie Brothers, Bonnie Raitt, and the other rock stars whose 1979 *No Nukes* concert and film galvanized baby-boomer sentiment against nuclear power. (Sample lyrics of the closing anthem: "Just give me the warm power of the sun . . . But won't you take all your atomic poison power away.")[87] Some of the blame might go to Jane Fonda, Michael Douglas, and the producers of the 1979 disaster film *The China Syndrome*, so named because the melted-down nuclear reactor core would supposedly sink through the Earth's crust all the way to China, after making "an area the size of Pennsylvania" uninhabitable. In a devilish coincidence, the Three Mile Island plant in central Pennsylvania suffered its partial meltdown two weeks after the movie's release, creating widespread panic and making the very idea of nuclear power as radioactive as its uranium fuel.

It's often said that with climate change, those who know the most are the most frightened, but with nuclear power, those who know the most are the least frightened.[88] As with oil tankers, cars, planes, buildings, and factories (chapter 12), engineers have learned from the accidents and near-misses and have progressively squeezed more safety out of nuclear reactors, reducing the risks of accidents and contamination far below those of fossil fuels. The advantage even extends to radioactivity, which is a natural property of the fly ash and flue gases emitted by burning coal.

Still, nuclear power is expensive, mainly because it must clear crippling regulatory hurdles while its competitors have been given easy passage. Also, in the United States, nuclear power plants are now being built, after a lengthy hiatus, by private companies using idiosyncratic designs, so they have not climbed the engineer's learning curve and settled on the best practices in design, fabrication, and construction. Sweden, France, and South Korea, in contrast, have built standardized reactors by the dozen and now enjoy cheap electricity with substantially lower carbon emissions. As Ivan Selin, former commissioner of the Nuclear Regulatory Commission, put it, "The French have two kinds of reactors and hundreds of kinds of cheese, whereas in the United States the figures are reversed."[89]

For nuclear power to play a transformative role in decarbonization it will eventually have to leap past the second-generation technology of light-water reactors. (The "first generation" consisted of prototypes from the 1950s and early 1960s.) Soon to come on line are a few Generation III reactors, which evolved from the current designs with improvements in safety and efficiency but so far have been plagued by financial and construction snafus. Generation IV reactors comprise a half-dozen new designs which promise to make nuclear plants a mass-produced commodity rather than finicky limited editions.[90] One type might be cranked out on an assembly line like jet engines, fitted into shipping containers, transported by rail, and installed on barges anchored offshore cities. This would allow them to clear the NIMBY hurdle, ride out storms or tsunamis, and be towed away at the end of their useful lives for decommissioning. Depending on the design, they could be buried and operated

underground, cooled by inert gas or molten salt that needn't be pressur-
ized, refueled continuously with a stream of pebbles rather than shut
down for the replacement of fuel rods, equipped to co-generate hydrogen
(the cleanest of fuels), and designed to shut themselves off without power
or human intervention if they overheat. Some would be fueled by relatively abundant thorium, and others by uranium extracted from seawater,
from dismantled nuclear weapons (the ultimate beating of swords into
plowshares), from the waste of existing reactors, or even from their own
waste—the closest we will ever get to a perpetual-motion machine, capable of powering the world for thousands of years. Even nuclear fusion,
long derided as the energy source that is "thirty years away and always
will be," really may be thirty years away (or less) this time.[91]

The benefits of advanced nuclear energy are incalculable. Most climate change efforts call for policy reforms (such as carbon pricing) which
remain contentious and will be hard to implement worldwide even in
the rosiest scenarios. An energy source that is cheaper, denser, and
cleaner than fossil fuels would sell itself, requiring no herculean political
will or international cooperation.[92] It would not just mitigate climate
change but furnish manifold other gifts. People in the developing world
could skip the middle rungs in the energy ladder, bringing their standard of living up to that of the West without choking on coal smoke.
Affordable desalination of seawater, an energy-ravenous process, could
irrigate farms, supply drinking water, and, by reducing the need for both
surface water and hydro power, allow dams to be dismantled, restoring
the flow of rivers to lakes and seas and revivifying entire ecosystems.
The team that brings clean and abundant energy to the world will benefit humanity more than all of history's saints, heroes, prophets, martyrs,
and laureates combined.

Breakthroughs in energy may come from startups founded by idealistic inventors, from the skunk works of energy companies, or from the
vanity projects of tech billionaires, especially if they have a diversified
portfolio of safe bets and crazy moonshots.[93] But research and development will also need a boost from governments, because these global public goods are too great a risk with too little reward for private companies.
Governments must play a role because, as Brand points out, "infrastructure is one of the things we hire governments to handle, especially energy infrastructure, which requires no end of legislation, bonds, rights
of way, regulations, subsidies, research, and public-private contracts
with detailed oversight."[94] This includes a regulatory environment that
is suited to 21st-century challenges rather than to 1970s-era technopho-

bia and nuclear dread. Some fourth-generation nuclear technologies are shovel-ready, but are trussed in regulatory green tape and may never see the light of day, at least not in the United States.[95] China, Russia, India, and Indonesia, which are hungry for energy, sick of smog, and free from American squeamishness and political gridlock, may take the lead.

Whoever does it, and whichever fuel they use, the success of deep decarbonization will hinge on technological progress. Why assume that the know-how of 2018 is the best the world can do? Decarbonization will need breakthroughs not just in nuclear power but on other technological frontiers: batteries to store the intermittent energy from renewables; Internet-like smart grids that distribute electricity from scattered sources to scattered users at scattered times; technologies that electrify and decarbonize industrial processes such as the production of cement, fertilizer, and steel; liquid biofuels for heavy trucks and planes that need dense, portable energy; and methods of capturing and storing CO_2.

$\sim$

The last of these is critical for a simple reason. Even if greenhouse gas emissions are halved by 2050 and zeroed by 2075, the world would still be on course for risky warming, because the CO_2 already emitted will remain in the atmosphere for a very long time. It's not enough to stop thickening the greenhouse; at some point we have to dismantle it.

The basic technology is more than a billion years old. Plants suck carbon out of the air as they use the energy in sunlight to combine CO_2 with H_2O and make sugars (like $C_6H_{12}O_6$), cellulose (a chain of $C_6H_{10}O_5$ units), and lignin (a chain of units like $C_{10}H_{14}O_4$); the latter two make up most of the biomass in wood and stems. The obvious way to remove CO_2 from the air, then, is to recruit as many carbon-hungry plants as we can to help us. We can do this by encouraging the transition from deforestation to reforestation and afforestation (planting new forests), by reversing tillage and wetland destruction, and by restoring coastal and marine habitats. And to reduce the amount of carbon that returns to the atmosphere when dead plants rot, we could encourage building with wood and other plant products, or cook the biomass into non-rotting charcoal and bury it as a soil amendment called biochar.[96]

Other ideas for carbon capture span a broad range of flakiness, at least by the standards of current technology. The more speculative end shades into geoengineering, and includes plans to disperse pulverized rock that takes up CO_2 as it weathers, to add alkali to clouds or the oceans to dissolve more CO_2 in water, and to fertilize the ocean with iron to accelerate photosynthesis by plankton.[97] The more proven end consists of technologies that can scrub CO_2 from the smokestacks of fossil fuel plants and pump it into nooks and crannies in the earth's crust. (Skimming the sparse 400 parts per million directly from the atmosphere is theoretically possible but prohibitively inefficient, though that could change if nuclear power became cheap enough.) The technologies can be retrofitted into existing factories and power plants, and though they are themselves energy-hungry, they could slash carbon emissions from the vast energy infrastructure that is already in place (resulting in so-called clean coal). The technologies can also be fitted onto gasification plants that convert coal into liquid fuels, which may still be needed for planes and heavy trucks. The geophysicist Daniel Schrag points out that the gasification process already has to separate CO_2 from the gas stream, so sequestering that CO_2 to protect the atmosphere is a modest incremental expense, and it would yield liquid fuel with a smaller carbon footprint than that of petroleum.[98] Better still, if the coal feedstock is supplemented with biomass (including grasses, agricultural waste, forest cuttings, municipal garbage, and perhaps someday genetically engineered plants or algae), it could be carbon-neutral. Best of all, if the feedstock consisted *exclusively* of biomass, it would be carbon-*negative*. The plants pull CO_2 out of the atmosphere, and when their biomass is used for energy (via combustion, fermentation, or gasification), the carbon capture process keeps it out. The combination, sometimes called BECCS—bioenergy with carbon capture and storage—has been called climate change's savior technology.[99]

Will any of this happen? The obstacles are unnerving; they include the world's growing thirst for energy, the convenience of fossil fuels with their vast infrastructure, the denial of the problem by energy corporations and the political right, the hostility to technological solutions from traditional Greens and the climate justice left, and the tragedy of the carbon commons. For all that, preventing climate change is an idea whose time has come. One indication is a trio of headlines that appeared in *Time* magazine within a three-week span in 2015: "China Shows It's Serious About Climate Change," "Walmart, McDonald's, and 79 Others Commit to Fight Global Warming," and "Americans' Denial of Climate Change Hits Record Low." In the same season the *New York Times* reported, "Poll Finds Global Consensus on a Need to Tackle Climate Change." In all but one of the forty countries surveyed (Pakistan), a majority of respondents were in favor of limiting greenhouse gas emissions, including 69 percent of the Americans.[100]

The global consensus is not just hot air. In December 2015, 195 countries signed a historic agreement that committed them to keeping the global temperature rise to "well below" 2°C (with a target of 1.5°C) and to setting aside $100 billion annually in climate mitigation financing for developing countries (which had been a sticking point in prior, unsuccessful attempts at a global consensus).[101] In October 2016, 115 of the signatories ratified the agreement, putting it into force. Most of the signatories submitted detailed plans on how they would pursue these goals through 2025, and all promised to update their plans every five years with stepped-up efforts. Without this ratcheting, the current plans are inadequate: they would allow the world's temperature to rise by 2.7°C, and would reduce the chance of a dangerous 4°C rise in 2100 by only 75 percent, which is still too close for comfort. But the public commitments, combined with contagious technological advances, could push the ratchet upward, in which case the Paris agreement would substantially reduce the likelihood of a 2°C rise and essentially eliminate the possibility of a 4°C rise.[102]

This game plan faced a setback in 2017 when Donald Trump, who had notoriously called climate change a Chinese hoax, announced that the United States would withdraw from the agreement. Even if the withdrawal takes place in November 2020 (the earliest possible date), the decarbonization driven by technology and economics will continue, and climate change policies will be advanced by cities, states, business and tech leaders, and the world's other countries, which have declared the deal "irreversible" and may pressure the United States to keep its word by imposing carbon tariffs on American exports and other sanctions.[103]

⁓

Even with fair winds and following seas, the effort needed to prevent climate change is immense, and we have no guarantee that the necessary transformations in technology and politics will be in place soon enough to slow down global warming before it causes extensive harm. This brings us to a last-ditch protective measure: lowering the world's temperature by reducing the amount of solar radiation that reaches the lower atmosphere and Earth's surface.[104] A fleet of airplanes could spray a fine mist of sulfates, calcite, or nanoparticles into the stratosphere, spreading a thin veil that would reflect back just enough sunlight to prevent dangerous warming.[105] This would mimic the effects of a volcanic eruption such as that of Mount Pinatubo in the Philippines in 1991, which spewed so much sulfur dioxide into the atmosphere that the planet cooled down by half a degree Celsius (about one degree Fahrenheit) for two years. Or a fleet of cloudships could spray a fine mist of seawater into the air. As the water evaporated, salt crystals would waft into the clouds and water vapor would condense around them, forming droplets that would whiten the clouds and reflect more sunlight back into space. These measures are relatively inexpensive, require no exotic new technologies, and could bring global temperatures down quickly. Other ideas for manipulating the atmosphere and oceans have been bruited about as well, though research on all of them is in its infancy.

The very idea of climate engineering sounds like the crazed scheme of a mad scientist, and it once was close to taboo. Critics see it as a Promethean folly that could have unintended consequences such as disrupting rainfall patterns and damaging the ozone layer. Since the effects of any measure applied to the entire planet are uneven from place to place, climate engineering raises the question of whose hand should be on the world's thermostat: as with a bickering couple, if one country lowered the temperature at the expense of another, it could set off a war. Once the world depended on climate engineering, then if for any reason it slacked off, temperatures in the carbon-soaked atmosphere would soar far more quickly than people could adapt. The mere mention of an escape hatch for the climate crisis creates a moral hazard, tempting countries to shirk their duty to reduce greenhouse gas emissions. And the accumulated CO_2 in the atmosphere would continue to dissolve in seawater, slowly turning the oceans into carbonic acid.

For all these reasons, no responsible person could maintain that we can just keep pumping carbon into the air and slather sunscreen onto the stratosphere to compensate. But in a 2013 book the physicist David Keith makes a case for a form of climate engineering that is *moderate, responsive,* and *temporary.* "Moderate" means that the amounts of sulfate or calcite would be just enough to reduce the rate of warming, not cancel it altogether; moderation is a virtue because small manipulations are less likely to bring unwelcome surprises. "Responsive" means that any manipulation would be careful, gradual, closely monitored, constantly adjusted, and, if indicated, halted altogether. And "temporary" means that the program would be designed only to give humanity breathing space until it eliminates greenhouse gas emissions and brings the CO_2 in the atmosphere back to preindustrial levels. In response to the fear that the world would become addicted to climate engineering forever, Keith remarks, "Is it plausible that we will not figure out how to pull, say, five gigatons of carbon per year out of the air by 2075? I don't buy it."[106]

Though Keith is among the world's foremost climate engineers, he cannot be accused of being carried away by innovation thrill. A similarly thoughtful case may be found in the journalist Oliver Morton's 2015 book *The Planet Remade,* which presents the historical, political, and moral dimensions of climate engineering alongside the technical state of the art. Morton shows that humanity has been disrupting global cycles of water, nitrogen, and carbon for more than a century, so it's too late to preserve a primeval Earth system. And given the enormity of the climate change problem, it's unwise to assume we will solve it quickly or easily. Research into how we might minimize the harm to millions of people before the solutions are completely in place only seems prudent, and Morton lays out scenarios of how a program of moderate and temporary climate engineering might be implemented even in a world that falls short of ideal global governance. The legal scholar Dan Kahan has shown that far from creating a moral hazard, providing information about climate engineering makes people *more* concerned about climate change and less biased by their political ideology.[107]

~

Despite a half-century of panic, humanity is not on an irrevocable path to ecological suicide. The fear of resource shortages is misconceived. So is the misanthropic environmentalism that sees modern humans as vile despoilers of a pristine planet. An enlightened environmentalism recognizes that humans need to use energy to lift themselves out of the poverty to which entropy and evolution consign them. It seeks the means to do so with the least harm to the planet and the living world. History suggests that this modern, pragmatic, and humanistic environmentalism can work. As the world gets richer and more tech-savvy, it dematerializes, decarbonizes, and densifies, sparing land and species. As people get richer and better educated, they care more about the environment, figure out ways to protect it, and are better able to pay the costs. Many parts of the environment are rebounding, emboldening us to deal with the admittedly severe problems that remain.

First among them is the emission of greenhouse gases and the threat they pose of dangerous climate change. People sometimes ask me whether I think that humanity will rise to the challenge or whether we will sit back and let disaster unfold. For what it's worth, I think we'll rise to the challenge, but it's vital to understand the nature of this optimism. The economist Paul Romer distinguishes between *complacent* optimism, the feeling of a child waiting for presents on Christmas morning, and *conditional* optimism, the feeling of a child who wants a treehouse and realizes that if he gets some wood and nails and persuades other kids to help him, he can build one.[108] We cannot be complacently optimistic about climate change, but we can be conditionally optimistic. We have some practicable ways to prevent the harms and we have the means to learn more. Problems are solvable. That does not mean that they will solve themselves, but it does mean that we can solve them *if* we sustain the benevolent forces of modernity that have allowed us to solve problems so far, including societal prosperity, wisely regulated markets, international governance, and investments in science and technology.

CHAPTER 11

PEACE

How deep do the currents of progress flow? Can they suddenly come to a halt or go into reverse? The history of violence provides an opportunity to confront these questions. In *The Better Angels of Our Nature* I showed that, as of the first decade of the 21st century, every objective measure of violence had been in decline. As I was writing it, reviewers warned me that it could all explode before the first copy hit the stores. (A war, possibly nuclear, between Iran and either Israel or the United States was the worry of the day.) Since the book was published in 2011, a cascade of bad news would seem to make it obsolete: civil war in Syria, atrocities in the Islamic State, terrorism in Western Europe, autocracy in Eastern Europe, shootings by police in the United States, and hate crimes and other outbursts of racism and misogyny from angry populists throughout the West.

But the same Availability and Negativity biases that made people incredulous about the possibility that violence had declined can make them quick to conclude that any decline has been reversed. In the next five chapters I will put the recent bad news in perspective by going back to the data. I'll plot the historical trajectories of several kinds of violence up to the present, including a reminder of the last data point available when *The Better Angels of Our Nature* went to press.[1] Seven years or so is an eyeblink in history, but it offers a crude indication of whether the book capitalized on a lucky instant or identified an ongoing trend. More important, I'll try to explain the trends in terms of deeper historical forces, placing them within the narrative of progress that is the subject of this book. (In doing so I'll introduce some new ideas on what those forces are.) I'll begin with the most extravagant form of violence, war.

~

For most of human history, war was the natural pastime of governments, peace a mere respite between wars.[2] This can be seen in figure 11-1, which plots the proportion of time over the last half-millennium that the great powers of the day were at war. (Great powers are the handful of states and empires that can project force beyond their borders, that treat each other as peers, and that collectively control a majority of the world's military resources.)[3] Wars between great powers, which include world wars, are the most intense forms of destruction our sorry species has managed to dream up, and they are responsible for a majority of the victims of all wars combined. The graph shows that at the dawn of the modern era the great powers were pretty much always at war. But nowadays they are never at war: the last one pitted the United States against China in Korea more than sixty years ago.

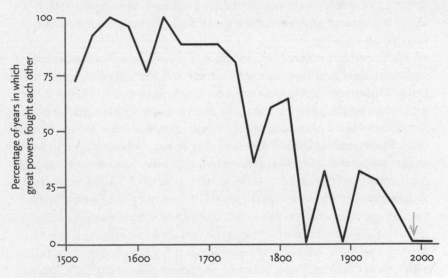

Figure 11-1: Great power war, 1500–2015

Source: Levy & Thompson 2011, updated for the 21st century. Percentage of years the great powers fought each other in wars, aggregated over 25-year periods, except for 2000–2015. The arrow points to 1975–1999, the last quarter-century plotted in fig. 5–12 of Pinker 2011.

The jagged decline of great power war conceals two trends that until recently went in opposite directions.[4] For 450 years, wars involving a great power became shorter and less frequent. But as their armies became better manned, trained, and armed, the wars that did take place became more lethal, culminating in the brief but stunningly destructive world wars. It was only after the second of these that all three measures of war—frequency, duration, and lethality—declined in tandem, and the world entered the period that has been called the Long Peace.

It's not just the great powers that have stopped fighting each other. War in the classic sense of an armed conflict between the uniformed armies of two nation-states appears to be obsolescent.[5] There have been no more than three in any year since 1945, none in most years since 1989, and none since the American-led invasion of Iraq in 2003, the longest stretch without an interstate war since the end of World War II.[6] Today, skirmishes between national armies kill dozens of people rather than the hundreds of thousands or millions who died in the all-out wars that nation-states have fought throughout history. The Long Peace has certainly been tested since 2011, such as in conflicts between Armenia and Azerbaijan, Russia and Ukraine, and the two Koreas, but in each case the belligerents backed down rather than escalating into all-out war. This doesn't, of course, mean that escalation to major war is impossible, just that it is considered extraordinary, something that nations try to avoid at (almost) all costs.

The geography of war also continues to shrink. In 2016 a peace agreement between the government of Colombia and Marxist FARC guerrillas ended the last active political armed conflict in the Western Hemisphere, and the last remnant of the Cold War. This is a momentous change from just decades before.[7] In Guatemala, El Salvador, and Peru, as in Colombia, leftist guerrillas battled American-backed governments, and in Nicaragua it was the other way around (American-backed contras battling a left-wing government), in conflicts that collectively killed more than 650,000 people.[8] The transition of an entire hemisphere to peace follows the path of other large regions of the world. Western Europe's bloody centuries of warfare, culminating in the two world wars, have given way to more than seven decades of peace. In East Asia, the wars of the mid-20th century took millions of lives—in Japan's conquests, the Chinese Civil War, and the wars in Korea and Vietnam. Yet despite serious political disputes, East and Southeast Asia today are almost entirely free from active interstate combat.

The world's wars are now concentrated almost exclusively in a zone stretching from Nigeria to Pakistan, an area containing less than a sixth of the world's population. Those wars are civil wars, which the Uppsala Conflict Data Program (UCDP) defines as an armed conflict between a government and an organized force which verifiably kills at least a thousand soldiers and civilians a year. Here we find some recent cause for discouragement. A precipitous decline in the number of civil wars after the end of the Cold War—from fourteen in 1990 to four in 2007—went back up to eleven in 2014 and 2015 and to twelve in 2016.[9] The flip is driven mainly by conflicts that have a radical Islamist group on one side (eight of the eleven in 2015, ten of the twelve in 2016); without them, there would have been no increase in the number of wars at all. Perhaps not coincidentally, two of the wars in 2014 and 2015 were fueled by another counter-Enlightenment ideology, Russian nationalism, which drove sep-aratist forces, backed by Vladimir Putin, to battle the government of Ukraine in two of its provinces.

The worst of the ongoing wars is in Syria, where the government of Bashar al-Assad has pulverized his country in an attempt to defeat a diverse set of rebel forces, Islamist and non-Islamist, with the assistance of Russia and Iran. The Syrian civil war, with 250,000 battle deaths as of 2016 (conservatively estimated), is responsible for most of the uptick in the global rate of war deaths shown in figure 11-2.[10]

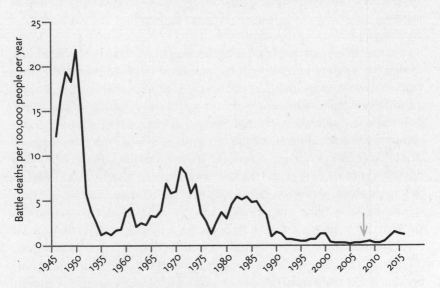

Figure 11-2: Battle deaths, 1946–2016

Sources: Adapted from Human Security Report Project 2007. For 1946–1988: *Peace Research Institute of Oslo Battle Deaths Dataset 1946–2008*, Lacina & Gleditsch 2005. For 1989–2015: *UCDP Battle-Related Deaths Dataset version 5.0*, Uppsala Conflict Data Program 2017, Melander, Pettersson, & Themnér 2016, updated with information from Therese Pettersson and Sam Taub of UCDP. World population figures: 1950–2016, US Census Bureau; 1946–1949, McEvedy & Jones 1978, with adjustments. The arrow points to 2008, the last year plotted in fig. 6–2 of Pinker 2011.

That uptick, however, comes at the end of a vertiginous six-decade plunge. World War II at its worst saw almost 300 battle deaths per 100,000 people per year; it is not shown in the graph because it would have scrunched the line for all subsequent years into a wrinkled carpet. In the postwar years, as the graph shows, the rate of deaths roller-coastered downward, cresting at 22 during the Korean War, 9 during the Vietnam War in the late 1960s and early 1970s, and 5 during the Iran-Iraq War in the mid-1980s, before bobbing along the floor at less than 0.5 between 2001 and 2011. It crept up to 1.5 in 2014 and subsided to 1.2 in 2016, the most recent year for which data are available.

Followers of the news in the mid-2010s might have expected the Syrian carnage to have erased all of the historic progress of the preceding decades. That's because they forget the many civil wars that ended without fanfare after 2009 (in Angola, Chad, India, Iran, Peru, and Sri Lanka) and also forget earlier ones with massive death tolls, such as the wars in Indochina (1946–54, 500,000 deaths), India (1946–48, a million deaths), China (1946–50, a million deaths), Sudan (1956–72, 500,000 deaths, and 1983–2002, a million deaths), Uganda (1971–78, 500,000 deaths), Ethiopia (1974–91, 750,000 deaths), Angola (1975–2002, a million deaths), and Mozambique (1981–92, 500,000 deaths).[11]

Searing images of desperate refugees from the Syrian civil war, many of them struggling to resettle in Europe, have led to the claim that the world now has more refugees than at any time in history. But this is another symptom of historical amnesia and the Availability bias. The political scientist Joshua Goldstein notes that today's four million Syrian refugees are outnumbered by the ten million displaced by the Bangladesh War of Independence in 1971, the fourteen million displaced by the partition of India in 1947, and the sixty million displaced by World War II in Europe alone, eras when the world's population was a fraction of what it is now. Quantifying this misery is by no means callous to the terrible suffering of today's victims. It honors the suffering of yesterday's victims, and it ensures that policymakers will act in their interests by working from an accurate understanding of the world. In particular, it should prevent them from drawing dangerous conclusions about "a world at war," which could tempt them to scrap global governance or return to a mythical "stability" of Cold War confrontation. "The world is not the problem," Goldstein notes; "Syria is the problem. . . . The policies and practices that ended wars [elsewhere] can with effort and intelligence end wars today in South Sudan, Yemen, and perhaps even Syria."[12]

Mass killings of unarmed civilians, also known as genocides, democides, or one-sided violence, can be as lethal as wars and often overlap with them. According to the historians Frank Chalk and Kurt Jonassohn,

"Genocide has been practiced in all regions of the world and during all periods in history."[13] During World War II, tens of millions of civilians were slaughtered by Hitler, Stalin, and imperial Japan, and in deliberate bombings of civilian areas by all sides (twice with nuclear weapons); at its peak the death rate was about 350 per 100,000 per year.[14] But contrary to the assertion that "the world has learned nothing from the Holocaust," the postwar period has seen nothing like the blood flood of the 1940s. Even within the postwar period, the rate of deaths in genocides has juddered down a steep sawtooth, as we see in two datasets shown in figure 11-3.

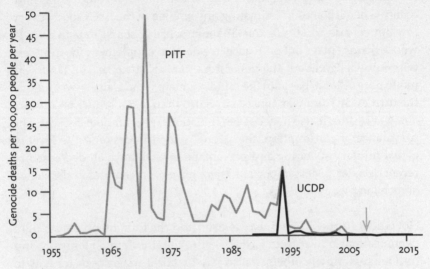

Figure 11-3: Genocide deaths, 1956–2016

Sources: PITF, 1955–2008: *Political Instability Task Force State Failure Problem Set*, 1955–2008, Marshall, Gurr, & Harff 2009; Center for Systemic Peace 2015. Calculations described in Pinker 2011, p. 338. UCDP, 1989–2016: *UCDP One-Sided Violence Dataset v. 2.5-2016*, Melander, Pettersson, & Themnér 2016; Uppsala Conflict Data Program 2017, "High fatality" estimates, updated with data provided by Sam Taub of UCDP, scaled by world population figures from US Census Bureau. The arrow points to 2008, the last year plotted in fig. 6–8 of Pinker 2011.

The peaks in the graph correspond to mass killings in the Indonesian anti-Communist "year of living dangerously" (1965–66, 700,000 deaths), the Chinese Cultural Revolution (1966–75, 600,000), Tutsis against Hutus in Burundi (1965–73, 140,000), the Bangladesh War of Independence (1971, 1.7 million), north-against-south violence in Sudan (1956–72, 500,000), Idi Amin's regime in Uganda (1972–79, 150,000), Pol Pot's regime in Cambodia (1975–79, 2.5 million), killings of political enemies in Vietnam (1965–75, 500,000), and more recent massacres in Bosnia (1992–95, 225,000), Rwanda

(1994, 700,000), and Darfur (2003–8, 373,000).[15] The barely perceptible swelling from 2014 to 2016 includes the atrocities that contribute to the impression that we are living in newly violent times: at least 4,500 Yazidis, Christians, and Shiite civilians killed by ISIS; 5,000 killed by Boko Haram in Nigeria, Cameroon, and Chad; and 1,750 killed by Muslim and Christian militias in the Central African Republic.[16] One can never use the word "fortunately" in connection with the killing of innocents, but the numbers in the 21st century are a fraction of those in earlier decades.

Of course, the numbers in a dataset cannot be interpreted as a direct readout of the underlying risk of war. The historical record is especially scanty when it comes to estimating any change in the likelihood of very rare but very destructive wars.[17] To make sense of sparse data in a world whose history plays out only once, we need to supplement the numbers with knowledge about the generators of war, since, as the UNESCO motto notes, "Wars begin in the minds of men." And indeed we find that the turn away from war consists in more than just a reduction in wars and war deaths; it also may be seen in nations' preparations for war. The prevalence of conscription, the size of armed forces, and the level of global military spending as a percentage of GDP have all decreased in recent decades.[18] Most important, there have been changes in the minds of men (and women).

〜

How did it happen? The Age of Reason and the Enlightenment brought denunciations of war from Pascal, Swift, Voltaire, Samuel Johnson, and the Quakers, among others. It also saw practical suggestions for how to reduce or even eliminate war, particularly Kant's famous essay "Perpetual Peace."[19] The spread of these ideas has been credited with the decline in great power wars in the 18th and 19th centuries and with several hiatuses in war during that interval.[20] But it was only after World War II that the pacifying forces identified by Kant and others were systematically put into place.

As we saw in chapter 1, many Enlightenment thinkers advanced the theory of gentle commerce, according to which international trade should make war less appealing. Sure enough, trade as a proportion of GDP shot up in the postwar era, and quantitative analyses have confirmed that trading countries are less likely to go to war, holding all else constant.[21]

Another brainchild of the Enlightenment is the theory that democratic government serves as a brake on glory-drunk leaders who would drag their countries into pointless wars. Starting in the 1970s, and accelerating after the fall of the Berlin Wall in 1989, more countries gave democracy a chance (chapter 14). While the categorical statement that no two democracies have ever gone to war is dubious, the data support a graded version of the Democratic Peace theory, in which pairs of countries that are more democratic are less likely to confront each other in militarized disputes.[22]

The Long Peace was also helped along by some realpolitik. The massive destructive powers of the American and Soviet armies (even without their nuclear weapons) made the Cold War superpowers think twice about confronting each other on the battlefield—which, to the world's surprise and relief, they never did.[23]

Yet the biggest single change in the international order is an idea we seldom appreciate today: *war is illegal*. For most of history, that was not the case. Might made right, war was the continuation of policy by other means, and to the victor went the spoils. If one country felt it had been wronged by another, it could declare war, conquer some territory as compensation, and expect the annexation to be recognized by the rest of the world. The reason that Arizona, California, Colorado, Nevada, New Mexico, and Utah are American states is that in 1846 the United States conquered them from Mexico in a war over unpaid debts. That cannot happen today: the world's nations have committed themselves to not waging war except in self-defense or with the approval of the United Nations Security Council. States are immortal, borders are grandfathered in, and any country that indulges in a war of conquest can expect opprobrium, not acquiescence, from the rest.

The legal scholars Oona Hathaway and Scott Shapiro argue that it's the outlawry of war that deserves much of the credit for the Long Peace. The idea that nations should agree to make war illegal was proposed by Kant in 1795. It was first agreed upon in the much-ridiculed 1928 Pact of Paris, also known as the Kellogg-Briand pact, but really became effective only with the founding of the United Nations in 1945. Since then, the conquest taboo has occasionally been enforced with a military response, such as when an international coalition reversed Iraq's conquest of Kuwait in 1990–91. More often the prohibition has functioned as a norm— "War is something that civilized nations just don't do"—backed by economic sanctions and symbolic punishments. Those penalties are effective to the extent that nations value their standing in the international community—a reminder of why we should cherish and strengthen that community in the face of threats from populist nationalism today.[24]

To be sure, the norm is sometimes honored in the breach, most recently in 2014, when Russia annexed Crimea. This would seem to confirm the cynical view that until we have a world government, international norms are toothless and will be flouted with impunity. Hathaway and Shapiro reply that laws within a country are broken, too, from parking violations to homicides, yet an imperfectly enforced law is better than no rule of law at all. The century before the Paris Peace Pact, they calculate, saw the equivalent of *eleven Crimea-sized annexations a year*, most of which stuck. But virtually every acre of land that was conquered after 1928 has been returned to the state that lost it. Frank Kellogg and Aristide Briand (the US secretary of state and the French foreign minister) may deserve the last laugh.

Hathaway and Shapiro point out that the outlawry of interstate war had a downside. As European empires vacated the colonial territories they had conquered, they often left behind weak states with fuzzy borders and no single recognized successor to govern them. The states often fell into civil war and intercommunal violence. Under the new international order, they were no longer legitimate targets of conquest by more effective powers, and hung on in semi-anarchy for years or decades.

The decline of interstate war was still a magnificent example of progress. Civil wars kill fewer people than interstate wars, and since the late 1980s civil wars have declined as well.[25] When the Cold War ended, the great powers became less interested in who won a civil war than in how to end it, and they supported UN peacekeeping forces and other international posses which inserted themselves between belligerents and, more often than not, really did keep the peace.[26] Also, as countries get richer, they become less vulnerable to civil war. Their governments can afford to provide services like health care, education, and policing and thus outcompete rebels for the allegiance of their citizens, and they can regain control of the frontier regions that warlords, mafias, and guerrillas (often the same people) stake out.[27] And since many wars are ignited by the mutual fear that unless a country attacks preemptively it will be annihilated by a preemptive attack (the game-theoretic scenario called a security dilemma or Hobbesian trap), the alighting of peace in a neighborhood, whatever its first cause, can be self-reinforcing. (Conversely, war can be contagious.)[28] That helps explain the shrinking geography of war, with most regions of the globe at peace.

~

Together with ideas and policies that reduce the incidence of war, there has been a change in values. The pacifying forces we have seen so far are, in a sense, technological: they are means by which the odds can be tilted in favor of peace if it's peace that people want. At least since the folk-song-and-Woodstock '60s, the idea that peace is inherently worthy has become second nature to Westerners. When military interventions have been launched they have been rationalized as regrettable but necessary measures to prevent greater violence. But not so long ago it was *war* that was considered worthy. War was glorious, thrilling, spiritual, manly, no-ble, heroic, altruistic—a cleansing purgative for the effeminacy, selfish-ness, consumerism, and hedonism of decadent bourgeois society.[29]

Today, the idea that it is inherently noble to kill and maim people and destroy their roads, bridges, farms, dwellings, schools, and hospitals strikes us as the raving of a madman. But during the 19th-century counter-Enlightenment, it all made sense. Romantic militarism became increasingly fashionable, not just among *Pickelhaube*-topped military of-ficers but among many artists and intellectuals. War "enlarges the mind of a people and raises their character," wrote Alexis de Tocqueville. It is "life itself," said Émile Zola; "the foundation of all the arts . . . [and] the high virtues and faculties of man," wrote John Ruskin.[30]

Romantic militarism sometimes merged with romantic nationalism, which exalted the language, culture, homeland, and racial makeup of an ethnic group—the ethos of blood and soil—and held that a nation could fulfill its destiny only as an ethnically cleansed sovereign state.[31] It drew strength from the muzzy notion that violent struggle is the life force of nature ("red in tooth and claw") and the engine of human progress. (This can be distinguished from the Enlightenment idea that the engine of human progress is problem-solving.) The valorization of struggle har-monized with Friedrich Hegel's theory of a dialectic in which historical forces bring forth a superior nation-state: wars are necessary, Hegel wrote, "for they save the state from social petrifaction and stagnation."[32] Marx adapted the idea to economic systems and prophesied that a pro-gression of violent class conflicts would climax in a communist utopia.[33]

But perhaps the biggest impetus to romantic militarism was declin-ism, the revulsion among intellectuals at the thought that ordinary peo-ple seemed to be enjoying their lives in peace and prosperity.[34] Cultural pessimism became particularly entrenched in Germany through the in-fluence of Schopenhauer, Nietzsche, Jacob Burckhardt, Georg Simmel, and Oswald Spengler, author in 1918–23 of *The Decline of the West*. (We will return to these ideas in chapter 23.) To this day, historians of World War I puzzle over why England and Germany, countries with a lot in common—Western, Christian, industrialized, affluent—would choose to hold a pointless bloodbath. The reasons are many and tangled, but insofar as they involve ideology, Germans before World War I "saw themselves as *outside* European or Western civilization," as Arthur Herman points out.[35] In particular, they thought they were bravely resisting the creep of a liberal, democratic, commercial culture that had been sapping the vitality of the West since the Enlightenment, with the complicity of Britain and the United States. Only from the ashes of a redemptive cataclysm, many thought, could a new heroic order arise. They got their wish for a cataclysm. After a second and even more horrific one, the romance had finally been drained from war, and peace became the stated goal of every Western and international institution. Human life has become more precious, while glory, honor, preeminence, manliness, heroism, and other symptoms of excess testosterone have been downgraded.

Many people refuse to believe that progress toward peace, however fitful, could even be possible. Human nature, they insist, includes an insatiable drive for conquest. (And not just human nature; some commentators project the megalomania of *Homo sapiens* males onto every form of intelligence, warning that we must not search for extraterrestrial life lest an advanced race of space aliens discovers our existence and comes over to subjugate us.) While a vision of world peace may have given John and Yoko some good songs, it is hopelessly naïve in the real world.

In fact, war may be just another obstacle an enlightened species learns to overcome, like pestilence, hunger, and poverty. Though conquest may be tempting over the short term, it's ultimately better to figure out how to get what you want without the costs of destructive conflict and the inherent hazards of living by the sword, namely that if you are a menace to others you have given them an incentive to destroy you first. Over the long run, a world in which all parties refrain from war is better for everyone. Inventions such as trade, democracy, economic development, peacekeeping forces, and international law and norms are tools that help build that world.

CHAPTER 12

SAFETY

The human body is a fragile thing. Even when people keep themselves fueled, functioning, and free of pathogens, they are vulnerable to "the thousand natural shocks that flesh is heir to." Our ancestors were easy pickings for predators like crocodiles and large cats. They were done in by the venom of snakes, spiders, insects, snails, and frogs. Trapped in the omnivore's dilemma, they could be poisoned by toxic ingredients in their expansive diets, including fish, beans, roots, seeds, and mushrooms. As they ventured up trees in pursuit of fruit and honey, their bodies obeyed Newton's law of universal gravitation and were liable to accelerate toward the ground at a rate of 9.8 meters per second. If they waded too far into lakes and rivers, the water could cut off their air supply. They played with fire and sometimes got burned. And they could be victims of malice aforethought: any technology that can fell an animal can fell a human rival.

Few people get eaten today, but every year tens of thousands die from snakebites, and other hazards continue to kill us in large numbers.[1] Accidents are the fourth-leading cause of death in the United States, after heart disease, cancer, and respiratory diseases. Worldwide, injuries account for about a tenth of all deaths, outnumbering the victims of AIDS, malaria, and tuberculosis combined, and are responsible for 11 percent of the years lost to death and disability.[2] Personal violence also takes a toll: it is among the top five hazards for young people in the United States and for all people in Latin America and sub-Saharan Africa.[3]

People have long given thought to the causes of danger and how they might be forfended. Perhaps the most stirring moment in Jewish religious observance is a prayer recited before the open Torah ark during the Days of Awe:

On Rosh Hashanah will be inscribed and on Yom Kippur will be sealed: . . . who will live and who will die; who will die at his allotted time and who before his time, who by water and who by fire, who by sword and who by beast, who by famine and who by thirst, who by earthquake and who by plague, who by strangling and who by stoning. . . . But repentance, prayer, and charity annul the severity of the decree.

Fortunately, our knowledge of how fatalities are caused has gone beyond divine inscription, and our means of preventing them have become more reliable than repentance, prayer, and charity. Human ingenuity has been vanquishing the major hazards of life, including every one enumerated in the prayer, and we are now living in the safest time in history.

In previous chapters we have seen how cognitive and moralistic biases work to damn the present and absolve the past. In this one we will see another way in which they conceal our progress. Though lethal injuries are a major scourge of human life, bringing the numbers down is not a sexy cause. The inventor of the highway guard rail did not get a Nobel Prize, nor are humanitarian awards given to designers of clearer prescription drug labels. Yet humanity has benefited tremendously from unsung efforts that have decimated the death toll from every kind of injury.

~

Who by sword. Let's begin with the category of injury that is the hardest to eliminate precisely because it is no accident, homicide. With the exception of the world wars, more people are killed in homicides than wars.[4] During the battle-scarred year of 2015 the ratio was around 4.5 to 1; more commonly it is 10 to 1 or higher. Homicides were an even greater threat to life in the past. In medieval Europe, lords massacred the serfs of their rivals, aristocrats and their retinues fought each other in duels, brigands and highwaymen murdered the victims of their robberies, and ordinary people stabbed each other over insults at the dinner table.[5]

But in a sweeping historical development that the German sociologist Norbert Elias called the Civilizing Process, Western Europeans, starting in the 14th century, began to resolve their disputes in less violent ways.[6] Elias credited the change to the emergence of centralized kingdoms out of the medieval patchwork of baronies and duchies, so that the endemic feuding, brigandage, and warlording were tamed by a "king's peace." Then, in the 19th century, criminal justice systems were further professionalized by municipal police forces and a more deliberative court system. Over those centuries Europe also developed an infrastructure of commerce, both physical, in the form of better roads and vehicles, and financial, in the form of currency and contracts. Gentle commerce proliferated, and the zero-sum plundering of land gave way to a positive-sum trade of goods and services. People became enmeshed in networks of commercial and occupational obligations laid out in legal and bureaucratic rules. Their norms for everyday conduct shifted from a macho culture of honor, in which affronts had to be answered with violence, to a gentlemanly culture of dignity, in which status was won by displays of propriety and self-control.

The historical criminologist Manuel Eisner has assembled datasets on homicide in Europe which put numbers to the narrative that Elias had published in 1939.[7] (Homicide rates are the most reliable indicator of violent crime across different times and places because a corpse is always hard to overlook, and rates of homicide correlate with rates of other violent crimes like robbery, assault, and rape.) Eisner argues that Elias's theory was on the right track, and not just in Europe. Whenever a government brings a frontier region under the rule of law and its people become integrated into a commercial society, rates of violence fall. In figure 12-1, I show Eisner's data for England, the Netherlands, and Italy, with updates through 2012; the curves for other Western European countries are similar. I have added lines for parts of the Americas in which law and order came later: colonial New England, followed by a region in the "Wild West," followed by Mexico, notorious for its violence today but far more violent in the past.

When I introduced the concept of progress I noted that no progressive trend is inexorable, and violent crime is a case in point. Starting in the 1960s, most Western democracies saw a boom in personal violence that erased a century of progress.[8] It was most dramatic in the United States, where the rate of homicide shot up by a factor of two and a half, and where urban and political life were upended by a widespread (and partly justified) fear of crime. Yet this reversal of progress has its own lessons for the nature of progress.

During the high-crime decades, most experts counseled that nothing could be done about violent crime. It was woven into the fabric of a violent American society, they said, and could not be controlled without solving the root causes of racism, poverty, and inequality. This version of historical pessimism may be called root-causism: the pseudo-profound idea that every social ill is a symptom of some deep moral sickness and can never be mitigated by simplistic treatments which fail to cure the gangrene at the core.[9] The problem with root-causism is not that real-world problems are simple but the opposite: they are more complex than a typical root-cause theory allows, especially when the theory is based on moralizing rather than data. So complex, in fact, that treating the symptoms may be the best way of dealing with the problem, because it does not require omniscience about the intricate tissue of actual causes. Indeed, by seeing what really does reduce the symptoms, one can test hypotheses about the causes, rather than just assuming them to be true.

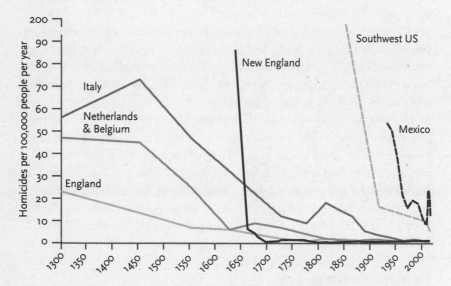

Figure 12-1: Homicide deaths, Western Europe, US, and Mexico, 1300–2015

Sources: England, Netherlands & Belgium, Italy, 1300–1994: Eisner 2003, plotted in fig. 3–3 of Pinker 2011. England, 2000–2014: UK Office for National Statistics. Italy and Netherlands, 2010–2012: United Nations Office on Drugs and Crime 2014. New England (New England, whites only, 1636–1790, and Vermont and New Hampshire, 1780–1890): Roth 2009, plotted in fig. 3–13 of Pinker 2011; 2006 and 2014 from FBI Uniform Crime Reports. Southwest US (Arizona, Nevada, and New Mexico), 1850 and 1914: Roth 2009, plotted in fig. 3–16 of Pinker 2011; 2006 and 2014 from FBI Uniform Crime Reports. Mexico: Carlos Vilalta, personal communication, originally from Instituto Nacional de Estadística y Geografía 2016 and Botello 2016, averaged over decades until 2010.

In the case of the 1960s crime explosion, even the facts at hand refuted the root-cause theory. That was the decade of civil rights, with racism in steep decline (chapter 15), and of an economic boom, with levels of inequality and unemployment for which we are nostalgic.[10] The 1930s, in contrast, was the decade of the Great Depression, Jim Crow laws, and monthly lynchings, yet the overall rate of violent crime plummeted. The root-cause theory was truly deracinated by a development that took everyone by surprise. Starting in 1992, the American homicide rate went into free fall during an era of steeply rising inequality, and then took another dive during the Great Recession beginning in 2007 (figure 12-2).[11] England, Canada, and most other industrialized countries also saw their homicide rates fall in the past two decades. (Conversely, in Venezuela during the Chávez-Maduro regime, inequality fell while homicide soared.)[12] Though numbers for the entire world exist only for this millennium and include heroic guesstimates for countries that are data deserts, the trend appears to be downward as well, from 8.8 homicides per 100,000 people in 2000 to 6.2 in 2012. That means there are 180,000 people walking around today who would have been murdered just in the last year if the global homicide rate had remained at its level of a dozen years before.[13]

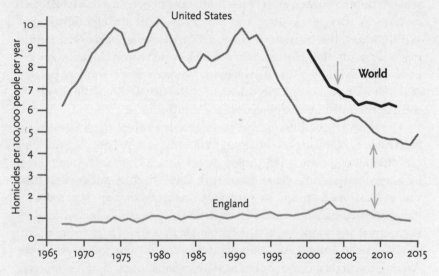

Figure 12-2: Homicide deaths, 1967–2015

Sources: United States: *FBI Uniform Crime Reports*, https://ucr.fbi.gov/, and Federal Bureau of Investigation 2016a. England (data include Wales): Office for National Statistics 2017. World, 2000: Krug et al. 2002. World, 2003–2011: United Nations Economic and Social Council 2014, fig. 1; the percentages were converted to homicide rates by setting the 2012 rate at 6.2, the estimate reported in United Nations Office on Drugs and Crime 2014, p. 12. The arrows point to the most recent years plotted in Pinker 2011 for the world (2004, fig. 3–9), US (2009, fig. 3–18), and England (2009, fig. 3–19).

Violent crime is a solvable problem. We may never get the homicide rate for the world down to the levels of Kuwait (0.4 per 100,000 per year), Iceland (0.3), or Singapore (0.2), let alone all the way to 0.[14] But in 2014, Eisner, in consultation with the World Health Organization, proposed a goal of reducing the rate of global homicide by 50 percent within thirty years.[15] The aspiration is not utopian but practical, based on two facts about the statistics of homicide.

The first is that the distribution of homicide is highly skewed at every level of granularity. The homicide rates of the most dangerous countries are several hundred times those of the safest, including Honduras (90.4 homicides per 100,000 per year), Venezuela (53.7), El Salvador (41.2), Jamaica (39.3), Lesotho (38), and South Africa (31).[16] Half of the world's homicides are committed in just twenty-three countries containing about a tenth of humanity, and a quarter are committed in just four: Brazil (25.2), Colombia (25.9), Mexico (12.9), and Venezuela. (The world's two murder zones—northern Latin America and southern sub-Saharan Africa—are distinct from its war zones, which stretch from Nigeria through the Middle East into Pakistan.) The lopsidedness continues down the fractal scale. Within a country, most of the homicides cluster in a few cities, such as Caracas (120 per 100,000) and San Pedro Sula (in Honduras, 187). Within cities, the homicides cluster in a few neighborhoods; within neighborhoods, they cluster in a few blocks; and within blocks, many are carried out by a few individuals.[17] In my hometown of Boston, 70 percent of the shootings take place in 5 percent of the city, and half the shootings were perpetrated by one percent of the youths.[18]

The other inspiration for the 50-30 goal is evident from figure 12-2: high rates of homicide can be brought down quickly. The most murderous affluent democracy, the United States, saw its homicide rate plunge by almost half in nine years; New York City's decline during that time was even steeper, around 75 percent.[19] Countries that are still more famous for their violence have also enjoyed steep declines, including Russia (from 19 per 100,000 in 2004 to 9.2 in 2012), South Africa (from 60.0 in 1995 to 31.2 in 2012), and Colombia (from 79.3 in 1991 to 25.9 in 2015).[20] Among the eighty-eight countries with reliable data, sixty-seven have seen a decline in the last fifteen years.[21] The unlucky ones (mostly in Latin America) have been ravaged by terrible increases, but even there, when leaders of cities and regions set their mind to reducing the bloodshed, they often succeed.[22] Figure 12-1 shows that Mexico, after suffering a reversal from 2007 to 2011 (entirely attributable to organized crime), enjoyed a reversal of the reversal by 2014, including an almost 90 percent drop from 2010 to 2012 in notorious Juárez.[23] Bogotá and Medellín saw declines by four-fifths in two decades, and São Paulo and the favelas of Rio de Janeiro saw declines by two-thirds.[24] Even the world's murder capital, San Pedro Sula, has seen homicide rates plunge by 62 percent in just *two years*.[25]

Now, combine the cockeyed distribution of violent crime with the proven possibility that high rates of violent crime can be brought down quickly, and the math is straightforward: a 50 percent reduction in thirty years is not just practicable but almost conservative.[26] And it's no statistical trick. The moral value of quantification is that it treats all lives as equally valuable, so actions that bring down the highest numbers of homicides prevent the greatest amount of human tragedy.

The lopsided skew of violent crime also points a flashing red arrow at the best way to reduce it.[27] Forget root causes. Stay close to the symptoms—the neighborhoods and individuals responsible for the biggest wedges of violence—and chip away at the incentives and opportunities that drive them.

It begins with law enforcement. As Thomas Hobbes argued during the Age of Reason, zones of anarchy are always violent.[28] It's not because everyone wants to prey on everyone else, but because in the absence of a government the threat of violence can be self-inflating. If even a few potential predators lurk in the region or could show up on short notice, people must adopt an aggressive posture to deter them. This deterrent is credible only if they advertise their resolve by retaliating against any affront and avenging any depredation, regardless of the cost. This "Hobbesian trap," as it is sometimes called, can easily set off cycles of feuding and vendetta: you have to be at least as violent as your adversaries lest you become their doormat. The largest category of homicide, and the one that varies the most across times and places, consists of confrontations between loosely acquainted young men over turf, reputation, or revenge. A disinterested third party with a monopoly on the legitimate use of force—that is, a state with a police force and judiciary—can nip this cycle in the bud. Not only does it disincentivize aggressors by the threat of punishment, but it reassures everyone else that the aggressors are disincentivized and thereby relieves them of the need for belligerent self-defense.

The most blatant evidence for the impact of law enforcement may be found in the sky-high rates of violence in the times and places where law enforcement is rudimentary, such as the upper left tips of the curves in figure 12-1. Equally persuasive is what happens when police go on strike: an eruption of looting and vigilantism.[29] But crime rates can also soar when law enforcement is merely ineffective—when it is so inept, corrupt, or overwhelmed that people know they can break the law with impunity. That was a contributor to the 1960s crime boom, when the judicial system was no match for a wave of baby boomers entering their crime-prone years, and it is a contributor to the high-crime regions of Latin America today.[30] Conversely, an expansion of policing and criminal punishment (though with a big overshoot in incarceration) explains a good part of the Great American Crime Decline of the 1990s.[31]

Here is Eisner's one-sentence summary of how to halve the homicide rate within three decades: "An effective rule of law, based on legitimate law enforcement, victim protection, swift and fair adjudication, moderate punishment, and humane prisons is critical to sustainable reductions in lethal violence."[32] The adjectives *effective, legitimate, swift, fair, moderate,* and *humane* differentiate his advice from the get-tough-on-crime rhetoric favored by right-wing politicians. The reasons were explained by Cesare Beccaria two hundred and fifty years ago. While the threat of ever-harsher punishments is both cheap and emotionally satisfying, it's not particularly effective, because scofflaws just treat them like rare accidents—horrible, yes, but a risk that comes with the job. Punishments that are predictable, even if less draconian, are likelier to be factored into day-to-day choices.

Together with the presence of law enforcement, the *legitimacy* of the regime appears to matter, because people not only respect legitimate authority themselves but factor in the degree to which they expect their potential adversaries to respect it. Eisner, together with the historian Randolph Roth, notes that crime often shoots up in decades in which people question their society and government, including the American Civil War, the 1960s, and post-Soviet Russia.[33]

Recent reviews of what does and doesn't work in crime prevention back up Eisner's advisory, particularly a massive meta-analysis by the sociologists Thomas Abt and Christopher Winship of 2,300 studies evaluating just about every policy, plan, program, project, initiative, intervention, nostrum, and gimmick that has been tried in recent decades.[34] They concluded that the single most effective tactic for reducing violent crime is *focused deterrence.* A "laser-like focus" must first be directed on the neighborhoods where crime is rampant or even just starting to creep up, with the "hot spots" identified by data gathered in real time. It must be further beamed at the individuals and gangs who are picking on victims or roaring for a fight. And it must deliver a simple and concrete message about the behavior that is expected of them, like "Stop shooting and we will help you, keep shooting and we will put you in prison." Getting the message through, and then enforcing it, depends on the cooperation of other members of the community—the store owners, preachers, coaches, probation officers, and relatives.

Also provably effective is cognitive behavioral therapy. This has nothing to do with psychoanalyzing an offender's childhood conflicts or propping his eyelids open while he retches to violent film clips like in *A Clockwork Orange*. It is a set of protocols designed to override the habits of thought and behavior that lead to criminal acts. Troublemakers are impulsive: they seize on sudden opportunities to steal or vandalize, and lash out at people who cross them, heedless of the long-term consequences.[35] These temptations can be counteracted with therapies that teach strategies of self-control. Troublemakers also have narcissistic and sociopathic thought patterns, such as that they are always in the right, that they are entitled to universal deference, that disagreements are personal insults, and that other people have no feelings or interests. Though they cannot be "cured" of these delusions, they can be trained to recognize and counteract them.[36] This swaggering mindset is amplified in a culture of honor, and it can be deconstructed in therapies of anger management and social-skills training as part of counseling for at-risk youth or programs to prevent recidivism.

Whether or not their impetuousness has been brought under control, potential miscreants can stay out of trouble simply because opportunities for instant gratification have been removed from their environments.[37] When cars are harder to steal, houses are harder to burgle, goods are harder to pilfer and fence, pedestrians carry more credit cards than cash, and dark alleys are lit and video-monitored, would-be criminals don't seek another outlet for their larcenous urges. The temptation passes, and a crime is not committed. Cheap consumer goods are another development that has turned weak-willed delinquents into law-abiding citizens despite themselves. Who nowadays would take the risk of breaking into an apartment just to steal a clock radio?

Together with anarchy, impulsiveness, and opportunity, a major trigger of criminal violence is contraband. Entrepreneurs in illegal goods and pastimes cannot file a lawsuit when they feel they have been swindled, or call the police when someone threatens them, so they have to protect their interests with a credible threat of violence. Violent crime exploded in the United States when alcohol was prohibited in the 1920s and when crack cocaine became popular in the late 1980s, and it is rampant in Latin American and Caribbean countries in which cocaine, heroin, and marijuana are trafficked today. Drug-fueled violence remains an unsolved international problem. Perhaps the ongoing decriminalization of marijuana, and in the future other drugs, will lift these industries out of their lawless underworld. In the meantime, Abt and Winship observe that "aggressive drug enforcement yields little anti-drug benefits and generally increases violence," while "drug courts and treatment have a long history of effectiveness."[38]

Any evidence-based reckoning is bound to pour cold water on programs that seemed promising in the theater of the imagination. Conspicuous by their absence from the list of what works are bold initiatives like slum clearance, gun buybacks, zero-tolerance policing, wilderness ordeals, three-strikes-and-you're-out mandatory sentencing, police-led drug awareness classes, and "scared straight" programs in which at-risk youths are exposed to squalid prisons and badass convicts. And perhaps most disappointing to those who hold strong opinions without needing evidence are the equivocal effects of gun legislation. Neither right-to-carry laws favored by the right, nor bans and restrictions favored by the left, have been shown to make much difference—though there is much we don't know, and political and practical impediments to finding out more.[39]

~

As I sought to explain various declines of violence in *The Better Angels of Our Nature* I put little stock in the idea that in the past "human life was cheap" and that over time it became more precious. It seemed woolly and untestable, almost circular, so I stuck to explanations that were closer to the phenomena, such as governance and trade. After sending in the manuscript, I had an experience that gave me second thoughts. To reward myself for completing that massive undertaking I decided to replace my rusty old car, and in the course of car shopping I bought the latest issue of *Car and Driver* magazine. The issue opened with an article called "Safety in Numbers: Traffic Deaths Fall to an All-Time Low," and it was illustrated with a graph that was instantly familiar: time on the x-axis, rate of death on the y-axis, and a line that snaked from the top left to the bottom right.[40] Between 1950 and 2009, the rate of death in traffic accidents fell *sixfold*. Staring up at me was yet another decline in violent death, but this time dominance and hatred had nothing to do with it. Some combination of forces had been working over the decades to reduce the risk of death from driving—as if, yes, life had become more precious. As society became richer, it spent more of its income, ingenuity, and moral passion on saving lives on the roads.

Later I learned that *Car and Driver* had been conservative. Had they plotted the dataset from its first year, 1921, it would have shown an almost *twenty-four-fold* reduction in the death rate. Figure 12-3 shows the full time line—though not even the full story, since for every person who died there were others who were crippled, disfigured, and racked with pain.

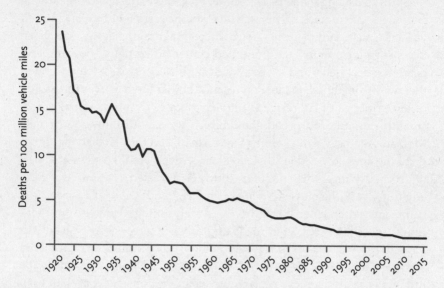

Figure 12-3: Motor vehicle accident deaths, US, 1921–2015

Sources: National Highway Traffic Safety Administration, accessed from http://www.informedforlife .org/demos/FCKeditor/UserFiles/File/TRAFFICFATALITIES(1899-2005).pdf, http://www-fars.nhtsa .dot.gov/Main/index.aspx, and https://crashstats.nhtsa.dot.gov/Api/Public/ViewPublication/812384.

The magazine graph was annotated with landmarks in auto safety which identified the technological, commercial, political, and moralistic forces at work. Over the short run they sometimes pushed against each other, but over the long run they collectively pulled the death rate down, down, down. At times there were moral crusades to reduce the carnage, with automobile manufacturers as the villains. In 1965 a young lawyer named Ralph Nader published *Unsafe at Any Speed*, a *j'accuse* of the industry for neglecting safety in automotive design. Soon after, the National Highway Traffic Safety Administration was established and legislation was passed requiring new cars to be equipped with a number of safety features. Yet the graph shows that steeper reductions came *before* the activism and the legislation, and the auto industry was sometimes ahead of its customers and regulators. A signpost in the graph pointing to 1956 notes, "Ford Motor Company offers the 'Lifeguard' package. . . . It includes seatbelts, a padded dash, padded visors, and a recessed steering-wheel hub designed to not turn drivers into a kebab during a collision. It is a sales failure." It took a decade for those features to become mandatory.

Sprinkled along the slope were other episodes of push and pull among engineers, consumers, corporate suits, and government bureaucrats. At various times, crumple zones, four-wheel dual braking systems, collapsible steering columns, high-mounted center brake lights, buzzing and garroting seat belts, and air bags and stability control systems wended their way from the lab to the showroom. Another lifesaver was the paving of long ribbons of countryside into divided, reflectored, guard-railed, smooth-curved, and broad-shouldered interstate highways. In 1980 Mothers Against Drunk Driving was formed, and they lobbied for higher drinking ages, lowered legal blood alcohol levels, and the stigmatization of drunk driving, which popular culture had treated as a source of comedy (such as in the movies *North by Northwest* and *Arthur*). Crash testing, traffic law enforcement, and driver education (together with unintentional boons like congested roads and economic recessions) saved still more lives. A *lot* of lives: since 1980, about 650,000 Americans have lived who would have died if traffic death rates had remained the same.[41] The numbers are all the more remarkable when we consider that with each passing decade, Americans drove more miles (55 billion in 1920, 458 billion in 1950, 1.5 trillion in 1980, and 3 trillion in 2013), so they were enjoying all the pleasures of leafy suburbs, soccer-playing children, seeing the USA in their Chevrolet, or just cruising down the streets, feeling out of sight, spending all their money on a Saturday night.[42] The additional miles driven did not eat up the safety gains: automobile deaths per capita (as opposed to per vehicle mile) peaked in 1937 at close to 30 per 100,000 per year, and have been in steady decline since the late 1970s, hitting 10.2 in 2014, the lowest rate since 1917.[43]

The progress in the number of motorists who arrive alive is not uniquely American. Fatality rates have sunk in other wealthy countries such as France, Australia, and of course safety-conscious Sweden. (I ended up buying a Volvo.) But it *can* be attributed to living in a wealthy country. Emerging nations like India, China, Brazil, and Nigeria have per capita traffic death rates that are double that of the United States and seven times that of Sweden.[44] Wealth buys life.

A decline in road deaths would be a dubious achievement if it left us more endangered than we were before the automobile was invented. But life before the car was not so safe either. The pictorial curator Otto Bettmann recounts contemporary accounts of city streets in the horse-drawn era:

"It takes more skill to cross Broadway . . . than to cross the Atlantic in a clamboat." . . . The engine of city mayhem was the horse. Underfed and nervous, this vital brute was often flogged to exhaustion by piti- less drivers, who exulted in pushing ahead "with utmost fury, defying law and delighting in destruction." Runaways were common. The havoc killed thousands of people. According to the National Safety Council, the horse-associated fatality rate was ten times the car- associated rate of modern times [in 1974, which is more than double the per capita rate today—SP].[45]

The Brooklyn Dodgers, before they moved to Los Angeles, had been named after the city's pedestrians, famous for their skill at darting out of the way of hurtling streetcars. (Not everyone in that era succeeded: my grandfather's sister was killed by a streetcar in Warsaw in the 1910s.) Like the lives of drivers and passengers, the lives of pedestrians have become more precious, thanks to lights, crosswalks, overpasses, traffic law enforcement, and the demise of hood ornaments, bumper bullets, and other chrome-plated weaponry. Figure 12-4 shows that walking the streets of America today is six times as safe as it was in 1927.

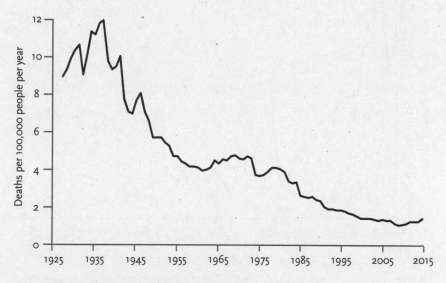

Figure 12-4: Pedestrian deaths, US, 1927–2015

Sources: National Highway Traffic Safety Administration. **For 1927–1984:** Federal Highway Adminis- tration 2003. **For 1985–1995:** National Center for Statistics and Analysis 1995. **For 1995–2005:** National Center for Statistics and Analysis 2006. **For 2005–2014:** National Center for Statistics and Analysis 2016. **For 2015:** National Center for Statistics and Analysis 2017.

The almost 5,000 pedestrians killed in 2014 is still a shocking toll (just compare it with the 44 killed by terrorists to much greater publicity), but it's better than the 15,500 who were mowed down in 1937, when the country had two-fifths as many people and far fewer cars. And the biggest salvation is to come. Within a decade of this writing, most new cars will be driven by computers rather than by slow-witted and scatterbrained humans. When robotic cars are ubiquitous, they could save more than a million lives a year, becoming one of the greatest gifts to human life since the invention of antibiotics.

A cliché in discussions of risk perception is that many people have a fear of flying but almost no one a fear of driving, despite the vastly greater safety of plane travel. But the overseers of air traffic safety are never satisfied. They scrutinize the black box and wreckage after every crash, and have steadily made an already safe mode of transportation even safer. Figure 12-5 shows that in 1970 the chance that an airline passenger would die in a plane crash was less than five in a million; by 2015 that small risk had fallen a hundredfold.

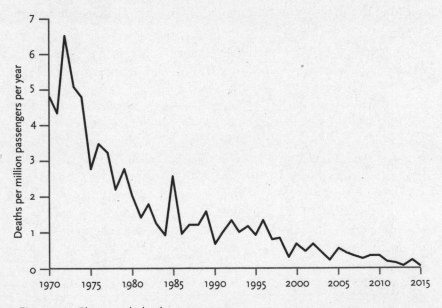

Figure 12-5: Plane crash deaths, 1970–2015

Source: Aviation Safety Network 2017. Data on the number of passengers are from World Bank 2016b.

～

Who by water and who by fire. Well before the invention of cars and planes, people were vulnerable to lethal dangers in their environments. The sociologist Robert Scott began his history of life in medieval Europe as follows: "On December 14, 1421, in the English city of Salisbury, a fourteen-year-old girl named Agnes suffered a grievous injury when a hot spit pierced her torso." (She was reportedly cured by a prayer to Saint Osmund.)[46] It was just one example of how the communities of medieval Europe were "very dangerous places." Infants and toddlers, who were left unattended while their parents worked, were especially vulnerable, as the historian Carole Rawcliffe explains:

The juxtaposition in dark, cramped surroundings of open hearths, straw bedding, rush-covered floors and naked flames posed a constant threat to curious infants. [Even at play] children were in danger because of ponds, agricultural or industrial implements, stacks of timber, unattended boats and loaded wagons, all of which appear with depressing frequency in coroners' reports as causes of death among the young.[47]

The *Encyclopedia of Children and Childhood in History and Society* notes that "to modern audiences, the image of a sow devouring a baby, which appears in Chaucer's 'The Knight's Tale,' borders on the bizarre, but it almost certainly reflected the common threat that animals posed to children."[48]

Adults were no safer. A Web site called *Everyday Life and Fatal Hazard in Sixteenth-Century England* (sometimes known as the Tudor Darwin Awards) posts monthly updates on the historians' analyses of coroners' reports. The causes of death include eating tainted mackerel, getting stuck while climbing through a window, being crushed by a stack of peat slabs, being strangled by a strap that hung baskets from one's shoulders, plunging off a cliff while hunting cormorants, and falling onto one's knife while slaughtering a pig.[49] In the absence of artificial lighting, anyone who ventured out after dark faced the risk of drowning in wells, rivers, ditches, moats, canals, and cesspools.

Today we don't worry about babies getting eaten by sows, but other hazards are still with us. After car crashes, the likeliest cause of accidental death consists of falls, followed by drownings and fires, followed by poisonings. We know this because epidemiologists and safety engineers tabulate accidental deaths with almost plane-wreckage attention to detail, classifying and sub-classifying them to determine which kill the most people and how the risks may be reduced. (The *International Classification of Diseases*, tenth revision, has codes for 153 kinds of falls alone, together with 39 exclusions.) As their advisories are translated into laws, building codes, inspection regimes, and best practices, the world becomes safer. Since the 1930s, the chance that Americans will fall to their deaths has declined by 72 percent, because they have been protected by railings, signage, window guards, grab bars, worker harnesses, safer flooring and ladders, and inspections. (Most of the remaining deaths are of frail, elderly people.) Figure 12-6 shows the fall of falling,[50] together with the trajectories of the other major risks of accidental death since 1903.

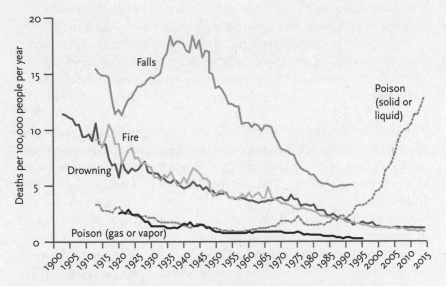

Figure 12-6: Deaths from falls, fire, drowning, and poison, US, 1903–2014

Source: National Safety Council 2016. Data for Fire, Drowning, and Poison (solid or liquid) are aggregated over 1903–1998 and 1999–2014 datasets. For 1999–2014, data for Poison (solid or liquid) include poisonings by gas or vapor. Data for Falls extend only to 1992 because of reporting artifacts in subsequent years (see note 50 for details).

The slopes for the liturgical categories of dying by fire and dying by water are almost identical, and the number of victims of each has declined by more than 90 percent. Fewer Americans drown today, thanks to lifejackets, lifeguards, fences around pools, instruction in swimming and lifesaving, and increased awareness of the vulnerability of small children, who can drown in bathtubs, toilets, even buckets.

Fewer are overcome by flames and smoke. In the 19th century, professional brigades were established to extinguish fires before they turned into conflagrations that could raze entire cities. In the middle of the 20th century, fire departments turned from just fighting fires to preventing them. The campaign was prompted by horrific blazes such as the 1942 Cocoanut Grove nightclub fire in Boston, which left 492 dead, and it was publicized with the help of heart-wrenching photos of firefighters carrying the lifeless bodies of small children out of smoldering houses. Fire was designated a nationwide moral emergency in reports from presidential commissions with titles like *America Burning*.[51] The campaign led to the now-ubiquitous sprinklers, smoke detectors, fire doors, fire escapes, fire drills, fire extinguishers, fire-retardant materials, and fire safety education mascots like Smokey the Bear and Sparky the Fire Dog. As a result, fire departments are putting themselves out of business. About 96 percent of their calls are for cardiac arrests and other medical emergencies, and most of the remainder are for small fires. (Contrary to a charming image, they don't rescue kittens from trees.) A typical firefighter will see just one burning building every other year.[52]

Fewer Americans are accidentally gassing themselves to death. One advance was a transition starting in the 1940s from toxic coal gas to non-toxic natural gas in household cooking and heating. Another was better design and maintenance of gas stoves and heaters so they wouldn't burn their fuel incompletely and spew carbon monoxide into the house. Starting in the 1970s, cars were equipped with catalytic converters, which had been designed to reduce air pollution but which also prevented them from becoming mobile gas chambers. And throughout the century people were increasingly reminded that it's a bad idea to run cars, generators, charcoal grills, and combustion heaters indoors or beneath windows.

Figure 12-6 shows an apparent exception to the conquest of accidents: the category called "Poison (solid or liquid)." The steep rise starting in the 1990s is anomalous in a society that is increasingly latched, alarmed, padded, guard-railed, and warning-stickered, and at first I could not understand why more Americans were apparently eating roach powder or drinking bleach. Then I realized that the category of accidental poisonings includes drug overdoses. (I should have recalled that Leonard Cohen's song based on the Yom Kippur prayer contains the lines "Who in her lonely slip / Who by barbiturate.") In 2013, 98 percent of the "Poison" deaths were from drugs (92 percent) or alcohol (6 percent), and almost all the others were from gases and vapors (mostly carbon monoxide). Household and occupational hazards like solvents, detergents, insecticides, and lighter fluid were responsible for less than a half of one percent of the poisoning deaths, and would scrape the bottom of figure 12-6.[53] Though small children still rummage under sinks, taste the offerings, and get rushed to poison control centers, few of them die.

So the single rising curve in figure 12-6 is not a counterexample to humanity's progress in reducing environmental hazards, though it certainly is a step backward with respect to a different kind of hazard, drug abuse. The curve begins to rise in the psychedelic 1960s, jerks up again during the crack cocaine epidemic of the 1980s, and blasts off during the far graver epidemic of opioid addiction in the 21st century. Starting in the 1990s, doctors overprescribed synthetic opioid painkillers like oxycodone, hydrocodone, and fentanyl, which are not just addictive but gateway drugs to heroin. Overdoses of both the legal and illegal opioids have become a major menace, killing more than 40,000 a year and lifting "poison" into the largest category of accidental death, exceeding even traffic accidents.[54]

Drug overdoses clearly are a different kind of phenomenon from car crashes, falls, fires, drownings, and gassings. People don't get addicted to carbon monoxide, or crave taller and taller ladders, so the kinds of mechanical safeguards that worked so well for environmental hazards will not be enough to end the opioid epidemic. Politicians and public health officials are coming to grips with the enormity of the problem, and countermeasures are being implemented: monitoring prescriptions, encouraging the use of safer analgesics, shaming or punishing pharma companies that recklessly promote the drugs, making the antidote naloxone more available, and treating addicts with opiate antagonists and cognitive behavior therapy.[55] A sign that the measures might be effective is that the number of overdoses of prescription opioids (though not of illicit heroin and fentanyl) peaked in 2010 and may be starting to come down.[56]

Also noteworthy is that opioid overdoses are largely an epidemic of the druggy Baby Boomer cohort reaching middle age. The peak age of poisoning deaths in 2011 was around fifty, up from the low forties in 2003, the late thirties in 1993, the early thirties in 1983, and the early twenties in 1973.[57] Do the subtractions and you find that in every decade it's the members of the generation born between 1953 and 1963 who are drugging themselves to death. Despite perennial panic about teenagers, today's kids are, relatively speaking, all right, or at least better. According to a major longitudinal study of teenagers called Monitoring the Future, high schoolers' use of alcohol, cigarettes, and drugs (other than marijuana and vaping) have dropped to the lowest levels since the survey began in 1976.[58]

~

With the shift from a manufacturing to a service economy, many social critics have expressed nostalgia for the era of factories, mines, and mills, probably because they never worked in one. On top of all the lethal hazards we've examined, industrial workplaces add countless others, because whatever a machine can do to its raw materials—sawing, crushing, baking, rendering, stamping, threshing, or butchering them—it can also do to the workers tending it. In 1892 President Benjamin Harrison noted that "American workmen are subjected to peril of life and limb as great as a soldier in time of war." Bettmann comments on some of the gruesome pictures and captions he collected from the era:

> The miner, it was said, "went down to work as to an open grave, not knowing when it might close on him." . . . Unprotected powershafts maimed and killed hoopskirted workers. . . . The circus stuntman and test pilot today enjoy greater life assurance than did the [railroad] brakeman of yesterday, whose work called for precarious leaps between bucking freight cars at the command of the locomotive's whistle. . . . Also subject to sudden death . . . were the train couplers, whose omnipresent hazard was loss of hands and fingers in the primitive link-and-pin devices. . . . Whether a worker was mutilated by a buzz saw, crushed by a beam, interred in a mine, or fell down a shaft, it was always "his own bad luck."[59]

"Bad luck" was a convenient explanation for employers, and until recently it was a part of a widespread fatalism about lethal accidents, which were commonly attributed to destiny or acts of God. (Today, safety engineers and public health researchers don't even use the word *accident*, since it implies a fickle finger of fate; the term of art is *unintentional injury*.) The first safety measures and insurance policies in the 18th and 19th centuries protected property, not people. As injuries and deaths started to increase unignorably during the Industrial Revolution, they were written off as "the price of progress," according to a nonhumanistic definition of "progress" that was not reckoned in human welfare. A railroad superintendent, justifying his refusal to put a roof over a loading platform, explained that "men are cheaper than shingles. . . . There's a dozen waiting when one drops out."[60] The inhuman pace of industrial production has been immortalized in cultural icons such as Charlie Chaplin on the assembly line in *Modern Times* and Lucille Ball in the chocolate factory in *I Love Lucy*.

Workplaces began to change in the late 19th century as the first labor unions organized, journalists took up the cause, and government agencies started to collect data quantifying the human toll.[61] Bettmann's comment on the lethality of work on trains was based on more than just pictures: in the 1890s, the annual death rate for trainmen was an astonishing 852 per 100,000, almost one percent a year. The carnage was reduced when an 1893 law mandated the use of air brakes and automatic couplers in all freight trains, the first federal law intended to improve workplace safety.

The safeguards spread to other occupations in the early decades of the 20th century, the Progressive Era. They were the result of agitation by reformers, labor unions, and muckraking journalists and novelists like Upton Sinclair.[62] The most effective reform was a simple change in the law brought over from Europe: employers' liability and workmen's compensation. Previously, injured workers or their survivors had to sue for compensation, usually unsuccessfully. Now, employers were required to compensate them at a fixed rate. The change appealed to management as much as to workers, since it made their costs more predictable and the workers more cooperative. Most important, it yoked the interests of management and labor: both had a stake in making workplaces safer, as did the insurers and government agencies that underwrote the compensation. Companies set up safety committees and safety departments, hired safety engineers, and implemented many protections, sometimes out of economic or humanitarian motives, sometimes as a response to public shaming after a well-publicized disaster, often under the duress of lawsuits and government regulations. The results are plain to see in figure 12-7.[63]

At almost 5,000 deaths in 2015, the number of workers killed on the job is still too high, but it's much better than the 20,000 deaths in 1929, when the population was less than two-fifths the size. Much of the savings is the result of the movement of the labor force from farms and factories to stores and offices. But much of it is a gift of the discovery that saving lives while producing the same number of widgets is a solvable engineering problem.

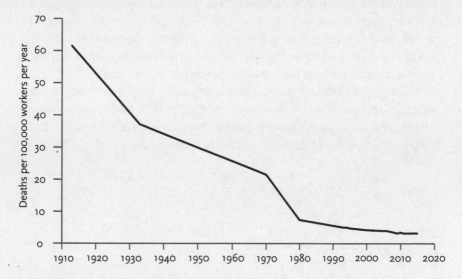

Figure 12-7: Occupational accident deaths, US, 1913–2015

Sources: Data are from different sources and may not be completely commensurable (see note 63 for details). **For 1913, 1933, and 1980:** Bureau of Labor Statistics, National Safety Council, and CDC National Institute for Occupational Safety and Health, respectively, cited in Centers for Disease Control 1999. **For 1970:** Occupational Safety and Health Administration, "Timeline of OSHA's 40 Year History," https://www.osha.gov/osha40/timeline.html. **For 1993–1994:** Bureau of Labor Statistics, cited in Pegula & Janocha 2013. **For 1995–2005:** National Center for Health Statistics 2014, table 38. **For 2006–2014:** Bureau of Labor Statistics 2016a. The latter data were reported as deaths per full-time-equivalent workers and are multiplied by .95 for rough commensurability with the preceding years, based on the year 2007, when the Census of Fatal Occupation Injuries reported rates both per worker (3.8) and per FTE (4.0).

Who by earthquake. Could the efforts of mortals even mitigate what lawyers call "acts of God"—the droughts, floods, wildfires, storms, volcanoes, avalanches, landslides, sinkholes, heat waves, cold snaps, meteor strikes, and yes, earthquakes that are the quintessentially uncontrollable catastrophes? The answer, shown in figure 12-8, is yes.

After the ironic 1910s, when the world was ravaged by a world war and an influenza pandemic but relatively spared from natural disasters, the rate of death from disasters has rapidly declined from its peak. It's not that with each passing decade the world has miraculously been blessed with fewer earthquakes, volcanoes, and meteors. It's that a richer and more technologically advanced society can prevent natural hazards from becoming human catastrophes. When an earthquake strikes, fewer people are crushed by collapsing masonry or burned in conflagrations.

When the rains stop, they can use water impounded in reservoirs. When the temperature soars or plummets, they stay in climate-controlled interiors. When a river floods its banks, their drinking water is safeguarded from human and industrial waste. The dams and levees that impound water for drinking and irrigation, when properly designed and built, make floods less likely in the first place. Early warning systems allow people to evacuate or take shelter before a cyclone makes landfall. Though geologists can't yet predict earthquakes, they can often predict volcanic eruptions, and can prepare the people who live along the Rim of Fire and other fault systems to take lifesaving precautions. And of course a richer world can rescue and treat its injured and quickly rebuild.

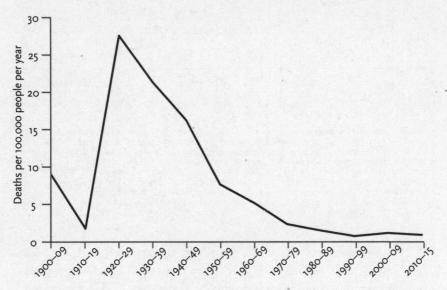

Figure 12-8: Natural disaster deaths, 1900–2015

Source: *Our World in Data*, Roser 2016q, based on data from EM-DAT, *The International Disaster Database*, www.emdat.be. The graph plots the sum of the death rates for Drought, Earthquake, Extreme temperature, Flood, Impact, Landslide, Mass movement (dry), Storm, Volcanic activity, and Wildfire (excluding Epidemics). In many decades a single disaster type dominates the numbers: droughts in the 1910s, 1920s, 1930s, and 1960s; floods in the 1930s and 1950s; earthquakes in the 1970s, 2000s, and 2010s.

It's the poorer countries today that are most vulnerable to natural hazards. A 2010 earthquake in Haiti killed more than 200,000 people, while a stronger one in Chile a few weeks later killed just 500. Haiti also loses ten times as many of its citizens to hurricanes as the richer Dominican Republic, the country with which it shares the island of Hispaniola. The good news is that as poorer countries get richer, they get safer (at least as long as economic development outpaces climate change). The annual death rate from natural disasters in low-income countries has come down from 0.7 per 100,000 in the 1970s to 0.2 today, which is lower than the rate for upper-middle-income countries in the 1970s. That's still higher than the rate for high-income countries today (0.05, down from 0.09), but it shows that rich and poor countries alike can make progress in defending themselves against a vengeful deity.[64]

And what about the very archetype of an act of God? The projectile that Zeus hurled down from Olympus? The standard idiom for an unpredictable date with death? The literal bolt from the blue? Figure 12-9 shows the history.

Yes, thanks to urbanization and to advances in weather prediction, safety education, medical treatment, and electrical systems, there has been a thirty-seven-fold decline since the turn of the 20th century in the chance that an American will be killed by a bolt of lightning.

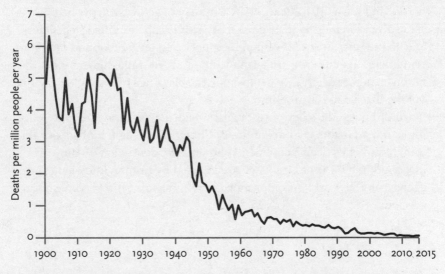

Figure 12-9: Lightning strike deaths, US, 1900–2015

Source: *Our World in Data*, Roser 2016q, based on data from National Oceanic and Atmospheric Administration, http://www.lightningsafety.noaa.gov/victims.shtml, and López & Holle 1998.

Humanity's conquest of everyday danger is a peculiarly unappreciated form of progress. (Some readers of a draft of this chapter wondered what it was even doing in a book on progress.) Though accidents kill more people than all but the worst wars, we seldom see them through a moral lens. As we say: Accidents will happen. Had we ever been confronted with the dilemma of whether a million deaths and tens of millions of injuries a year was a price worth paying for the convenience of driving our own cars at enjoyable speeds, few would have argued that it was. Yet that is the monstrous choice we tacitly made, because the dilemma was never put to us in those terms.[65] Now and again a hazard is moralized and a crusade against it is mounted, particularly if a disaster makes the news and a villain can be fingered (a greedy factory owner, a negligent public official). But soon it recedes back into the lottery of life.

Just as people tend not to see accidents as atrocities (at least when they are not the victims), they don't see gains in safety as moral triumphs, if they are aware of them at all. Yet the sparing of millions of lives, and the reduction of infirmity, disfigurement, and suffering on a massive scale, deserve our gratitude and demand an explanation. That is true even of murder, the most moralized of acts, whose rate has plummeted for reasons that defy standard narratives.

Like other forms of progress, the ascent of safety was led by some heroes, but it was also advanced by a motley of actors who pushed in the same direction inch by inch: grassroots activists, paternalistic legislators, and an unsung cadre of inventors, engineers, policy wonks, and number-crunchers. Though we sometimes chafe at the false alarms and the nanny-state intrusions, we get to enjoy the blessings of technology without the threats to life and limb.

And though the story of seat belts, smoke alarms, and hot-spot policing is not a customary part of the Enlightenment saga, it plays out the Enlightenment's deepest themes. Who will live and who will die are not inscribed in a Book of Life. They are affected by human knowledge and agency, as the world becomes more intelligible and life becomes more precious.

CHAPTER 13

TERRORISM

W hen I wrote in the preceding chapter that we are living in the safest time in history, I was aware of the incredulity those words would evoke. In recent years, highly publicized terrorist attacks and rampage killings have set the world on edge and fostered an illusion that we live in newly dangerous times. In 2016, a majority of Americans named terrorism as the most important issue facing the country, said they were worried that they or a family member would be a victim, and identified ISIS as a threat to the existence or survival of the United States.[1] The fear has addled not just ordinary citizens trying to get a pollster off the phone but public intellectuals, especially cultural pessimists perennially hungry for signs that Western civilization is (as always) on the verge of collapse. The political philosopher John Gray, an avowed progressophobe, has described the contemporary societies of Western Europe as "terrains of violent conflict" in which "peace and war [are] fatally blurred."[2]

But yes, all this is an illusion. Terrorism is a unique hazard because it combines major dread with minor harm. I will not count trends in terrorism as an example of progress, since they don't show the long-term decline we've seen for disease, hunger, poverty, war, violent crime, and accidents. But I will show that terrorism is a distraction in our assessment of progress, and, in a way, a backhanded tribute to that progress.

Gray dismissed actual data on violence as "amulets" and "sorcery." The following table shows why he needed this ideological innumeracy to prosecute his jeremiad. It shows the number of victims of four categories of killing—terrorism, war, homicide, and accidents—together with the total of all deaths, in the most recent year for which data are available (2015 or earlier). A graph is impossible, because swatches for the terrorism numbers would be smaller than a pixel.

Table 13-1: Deaths from Terrorism, War, Homicide, and Accidents

	US	Western Europe	World
Terrorism	44	175	38,422
War	28	5	97,496
Homicide	15,696	3,962	437,000
Motor vehicle accidents	35,398	19,219	1,250,000
All accidents	136,053	126,482	5,000,000
All deaths	2,626,418	3,887,598	56,400,000

"Western Europe" is defined as in the Global Terrorism Database, comprising 24 countries and a 2014 population of 418,245,997 (Statistics Times 2015). I omit Andorra, Corsica, Gibraltar, Luxembourg, and the Isle of Man.
Sources: Terrorism (2015): National Consortium for the Study of Terrorism and Responses to Terrorism 2016. **War, US and Western Europe (UK + NATO) (2015):** icasualties.org, http://icasualties.org. **War, World (2015):** *UCDP Battle-Related Deaths Dataset*, Uppsala Conflict Data Program 2017. **Homicide, US (2015):** Federal Bureau of Investigation 2016a. **Homicide, Western Europe and World (2012 or most recent):** United Nations Office on Drugs and Crime 2013. Data for Norway exclude the Utøya terrorist attack. **Motor vehicle accidents, All accidents, and All deaths, US (2014):** Kochanek et al. 2016, table 10. **Motor vehicle accidents, Western Europe (2013):** World Health Organization 2016c. **All accidents, Western Europe (2014 or most recent):** World Health Organization 2015a. **Motor vehicle accidents and All accidents, World (2012):** World Health Organization 2014. **All deaths, Western Europe (2012 or most recent):** World Health Organization 2017a. **All deaths, World (2015):** World Health Organization 2017c.

Start with the United States. What jumps out of the table is the tiny number of deaths in 2015 caused by terrorism compared with those from hazards that inspire far less anguish or none at all. (In 2014 the terrorist death toll was even lower, at 19.) Even the estimate of 44 is generous: it comes from the Global Terrorism Database, which counts hate crimes and most rampage shootings as examples of "terrorism." The toll is comparable to the number of military fatalities in Afghanistan and Iraq (28 in 2015, 58 in 2014), which, consistent with the age-old devaluing of the lives of soldiers, received a fraction of the news coverage. The next rows down reveal that in 2015 an American was more than 350 times as likely to be killed in a police-blotter homicide as in a terrorist attack, 800 times as likely to be killed in a car crash, and 3,000 times as likely to die in an accident of any kind. (Among the categories of accident that typically kill more than 44 people in a given year are "Lightning," "Contact with hot tap water," "Contact with hornets, wasps, and bees," "Bitten or struck by mammals other than dogs," "Drowning and submersion while in or falling into bathtub," and "Ignition or melting of clothing and apparel other than nightwear.")[3]

In Western Europe, the relative danger of terrorism was higher than in the United States. In part this is because 2015 was an *annus horribilis* for terrorism in that region, including the deadly attacks on the *Charlie Hebdo* offices, the Bataclan theater, and other targets in Paris and its surroundings. (In 2014, just 5 people were killed.) But the relatively higher terrorism risk is also a sign of how much safer Europe is in every other way. Western Europeans are less murderous than Americans (with about a quarter their homicide rate) and also less car-crazy, so fewer die on the road.[4] Even with these factors tipping the scale toward terrorism, a Western European in 2015 was more than 20 times as likely to die in one of their (relatively rare) homicides as in a terrorist attack, more than 100 times as likely to die in a car crash, and more than 700 times as likely to be crushed, poisoned, burned, asphyxiated, or otherwise killed in an accident.

The third column shows that for all the recent anguish about terrorism in the West, we have it easy compared with other parts of the world. Though the United States and Western Europe contain about a tenth of the world's population, in 2015 they suffered one-half of one percent of the terrorist deaths. That's not because terrorism is a major cause of death elsewhere. It's because terrorism, as it is now defined, is largely a phenomenon of war, and wars no longer take place in the United States or Western Europe. In the years since the attacks of September 11, 2001, violence that used to be called "insurgency" or "guerrilla warfare" is now often classified as "terrorism."[5] (The Global Terrorism Database, incredibly, does not classify any deaths in Vietnam in the last five years of the war there as "terrorism.")[6] A majority of the world's terrorist deaths take place in zones of civil war (including 8,831 in Iraq, 6,208 in Afghanistan, 5,288 in Nigeria, 3,916 in Syria, 1,606 in Pakistan, and 689 in Libya), and many of these are double-counted as war deaths, because "terrorism" during a civil war is simply a war crime—a deliberate attack on civilians—committed by a group other than the government. (Excluding these six civil war zones, the terrorism death count for 2015 was 11,884.) Yet even with the double counting of terrorism and war during the 21st century's worst year for war deaths, a global citizen was 11 times as likely to have died in a homicide as in a terrorist attack, more than 30 times as likely to have died in a car crash, and more than 125 times as likely to have died in an accident of any kind.

Has terrorism, whatever its toll, increased over time? The historical trends are elusive. Because "terrorism" is an elastic category, the trend lines look different depending on whether a dataset includes civil war crimes, multiple murders (which include robberies or mafia hits in which several victims are shot), or suicidal rampages in which the killer ranted about some political grievance beforehand. (The Global Terrorism Database, for example, includes the 1999 Columbine school massacre but not the 2012 Sandy Hook school massacre.) Also, mass killings are media-driven spectacles, in which coverage inspires copycats, so they can yo-yo up and down as one event inspires another until the novelty wears off for a while.[7] In the United States, the number of "active shooter incidents" (public rampage killings with guns) has wobbled with an upward trend since 2000, though the number of "mass murders" (four or more deaths in an incident) shows no systematic change (if anything, it shows a slight decline) from 1976 to 2011.[8] The per capita death rate from "terrorism incidents" is shown in figure 13-1, together with the messy trends for Western Europe and the world.

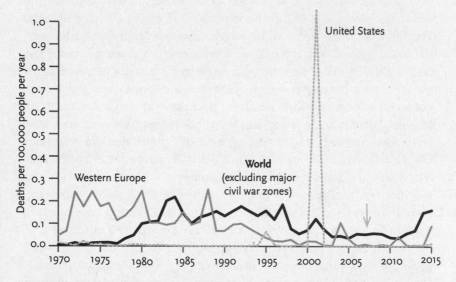

Figure 13-1: Terrorism deaths, 1970–2015

Sources: "Global Terrorism Database," National Consortium for the Study of Terrorism and Responses to Terrorism 2016, https://www.start.umd.edu/gtd/. The rate for the world excludes deaths in Afghanistan after 2001, Iraq after 2003, Pakistan after 2004, Nigeria after 2009, Syria after 2011, and Libya after 2014. Population estimates for the world and Western Europe are from the United Nation's 2015 Revision of World Population Prospects (https://esa.un.org/unpd/wpp/); estimates for the United States are from US Census Bureau 2017. The vertical arrow points to 2007, the last year plotted in figs. 6–9, 6–10, and 6–11 in Pinker 2011.

The death rate for American terrorism for the year 2001, which includes the 3,000 deaths from the 9/11 attacks, dominates the graph. Elsewhere we see a bump for the Oklahoma City bombing in 1995 (165 deaths) and barely perceptible wrinkles in other years.[9] Excluding 9/11 and Oklahoma, about twice as many Americans have been killed since 1990 by right-wing extremists as by Islamist terror groups.[10] The line for Western Europe shows that the rise in 2015 came after a decade of relative quiescence, and is not even the worst that Western Europe has seen: the rate of killing was higher in the 1970s and 1980s, when Marxist and secessionist groups (including the Irish Republican Army and the Basque ETA movement) carried out regular bombings and shootings. The line for the world as a whole (excluding recent deaths in major war zones, which we examined in the chapter on war) contains a spiky plateau for the 1980s and 1990s, a fall after the end of the Cold War, and a recent rise to a level that still falls below that of the earlier decades. So the historical trends, like the current numbers, belie the fear that we are living in newly dangerous times, particularly in the West.

~

Though terrorism poses a minuscule danger compared with other risks, it creates outsize panic and hysteria because that is what it is designed to do. Modern terrorism is a by-product of the vast reach of the media.[11] A group or an individual seeks a slice of the world's attention by the one guaranteed means of attracting it: killing innocent people, especially in circumstances in which readers of the news can imagine themselves. News media gobble the bait and give the atrocities saturation coverage. The Availability heuristic kicks in and people become stricken with a fear that is unrelated to the level of danger.

It's not just the salience of a horrific event that stokes the terror. Our emotions are far more engaged when the cause of a tragedy is malevolent intent rather than accidental misfortune.[12] (I confess that as a frequent visitor to London, I was far more upset when I read the headline RUSSELL SQUARE "TERROR" KNIFE ATTACK LEAVES WOMAN DEAD than when I read RENOWNED ART COLLECTOR DIES AFTER BEING HIT BY BUS IN OXFORD STREET TRAGEDY.) Something is uniquely unsettling about the thought of a human being who wants to kill you, and for a good evolutionary reason. Accidental causes of death don't *try* to do you in, and they don't care how you react, whereas human malefactors deploy their intelligence to outsmart you, and vice versa.[13]

Given that terrorists are not mindless hazards but human agents with goals, could it be *rational* to worry about them despite the small amount of damage they do? After all, we are justly outraged by despots who execute dissidents, even though the number of their victims may be as small as those of terrorism. The difference is that despotic violence has strategic effects that are disproportionate to the body count: it eliminates the most potent threats to the regime, and it deters the rest of the population from replacing them. Terrorist violence, almost by definition, strikes victims at random. The objective significance of the threat, then, beyond the immediate damage, depends on what the scattershot killing is designed to accomplish.

With many terrorists, the goal is little more than publicity itself. The legal scholar Adam Lankford has analyzed the motives of the overlapping categories of suicide terrorists, rampage shooters, and hate crime killers, including both the self-radicalized lone wolves and the bomb fodder recruited by terrorist masterminds.[14] The killers tend to be loners and losers, many with untreated mental illness, who are consumed with resentment and fantasize about revenge and recognition. Some fused their bitterness with Islamist ideology, others with a nebulous cause such as "starting a race war" or "a revolution against the federal government, taxes, and anti-gun laws." Killing a lot of people offered them the chance to be a somebody, even if only in the anticipation, and going out in a blaze of glory meant that they didn't have to deal with the irksome aftermath of being a mass murderer. The promise of paradise, and an ideology that rationalizes how the massacre serves a greater good, makes the posthumous fame all the more inviting.

Other terrorists belong to militant groups that seek to call attention to their cause, to extort a government to change its policies, to provoke it into an extreme response that might recruit new sympathizers or create a zone of chaos for them to exploit, or to undermine the government by spreading the impression that it cannot protect its own citizens. Before we conclude that they "pose a threat to the existence or survival of the United States," we should bear in mind how weak the tactic actually is.[15] The historian Yuval Harari notes that terrorism is the opposite of military action, which tries to damage the enemy's ability to retaliate and prevail.[16] When Japan attacked Pearl Harbor in 1941, it left the United States without a fleet to send to Southeast Asia in response. It would have been mad for Japan to have opted for terrorism, say, by torpedoing a passenger ship to provoke the United States into responding with an intact navy. From their position of weakness, Harari notes, what terrorists seek to accomplish is not damage but theater. The image that most people retain from 9/11 is not Al Qaeda's attack on the Pentagon—which actually destroyed part of the enemy's military headquarters and killed commanders and analysts—but its attack on the totemic World Trade Center, which killed brokers, accountants, and other civilians.

Though terrorists hope for the best, their small-scale violence almost never gets them what they want. Separate surveys by the political scientists Max Abrahms, Audrey Cronin, and Virginia Page Fortna of hundreds of terrorist movements active since the 1960s show that they all were extinguished or faded away without attaining their strategic goals.[17]

Indeed, the rise of terrorism in public awareness is not a sign of how dangerous the world has become but the opposite. The political scientist Robert Jervis observes that the placement of terrorism at the top of the list of threats "in part stems from a security environment that is remarkably benign."[18] It is not only interstate war that has become rare; so has the use of political violence in the domestic arena. Harari points out that in the Middle Ages, every sector of society retained a private militia—aristocrats, guilds, towns, even churches and monasteries—and they secured their interests by force: "If in 1150 a few Muslim extremists had murdered a handful of civilians in Jerusalem, demanding that the Crusaders leave the Holy Land, the reaction would have been ridicule rather than terror. If you wanted to be taken seriously, you should have at least gained control of a fortified castle or two." As modern states have successfully claimed a monopoly on force, driving down the rate of killing within their borders, they opened a niche for terrorism:

> The state has stressed so many times that it will not tolerate political violence within its borders that it has no alternative but to see any act of terrorism as intolerable. The citizens, for their part, have become used to zero political violence, so the theatre of terror incites in them visceral fears of anarchy, making them feel as if the social order is about to collapse. After centuries of bloody struggles, we have crawled out of the black hole of violence, but we feel that the black hole is still there, patiently waiting to swallow us again. A few gruesome atrocities and we imagine that we are falling back in.[19]

As states try to carry out the impossible mandate of protecting their citizens from all political violence everywhere and all the time, they are tempted to respond with theater of their own. The most damaging effect of terrorism is countries' overreaction to it, the case in point being the American-led invasions of Afghanistan and Iraq following 9/11.

Instead, countries could deal with terrorism by deploying their greatest advantage: knowledge and analysis, not least knowledge of the numbers. The uppermost goal should be to make sure the numbers stay small by securing weapons of mass destruction (chapter 19). Ideologies that justify violence against innocents, such as militant religions, nationalism, and Marxism, can be countered with better systems of value and belief (chapter 23). The media can examine their essential role in the show business of terrorism by calibrating their coverage to the objective dangers and giving more thought to the perverse incentives they have set up. (Lankford, together with the sociologist Eric Madfis, has recommended a policy for rampage shootings of "Don't Name Them, Don't Show Them, but Report Everything Else," based on a policy for juvenile shooters already in effect in Canada and on other strategies of calculated media self-restraint.)[20] Governments can step up their intelligence and clandestine actions against networks of terrorism and their financial tributaries. And people could be encouraged to keep calm and carry on, as the British wartime poster famously urged during a time of much greater peril.

Over the long run, terrorist movements sputter out as their small-scale violence fails to achieve their strategic goals, even as it causes local misery and fear.[21] It happened to the anarchist movements at the turn of the 20th century (after many bombings and assassinations), it happened to the Marxist and secessionist groups in the second half of the 20th century, and it will almost certainly happen to ISIS in the 21st. We may never drive the already low numbers of terrorist casualties to zero, but we can remember that terror about terrorism is a sign not of how dangerous our society has become, but of how safe.

CHAPTER 14

DEMOCRACY

Since the first governments appeared around five thousand years ago, humanity has tried to steer a course between the violence of anarchy and the violence of tyranny. In the absence of a government or powerful neighbors, tribal peoples tend to fall into cycles of raiding and feuding, with death rates exceeding those of modern societies, even including their most violent eras.[1] Early governments pacified the people they ruled, reducing internecine violence, but imposed a reign of terror that included slavery, harems, human sacrifice, summary executions, and the torture and mutilation of dissidents and deviants.[2] (The Bible has no shortage of examples.) Despotism has persisted through history not just because being a despot is nice work if you can get it, but because from the people's standpoint the alternative was often worse. Matthew White, who calls himself a necrometrician, has estimated the death tolls of the hundred bloodiest episodes in 2,500 years of human history. After looking for patterns in the list, he reported this one as his first:

> Chaos is deadlier than tyranny. More of these multicides result from the breakdown of authority rather than the exercise of authority. In comparison to a handful of dictators such as Idi Amin and Saddam Hussein who exercised their absolute power to kill hundreds of thousands, I found more and deadlier upheavals like the Time of Troubles [in 17th-century Russia], the Chinese Civil War [1926–37, 1945–49], and the Mexican Revolution [1910–20] where no one exercised enough control to stop the death of millions.[3]

One can think of *democracy* as a form of government that threads the needle, exerting just enough force to prevent people from preying on each other without preying on the people itself. A good democratic government allows people to pursue their lives in safety, protected from the violence of anarchy, and in freedom, protected from the violence of tyranny. For that reason alone, democracy is a major contributor to human flourishing. But it's not the only reason: democracies also have higher rates of economic growth, fewer wars and genocides, healthier and better-educated citizens, and virtually no famines.[4] If the world has become more democratic over time, that is progress.

In fact the world *has* become more democratic, though not in a steadily rising tide. The political scientist Samuel Huntington organized the history of democratization into three waves.[5] The first swelled in the 19th century, when that great Enlightenment experiment, American constitutional democracy with its checks on government power, seemed to be working. The experiment, with local variations, was emulated by a number of countries, mainly in Western Europe, cresting at twenty-nine in 1922. The first wave was pushed back by the rise of fascism, and by 1942 had ebbed to just twelve countries. With the defeat of fascism in World War II, a second wave gathered force as colonies gained independence from their European overlords, pushing the number of recognized democracies up to thirty-six by 1962. Still, European democracies were sandwiched between Soviet-dominated dictatorships to the east and fascist dictatorships in Portugal and Spain to the southwest. And the second wave was soon pushed back by military juntas in Greece and Latin America, authoritarian regimes in Asia, and Communist takeovers in Africa, the Middle East, and Southeast Asia.[6] By the mid-1970s the prospects for democracy looked bleak. The West German chancellor Willy Brandt lamented that "Western Europe has only 20 or 30 more years of democracy left in it; after that it will slide, engineless and rudderless, under the surrounding sea of dictatorship." The American senator and social scientist Daniel Patrick Moynihan agreed, writing that "liberal democracy on the American model increasingly tends to the condition of monarchy in the 19th century: a holdover form of government, one which persists in isolated or peculiar places here and there, and may even serve well enough for special circumstances, but which has simply no relevance to the future. It is where the world was, not where it is going."[7]

Before the ink was dry on these lamentations, democratization's third wave—more like a tsunami—erupted. Military and fascist governments fell in southern Europe (Greece and Portugal in 1974, Spain in 1975), Latin America (including Argentina in 1983, Brazil in 1985, and Chile in 1990), and Asia (including Taiwan and the Philippines around 1986, South Korea around 1987, and Indonesia in 1998). The Berlin Wall was torn down in 1989, freeing the nations of Eastern Europe to establish democratic governments, and communism imploded in the Soviet Union in 1991, clearing space for Russia and most of the other republics to make the transition. Some African countries threw off their strongmen, and the last European colonies to gain independence, mostly in the Caribbean and Oceania, opted for democracy as their first form of government. In 1989 the political scientist Francis Fukuyama published a famous essay in which he proposed that liberal democracy represented "the end of history," not because nothing would ever happen again but because the world was coming to a consensus over the humanly best form of governance and no longer had to fight over it.[8]

Fukuyama coined a runaway meme: in the decades since his essay appeared, books and articles have announced "the end of" nature, science, faith, poverty, reason, money, men, lawyers, illness, the free market, and sex. But Fukuyama also became a punching bag as editorialists, commenting on the latest bit of bad news, gleefully announced "the return of history" and the rise of alternatives to democracy such as theocracy in the Muslim world and authoritarian capitalism in China. Democracies themselves appeared to be backsliding into authoritarianism with populist victories in Poland and Hungary and power grabs by Recep Erdogan in Turkey and Vladimir Putin in Russia (the return of the sultan and the czar). Historical pessimists, with their customary schadenfreude, announced that the third wave of democratization had given way to an "undertow," "recession," "erosion," "rollback," or "meltdown."[9] Democratization, they said, was a conceit of Westerners projecting their tastes onto the rest of the world, whereas authoritarianism seemed to suit most of humanity just fine.

Could recent history really imply that people are happy to be brutalized by their governments? The very idea is doubtful for two reasons. Most obviously, in a country that is not democratic, how could you tell? The pent-up demand for democracy might be enormous, but no one dares express it lest they be jailed or shot. The other is the headline fallacy: crackdowns make the news more often than liberalizations, and the Availability bias could make us forget about all the boring countries that become democratic bit by bit.

As always, the only way to know which way the world is going is to quantify. This raises the question of what counts as a "democracy," a word that has developed such an aura of goodness as to have become almost meaningless. A good rule of thumb is that any country that has the word "democratic" in its official name, like the Democratic People's Republic of

Korea (a.k.a. North Korea) or the German Democratic Republic (a.k.a. East Germany), isn't one. Nor is it helpful to ask the citizens of undemocratic states what they think the word means: almost half think it means "The army takes over when the government is incompetent" or "Religious leaders ultimately interpret the laws."[10] Ratings by experts have a related problem when their checklists embrace a hodgepodge of good things such as "freedom from socioeconomic inequalities" and "freedom from war."[11] Yet another complication is that countries vary continuously in the different components of democracy such as freedom of speech, the openness of the political process, and the constraints on its leaders' power, so any tally that dichotomizes nations into "democracies" and "autocracies" will fluctuate from year to year depending on arbitrary choices about where to place the countries that hover near the boundary (a problem exacerbated when the raters' standards rise over time, a phenomenon we will return to).[12] The Polity Project deals with these obstacles by using a fixed set of criteria to assign a score between –10 and 10 to every country in every year indicating how autocratic or democratic it is, focusing on citizens' ability to express political preferences, constraints on the power of the executive, and a guarantee of civil liberties.[13] The sum for the world since 1800, spanning the three waves of democratization, is shown in figure 14-1.

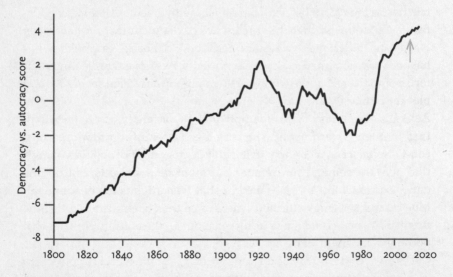

Figure 14-1: Democracy versus autocracy, 1800–2015

Source: *HumanProgress*, http://humanprogress.org/f1/2560, based on *Polity IV Annual Time-Series, 1800–2015*, Marshall, Gurr, & Jaggers 2016. Scores are summed over sovereign states with a population greater than 500,000, and range from –10 for a complete autocracy to 10 for a perfect democracy. The arrow points to 2008, the last year plotted in fig. 5–23 of Pinker 2011.

The graph shows that the third wave of democratization is far from over, let alone ebbing, even if it has not continued to surge at the rate of the years surrounding the fall of the Berlin Wall in 1989. At that time, the world had 52 democracies (defined by the Polity Project as countries with a score of 6 or higher on their scale), up from 31 in 1971. After swelling in the 1990s, this third wave spilled into the 21st century in a rainbow of "color revolutions" including Croatia (2000), Serbia (2000), Georgia (2003), Ukraine (2004), and Kyrgyzstan (2005), bringing the total at the start of the Obama presidency in 2009 to 87.[14] Belying the image of a rollback or meltdown under his watch, the number continued to grow. As of 2015, the most recent year in the dataset, the total stood at 103. The Nobel Peace Prize was awarded that year to a coalition of organizations in Tunisia that solidified a transition to democracy, a success story from the Arab Spring of 2011. It also saw transitions to democracy in Myanmar and Burkina Faso, and positive movements in five other countries, including Nigeria and Sri Lanka. The world's 103 democracies in 2015 embraced 56 percent of the world's population, and if we add the 17 countries that were more democratic than autocratic, we get a total of *two-thirds* of the world's population living in free or relatively free societies, compared with less than two-fifths in 1950, a fifth in 1900, seven percent in 1850, and one percent in 1816. Of the people living in the 60 nondemocratic countries today (20 full autocracies, 40 more autocratic than democratic), four-fifths reside in a single country, China.[15]

Though history has not ended, Fukuyama had a point: democracy has proved to be more attractive than its eulogizers acknowledge.[16] After the first wave of democratization broke, there were theories "explaining" how democracy could never take root in Catholic, non-Western, Asian, Muslim, poor, or ethnically diverse countries, each refuted in turn. It is true that stable, top-shelf democracy is likelier to be found in countries that are richer and more highly educated.[17] But governments that are more democratic than not are a motley collection: they are entrenched in most of Latin America, in floridly multiethnic India, in Muslim Malaysia, Indonesia, Niger, and Kosovo, in fourteen countries in sub-Saharan Africa (including Namibia, Senegal, and Benin), and in poor countries elsewhere such as Nepal, Timor-Leste, and most of the Caribbean.[18]

Even the autocracies of Russia and China, which show few signs of liberalizing, are incomparably less repressive than the regimes of Stalin, Brezhnev, and Mao.[19] Johan Norberg summarizes life in China: "The Chinese people today can move almost however they like, buy a home, choose an education, pick a job, start a business, belong to a church (as long as they are Buddhists, Taoist, Muslims, Catholics or Protestants), dress as they like, marry whom they like, be openly gay without ending up in a labor camp, travel abroad freely, and even criticize aspects of the Party's policy (though not its right to rule unopposed). Even 'not free' is not what it used to be."[20]

~

Why has the tide of democratization repeatedly exceeded expectations? The various backslidings, reversals, and black holes for democracy have led to theories which posit onerous prerequisites and an agonizing ordeal of democratization. (This serves as a convenient pretext for dictators to insist that their countries are not ready for it, like the revolutionary leader in Woody Allen's *Bananas* who upon taking power announces, "These people are peasants. They are too ignorant to vote.") The awe is reinforced by a civics-class idealization of democracy in which an informed populace deliberates about the common good and carefully selects leaders who carry out their preference.

By that standard, the number of democracies in the world is zero in the past, zero in the present, and almost certainly zero in the future. Political scientists are repeatedly astonished by the shallowness and incoherence of people's political beliefs, and by the tenuous connection of their preferences to their votes and to the behavior of their representatives.[21] Most voters are ignorant not just of current policy options but of basic facts, such as what the major branches of government are, who the United States fought in World War II, and which countries have used nuclear weapons. Their opinions flip depending on how a question is worded: they say that the government spends too much on "welfare" but too little on "assistance to the poor," and that it should "use military force" but not "go to war." When they do formulate a preference, they commonly vote for a candidate with the opposite one. But it hardly matters, because once in office politicians vote the positions of their party regardless of the opinions of their constituents.

Nor does voting even provide much of a feedback signal about a government's performance. Voters punish incumbents for recent events over which they have dubious control, such as macroeconomic swings and terrorist strikes, or no control at all, such as droughts, floods, even shark attacks. Many political scientists have concluded that most people correctly recognize that their votes are astronomically unlikely to affect the outcome of an election, and so they prioritize work, family, and leisure over educating themselves about politics and calibrating their votes. They use the franchise as a form of self-expression: they vote for candidates who they think are like them and stand for their kind of people.

So despite the widespread belief that elections are the quintessence of democracy, they are only one of the mechanisms by which a government is held responsible to those it governs, and not always a constructive one. When an election is a contest between aspiring despots, rival factions fear the worst if the other side wins and try to intimidate each other from the ballot box. Also, autocrats can learn to use elections to their advantage. The latest fashion in dictatorship has been called the competitive, electoral, kleptocratic, statist, or patronal authoritarian regime.[22] (Putin's Russia is the prototype.) The incumbents use the formidable resources of the state to harass the opposition, set up fake opposition parties, use state-controlled media to spread congenial narratives, manipulate electoral rules, tilt voter registration, and jigger the elections themselves. (Patronal authoritarians, for all that, are not invulnerable—the color revolutions sent several of them packing.)

If neither voters nor elected leaders can be counted on to uphold the ideals of democracy, why should this form of government work so not-badly—the worst form of government except all the others that have been tried, as Churchill famously put it? In his 1945 book *The Open Society and Its Enemies*, the philosopher Karl Popper argued that democracy should be understood not as the answer to the question "Who should rule?" (namely, "The People"), but as a solution to the problem of how to dismiss bad leadership without bloodshed.[23] The political scientist John Mueller broadens the idea from a binary Judgment Day to continuous day-to-day feedback. Democracy, he suggests, is essentially based on giving people the freedom to complain: "It comes about when the people effectively agree not to use violence to replace the leadership, and the leadership leaves them free to try to dislodge it by any other means."[24] He explains how this can work:

If citizens have the right to complain, to petition, to organize, to protest, to demonstrate, to strike, to threaten to emigrate or secede, to shout, to publish, to export their funds, to express a lack of confidence, and to wheedle in back corridors, government will tend to respond to the sounds of the shouters and the importunings of the wheedlers: that is, it will necessarily become responsive—pay attention—whether there are elections or not.[25]

Women's suffrage is an example: by definition, they could not vote to grant themselves the vote, but they got it by other means.

The contrast between the messy reality of democracy and the civics-class ideal leads to perennial disillusionment. John Kenneth Galbraith once advised that if you ever want a lucrative book contract, just propose to write *The Crisis of American Democracy*. Reviewing the history, Mueller concludes that "inequality, disagreement, apathy, and ignorance seem to be normal, not abnormal, in a democracy, and to a considerable degree the beauty of the form is that it works despite these qualities—or, in some important respects, because of them."[26]

In this minimalist conception, democracy is not a particularly abstruse or demanding form of government. Its main prerequisite is that a government be competent enough to protect people from anarchic violence so they don't fall prey to, or even welcome, the first strongman who promises he can do the job. (Chaos is deadlier than tyranny.) That's one reason why democracy has trouble getting a toehold in extremely poor countries with weak governments, such as in sub-Saharan Africa, and in countries whose government has been decapitated, such as Afghanistan and Iraq following the American-led invasions. As the political scientists Steven Levitsky and Lucan Way point out, "State failure brings violence and instability; it almost never brings democratization."[27]

Ideas matter, too. For democracy to take root, influential people (particularly people with guns) have to think that it is better than alternatives such as theocracy, the divine right of kings, colonial paternalism, the dictatorship of the proletariat (in practice, its "revolutionary vanguard"), or authoritarian rule by a charismatic leader who directly embodies the will of the people. This helps explain other patterns in the annals of democratization, such as why democracy is less likely to take root in countries with less education, in countries that are remote from Western influence (such as in Central Asia), and in countries whose regimes were born of violent, ideologically driven revolutions (such as China, Cuba, Iran, North Korea, and Vietnam).[28] Conversely, as people recognize that democracies are relatively nice places to live, the idea of democracy can become contagious and the number can increase over time.

～

The freedom to complain rests on an assurance that the government won't punish or silence the complainer. The front line in democratization, then, is constraining the government from abusing its monopoly on force to brutalize its uppity citizens.

A series of international agreements beginning with the Universal Declaration of Human Rights of 1948 drew red lines around thuggish governmental tactics, particularly torture, extrajudicial killings, the imprisonment of dissidents, and the ugly transitive verb coined during the Argentinian military regime of 1976–83, *to disappear someone.* These red lines are not the same as electoral democracy, since a majority of voters may be indifferent to government brutality as long as it isn't directed at them. In practice, democratic countries do show greater respect for human rights.[29] But the world also has some benevolent autocracies, like Singapore, and some repressive democracies, like Pakistan. This leads to a key question about whether the waves of democratization are really a form of progress. Has the rise in democracy brought a rise in human rights, or are dictators just using elections and other democratic trappings to cover their abuses with a smiley-face?

The US State Department, Amnesty International, and other organizations have monitored violations of human rights over the decades. If one were to look at their numbers since the 1970s, it would appear that governments are as repressive as ever—despite the spread of democracy, human rights norms, international criminal courts, and the watchdog organizations themselves. This has led to pronouncements (delivered with alarm by rights activists and with glee by cultural pessimists) that we have reached "the endtimes of human rights," "the twilight of human rights law," and, of course, "the post–human rights world."[30]

But progress has a way of covering its tracks. As our moral standards rise over the years, we become alert to harms that would have gone unnoticed in the past. Moreover, activist organizations feel they must always cry "crisis" to keep the heat up (though the strategy can backfire, implying that decades of activism have been a waste of time). The political scientist Kathryn Sikkink calls this the information paradox: as human rights watchdogs admirably look harder for abuse, look in more places for abuse, and classify more acts as abuse, they find more of it—but if we don't compensate for their keener powers of detection, we can be misled into thinking that there is more abuse to detect.[31]

The political scientist Christopher Fariss has cut this knot with a mathematical model that compensates for more dogged reporting over time and estimates the actual amount of human rights abuse in the world. Figure 14-2 shows his scores for four countries from 1949 to 2014 and for the world as a whole. The graph displays numbers spat out by a mathematical model, so we should not take the exact values too seriously, but they do indicate differences and trends. The top line is for a country that represents a gold standard for human rights. As with most measures of human flourishing, it is Scandinavian, in this case Norway, and it started high and has grown higher. We see diverging lines for the two Koreas: North, which started low and sank even lower, and South, which rose from a right-wing autocracy during the Cold War into positive territory today. In China, human rights hit bottom during the Cultural Revolution, shot up after the death of Mao, and crested during the 1980s democracy movement before the government cracked down after the Tiananmen Square protests, though they are still well above the Maoist-era lowlands. But the most significant curve is the one for the world as a whole: for all its setbacks, the arc of human rights bends upward.

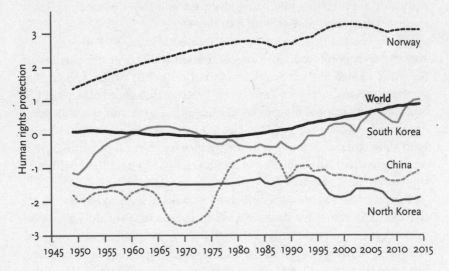

Figure 14-2: Human rights, 1949–2014

Source: *Our World in Data*, Roser 2016i, graphing an index devised by Fariss 2014, which estimates protection from torture, extrajudicial killing, political imprisonment, and disappearances. "0" is the mean over all countries and years; the units are standard deviations.

～

How does the curtailment of government power unfold in real time? An unusually clear window into the machinery of human progress is the fate of the ultimate exercise of violence by the state: deliberately killing its citizens.

Capital punishment was once ubiquitous among countries, and it was applied to hundreds of misdemeanors in gruesome public spectacles of torture and humiliation.[32] (The crucifixion of Jesus together with two common thieves is as good a reminder as any.) After the Enlightenment,

European countries stopped executing people for any but the most heinous crimes: by the middle of the 19th century, Britain had reduced the number of capital offenses from 222 to 4. And the countries looked for methods of execution such as drop hanging that were as humane as such a gruesome practice could pretend to be. After World War II, when the Universal Declaration of Human Rights inaugurated a second humanitarian revolution, capital punishment was abolished altogether in country after country, and in Europe today it lingers only in Belarus.

The abolition of capital punishment has gone global (figure 14-3), and today the death penalty is on death row.[33] In the last three decades, two or three countries have abolished it every year, and less than a fifth of the world's nations continue to execute people. (While ninety countries retain capital punishment in their law books, most have not put anyone to death in at least a decade.) The UN Special Rapporteur on executions, Christof Heyns, points out that if the current rate of abolition continues (not that he's prophesying it will), capital punishment will vanish from the face of the earth by 2026.[34]

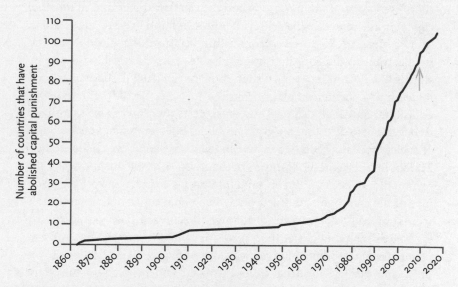

Figure 14-3: Death penalty abolitions, 1863–2016

Source: "Capital Punishment by Country: Abolition Chronology," *Wikipedia*, retrieved Aug. 15, 2016. Several European countries abolished the death penalty in their mainland earlier than indicated here, but the time line records the last abolition in any territory under their jurisdiction. The arrow points to 2008, the last year plotted in fig. 4–3 of Pinker 2011.

The top five countries that still execute people in significant numbers form an unlikely club: China and Iran (more than a thousand apiece annually), Pakistan, Saudi Arabia, and the United States. As in other areas of human flourishing (such as crime, war, health, longevity, accidents, and education), the United States is a laggard among wealthy democracies. This American exceptionalism illuminates the tortuous path by which moral progress proceeds from philosophical arguments to facts on the ground. It also showcases the tension between the two conceptions of democracy we have been examining: a form of government whose power to inflict violence on its citizens is sharply circumscribed, and a form of government that carries out the will of the majority of its people. The reason the United States is a death-penalty outlier is that it is, in one sense, *too* democratic.

In his history of the abolition of capital punishment in Europe, the legal scholar Andrew Hammel points out that in most times and places the death penalty strikes people as perfectly just: if you take a life, you deserve to lose your own.[35] It was only with the Enlightenment that forceful arguments against the death penalty began to appear.[36] One argument was that the state's mandate to exercise violence may not breach the sacred zone of human life. Another was that the deterrent effect of capital punishment can be achieved with surer and less brutal penalties.

The ideas trickled down from a thin stratum of philosophers and intellectuals to the educated upper classes, particularly liberal professionals like doctors, lawyers, writers, and journalists. Abolition was soon folded into a portfolio of other progressive causes, including mandatory education, universal suffrage, and workers' rights. It was also sacralized under the halo of "human rights" and held out as a symbol of "the kind of society we choose to live in and the kind of people we choose to be." The abolitionist elites in Europe got their way over the misgivings of the common man because European democracies did not convert the opinions of the common man into policy. The penal codes of their countries were drafted by committees of renowned scholars, passed into law by legislators who thought of themselves as a natural aristocracy, and implemented by appointed judges who were lifelong civil servants. It was only after a couple of decades had elapsed and people saw that their country had not fallen into chaos—at which point it would have taken a concerted effort to *reintroduce* capital punishment—that the populace came around to seeing it as unnecessary.

But the United States, for better or worse, is closer to having government by the people for the people. Other than for a few federal crimes like terrorism and treason, the death penalty is decided upon by individual states, voted on by legislators who are close to their constituents, and in many states sought and approved by prosecutors and judges who have to stand for reelection. Southern states have a longstanding culture of honor, with its ethos of justified retaliation, and not surprisingly, American executions are concentrated in a handful of Southern states, mainly Texas, Georgia, and Missouri—indeed, in a handful of *counties* in those states.[37]

Yet the United States, too, has been swept by the historical current, and capital punishment is on the way out despite its continuing popular appeal (with 61 percent in favor in 2015).[38] Seven states have repealed the death penalty in the past decade, an additional sixteen have moratoria, and thirty have not executed anyone in five years. Even Texas executed only seven prisoners in 2016, compared with forty in 2000. Figure 14-4 shows the steady decline of the use of the death penalty in the United States, with what may be a final slide to zero visible in the rightmost segment. And true to the pattern in Europe, as the practice becomes obsolescent, public opinion straggles behind: in 2016, popular support for the death penalty slipped just below 50 percent for the first time in almost fifty years.[39]

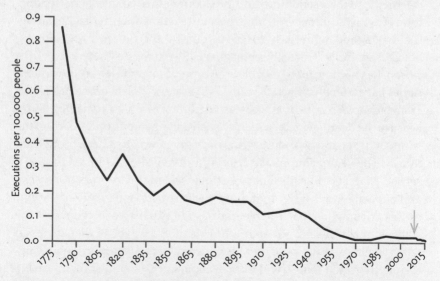

Figure 14-4: Executions, US, 1780–2016

Sources: Death Penalty Information Center 2017. Population estimates from US Census Bureau 2017. The arrow points to 2010, the last year plotted in fig. 4–4 of Pinker 2011.

How can the United States be doing away with capital punishment almost despite itself? Here we see another path along which moral prog-

ress can take place. Though the American political system is more populist than those of its Western peers, it still falls short of being a direct
participatory democracy like ancient Athens (which, pointedly, put Socrates to death). With the historical expansion of sympathy and reason,
even the staunchest fans of capital punishment have lost their stomach
for lynch mobs, hanging judges, and rowdy public executions, and insist
that the practice be carried out with a modicum of dignity and care. That
requires an intricate apparatus of death and a team of mechanics to run
and repair it. As the machine wears out and the mechanics refuse to
maintain it, it becomes increasingly unwieldy and invites being
scrapped.[40] The American death penalty is not so much being abolished
as falling apart, piece by piece.

First, advances in forensic science, particularly DNA fingerprinting,
have shown that innocent people have almost certainly been put to
death, a scenario that unnerves even ardent supporters of the death penalty. Second, the grisly business of snuffing out a life has evolved from
the gory sadism of crucifixion and disembowelment, to the quick but
still graphic ropes, bullets, and blades, to the invisible agents of gas and
electricity, to the pseudo-medical procedure of lethal injection. But doctors refuse to administer it, pharmaceutical companies refuse to supply
the drugs, and witnesses are disturbed by the death throes during
botched attempts. Third, the chief alternative to the death penalty, life in
prison, has become more reliable as escape-proof and riot-proof peniten-
tiaries have been perfected. Fourth, as the rate of violent crime has plummeted (chapter 12), people feel less need for draconian remedies. Fifth,
because the death penalty is seen as such a momentous undertaking, the
summary executions of earlier eras have given way to a drawn-out legal
ordeal. The sentencing phase after a guilty verdict is tantamount to a
second trial, and a death sentence triggers a lengthy process of reviews
and appeals—so lengthy that most death-row prisoners die of natural
causes. Meanwhile, the billable hours from expensive lawyers cost the
state eight times as much as life in prison. Sixth, social disparities in
death sentences, with poor and black defendants disproportionately being put to death ("Those without the capital get the punishment"), have
weighed increasingly on the nation's conscience. Finally, the Supreme
Court, which is repeatedly tasked with formulating a consistent rationale
for this crazy quilt, has struggled to rationalize the practice, and has
chipped away at it piece by piece. In recent years it has ruled that states
may not execute juveniles, people with intellectual disabilities, or perpetrators of crimes other than murder, and it came close to ruling against

the hit-and-miss method of lethal injection. Court watchers believe it is only a matter of time before the Justices are forced to confront the caprice of the whole macabre practice head on, invoke "evolving standards of decency," and strike it down as a violation of the Eighth Amendment's prohibition of cruel and unusual punishment once and for all.

The uncanny assemblage of scientific, institutional, legal, and social forces all pushing to strip government of its power to kill makes it seem as if there really is a mysterious arc bending toward justice. More prosaically, we are seeing a moral principle—Life is sacred, so killing is onerous—become distributed across a wide range of actors and institutions that have to cooperate to make the death penalty possible. As these actors and institutions implement the principle more consistently and thoroughly, they inexorably push the country away from the impulse to avenge a life with a life. The pathways are manifold and tortuous, the effects are slow and then sudden, but in the fullness of time an idea from the Enlightenment can transform the world.

CHAPTER 15

EQUAL RIGHTS

Humans are liable to treat entire categories of other humans as means to an end or nuisances to be cast aside. Coalitions bound by race or creed seek to dominate rival coalitions. Men try to control the labor, freedom, and sexuality of women.[1] People translate their discomfort with sexual nonconformity into moralistic condemnation.[2] We call these phenomena racism, sexism, and homophobia, and they have been rampant, to varying degrees, in most cultures throughout history. The disavowal of these evils is a large part of what we call civil rights or equal rights. The historical expansion of these rights—the stories of Selma, Seneca Falls, and Stonewall—is a stirring chapter in the story of human progress.[3]

The rights of racial minorities, women, and gay people continue to advance, each recently emblazoned on a milestone. The year 2017 saw the completion of two terms in office by the first African American president, an achievement movingly captured by First Lady Michelle Obama in a speech at the Democratic National Convention in 2016: "I wake up every morning in a house that was built by slaves, and I watch my daughters, two beautiful, intelligent black young women, playing with their dogs on the White House lawn." Barack Obama was succeeded by the first woman nominee of a major party in a presidential election, less than a century after American women were even allowed to vote; she won a solid plurality of the popular vote and would have been president were it not for peculiarities of the Electoral College system and other quirks of that election year. In a parallel universe very similar to this one until November 8, 2016, the world's three most influential nations (the United States, the United Kingdom, and Germany) are all led by women.[4] And in 2015, just a dozen years after it ruled that homosexual activity may not be criminalized, the Supreme Court guaranteed the right of marriage to same-sex couples.

But it's in the nature of progress that it erases its tracks, and its champions fixate on the remaining injustices and forget how far we have come. An axiom of progressive opinion, especially in universities, is that we continue to live in a deeply racist, sexist, and homophobic society— which would imply that progressivism is a waste of time, having accomplished nothing after decades of struggle.

Like other forms of progressophobia, the denial of advances in rights has been abetted by sensational headlines. A string of highly publicized killings by American police officers of unarmed African American suspects, some of them caught on smartphone videos, has led to a sense that the country is suffering an epidemic of racist attacks by police on black men. Media coverage of athletes who have assaulted their wives or girlfriends, and of episodes of rape on college campuses, has suggested to many that we are undergoing a surge of violence against women. And one of the most heinous crimes in American history took place in 2016 when Omar Mateen opened fire at a gay nightclub in Orlando, killing forty-nine people and wounding another fifty-three.

The belief in an absence of progress has been fortified by the recent history of the universe we do live in, where Donald Trump rather than Hillary Clinton was the beneficiary of the American electoral system in 2016. During his campaign, Trump uttered misogynistic, anti-Hispanic, and anti-Muslim insults that were well outside the norms of American political discourse, and the rowdy followers he encouraged at his rallies were even more offensive. Some commentators worried that his victory represented a turning point in the nation's progress toward equality and rights, or that it uncovered the ugly truth that we had never made progress in the first place.

The goal of this chapter is to plumb the depths of the current that carries equal rights along. Is it an illusion, a turbulent whirlpool atop a stagnant pond? Does it easily change direction and flow backwards? Or does justice roll on like a river, righteousness like a mighty stream?[5] I'll end with a coda about progress in the rights of the most easily victimized sector of humanity, children.

~

By now you should be skeptical about reading history from the headlines, and that applies to the recent assaults on equal rights. The data suggest that the number of police shootings has *decreased*, not increased, in recent decades (even as the ones that do occur are captured on video), and three independent analyses have found that a black suspect is no more likely than a white suspect to be killed by the police.[6] (American police shoot too many people, but it's not primarily a racial issue.) A spate of news about rape cannot tell us whether there is now more violence against women, a bad thing, or whether we now care more about violence against women, a good thing. And to this day it is unclear whether the Orlando nightclub massacre was committed out of homophobia, sympathy for ISIS, or the drive for posthumous notoriety that motivates most rampage shooters.

Better first drafts of history can be gleaned from data on values and from vital statistics. The Pew Research Center has probed Americans' opinions on race, gender, and sexual orientation over the past quarter century, and has reported that these attitudes have undergone a "fundamental shift" toward tolerance and respect of rights, with formerly widespread prejudices sinking into oblivion.[7] The shift is visible in figure 15-1, which plots reactions to three survey statements that are representative of many others.

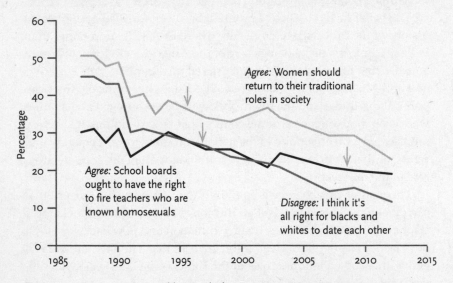

Figure 15-1: Racist, sexist, and homophobic opinions, US, 1987–2012

Source: Pew Research Center 2012b. The arrows point to the most recent years plotted in Pinker 2011 for similar questions: Blacks, 1997 (fig. 7–7); Women, 1995 (fig. 7–11); Homosexuals, 2009 (fig. 7–24).

Other surveys show the same shifts.[8] Not only has the American population become more liberal, but each generational cohort is more liberal than the one born before it.[9] As we will see, people tend to carry their values with them as they age, so the Millennials (those born after 1980), who are even less prejudiced than the national average, tell us which way the country is going.[10]

Of course one can wonder whether figure 15-1 displays a decline in prejudice or simply a decline in the social acceptability of prejudice, with fewer people willing to confess their disreputable attitudes to a pollster. The problem has long haunted social scientists, but recently the economist Seth Stephens-Davidowitz has discovered an indicator of attitudes that is the closest we've come to a digital truth serum.[11] In the privacy of their keyboards and screens, people query Google with every curiosity, anxiety, and guilty pleasure you can imagine, together with many you can't imagine. (Common searches include "How to make my penis bigger" and "My vagina smells like fish.") Google has amassed big data on the strings that people search for in different months and regions (though not the identity of the searchers), together with tools for analyzing them. Stephens-Davidowitz discovered that searches for the word *nigger* (mostly in pursuit of racist jokes) correlate with other indicators of racial prejudice across regions, such as vote totals for Barack Obama in 2008 that were lower than expected for a Democrat.[12] He suggests that these searches can serve as an unobtrusive indicator of private racism.

Let's use them to track recent trends in racism, and while we're at it, private sexism and homophobia as well. Well into my adolescence, jokes featuring dumb Poles, ditzy dames, and lisping, limp-wristed homosexuals were common in network television and newspaper comics. Today they are taboo in mainstream media. But do bigoted jokes remain a private indulgence, or have private attitudes changed so much that people feel offended, sullied, or bored by them? Figure 15-2 shows the results. The curves suggest that Americans are not just more abashed about confessing to prejudice than they used to be; they privately don't find it as amusing.[13] And contrary to the fear that the rise of Trump reflects (or emboldens) prejudice, the curves continue their decline through his period of notoriety in 2015–2016 and inauguration in early 2017.

Stephens-Davidowitz has pointed out to me that these curves probably *underestimate* the decline in prejudice because of a shift in who's Googling. When the records began in 2004, Googlers were mostly young and urban. Older and rural people tend to be latecomers to technology, and if they are the ones who are likelier to search for the offensive terms, that would inflate the proportion in later years and conceal the extent of the decline in bigotry. Google doesn't record the searchers' ages or levels of education, but it does record where the searches come from. In response to my query, Stephens-Davidowitz confirmed that bigoted searches tended to come from regions with older and less-educated populations. Compared with the country as a whole, retirement communities are seven times as likely to search for "nigger jokes" and thirty times as likely to search for "fag jokes." ("Google AdWords," he told me apologetically, "doesn't give data on 'bitch jokes.'") Stephens-Davidowitz also got his hands on a trove of search data from AOL, which, unlike Google, tracks the searches made by individuals (though not, of course, their identities). These threads confirmed that racists may be a dwindling breed: someone who searches for "nigger" is likely to search for other topics that appeal to senior citizens, such as "social security" and "Frank Sinatra." The main exception was a sliver of teenagers who also searched for bestiality, decapitation videos, and child pornography—anything you're not supposed to search for. But aside from these transgressive youths (and there have always been transgressive youths), private prejudice is declining with time *and* declining with youth, which means that we can expect it to decline still further as aging bigots cede the stage to less prejudiced cohorts.

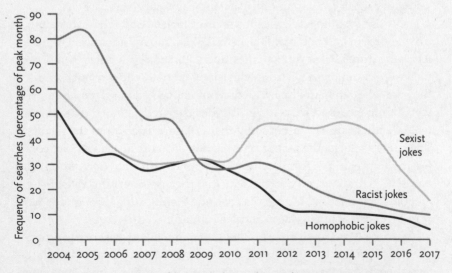

Figure 15-2: Racist, sexist, and homophobic Web searches, US, 2004–2017

Source: Google Trends (www.google.com/trends), searches for "nigger jokes," "bitch jokes," and "fag jokes," United States, 2004–2017, relative to total search volume. Data (accessed Jan. 22, 2017) are by month, expressed as a percentage of the peak month for each search term, then averaged over the months of each year, and smoothed.

Until they do, these older and less-educated people (mainly white men) may not respect the benign taboos on racism, sexism, and homophobia that have become second nature to the mainstream, and may even dismiss them as "political correctness." Today they can find each other on the Internet and coalesce under a demagogue. As we will see in chapter 20, Trump's success, like that of right-wing populists in other Western countries, is better understood as the mobilization of an aggrieved and shrinking demographic in a polarized political landscape than as the sudden reversal of a century-long movement toward equal rights.

~

Progress in equal rights may be seen not just in political milestones and opinion bellwethers but in data on people's lives. Among African Americans, the poverty rate fell from 55 percent in 1960 to 27.6 percent in 2011.[14] Life expectancy rose from 33 in 1900 (17.6 years below that of whites) to 75.6 years in 2015 (less than 3 years below whites).[15] African Americans who make it to 65 have *longer* lives ahead of them than white Americans of the same age. The rate of illiteracy fell among African Americans from 45 percent in 1900 to effectively zero percent today.[16] As we will see in the next chapter, the racial gap in children's readiness for school has been shrinking. As we will see in chapter 18, so has the racial gap in happiness.[17]

Racist violence against African Americans, once a regular occurrence in night raids and lynchings (three a week at the turn of the 20th century), plummeted in the 20th century, and has fallen further since the FBI started amalgamating reports on hate crimes in 1996, as figure 15-3 shows. (Very few of these crimes are homicides, in most years one or zero.)[18] The slight uptick in 2015 (the most recent year available) cannot be blamed on Trump, since it parallels the uptick in violent crime that year (see figure 12-2), and hate crimes track rates of overall lawlessness more closely than they do remarks by politicians.[19]

Figure 15-3 shows that hate crimes against Asian, Jewish, and white targets have declined as well. And despite claims that Islamophobia has become rampant in America, hate crimes targeting Muslims have shown little change other than a one-time rise following 9/11 and upticks following other Islamist terror attacks, such as the ones in Paris and San Bernardino in 2015.[20] At the time of this writing, FBI data from 2016 are not available, so it's premature to accept the widespread claims of a Trumpist surge in hate crimes that year. The claims come from advocacy organizations, whose funding depends on whipping up fear, rather than disinterested recordkeepers; some of the incidents were ironic hoaxes, and many were boorish outbursts rather than actual crimes.[21] Aside from post-terrorist and crime-related blips, the trend in hate crimes is downward.

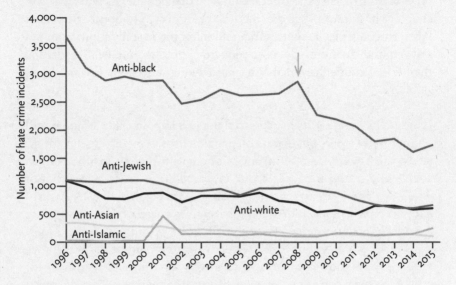

Figure 15-3: Hate crimes, US, 1996–2015

Source: Federal Bureau of Investigation 2016b. The arrow points to 2008, the last year plotted in fig. 7–4 of Pinker 2011.

Women's status, too, is ascendant. As recently as my childhood, American women in most states could not take out a loan or credit card in their own names, had to look for jobs in the HELP WANTED—FEMALE section of the classified ads, and could not press charges of rape against their husbands.[22] Today, women make up 47 percent of the labor force and a majority of university students.[23] Violence against women is best measured by victimization surveys, because they circumvent the problem of underreporting to the police; these instruments show that rates of rape and violence against wives and girlfriends have been sinking for decades and are now at a quarter or less of their peaks in the past (figure 15-4).[24] Too many of these crimes still take place, but we should be encouraged by the fact that a heightened concern about violence against women is not futile moralizing but has brought measurable progress—which means that continuing this concern can lead to greater progress still.

No form of progress is inevitable, but the historical erosion of racism, sexism, and homophobia are more than a change in fashion. As we will see, it seems to be pushed along by the tide of modernity. In a cosmopolitan society, people rub shoulders, do business, and find themselves in the same boat with other kinds of people, and that tends to make them more sympathetic to one another.[25] Also, as people are forced to justify the way they treat other people, rather than dominating them out of instinctive, religious, or historical inertia, any justification for prejudicial treatment will crumble under scrutiny.[26] Racial segregation, male-only suffrage, and the criminalization of homosexuality are literally indefensible: people tried to defend them in their times, and they lost the argument.

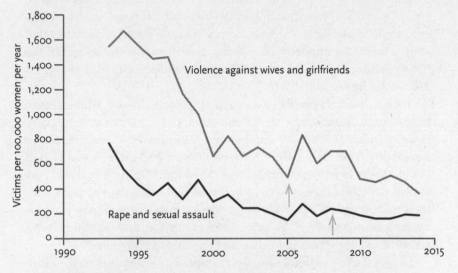

Figure 15-4: Rape and domestic violence, US, 1993–2014

Sources: US Bureau of Justice Statistics, *National Crime Victimization Survey,* Victimization Analysis Tool, http://www.bjs.gov/index.cfm?ty=nvat, with additional data provided by Jennifer Truman of BJS. The gray line represents "Intimate partner violence" with female victims. The arrows point to 2005, the last year plotted in fig. 7–13, and 2008, the last year plotted in fig. 7–10, of Pinker 2011.

These forces can prevail over the long term even against the tug of populist backlash. The global momentum toward abolition of the death penalty (chapter 14), despite its perennial popular appeal, offers a lesson in the messy ways of progress. As indefensible or unworkable ideas fall by the wayside, they are removed from the pool of thinkable options, even among those who like to think that they think the unthinkable, and the political fringe is dragged forward despite itself. That's why even in the most regressive political movement in recent American history there were no calls for reinstating Jim Crow laws, ending women's suffrage, or recriminalizing homosexuality.

~

Racial and ethnic prejudice is declining not just in the West but worldwide. In 1950, almost half the world's countries had laws that discriminated against ethnic or racial minorities (including, of course, the United
States). By 2003 fewer than a fifth did, and they were outnumbered by
countries with affirmative action policies that *favored* disadvantaged minorities.[27] A huge 2008 survey by the World Public Opinion poll of
twenty-one developed and developing nations found that in every one,
large majorities of respondents (around 90 percent on average) say that
it's important for people of different races, ethnicities, and religions to be
treated equally.[28] Notwithstanding the habitual self-flagellation by Western intellectuals about Western racism, it's non-Western countries that
are the least tolerant. But even in India, the country at the bottom of the
list, 59 percent of the respondents affirmed racial equality, and 76 percent
affirmed religious equality.[29]

With women's rights, too, the progress is global. In 1900, women could
vote in only one country, New Zealand. Today they can vote in every
country in which men can vote but one, Vatican City. Women make up
almost 40 percent of the labor force worldwide and more than a fifth of
the members of national parliaments. The World Opinion Poll and Pew
Global Attitudes Project have each found that more than 85 percent of
their respondents believe in full equality for men and women, with rates
ranging from 60 percent in India, to 88 percent in six Muslim-majority
countries, to 98 percent in Mexico and the United Kingdom.[30]

In 1993 the UN General Assembly adopted a Declaration on the Elimination of Violence Against Women. Since then most countries have implemented laws and public-awareness campaigns to reduce rape, forced
marriage, child marriage, genital mutilation, honor killings, domestic
violence, and wartime atrocities. Though some of these measures are
toothless, there are grounds for optimism over the long term. Global
shaming campaigns, even when they start out as purely aspirational,
have in the past led to dramatic reductions in slavery, dueling, whaling,
foot-binding, piracy, privateering, chemical warfare, apartheid, and atmospheric nuclear testing.[31] Female genital mutilation is an example:
though still practiced in twenty-nine African countries (together with
Indonesia, Iraq, India, Pakistan, and Yemen), a majority of both men and
women in those countries believe it should stop, and over the past thirty
years rates have fallen by a third.[32] In 2016 the Pan-African Parliament,
working with the UN Population Fund, endorsed a ban on the practice,
together with child marriage.[33]

Gay rights is another idea whose time has come. Homosexual acts used to be a criminal offense in almost every country in the world.[34] The first arguments that behavior between consenting adults is no one else's business were formulated during the Enlightenment by Montesquieu, Voltaire, Beccaria, and Bentham. A smattering of countries decriminalized homosexuality soon thereafter, and the number shot up with the gay rights revolution of the 1970s. Though homosexuality is still a crime in more than seventy countries (and a capital crime in eleven Islamic ones), and despite backsliding in Russia and several African countries, the global trend, encouraged by the UN and every human rights organization, continues toward liberalization.[35] Figure 15-5 shows the time line: in the past six years, an additional eight countries have stricken homosexuality from their criminal codes.

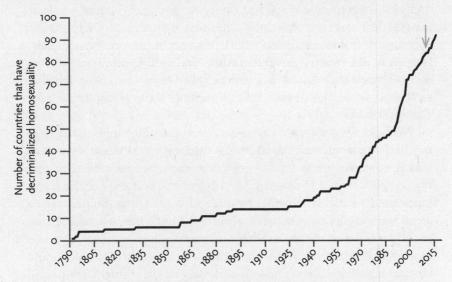

Figure 15-5: Decriminalization of homosexuality, 1791–2016

Sources: Ottosson 2006, 2009. Dates for an additional sixteen countries were obtained from "LGBT Rights by Country or Territory," *Wikipedia*, retrieved July 31, 2016. Dates for an additional thirty-six countries that currently allow homosexuality are not listed in either source. The arrow points to 2009, the last year plotted in fig. 7–23 of Pinker 2011.

The worldwide progress against racism, sexism, and homophobia, even with its bumpiness and setbacks, can feel like an overarching sweep. Martin Luther King Jr. famously quoted the abolitionist Theodore Parker's image of an arc bending toward justice. Parker confessed that he could not complete the arc by sight but could "divine it by conscience."

Is there a more objective way of determining whether there is a historical arc toward justice, and if so, what bends it?

One view of the moral arc is provided by the World Values Survey, which has polled 150,000 people in more than ninety-five countries containing almost 90 percent of the world's population over a span of several decades. In his book *Freedom Rising,* the political scientist Christian Welzel (building on a collaboration with Ron Inglehart, Pippa Norris, and others) has proposed that the process of modernization has stimulated the rise of "emancipative values."[36] As societies shift from agrarian to industrial to informational, their citizens become less anxious about fending off enemies and other existential threats and more eager to express their ideals and to pursue opportunities in life. This shifts their values toward greater freedom for themselves and others. The transition is consistent with the psychologist Abraham Maslow's theory of a hierarchy of needs from survival and safety to belonging, esteem, and self-actualization (and with Brecht's "Grub first, then ethics"). People begin to prioritize freedom over security, diversity over uniformity, autonomy over authority, creativity over discipline, and individuality over conformity. Emancipative values may also be called liberal values, in the classical sense related to "liberty" and "liberation" (rather than the sense of political leftism).

Welzel derived a way to capture a commitment to emancipative values in a single number, based on his discovery that the answers to a cluster of survey items tend to correlate across people, countries, and regions of the world with a common history and culture. The items embrace gender equality (whether people feel that women should have an equal right to jobs, political leadership, and a university education), personal choice (whether they feel that divorce, homosexuality, and abortion may be justified), political voice (whether they believe that people should be guaranteed freedom of speech and a say in government, communities, and the workplace), and childrearing philosophy (whether they feel that children should be encouraged to be obedient or independent and imaginative). The correlations among these items are far from perfect— abortion, in particular, divides people who agree on much else—but they tend to go together and collectively predict many things about a country.

Before we look at historical changes in values, we have to keep in mind that the passage of time doesn't simply flip the pages of the calendar. As time goes by, people get older, and eventually they die and are replaced by a new generation. Any secular (in the sense of historical or long-term) change in human behavior, then, can take place for three reasons.[37] The trend can be a Period Effect: a change in the times, the zeitgeist, or the national mood that lifts or lowers all the boats. It can be an Age (or Life Cycle) Effect: people change as they grow from mewling infant to whining schoolboy to sighing lover to round-bellied justice, and so on. Since there are booms and busts in a nation's birthrate, the population average will automatically change with the changing proportion of young, middle-aged, and old people, even if the prevailing values at each age are the same. Finally, the trend can be a Cohort (or Generational) Effect: people born at a certain time may be stamped with traits they carry through their lives, and the average for the population will reflect the changing mixture of cohorts as one generation exits the stage and another enters. It's impossible to disentangle the effects of age, period, and cohort perfectly, because as one period transitions into the next, each cohort gets older. But by measuring a trait across a population in several periods, and separating the data from the different cohorts in each one, one can make reasonable inferences about the three kinds of change.

Let's first look at the history of the most developed nations, such as those of North America, Western Europe, and Japan. Figure 15-6 shows the trajectory of emancipative values over a span of a century. It plots survey data collected from adults (ranging in age from eighteen to eighty-five), at two periods (1980 and 2005), representing cohorts born between 1895 and 1980. (American cohorts are commonly divided into the GI Generation, born between 1900 and 1924; the Silent Generation, 1925–45; the Baby Boomers, 1946–64; Generation X, 1965–79; and the Millennials, 1980–2000.) The cohorts are arranged along the horizontal axis by birth year; each of the two testing years is plotted on a line. (Data from 2011 to 2014, which extend the series to late Millennials born through 1996, are similar to those of 2005.)

The graph displays a historical trend that is seldom appreciated in the hurly-burly of political debate: for all the talk about right-wing backlashes and angry white men, the values of Western countries have been getting steadily more liberal (which, as we will see, is one of the reasons those men are so angry).[38] The line for 2005 is higher than the line for 1980 (showing that everyone got more liberal over time), and both curves rise from left to right (showing that younger generations in both periods were more liberal than older generations). The rises are substantial: about three-quarters of a standard deviation apiece for the twenty-five years of passing time and for each twenty-five-year generation. (The rises are also unappreciated: a 2016 Ipsos poll showed that in almost every developed country, people think their compatriots are more socially conservative than they really are.)[39] A critical discovery displayed in the graph is that the liberalization does *not* reflect a growing bulge of liberal young people who will backslide into conservatism as they get older. If that were true, the two curves would sit side by side instead of one floating above the other, and a vertical line representing a given cohort would impale the 2005 curve at a *lower* value, reflecting conservative old age, rather than the higher value we see, reflecting the more liberal zeitgeist. Young people take their emancipative values with them as they age, a finding we'll return to when we ponder the future of progress in chapter 20.[40]

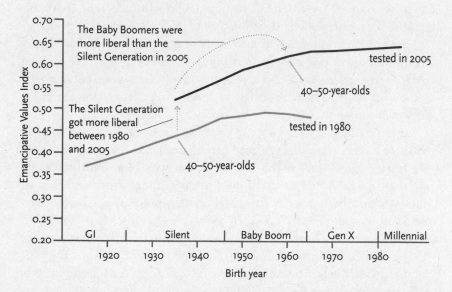

Figure 15-6: Liberal values across time and generations, developed countries, 1980–2005

Source: Welzel 2013, fig. 4.1. World Values Survey data are from Australia, Canada, France, West Germany, Italy, Japan, the Netherlands, Norway, Sweden, the United Kingdom, and the United States (each country weighted equally).

The liberalization trends shown in figure 15-6 come from the Prius-driving, chai-sipping, kale-eating populations of post-industrial Western countries. What about the rest of humanity? Welzel grouped the ninety-five countries in the World Values Survey into ten zones with similar histories and cultures. He also took advantage of the absence of a life-cycle effect to extrapolate emancipative values backwards: the values of a sixty-year-old in 2000, adjusted for the effects of forty years of liberalization in his or her country as a whole, provide a good estimate of the values of a twenty-year-old in 1960. Figure 15-7 shows the trends in liberal values for the different parts of the world over a span of almost fifty years, combining the effects of the changing zeitgeist in each country (like the jump between curves in figure 15-6) with the changing cohorts (the rise along each curve).

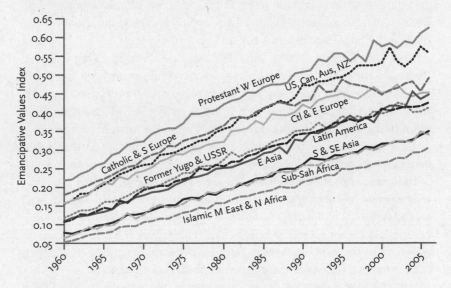

Figure 15-7: Liberal values across time (extrapolated), world's culture zones, 1960–2006

Source: World Values Survey, as analyzed in Welzel 2013, fig. 4.4, updated with data provided by Welzel. Emancipative value estimates for each country in each year are calculated for a hypothetical sample of a fixed age, based on each respondent's birth cohort, the year of testing, and a country-specific period effect. The labels are geographic mnemonics for Welzel's "culture zones" and do not literally apply to every country in a zone. I have renamed some of the zones: Protestant Western Europe corresponds to Welzel's "Reformed West." US, Canada, Australia, New Zealand = "New West." Catholic & Southern Europe = "Old West." Central & Eastern Europe = "Returned West." East Asia = "Sinic East." Former Yugoslavia & USSR = "Orthodox East." South & Southeast Asia = "Indic East." Countries in each zone are weighted equally.

The graph, unsurprisingly, reveals that differences across the world's culture zones are substantial. The Protestant countries of Western Europe, such as the Netherlands, Scandinavia, and the United Kingdom, are the world's most liberal, followed by the United States and other wealthy English-speaking countries, then Catholic and Southern Europe, then the former Communist countries of central Europe. Latin America, the industrialized countries of East Asia, and the former republics of the Soviet Union and Yugoslavia are more socially conservative, followed by South and Southeast Asia and sub-Saharan Africa. The world's most illiberal region is the Islamic Middle East.

What is surprising, though, is that *in every part of the world, people have become more liberal.* A lot more liberal: young Muslims in the Middle East, the world's most conservative culture, have values today that are comparable to those of young people in Western Europe, the world's most liberal culture, in the early 1960s. Though in every culture both the zeitgeist and the generations became more liberal, in some, like the Islamic Middle East, the liberalization was driven mainly by the generational turnover, and it played an obvious role in the Arab Spring.[41]

Can we identify the causes that differentiate the world's regions and liberalize them all over time? Many society-wide traits correlate with emancipative values, and—in a problem we encounter repeatedly—they tend to correlate with each other, a nuisance for social scientists who want to distinguish causation from correlation.[42] Prosperity (measured as GDP per capita) correlates with emancipative values, presumably because as people become healthier and more secure they can experiment with liberalizing their societies. The data show that more liberal countries are also, on average, better educated, more urban, less fecund, less inbred (with fewer marriages among cousins), more peaceful, more democratic, less corrupt, and less crime- and coup-ridden.[43] Their economies, now and in the past, tend to be built on networks of commerce rather than large plantations or the extraction of oil and minerals.

Yet the single best predictor of emancipative values is the World Bank's Knowledge Index, which combines per capita measures of education (adult literacy and enrollment in high schools and colleges), information access (telephones, computers, and Internet users), scientific and technological productivity (researchers, patents, and journal articles), and institutional integrity (rule of law, regulatory quality, and open economies).[44] Welzel found that the Knowledge Index accounts for *seventy percent* of the variation in emancipative values across countries, making it a far better predictor than GDP.[45] The statistical result vindicates a key insight of the Enlightenment: knowledge and sound institutions lead to moral progress.

⌒

Any tour of progress in rights must look at the most vulnerable sector of humanity, children, who cannot agitate for their own interests but depend upon the compassion of others. We've already seen that children the world over have become better off: they are less likely to enter the world motherless, die before their fifth birthday, or grow up stunted for lack of food. Here we'll see that in addition to escaping these natural assaults, children are increasingly escaping human-made ones: they are safer than they were before, and likelier to enjoy a true childhood.

The well-being of children is yet another case in which lurid head-lines terrify news readers even as they have less to be terrified about. Media reports of school shootings, abductions, bullying, cyberbullying, sexting, date rape, and sexual and physical abuse make it seem as if children are living in increasingly perilous times. The data say other-wise. Teenagers' retreat from dangerous drugs, mentioned in chapter 12, is just one example. In a 2014 review of the literature on violence against children in the United States, the sociologist David Finkelhor and his colleagues reported, "Of 50 trends in exposure examined, there were 27 significant declines and no significant increases between 2003 and 2011. Declines were particularly large for assault victimization, bullying, and sexual victimization."[46] Three of those trends are shown in figure 15-8.

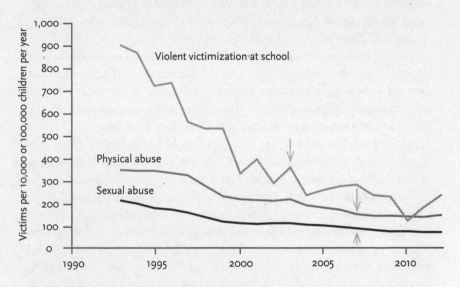

Figure 15-8: Victimization of children, US, 1993–2012

Sources: Physical abuse and **Sexual abuse** (mainly by caregivers): National Child Abuse and Neglect Data System, http://www.ndacan.cornell.edu/, analyzed by Finkelhor 2014; Finkelhor et al. 2014. **Violent victimization at school:** US Bureau of Justice Statistics, *National Crime Victimization Survey*, Victim-ization Analysis Tool, http://www.bjs.gov/index.cfm?ty=nvat. Rates for physical and sexual abuse are per 100,000 children younger than 18. Rates for violent victimization at school are per 10,000 children aged 12–17. The arrows point to 2003 and 2007, the last years plotted in fig. 7–22 and fig. 7–20 in Pinker 2011, respectively.

Another declining form of violence against children is corporal punishment—the spanking, smacking, paddling, birching, tanning, hid-ing, thrashing, and other crude methods of behavior modification that parents and teachers have inflicted on helpless children at least since the 7th-century-BCE advisory "Spare the rod and spoil the child." Corporal punishment has been condemned in several United Nations resolutions and has been outlawed in more than half the world's countries. The United States, once again, is an outlier among advanced democracies in allowing children to be paddled in schools, but even here, approval of all forms of corporal punishment is in slow but steady decline.[47]

Nine-year-old Oliver Twist's stint at picking oakum out of tarry ropes in an English workhouse is a fictional glimpse at one of the most widespread abuses of children, child labor. Together with Dickens's novel, Elizabeth Barrett Browning's 1843 poem "The Cry of the Children" and many journalistic exposés awakened 19th-century readers to the horrific conditions under which children were forced to work in that era. Small children stood on boxes to tend dangerous machinery in mills, mines, and canneries, breathing air thick with cotton or coal dust, kept awake by splashes of cold water in their faces, collapsing into sleep after exhausting shifts with food still in their mouths.

But the cruelties of child labor did not begin in Victorian factories.[48] Children have always been set to work as farmhands and domestics, and they were commonly hired out as servants to other people or as laborers in cottage industries, often from the age when they could walk. In the 17th century, for example, children put to work in a kitchen would crank a spit with a slab of meat for hours, protected from the fire only by a bale of wet hay.[49] No one thought of child labor as exploitation; it was a form of moral education, protecting children from idleness and sloth.

Starting with influential treatises by John Locke in 1693 and Jean-Jacques Rousseau in 1762, childhood was reconceptualized.[50] A carefree youth was now considered a human birthright. Play was an essential form of learning, and the early years of life shaped the adult and determined the future of society. In the decades around the turn of the 20th century, childhood was "sacralized," as the economist Viviana Zelizer has put it, and children achieved their current status as "economically worthless, emotionally priceless."[51] Under pressure from children's advocates, and helped along by affluence, smaller families, an expanding circle of sympathy, and an increasing premium on education, Western societies gradually did away with child labor. A snapshot of these forces pushing in the same direction may be found in an advertisement for tractors in a 1921 issue of the magazine *Successful Farming* entitled "Keep the Boy in School":

> The pressure of urgent Spring work is often the cause of keeping the boy out of school for several months. It may seem necessary—but it isn't fair to the boy! You are placing a life handicap in his path if you deprive him

of education. In this age, education is becoming more and more essential for success and prestige in all walks of life, including farming.

Should you feel that your own education was neglected, through no fault of yours, then you naturally will want your children to enjoy the benefits of a real education—to have some things you may have missed.

With the help of a Case Kerosene Tractor it is possible for one man to do more work, in a given time, than a good man and an industrious boy, together, working with horses. By investing in a Case Tractor and Ground Detour Plow and Harrow outfit now, your boy can get his schooling without interruption, and the Spring work will not suffer by his absence.

Keep the boy in school—and let a Case Kerosene Tractor take his place in the field. You'll never regret either investment.[52]

In many countries the coup de grâce was legislation that made schooling compulsory and thus made child laborers conspicuously illegal. Figure 15-9 shows that the proportion of children in the labor force in England was halved between 1850 and 1910, before child labor was outlawed altogether in 1918, and the United States followed a similar trajectory.

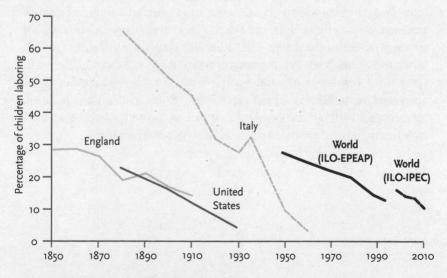

Figure 15-9: Child labor, 1850–2012

Sources: *Our World in Data*, Ortiz-Ospina & Roser 2016a, and the following. **England:** Percentage of children aged 10–14 recorded as working, Cunningham 1996. **United States:** Whaples 2005. **Italy:** Child work incidence, ages 10–14, Toniolo & Vecchi 2007. **World ILO-EPEAP** (International Labour Organization Programme on Estimates and Projections of the Economically Active Population): Child Labor, ages 10–14, Basu 1999. **World ILO-IPEC** (International Labour Organization International Programme on the Elimination of Child Labour): Child Labor, ages 5–17, International Labour Organization 2013.

The graph also shows the precipitous decline in Italy, together with two recent time series for the world. The lines are not commensurable because of differences in the age ranges and definitions of "child labor," but they show the same trend: downward. In 2012, 16.7 percent of the world's children worked an hour a week or more, 10.6 percent engaged in objectionable "child labor" (long hours or tender age), and 5.4 percent engaged in hazardous work—far too many, but less than half the rate of just a dozen years before. Child labor, now as always, is concentrated not in manufacturing but in agriculture, forestry, and fishing, and it goes with national poverty, as both cause and effect: the poorer the country, the larger the percentage of its children who work.[53] As wages rise, or when governments pay parents to send their children to school, child labor plummets, which suggests that poor parents send their children to work out of desperation rather than greed.[54]

As with other crimes and tragedies of the human condition, progress in ending child labor has been powered both by the global rise of affluence and by humanistic moral campaigns. In 1999, 180 countries ratified the Worst Forms of Child Labour Convention. The "worst forms" that were banned include hazardous labor and the exploitation of children in slavery, human trafficking, debt bondage, prostitution, pornography, drug trafficking, and war. Though the International Labour Organization's target of eliminating the worst forms by 2016 was not met, the momentum is unmistakable. The cause was symbolically ratified in 2014 when the Nobel Peace Prize was awarded to Kailash Satyarthi, the activist against child labor, who had been instrumental in the adoption of the 1999 resolution. He shared the prize with Malala Yousafzai, the heroic advocate for girls' education. And that brings us to yet another advance in human flourishing, the expansion of access to knowledge.

CHAPTER 16

KNOWLEDGE

Homo sapiens, "knowing man," is the species that uses information to resist the rot of entropy and the burdens of evolution. Humans everywhere acquire knowledge about their landscape, its flora and fauna, the tools and weapons that can subdue them, and the networks and norms that entangle them with kin, allies, and enemies. They accumulate and share that knowledge with the use of language, gesture, and face-to-face tutelage.[1]

At a few times in history, people have hit on technologies that multiply, indeed, exponentiate, the growth of knowledge, such as writing, printing, and electronic media. The supernova of knowledge continuously redefines what it means to be human. Our understanding of who we are, where we came from, how the world works, and what matters in life depends on partaking of the vast and ever-expanding store of knowledge. Though unlettered hunters, herders, and peasants are fully human, anthropologists often comment on their orientation to the present, the local, the physical.[2] To be aware of one's country and its history, of the diversity of customs and beliefs across the globe and through the ages, of the blunders and triumphs of past civilizations, of the microcosms of cells and atoms and the macrocosms of planets and galaxies, of the ethereal reality of number and logic and pattern—such awareness truly lifts us to a higher plane of consciousness. It is a gift of belonging to a brainy species with a long history.

It's been a long time since our culture's store of knowledge could be passed along by storytelling and apprenticeship. Formal schools are millennia old; I grew up with the Talmudic story of the 1st-century Rabbi Hillel who as a young man nearly froze to death after he climbed onto the roof of a school whose tuition he could not afford so that he could eavesdrop on lessons through the skylight. At various times, schools have been charged with instilling practical, religious, or patriotic wisdom in the young, but the Enlightenment, with its apotheosis of knowledge, would broaden their remit. "With the coming of the modern age," the educational theorist George Counts observes, "formal education assumed a significance far in excess of anything that the world had yet seen. The school, which had been a minor social agency in most of the societies of the past, directly affecting the lives of but a small fraction of the population, expanded horizontally and vertically until it took its place along with the state, the church, the family and property as one of society's most powerful institutions."[3] Today, education is compulsory in most countries, and it is recognized as a fundamental human right by the 170 members of the United Nations that signed the 1966 International Covenant on Economic, Social and Cultural Rights.[4]

The mind-altering effects of education extend to every sphere of life, in ways that range from the obvious to the spooky. At the obvious end of the range, we saw in chapter 6 that a little knowledge about sanitation, nutrition, and safe sex can go a long way toward improving health and extending life. Also obvious is that literacy and numeracy are the foundations of modern wealth creation. In the developing world a young woman can't even work as a household servant if she is unable to read a note or count out supplies, and higher rungs of the occupational ladder require ever-increasing abilities to understand technical material. The first countries that made the Great Escape from universal poverty in the 19th century, and the countries that have grown the fastest ever since, are the countries that educated their children most intensely.[5]

As with every question in social science, correlation is not causation. Do better-educated countries get richer, or can richer countries afford more education? One way to cut the knot is to take advantage of the fact that a cause must precede its effect. Studies that assess education at Time 1 and wealth at Time 2, holding all else constant, suggest that investing in education really does make countries richer. At least it does if the education is secular and rationalistic. Until the 20th century, Spain was an economic laggard among Western countries, even though Spaniards were highly schooled, because Spanish education was controlled by the Catholic Church, and "the children of the masses received only oral instruction in the Creed, the catechism, and a few simple manual skills. . . . Science, mathematics, political economy, and secular history were considered too controversial for anyone but trained theologians."[6] Clerical meddling has similarly been blamed for the economic lag of parts of the Arab world today.[7]

At the more spiritual end of the range, education brings gifts that go well beyond practical know-how and economic growth: better education today makes a country more democratic and peaceful tomorrow.[8] The wide-ranging effects of education make it hard to discern the intervening links in the causal chain from formal schooling to social harmony. Some of the links may simply be demographic and economic. Better-educated girls grow up to have fewer babies, and so are less likely to beget youth bulges with their surfeit of troublemaking young men.[9] And better-educated countries are richer, and as we saw in chapters 11 and 14, richer countries tend to be more peaceful and democratic.

But some of the causal pathways vindicate the values of the Enlightenment. So much changes when you get an education! You unlearn dangerous superstitions, such as that leaders rule by divine right, or that people who don't look like you are less than human. You learn that there are other cultures that are as tied to their ways of life as you are to yours, and for no better or worse reason. You learn that charismatic saviors have led their countries to disaster. You learn that your own convictions, no matter how heartfelt or popular, may be mistaken. You learn that there are better and worse ways to live, and that other people and other cultures may know things that you don't. Not least, you learn that there are ways of resolving conflicts without violence. All these epiphanies militate against knuckling under the rule of an autocrat or joining a crusade to subdue and kill your neighbors. Of course, none of this wisdom is guaranteed, particularly when authorities promulgate their own dogmas, alternative facts, and conspiracy theories—and, in a backhanded compliment to the power of knowledge, stifle the people and ideas that might discredit them.

Studies of the effects of education confirm that educated people really are more enlightened. They are less racist, sexist, xenophobic, homophobic, and authoritarian.[10] They place a higher value on imagination, independence, and free speech.[11] They are more likely to vote, volunteer, express political views, and belong to civic associations such as unions, political parties, and religious and community organizations.[12] They are also likelier to trust their fellow citizens—a prime ingredient of the precious elixir called social capital which gives people the confidence to contract, invest, and obey the law without fearing that they are chumps who will be shafted by everyone else.[13]

For all these reasons, the growth of education—and its first dividend, literacy—is a flagship of human progress. And as with so many other dimensions of progress, we see a familiar narrative: until the Enlightenment, almost everyone was abject; then, a few countries started to pull away from the pack; recently, the rest of the world has been catching up; soon, the bounty will be near-universal. Figure 16-1 shows that before the 17th century, literacy was the privilege of a small elite in Western Europe, less than an eighth of the population, and that was true for the world as a whole well into the 19th century. The world's literacy rate doubled in the next century and quadrupled in the century after that, so now 83 percent of the world is literate. Even that figure understates the literatization of the world, because the illiterate fifth is mostly middle-aged or elderly. In many Middle Eastern and North African countries, more than three-quarters of the people over sixty-five are illiterate, whereas the rate for those in their teens and twenties is in the single digits.[14] The literacy rate for young adults (aged fifteen to twenty-four) in 2010 was 91 percent—about the same as for the entire population of the United States in 1910.[15] Not surprisingly, the lowest rates of literacy are found in the world's poorest and most war-torn countries, such as South Sudan (32 percent), Central African Republic (37 percent), and Afghanistan (38 percent).[16]

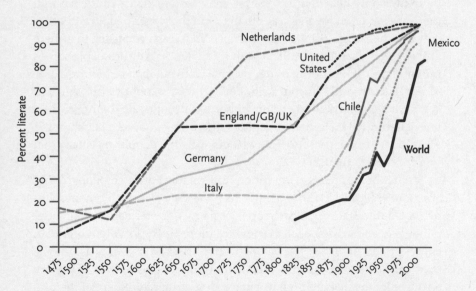

Figure 16-1: Literacy, 1475–2010

Source: *Our World in Data*, Roser & Ortiz-Ospina 2016b, including data from the following. **Before 1800:** Buringh & van Zanden 2009. **World:** van Zanden et al. 2014. **US:** National Center for Education Statistics. **After 2000:** Central Intelligence Agency 2016.

Literacy is the foundation for the rest of education, and figure 16-2 shows the world's progress in sending children to school.[17] The time line is familiar: in 1820, more than 80 percent of the world was unschooled; by 1900, a large majority of Western Europe and the Anglosphere had the benefit of a basic education; today, that's true of more than 80 percent of the world. The least fortunate region, sub-Saharan Africa, has a rate comparable to that of the world in 1980, Latin America in 1970, East Asia in the 1960s, Eastern Europe in 1930, and Western Europe in 1880. According to current projections, by the middle of this century, only five countries will have more than a fifth of their population uneducated, and by the end of the century the worldwide proportion will fall to zero.[18]

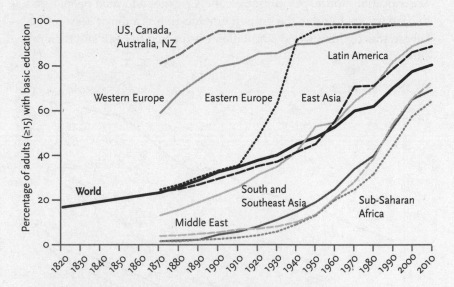

Figure 16-2: Basic education, 1820–2010

Source: *Our World in Data,* Roser & Ortiz–Ospina 2018, based on data from van Zanden et al. 2014. The graphs indicate the share of the population aged 15 or older that had completed at least a year of education (more in later eras); see van Leeuwen & van Leewen-Li 2014, pp. 88–93.

"Of making many books there is no end; and much study is a weariness of the flesh."[19] Unlike measures of well-being that have a natural floor of zero, like war and disease, or a natural ceiling of a hundred percent, like nutrition and literacy, the quest for knowledge is unbounded. Not only does knowledge itself expand indefinitely, but the premium for knowledge in an economy that is driven by technology has been soaring.[20] While global rates of literacy and basic education are converging to their natural ceiling, the number of years of schooling, extending into tertiary and postgraduate education in colleges and universities, continues to grow in every country. In 1920, just 28 percent of American teenagers between fourteen and seventeen were in high school; by 1930, the proportion had grown to almost half, and by 2011, 80 percent graduated, of whom almost 70 percent went on to college.[21] In 1940, less than 5 percent of Americans held a bachelor's degree; by 2015, almost a third did.[22] Figure 16-3 shows the parallel trajectories of the length of schooling in a sample of countries, with recent highs ranging from four years in Sierra Leone to thirteen years (some college) in the United States. According to one projection, by the end of the century more than 90 percent of the world's population will have some secondary education, and 40 percent some college.[23] Since educated people tend to have fewer children, the growth of education is a major reason that, later in this century, world population is expected to peak and then decline (figure 10-1).

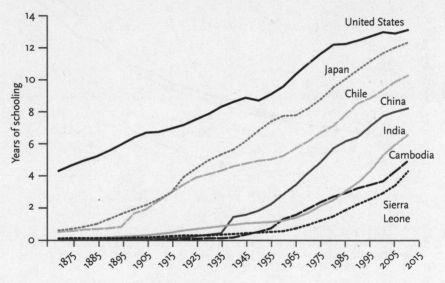

Figure 16-3: Years of schooling, 1870–2010

Source: *Our World in Data,* Roser & Ortiz-Ospina 2016a, based on data from Lee & Lee 2016. Data are for the population aged 15–64.

Though we see little or no global convergence in the length of formal schooling, an ongoing revolution in the dissemination of knowledge makes the gap less relevant. Most of the world's knowledge is now online rather than locked in libraries (much of it free), and massive open online courses (MOOCs) and other forms of distance learning are becoming available to anyone with a smartphone.

Other disparities in education are shrinking as well. In the United States, measures of school readiness among low-income, Hispanic, and African American children increased substantially between 1998 and 2010, possibly because free preschool programs are more widely available, and because poor families today have more books, computers, and Internet access and the parents spend more time interacting with their children.[24]

Even more consequentially, the ultimate form of sex discrimination—keeping girls out of school—is in decline. The change is consequential not just because women make up half the population, so educating them doubles the size of the skill pool, but because the hand that rocks the cradle rules the world. When girls are educated, they are healthier, have fewer and healthier children, and are more productive—and so are their countries.[25] It took the West centuries to figure out that educating the whole population, not just the half with testicles, was a good idea: the line for England in figure 16-4 shows that Englishwomen did not become/ as literate as Englishmen until 1885. The world as a whole caught on even later but quickly made up for lost time, going from teaching only two-thirds as many girls as boys to read in 1975 to teaching them in equal numbers in 2014. The United Nations has announced that the world has

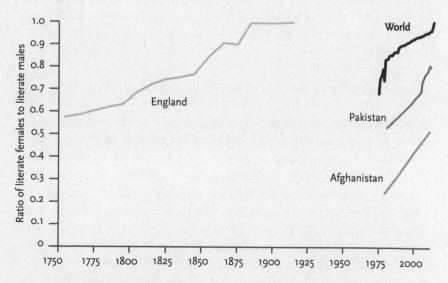

Figure 16-4: Female literacy, 1750–2014

Sources: England (all adults): Clark 2007, p. 179. World, Pakistan, & Afghanistan (ages 15–24): Human-Progress, http://www.humanprogress.org/f1/2101, based on data from UNESCO Institute for Statistics, summarized in World Bank 2016f. Data for the world are averaged over slightly different sets of countries in different years.

met the 2015 Millennium Development Goal of achieving gender parity in primary, secondary, and tertiary education.[26]

The other two lines tell their own story. The country with the worst gender ratio for literacy is Afghanistan. Not only is Afghanistan near the bottom in almost every measure of human development (including its overall literacy rate, which in 2011 stood at an abysmal .52), but from 1996 to 2001 it was under the control of the Taliban, the Islamic fundamentalist movement that, among other atrocities, forbade girls and women from attending school. The Taliban has continued to intimidate girls from getting an education in the regions of Afghanistan and neighboring Pakistan it controls. Starting in 2009 the twelve-year-old Malala Yousafzai, whose family ran a chain of schools in the Swat district of Pakistan, publicly spoke out for girls' right to an education. On a day that will live in infamy, October 9, 2012, a Taliban gunman boarded her school bus and shot her in the head. She survived to become the youngest winner of the Nobel Peace Prize and one of the world's most admired women. Yet even in these benighted parts of the world, progress can be seen.[27] In the past three decades the literacy gender ratio has doubled in Afghanistan and increased by half in Pakistan, whose ratio now matches that for the world in 1980 and for England in 1850. Nothing is certain, but the global tide of activism, economic development, common sense, and common decency are likely to push the ratio to its natural ceiling.

~

Could the world be getting not just more literate and knowledgeable but actually smarter? Might people be increasingly adept at learning new skills, grasping abstract ideas, and solving unforeseen problems? Amazingly, the answer is yes. Intelligence Quotient (IQ) scores have been rising for more than a century, in every part of the world, at a rate of about three IQ points (a fifth of a standard deviation) per decade. When the philosopher James Flynn first brought this phenomenon to psychologists' attention in 1984, many thought it must have been a mistake or trick.[28] For one thing, we know that intelligence is highly heritable, and the world has not engaged in a massive eugenics project in which smarter people have had more babies generation after generation.[29] Nor have people been marrying outside their clan and tribe (thus avoiding inbreeding and increasing hybrid vigor) in great enough numbers for a long enough time to explain the rise.[30] Also, it beggars belief to think that an average person of 1910, if he or she had entered a time machine and materialized today, would be borderline retarded by our standards, while if Joe and Jane Average made the reverse journey, they would outsmart 98 percent of the befrocked and bewhiskered Edwardians who greeted them as they emerged. Yet surprising as it is, the Flynn effect is no longer in doubt, and it has recently been confirmed in a meta-analysis of 271 samples from thirty-one countries with four million people.[31] Figure 16-5 plots the "secular rise in IQ scores," as psychologists call it (*secular* in the sense of long-term rather than irreligious).

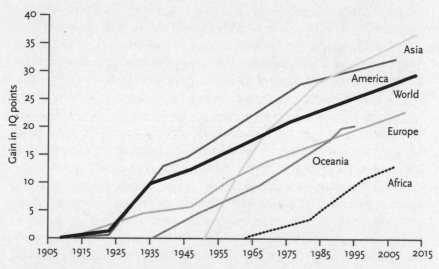

Figure 16-5: IQ gains, 1909–2013

Source: Pietschnig & Voracek 2015, supplemental online material. The lines display changes in IQ measured by different tests starting at different times and cannot be compared with one another.

Note that each line plots the *change* in IQ scores in a continent relative to the average score in the earliest year for which data are available, which is arbitrarily set to 0 because the tests and periods for the different continents are not directly commensurable. We cannot read the graph as we did the previous ones and infer, for example, that the IQ of Africa in 2007 is equivalent to the IQ of Australia and New Zealand in 1970. Not surprisingly, the rise in IQ scores obeys Stein's Law: Things that can't go on forever don't. The Flynn effect is now petering out in some of the countries in which it has been going on the longest.[32]

Though it's not easy to pinpoint the causes of the rise in IQ scores, it's no paradox that a heritable trait can be boosted by changes in the environment. That's what happened with height, a trait that also is highly heritable and has increased over the decades, and for some of the same reasons: better nutrition and less disease. Brains are greedy organs, consuming about a fifth of the body's energy, and they are made of fats and proteins that are demanding for the body to produce. Fighting off infections is metabolically expensive, and the immune system of a sick child may commandeer resources that would otherwise go to brain development. Also helping with brain development is a cleaner environment, with lower levels of lead and other toxins. Food, health, and environmental quality are among the perquisites of a richer society, and not surprisingly, the Flynn effect is correlated with increases in GDP per capita.[33]

But nutrition and health can explain only a part of the Flynn effect.[34] For one thing, their benefits should be concentrated in pulling up the lower half of the bell curve of IQ scores, populated by the duller people who had been held back by poor food and health. (After all, past a certain point, additional food makes people fatter, not smarter.) Indeed, in some times and places the Flynn effect *is* concentrated in the lower half, bringing the duller closer to the average. But in other times and places the entire curve crept rightward: the smart got smarter too, even though they started out healthy and well-fed. Second, improvements in health and nutrition should affect children most of all, and then the adults they grow into. But the Flynn effect is stronger for adults than for children, suggesting that experiences on the way to adulthood, not just biological constitution in early childhood, have pushed IQ scores higher. (The most obvious of these experiences is education.) Also, while IQ has risen over the decades, and nutrition, health, and height have risen over the decades, their various ascents and plateaus don't track each other particularly closely.

But the main reason that health and nutrition aren't enough to explain the IQ rise is that what has risen over time is not overall brainpower. The Flynn effect is not an increase in g, the general intelligence factor that underlies every subtype of intelligence (verbal, spatial, mathematical, memory, and so on) and is the aspect of intelligence most directly affected by the genes.[35] While overall IQ has risen, and scores on each intelligence subtest have risen, some subtest scores have risen more rapidly than others in a pattern different from the pattern linked to the genes. That's another reason the Flynn effect does not cast doubt on the high heritability of IQ.

So which kinds of intellectual performance have been pushed upward by the better environments of recent decades? Surprisingly, the steepest gains have not been found in the concrete skills that are directly taught in school, such as general knowledge, arithmetic, and vocabulary. They have been found in the abstract, fluid kinds of intelligence, the ones tapped by similarity questions ("What do an hour and a year have in common?"), analogies ("BIRD is to EGG as TREE is to what?"), and visual matrices (where the test-taker has to choose a complex geometric figure that fits into a rule-governed sequence). What has increased the most, then, is an analytic mindset: putting concepts into abstract categories (an hour and a year are "units of time"), mentally dissecting objects into their parts and relationships rather than absorbing them as wholes, and placing oneself in a hypothetical world defined by certain rules and exploring its logical implications while setting aside everyday experience ("Suppose that in Country X everything is made of plastic. Are the ovens made of plastic?").[36] An analytic mindset is inculcated by formal schooling, even if a teacher never singles it out in a lesson, as long as the curriculum requires understanding and reasoning rather than rote memorization (and that has been the trend in education since the early decades of the 20th century).[37] Outside the schoolhouse, analytic thinking is encouraged by a culture that trades in visual symbols (subway maps, digital displays), analytic tools (spreadsheets, stock reports), and academic concepts that trickle down into common parlance (*supply and demand, on average, human rights, win-win, correlation versus causation, false positive*).

Does the Flynn effect matter in the real world? Almost certainly. A high IQ is not just a number that you can brag about in a bar or that gets you into Mensa; it is a tailwind in life.[38] People with high scores on intelligence tests get better jobs, perform better in their jobs, enjoy better health and longer lives, are less likely to get into trouble with the law, and have a greater number of noteworthy accomplishments like starting companies, earning patents, and creating respected works of art—all holding socioeconomic status constant. (The myth, still popular among leftist intellectuals, that IQ doesn't exist or cannot be reliably measured was refuted decades ago.) We don't know whether these bonuses come from g alone or also from the Flynn component of intelligence, but the answer is probably both. Flynn has speculated, and I agree, that abstract reasoning can even hone the moral sense. The cognitive act of extricating oneself from the particulars of one's life and pondering "There but for fortune go I" or "What would the world be like if everyone did this?" can be a gateway to compassion and ethics.[39]

Since intelligence brings good things, and intelligence has been increasing, can we see a dividend from increasing intelligence in improvements to the world? Some skeptics (including, at the outset, Flynn himself) doubted whether the 20th century really produced more brilliant ideas than the ages of Hume, Goethe, and Darwin.[40] Then again, the geniuses of the past had the advantage of exploring virgin territory. Once someone discovers the analytic-synthetic distinction or the theory of natural selection, no one can ever discover it again. Today the intellectual landscape is well trodden, and it's harder for a solitary genius to tower above the crowd of hypereducated and networked thinkers who are mapping every nook and cranny. Still, there have been some signs of a smarter populace, such as the fact that the world's top-ranked chess and bridge players have been getting younger. And no one can second-guess the warp speed of advances in science and technology of the past half-century.

Most dramatically, an increase in one kind of abstract intelligence is visible all over the world: mastery of digital technology. Cyberspace is the ultimate abstract realm, in which goals are achieved not by pushing matter around in space but by manipulating intangible symbols and patterns. When people were first confronted with digital interfaces in the 1970s, like videocassette recorders and ticket machines in new subway systems, they were baffled. It was a running joke of the 1980s that most VCRs eternally flashed "12:00" at owners who couldn't figure out how to set the time. But Generation X and the Millennials have famously thrived in the digital realm. (In one cartoon of the new millennium, a father says to his young boy, "Son, your mother and I have bought software to control what you see on the Internet. Um . . . Could you install it for us?") The developing world has thrived in that realm as well, often leapfrogging the West in its adoption of smartphones and of applications for them such as mobile banking, education, and real-time market updates.[41]

Could the Flynn effect help explain the other rises in well-being we have seen in these chapters? An analysis by the economist R. W. Hafer suggests it could. Holding all the usual confounding variables constant—education, GDP, government spending, even a country's religious makeup and its history of colonization—he found that a country's average IQ predicted its subsequent growth in GDP per capita, together with growth in noneconomic measures of well-being like longevity and leisure time. An 11-point increase in IQ, he estimated, would accelerate a country's growth rate enough to double well-being in just nineteen years rather than twenty-seven. Policies that hurry the Flynn effect along, namely investments in health, nutrition, and education, could make a country richer, better governed, and happier down the road.[42]

~

What's good for humanity is not always good for social science, and it may be impossible to unsnarl the bundle of correlations among all the ways that life has improved and trace the causal arrows with certainty. But let's stop fretting for a moment about how hard it is to disentangle the strands and instead take note of their common direction. The very fact that so many dimensions of well-being are correlated across countries and decades suggests there may be a coherent phenomenon lurking beneath them—what statisticians call a general factor, a principal component, or a hidden, latent, or intervening variable.[43] We even have a name for that factor: progress.

No one has calculated this vector of progress underlying all the dimensions of human flourishing, but the United Nations Development Programme, inspired by the economists Mahbub ul Haq and Amartya Sen, offers a Human Development Index that is a composite of three of the major ones: life expectancy, GDP per capita, and education (being healthy, wealthy, and wise).[44] With this chapter we have now examined all of these goods, and it's an appropriate point to step back and take in the history of quantifiable human progress before we turn to its more qualitative aspects in the next two chapters.

Two economists have developed their own versions of a human development index that can be estimated retroactively into the 19th century, each of which aggregates measures of longevity, income, and education in different ways. Leandro Prados de la Escosura's Historical Index of Human Development, which goes back to 1870, averages the three measures with a geometric rather than an arithmetic mean (so that an extreme value on one measure cannot swamp the other two), and transforms the longevity and education measures to compensate for diminishing returns at their high end. Auke Rijpma of the "How Was Life?" project (whose data have appeared in a number of graphs in this book) developed a Well-Being Composite that goes back to 1820; together with the big three, it throws in measures of height (a proxy for health), democracy, homicide, income inequality, and biodiversity. (The latter two are the only ones that don't systematically improve over the

past two centuries.) The grades for the world on these two report cards
are shown in figure 16-6.

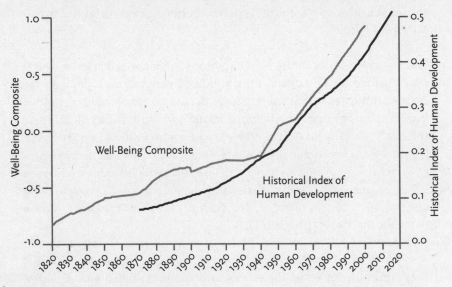

Figure 16-6: Global well-being, 1820–2015

Sources: Historical Index of Human Development: Prados de la Escosura 2015, 0–1 scale, available at
Our World in Data, Roser 2016h. **Well-Being Composite:** Rijpma 2014, p. 259, standard deviation scale
over country-decades.

To behold this graph is to apprehend human progress at a glance.
And packed into the lines are two vital subplots. One is that although the
world remains highly unequal, every region has been improving, and
the worst-off parts of the world today are better off than the best-off parts
not long ago.[45] (If we divide the world into the West and the Rest, we find
that the Rest in 2007 had reached the level of the West in 1950.) The other
is that while almost every indicator of human well-being correlates with
wealth, the lines don't just reflect a wealthier world: longevity, health,
and knowledge have increased even in many of the times and places
where wealth has not.[46] The fact that all aspects of human flourishing
tend to improve over the long run even when they are not in perfect sync
vindicates the idea that there is such a thing as progress.

CHAPTER 17

QUALITY OF LIFE

Though only the callous would deny that the conquests of disease, hunger, and illiteracy are stupendous achievements, one can still wonder whether continuous improvements in the kinds of things that economists measure should count as genuine progress. Once basic needs are satisfied, doesn't additional affluence just encourage people to indulge in shallow consumerism? And weren't increases in health and literacy trumpeted by the Five-Year Planners in the Soviet Union, China, and Cuba, all of which were rather grim places to live? People can be healthy, solvent, and literate and still not lead rich and meaningful lives.

Some of these reservations have already been answered. We've seen that totalitarianism, the main impediment to the good life in communist so-called utopias, has been receding. We've also seen that a major dimension of flourishing that is not captured by the standard metrics—the rights of women, children, and minorities—is on a steady rise. This chapter is about a broader cultural pessimism: the worry that all that extra healthy life span and income may not have increased human flourishing after all if they just consign people to a rat race of frenzied careerism, hollow consumption, mindless entertainment, and soul-deadening anomie.

To be sure, one can object to the objection, which comes from a long tradition of cultural and religious elites sneering at the supposedly empty lives of the bourgeoisie and proletariat. Cultural criticism can be a thinly disguised snobbery that shades into misanthropy. In *The Intellectuals and the Masses*, the critic John Carey shows how the British literary intelligentsia in the first decades of the 20th century harbored a contempt for the common person which bordered on the genocidal.[1] In practice, "consumerism" often means "consumption by the other guy," since the elites who condemn it tend themselves to be conspicuous consumers of exorbitant luxuries like hardcover books, good food and wine, live artistic performances, overseas travel, and Ivy-class education for their children. If more people can afford *their* preferred luxuries, even if they are frivolous by the lights of their cultural betters, that has to be counted as a good thing. In an old joke, a soapbox orator addresses a crowd on the glories of communism: "Come the revolution, everyone will eat strawberries and cream!" A man at the front whimpers, "But I don't like strawberries and cream." The speaker thunders, "Come the revolution, you *will* like strawberries and cream!"[2]

In *Development as Freedom*, Amartya Sen sidesteps this trap by proposing that the ultimate goal of development is to enable people to make choices: strawberries and cream for those who want them. The philosopher Martha Nussbaum has taken the idea a step further and laid out a set of "fundamental capabilities" that all people should be given the opportunity to exercise.[3] One can think of them as the justifiable sources of satisfaction and fulfillment that human nature makes available to us. Her list begins with capabilities that, as we have seen, the modern world increasingly allows people to realize: longevity, health, safety, literacy, knowledge, free expression, and political participation. It goes on to include aesthetic experience, recreation and play, enjoyment of nature, emotional attachments, social affiliations, and opportunities to reflect on and engage in one's own conception of the good life.

In this chapter I'll show how modernity is increasingly allowing people to exercise these capabilities, too—that life is getting better even beyond the standard economists' metrics like longevity and wealth. Admittedly, many people still don't like strawberries and cream, and they may exercise one capability—enjoying their freedom to watch television and play video games—to forgo others, such as aesthetic appreciation and enjoyment of nature. (When Dorothy Parker was challenged to use the word *horticulture* in a sentence, she answered, "You can lead a horticulture, but you can't make her think.") But an expansive cafeteria of opportunities to enjoy the aesthetic, intellectual, social, cultural, and natural delights of the world, regardless of which ones people put on their trays, is the ultimate form of progress.

～

Time is what life is made of, and one metric of progress is a reduction in the time people must devote to keeping themselves alive at the expense of the other, more enjoyable things in life. "In the sweat of thy face shalt thou eat bread," said the ever-merciful God as he exiled Adam and Eve from Eden, and for most people throughout history, sweat they did.

Farming is a sunup-to-sundown occupation, and though foragers hunt and gather just a few hours a day, they spend many more hours processing the food (for example, smashing rock-hard nuts), in addition to gathering firewood, carrying water, and laboring at other chores. The San of the Kalahari, once called "the original affluent society," turn out to work at least eight hours a day, six to seven days a week, on food alone.[4]

The 60-hour workweek of Bob Cratchit, with only one day off a year (Christmas, of course), was in fact lenient by the standards of his era. Figure 17-1 shows that in 1870 Western Europeans worked an average of 66 hours a week (the Belgians worked 72), while Americans worked 62 hours. Over the past century and a half, workers have increasingly been emancipated from their wage slavery, more dramatically in social-democratic Western Europe (where they now work 28 fewer hours a week) than in the go-getter United States (where they work 22 fewer hours).[5] As late as the 1950s, my paternal grandfather worked behind the cheese counter in an unheated Montreal market day and night, seven days a week, afraid to ask for shorter hours lest he be replaced. When my young parents protested on his behalf, he was given sporadic days off (which the owner no doubt perceived, like Scrooge, as "a poor excuse for picking a man's pocket"), until better labor-law enforcement gave him a predictable six-day workweek.

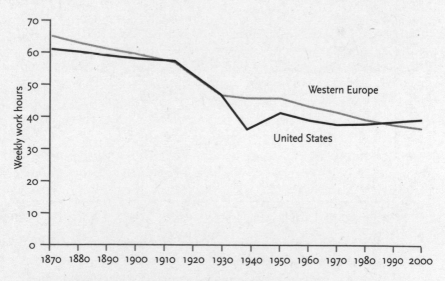

Figure 17-1: Work hours, Western Europe and US, 1870–2000

Source: Roser 2016t, based on data from Huberman & Minns 2007 on full-time production workers (both sexes) in nonagricultural activities.

Though a lucky few of us are paid to exercise our fundamental capabilities and willingly put in Victorian hours, most workers are grateful for the two dozen extra hours a week they have available to fulfill themselves in other ways. (On his hard-won day off, my grandfather would read the Yiddish papers, dress up in a jacket, tie, and fedora, and visit his sisters or my family.)

Likewise, though many of my fellow professors end their careers carried out of their offices feet first, workers in many other jobs are happy to spend their golden years reading, taking courses, seeing the national parks in a Winnebago, or dandling Vera, Chuck, and Dave in a cottage on the Isle of Wight. This, too, is a gift of modernity. As Morgan Housel notes, "We constantly worry about the looming 'retirement funding crisis' in America without realizing that the entire concept of retirement is unique to the last five decades. It wasn't long ago that the average American man had two stages of life: work and death. . . . Think of it this way: The average American now retires at age 62. One hundred years ago, the average American died at age 51."[6] Figure 17-2 shows that in 1880, almost 80 percent of American men of what we now consider retirement age were still in the workforce, and that by 1990 the proportion had fallen to less than 20 percent.

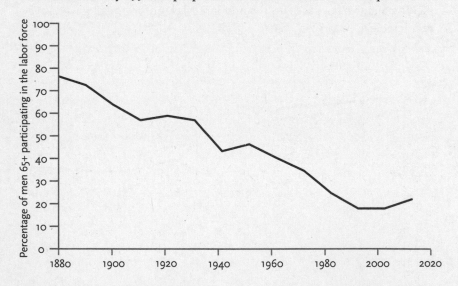

Figure 17-2: Retirement, US, 1880–2010

Source: Housel 2013, based on data from the Bureau of Labor Statistics, and Costa 1998.

Rather than looking forward to retirement, people used to dread the injury or frailty that would keep them from work and send them to the almshouse—the "haunting fear in the winter of life," as it was known.[7] Even after the Social Security Act of 1935 protected the elderly from utter destitution, poverty was a common end to a working life, and I grew up with the image (possibly an urban legend) of pensioners who subsisted on dog food. But with stronger public and private safety nets in place, senior citizens today are richer than people of working age: the poverty rate for people over 65 plunged from 35 percent in 1960 to less than 10 percent in 2011, well below the national rate of 15 percent.[8]

Thanks to the labor movement, legislation, and increased worker productivity, another once-crazy pipe dream has become a reality: paid vacations. Today an average American worker with five years on the job receives 22 days of paid time off a year (compared with 16 days in 1970), and that is miserly by the standards of Western Europe.[9] The combination of a shorter workweek, more paid time off, and a longer retirement means that the fraction of a person's life that is taken up by work has fallen by a quarter just since 1960.[10] Trends for the developing world vary by country, but as these countries get richer they are likely to follow those of the West.[11]

There is yet another way in which thick tranches of life have been freed up for people to pursue higher callings. In chapter 9 we saw that appliances such as refrigerators, vacuum cleaners, washing machines, and microwave ovens have become common or universal, even among the American poor. In 1919, an average American wage earner had to work 1,800 hours to pay for a refrigerator; in 2014, he or she had to work fewer than 24 hours (and the new fridge was frost-free and came with an icemaker).[12] Mindless consumerism? Not when you remember that food, clothing, and shelter are the three necessities of life, that entropy degrades all three, and that the time it takes to keep them usable is time that could be devoted to other pursuits. Electricity, running water, and appliances (or as they used to be called, "labor-saving devices") give us that time back—the many hours our grandmothers spent pumping, canning, churning, pickling, curing, sweeping, waxing, scrubbing, wringing, sudsing, drying, stitching, mending, knitting, darning, and, as they used to remind us, "slaving over a hot stove, working our fingers to the bone." Figure 17-3 shows that as utilities and appliances penetrated American households during the 20th century, the amount of life that people lost to housework—which, not surprisingly, people say is their least favorite way to spend their time—fell almost fourfold, from 58 hours a week in 1900 to 15.5 hours in 2011.[13] Time spent on laundry alone fell from 11.5 hours a week in 1920 to 1.5 in 2014.[14] For returning "washday" to our lives, Hans

Rosling suggests, the washing machine deserves to be called the greatest invention of the Industrial Revolution.[15]

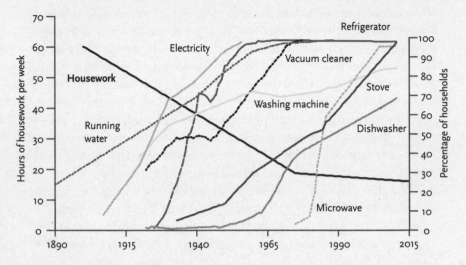

Figure 17-3: Utilities, appliances, and housework, US, 1900–2015

Sources: Before 2005: Greenwood, Seshadri, & Yorukoglu 2005. Appliances, 2005 and 2011: US Census Bureau, Siebens 2013. Housework, 2015: *Our World in Data*, Roser 2016t, based on the American Time Use Survey, Bureau of Labor Statistics 2016b.

As a feminist-era husband I can truthfully use the first-person plural in celebrating this gain. But in most times and places housework is gendered, so the liberation of humankind from household labor is in practice the liberation of *women* from household labor. Perhaps the liberation of women in general. Arguments for the equality of women go back to Mary Astell's 1700 treatise and are irrefutable, so why did they take centuries to catch on? In a 1912 interview in *Good Housekeeping* magazine, Thomas Edison prophesied one of the great social transformations of the 20th century:

> The housewife of the future will be neither a slave to servants nor herself a drudge. She will give less attention to the home, because the home will need less; she will be rather a domestic engineer than a domestic laborer, with the greatest of all handmaidens, electricity, at her service. This and other mechanical forces will so revolutionize the woman's world that a large portion of the aggregate of woman's energy will be conserved for use in broader, more constructive fields.[16]

Time is not the only life-enriching resource granted to us by technology. Another is light. Light is so empowering that it serves as the metaphor of choice for a superior intellectual and spiritual state: *enlightenment*. In the natural world we are plunged into darkness for half of our existence, but human-made light allows us to take back the night for reading, moving about, seeing people's faces, and otherwise engaging with our surroundings. The economist William Nordhaus has cited the plunging price (and hence the soaring availability) of this universally treasured resource as an emblem of progress. Figure 17-4 shows that the inflation-adjusted price of a million lumen-hours of light (about what you would need to read for two and a half hours a day for a year) has fallen *twelve thousandfold* since the Middle Ages (once called the Dark Ages), from around £35,500 in 1300 to less than £3 today. These days (and nights), if you aren't reading, conversing, getting out, or otherwise edifying yourself, it's not because you can't afford the light.

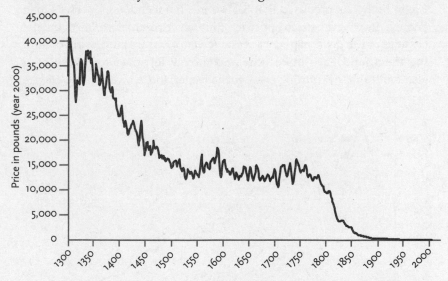

Figure 17-4: Cost of light, England, 1300–2006

Source: *Our World in Data*, Roser 2016d, based on data from Fouquet & Pearson 2012. Cost of one million lumen-hours (about 833 hours from an 80-watt incandescent bulb), in pounds sterling (inflation-adjusted to the year 2000).

The plunging cash value of artificial light actually understates the progress, because, as Adam Smith pointed out, "The real price of every thing . . . is the toil and trouble of acquiring it."[17] Nordhaus estimated how many hours a person would have to work to earn an hour of light to read by at different times in history.[18] A Babylonian in 1750 BCE would have had to labor fifty hours to spend one hour reading his cuneiform tablets by a sesame-oil lamp. In 1800, an Englishman had to toil for six hours to burn a tallow candle for an hour. (Imagine planning your family budget around that—you might settle for darkness.) In 1880, you'd need to work fifteen minutes to burn a kerosene lamp for an hour; in 1950, eight seconds for the same hour from an incandescent bulb; and in 1994, a *half-second* for the same hour from a compact fluorescent bulb—a 43,000-fold leap in affordability in two centuries. And the progress wasn't finished: Nordhaus published his article before LED bulbs flooded the market. Soon, cheap, solar-powered LED lamps will transform the lives of the more than one billion people without access to electricity, allowing them to read the news or do their homework without huddling around an oil drum filled with burning garbage.

The declining proportion of our lives we have to forfeit for light, appliances, and food may be part of a general law. The technology expert Kevin Kelly has proposed that "over time, if a technology persists long enough, its costs begin to approach (but never reach) zero."[19] As the necessities of life get cheaper, we waste fewer of our waking hours obtaining them, and have more time and money left over for everything else—and the "everything else" gets cheaper, too, so we can experience

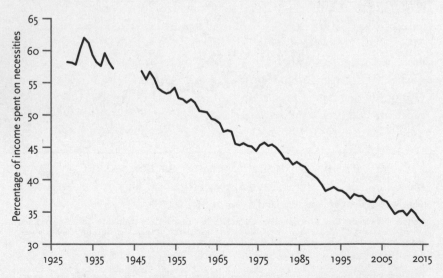

Figure 17-5: Spending on necessities, US, 1929–2016

Source: *HumanProgress*, http://humanprogress.org/static/1937, adapted from a graph by Mark Perry, using data from the Bureau of Economic Analysis, https://www.bea.gov/iTable/index_nipa.cfm. Proportion of disposable income spent on food at home, cars, clothing, household furnishings, housing, utilities, and gasoline. Data from 1941 to 1946 are omitted because they are distorted by rationing and soldiers' salaries during World War II.

more of them. Figure 17-5 shows that in 1929 Americans spent more than 60 percent of their disposable income on necessities; by 2016 that had fallen to a third.

What are people doing with that extra time and money? Are they truly enriching their lives, or are they just buying more golf clubs and designer handbags? Though it's presumptuous to pass judgment on how people choose to spend their days, we can focus on the pursuits that almost everyone would agree are constituents of a good life: connecting with loved ones and friends, experiencing the richness of the natural and cultural worlds, and having access to the fruits of intellectual and artistic creativity.

With the rise of two-career couples, overscheduled kids, and digital devices, there is a widespread belief (and recurring media panic) that families are caught in a time crunch that's killing the family dinner. (Both Al Gore and Dan Quayle lamented its demise in the run-up to the 2000 presidential election—and that was before smartphones and social media.) But the new tugs and distractions have to be weighed against the 24 extra hours that modernity has granted to breadwinners every week and the 42 extra hours it has granted to homemakers. Though people increasingly complain about how crazy-busy they are ("yuppie kvetching," as one team of economists put it), a different picture emerges when they are asked to keep track of their time. In 2015, men reported 42 hours of leisure per week, around 10 more than their counterparts did fifty years earlier, and women reported 36 hours, more than 6 hours more (figure 17-6).[20] (To be fair, the yuppies might have something to kvetch about: less-educated people reported having more leisure, and this inequality-in-reverse has grown over these fifty years.) Similar trends have been reported in Western Europe.[21]

Nor are Americans consistently feeling more harried. A review by the sociologist John Robinson shows some ups and downs between 1965 and 2010 in the percentage who say they feel "always rushed" (with a low of 18 percent in 1976 and a high of 35 percent in 1998), but no consistent trend over forty-five years.[22] And at the end of the day, the family dinner is alive and well. Several studies and polls agree that the number of dinners families have together changed little from 1960 through 2014, despite the iPhones, PlayStations, and Facebook accounts.[23] Indeed, over the course of the 20th century, typical American parents spent more time, not less, with their children.[24] In 1924, only 45 percent of mothers spent two or more hours a day with their children

(7 percent spent *no* time with them), and only 60 percent of fathers spent at least an hour a day with them. By 1999, the proportions had risen to 71 and 83 percent.[25] In fact, single and working mothers today spend more time with their children than stay-at-home married mothers did in 1965.[26] (An increase in hours spent caring for children is the main reason for the dip in leisure time visible in figure 17-6.)[27] But time-use studies are no match for Norman Rockwell and *Leave It to Beaver*, and many people misremember the mid-20th century as a golden age of family togetherness.

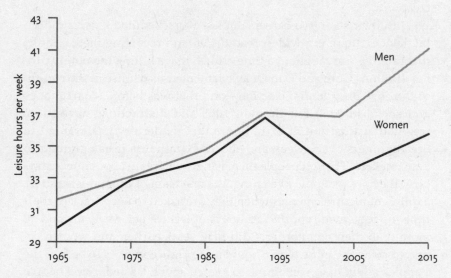

Figure 17-6: Leisure time, US, 1965–2015

Sources: 1965–2003: Aguiar & Hurst 2007, table III, Leisure Measure 1. 2015: American Time Use Survey, Bureau of Labor Statistics 2016c, summing Leisure and Sports, Lawn and Garden Care, and Volunteering for commensurability with Aguiar & Hurst's Measure 1.

Electronic media are commonly cited as a threat to human relationships, and certainly Facebook friends are a poor substitute for face-to-face contact with flesh-and-blood companions.[28] Yet overall, electronic technology has been a priceless gift to human closeness. A century ago, if family members moved to a distant city, one might never hear their voices or see their faces again. Grandchildren grew up without their grandparents laying eyes on them. Couples separated by study, work, or war would reread a letter dozens of times and tumble into despair if the next one was late, not knowing whether the postal service had lost it or whether the lover was angry, faithless, or dead (an agony re-counted in songs like the Marvelettes' and Beatles' "Please Mr. Postman" and Simon & Garfunkel's "Why Don't You Write Me?"). Even when long-distance telephony allowed people to reach out and touch someone, the exorbitant cost put a strain on intimacy. People of my generation remember the awkwardness of speed-talking on a pay phone while feeding it quarters between bongs, or the breakneck sprint when called to the family phone ("IT'S LONG DISTANCE!!!"), or the sinking feeling of the rent money evaporating as a pleasant conversation unfolded. "Only connect," advised E. M. Forster, and electronic technology is allowing us to connect as never before. Today, almost half of the world's population has Internet access, and three-quarters have access to a mobile phone. The marginal cost of a long-distance conversation is essentially zero, and the conversants can now see as well as hear each other.

And speaking of seeing, the plunging cost of photography is another gift to the richness of experience. In past eras people had only a mental image to remind them of a family member, living or dead. Today, like billions of others, I get a wave of gratitude for my blessings several times a day as my eyes alight on a photo of my loved ones. Affordable photography also allows the high points of life to be lived many times, not just once: the precious occasions, the stunning sights, the long-gone cityscapes, the elderly in their prime, the grown-ups as children, the children as babies.

Even in the future, when we have 3-D holographic surround-sound virtual reality with haptic exoskeleton gloves, we will still want to be within touching distance of the people we love, so the shrinking cost of transportation is another boon to humanity. Trains, buses, and cars have multiplied the opportunities for us to get together, and the remarkable democratization of plane travel has removed the barriers of distance and oceans. The term *jet set* for chic celebrities is an anachronism from the 1960s, when no more than a fifth of Americans had ever flown in a plane. Despite soaring fuel costs, the real price of plane travel in the United States has fallen by more than half since the late 1970s, when the airlines were deregulated (figure 17-7). In 1974, it cost $1,442 (in 2011 dollars) to fly from New York to Los Angeles; today it can be done for less than $300. As prices fell, more people flew: in 2000 more than half of Americans took at least one round-trip flight. You might have to spread-eagle while a guard slides a wand up your crotch, you may have an elbow in your ribs and a seatback in your chin, but long-distance lovers get to see each other, and if your mother gets sick you can be there the next day.

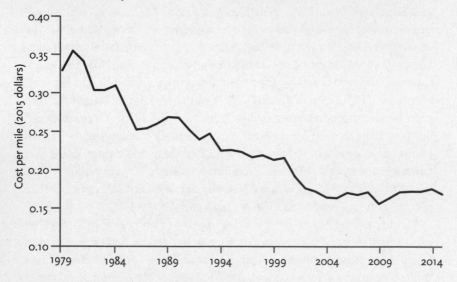

Figure 17-7: Cost of air travel, US, 1979–2015

Source: Thompson 2013, updated with data from Airlines for America, http://airlines.org/dataset/annual -round-trip-fares-and-fees-domestic/. Domestic travel, excluding checked baggage fees (which would raise the average cost for baggage-checking passengers by about a half-cent per mile since 2008).

Affordable transportation does more than reunite people. It also allows them to sample the phantasmagoria of Planet Earth. This is the pastime that we exalt as "travel" when we do it and revile as "tourism" when someone else does it, but it surely has to count as one of the things that make life worth living. To see the Grand Canyon, New York, the Aurora Borealis, Jerusalem—these are not just sensuous pleasures but experiences that widen the scope of our consciousness, allowing us to take in the vastness of space, time, nature, and human initiative. Though we bristle at the motor coaches and tour guides, the selfie-shooting throngs in their tacky shorts, we must concede that life is better when people can expand their awareness of our planet and species rather than being imprisoned within walking distance of their place of birth. With the rise of disposable income and the declining cost of plane travel, more people have been exploring the world, as we see in figure 17-8.

And no, the travelers aren't just lining up for wax museums and rides at Disney World. The number of areas in the world that are protected from development and economic exploitation exceeds 160,000 and increases daily. As we saw in figure 10-6, far more of the natural world is being set aside in nature preserves.

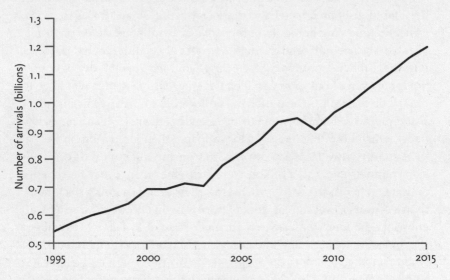

Figure 17-8: International tourism, 1995–2015

Source: World Bank 2016e, based on data from the World Tourism Organization, *Yearbook of Tourism Statistics.*

Another way in which the scope of our aesthetic experience has been magnified is food. The late 19th-century American diet consisted mainly of pork and starch.[29] Before refrigeration and motorized transport, most fruits and vegetables would have spoiled before they reached a consumer, so farmers grew nonperishables like turnips, beans, and potatoes. Apples were the only fruit, most of which went into cider. (As recently as the 1970s, Florida souvenir shops sold bags of oranges for tourists to take home as gifts.) The American diet was called "white bread" and "meat-and-potatoes" for good reason. Adventurous cooks might whip up some Spam fritters, mock apple pie made from Ritz crackers, or "Perfection Salad" (coleslaw in lemon Jell-O). New cuisines introduced by immigrants were so exotic that they became the butt of jokes, including Italian ("Mamma mia, that's a spicy meatball!"), Mexican ("Solves the gas shortage"), Chinese ("An hour later and you're hungry again"), and Japanese ("Bait, not food"). Today, even small towns and shopping mall food courts offer a cosmopolitan menu, sometimes with all these cuisines plus Greek, Thai, Indian, Vietnamese, and Middle Eastern. Grocers have broadened their offerings as well, from a few hundred items in the 1920s to 2,200 in the 1950s, 17,500 in the 1980s, and 39,500 in 2015.[30]

Last but not least, access to the finest products of the human mind has been fabulously broadened and democratized. It's hard for us to reconstruct the gnawing boredom of the isolated rural households of yesteryear.[31] In the late 19th century there was not only no Internet but no radio, television, movies, or musical recordings, and for the majority of households not even a book or newspaper. For entertainment, men would go to the saloon to drink.[32] The writer and editor William Dean Howells (1837–1920) entertained himself as a boy by rereading the pages of an old newspaper which his father had used to wallpaper their Ohio cabin.

A country-dweller today can choose from among hundreds of television channels and half a billion Web sites, embracing every newspaper and magazine in the world (including their archives going back more than a century), every great work of literature that is out of copyright, an encyclopedia more than seventy times the size of *Britannica* with about the same level of accuracy, and every classic work of art and music.[33] He could fact-check rumors on *Snopes,* teach himself math and science at *Khan Academy,* build his word power with the *American Heritage Dictionary,* enlighten himself with the *Stanford Encyclopedia of Philosophy,* and watch lectures by the world's great scholars, writers, and critics, many long dead. Today an impoverished Hillel would not have to pass out from cold while eavesdropping on lessons through the skylight of a schoolhouse.

Even for wealthy Western urbanites, who always had the run of the palaces of culture, access to arts and letters has expanded tremendously. When I was a student, a movie buff had to wait years for a classic film to be shown at a local repertory theater or on late-night television, if it was shown at all; today it can be streamed on demand. I can listen to any of thousands of songs while jogging, washing the dishes, or waiting in line at the Registry of Motor Vehicles. With a few keystrokes, I could lose myself in the complete works of Caravaggio, the original trailer for *Rashomon,* Dylan Thomas reciting "And Death Shall Have No Dominion," Eleanor Roosevelt reading aloud the Universal Declaration of Human Rights, Maria Callas singing "O mio babbino caro," Billie Holiday singing "My Man Don't Love Me," and Solomon Linda singing "Mbube"— experiences I could not have had for love or money just a few years ago. Cheap hi-fi headphones and, soon, cardboard virtual-reality glasses enhance the aesthetic experience well beyond the tinny speakers and muddy black-and-white reproductions of my youth. And those who like paper can buy a used copy of Doris Lessing's *The Golden Notebook,* Vladimir Nabokov's *Pale Fire,* or Wole Soyinka's *Aké: The Years of Childhood* for a dollar apiece.

A combination of Internet technology and crowdsourcing from thousands of volunteers has led to flabbergasting access to the great works of humankind. There can be no question of which was the greatest era for culture; the answer has to be today, until it is superseded by tomorrow. The answer does not depend on invidious comparisons of the quality of the works of today and those of the past (which we are in no position to make, just as many of the great works of the past were not appreciated in their time). It follows from our ceaseless creativity and our fantastically cumulative cultural memory. We have, at our fingertips, virtually all the works of genius prior to our time, together with those of our own time, whereas the people who lived before our time had neither. Better still, the world's cultural patrimony is now available not just to the rich and well-located but to anyone who is connected to the vast web of knowledge, which means most of humanity and soon all of it.

CHAPTER 18

HAPPINESS

B ut are we any happier? If we have a shred of cosmic gratitude, we ought to be. An American in 2015, compared with his or her counterpart a half-century earlier, will live nine years longer, have had three more years of education, earn an additional $33,000 a year per family member (only a third of which, rather than half, will go to necessities), and have an additional eight hours a week of leisure. He or she can spend that leisure time reading on the Web, listening to music on a smartphone, streaming movies on high-definition TV, Skyping with friends and relatives, or dining on Thai food instead of Spam fritters.

But if popular impressions are a guide, today's Americans are not one and a half times happier (as they would be if happiness tracked income), or a third happier (if it tracked education), or even an eighth happier (if it tracked longevity). People seem to bitch, moan, whine, carp, and kvetch as much as ever, and the proportion of Americans who tell pollsters that they are happy has remained steady for decades. Popular culture has noticed the ingratitude in the Internet meme and Twitter hashtag #first worldproblems and in a monologue by the comedian Louis C.K. known as "Everything's Amazing and Nobody's Happy":

> When I read things like, "The foundations of capitalism are shattering," I'm like, maybe we need some time where we're walking around with a donkey with pots clanging on the sides. . . . 'Cause now we live in an amazing world, and it's wasted on the crappiest generation of spoiled idiots. . . . Flying is the worst one, because people come back from flights, and they tell you their story. . . . They're like, "It was the worst day of my life. . . . We get on the plane and they made us sit there on the runway for forty minutes." . . . Oh really, then what happened next? Did you *fly* through the *air*, incredibly, like a bird? Did you soar

into the clouds, impossibly? Did you partake in the *miracle* of human
flight, and then land softly on giant tires that you couldn't even conceive how they fuckin' put air in them? . . . You're sitting in a *chair* in
the *sky*. You're like a Greek myth right now! . . . People say there's
delays? . . . Air travel's too slow? New York to California in five hours.
That used to take thirty years! And a bunch of you would die on the
way there, and you'd get shot in the neck with an arrow, and the other
passengers would just bury you and put a stick there with your hat on
it and keep walking. . . . The Wright Brothers would kick us all in the
[crotch] if they knew.[1]

Writing in 1999, John Mueller summed up the common understanding of modernity at the time: "People seem simply to have taken the remarkable economic improvement in stride and have deftly found new
concerns to get upset about. In an important sense, then, things never get
better."[2] The understanding was based on more than just impressions of
American malaise. In 1973 the economist Richard Easterlin identified a
paradox that has since been named for him.[3] Though in comparisons
within a country richer people are happier, in comparisons *across* countries the richer ones appeared to be no happier than poorer ones. And in
comparisons *over time*, people did not appear to get happier as their
countries got richer.

The Easterlin paradox was explained with two theories from psychology. According to the theory of the hedonic treadmill, people adapt to
changes in their fortunes, like eyes adapting to light or darkness, and
quickly return to a genetically determined baseline.[4] According to the
theory of social comparison (or reference groups, status anxiety, or relative deprivation, which we examined in chapter 9), people's happiness is
determined by how well they think they are doing relative to their compatriots, so as the country as a whole gets richer, no one feels happier—
indeed, if their country becomes more unequal, then even if they get
richer they may feel worse.[5]

If, in this sense, things never get better, one can wonder whether all
that economic, medical, and technological so-called progress was worth
it. Many argue that it was not. We have been spiritually impoverished,
they say, by the rise of individualism, materialism, consumerism, and
decadent wealth, and by the erosion of traditional communities with
their hearty social bonds and their sense of meaning and purpose bestowed by religion. That is why, one often reads, depression, anxiety,
loneliness, and suicide have been soaring, and why Sweden, that secular

paradise, has a famously high rate of suicide. In 2016 the activist George Monbiot prosecuted the cultural pessimist's time-honored campaign against modernity in an op-ed entitled "Neoliberalism Is Creating Loneliness. That's What's Wrenching Society Apart." The tag line was, "Epidemics of mental illness are crushing the minds and bodies of millions. It's time to ask where we are heading and why." The article itself warned, "The latest, catastrophic figures for children's mental health in England reflect a global crisis."[6]

If all those extra years of life and health, all that additional knowledge and leisure and breadth of experience, all those advances in peace and safety and democracy and rights, have really left us no happier but just lonelier and more suicidal, it would be history's greatest joke on humanity. But before we start walking around with a donkey with pots clanging on the sides, we had better take a closer look at the facts about human happiness.

～

At least since the Axial Age, thinkers have deliberated about what makes for a good life, and today happiness has become a major topic in social science.[7] Some intellectuals are incredulous, even offended, that happiness has become a subject for economists rather than just poets, essayists, and philosophers. But the approaches are not opposed. Social scientists often begin their studies of happiness with ideas that were first conceived by artists and philosophers, and they can pose questions about historical and global patterns that cannot be answered by solitary reflection, no matter how insightful. That is especially true for the question of whether progress has left people happier. To answer it, we must first assuage the critics' incredulity over the possibility that happiness can even be measured.

Artists, philosophers, and social scientists agree that well-being is not a single dimension. People can be better off in some ways and worse off in others. Let's distinguish the major ones.

We can begin with objective aspects of well-being: the gifts we deem intrinsically worthwhile whether or not their possessors appreciate them. At the top of that list is life itself; also on it are health, education, freedom, and leisure. That is the mindset behind Louis C.K.'s social criticism and, in part, behind Amartya Sen's and Martha Nussbaum's conceptions of fundamental human capabilities.[8] In this sense we can say that people who live long, healthy, and stimulating lives are truly better off even if they have a morose temperament or are in a bad mood or are spoiled idiots and fail to count their blessings. One rationale for this apparent paternalism is that life, health, and freedom are prerequisites to everything else, including the very act of pondering what is worthwhile in life, and so they are worthy by their very nature. Another is that the people who have the luxury of failing to appreciate their good fortune make up a biased sample of lucky survivors. If we could canvass the souls of the dead children and mothers and the victims of war and starvation and disease, or if we went back in time and gave them a choice between proceeding with their lives in a premodern or modern world, we might uncover an appreciation of modernity that is more commensurate with its objective benefits. These dimensions of well-being have been the topics of the preceding chapters, and the verdict on whether they have improved over time is in.

Among these intrinsic goods is freedom or autonomy: the availability of options to lead a good life (positive freedom) and the absence of coercion that prevents a person from choosing among them (negative freedom). Sen gave a shout-out to this value in the title of his book on the ultimate goal of the development of nations: *Development as Freedom*. Positive freedom is related to the economist's notion of utility (what people want; what they spend their wealth on), and negative freedom to the political scientist's notions of democracy and human rights. As I mentioned, freedom (together with life and reason) is a prerequisite to the very act of evaluating what is good in life. Unless we are impotently lamenting or celebrating our fate, then whenever we assess our condition we are presupposing that people in the past could have chosen otherwise. And when we ask where we should be heading, we presuppose that we have choices about what to pursue. For these reasons, freedom itself is inherently worthy.

In theory, freedom is independent of happiness. People can surrender to fatal attractions, crave pleasures that are bad for them, regret a choice the morning after, or ignore advice to be careful what they wish for.[9] In practice, freedom and the other good things in life go together. Whether assessed objectively through a democracy index for a country as a whole, or subjectively through people's ratings of whether they feel they have "free choice and control over their lives," the level of happiness in a country is correlated with the level of freedom.[10] Also, people single out freedom as a component of a *meaningful* life, whether or not it leads to a happy life.[11] Like Frank Sinatra, they may have regrets, they may take blows, but they do it their way. People can even value autonomy *over* happiness: many who have gone through a painful divorce, for example, would still not choose to return to a time when their parents would have arranged their marriages.

What about happiness itself? How can a scientist measure something as subjective as well-being? The best way to find out how happy people are is to ask them. Who could be a better judge? An old *Saturday Night Live* skit has Gilda Radner in a postcoital conversation with a nervous lover (played by Chevy Chase) who is worried she didn't have an orgasm, and she consoles him by saying, "Sometimes I do and I don't even know it." We laugh because when it comes to subjective experience, the experiencer herself is the ultimate authority. But we don't have to take people's word for it: self-reports of well-being turn out to correlate with everything else we think of as indicating happiness, including smiles, a buoyant demeanor, activity in the parts of the brain that respond to cute babies, and, Gilda and Chevy notwithstanding, judgments by other people.[12]

Happiness has two sides, an experiential or emotional side, and an evaluative or cognitive side.[13] The experiential component consists of a balance between positive emotions like elation, joy, pride, and delight, and negative emotions like worry, anger, and sadness. Scientists can sample these experiences in real time by having people wear a beeper that goes off at random times and prompts them to indicate how they are feeling. The ultimate measure of happiness would consist of a lifetime integral or weighted sum of how happy people are feeling and how long they feel that way. Though experience sampling is the most direct way of assessing subjective well-being, it's laborious and expensive, and there are no good datasets that compare people in different countries or track them over the years. The next best thing is to ask people how they are feeling at the time, or how they remember having felt during the day or week before.

This brings us to the other side of well-being, people's *evaluations* of how they are living their lives. People can be asked to reflect on how satisfied they feel "these days" or "as a whole" or "taking all things together," or to render the almost philosophical judgment of where they stand on a ten-rung ladder ranging from "the worst possible life for you" to "the best possible life for you." People find these questions hard (not surprisingly, since they *are* hard), and their responses may be warped by the weather, their current mood, and what they were asked about immediately beforehand (with questions to college students about their dating life, or to anyone about politics, having a reliably depressive effect). Social scientists have become resigned to the fact that happiness, satisfaction, and best-versus-worst-possible life are blurred in people's minds and that it's often easiest just to average them together.[14]

Emotions and evaluations are, of course, related, though imperfectly: an abundance of happiness makes for a better life, but an absence of worry and sadness does not.[15] And this brings us to the final dimension of a good life, meaning and purpose. This is the quality that, together with happiness, goes into Aristotle's ideal of *eudaemonia* or "good spirit."[16] Happiness isn't everything. We can make choices that leave us unhappy in the short term but fulfilled over the course of a life, such as raising a child, writing a book, or fighting for a worthy cause.

Though no mortal can stipulate what *really* makes a life meaningful, the psychologist Roy Baumeister and his colleagues probed for what makes people *feel* their lives are meaningful. The respondents separately rated how happy and how meaningful their lives were, and they answered a long list of questions about their thoughts, activities, and circumstances. The results suggest that many of the things that make people happy also make their lives meaningful, such as being connected to others, feeling productive, and not being alone or bored. But other things can make lives happier while leaving them no more meaningful or even less so.

People who lead happy but not necessarily meaningful lives have all their needs satisfied: they are healthy, have enough money, and feel good a lot of the time. People who lead meaningful lives may enjoy none of these boons. Happy people live in the present; those with meaningful lives have a narrative about their past and a plan for the future. Those with happy but meaningless lives are takers and beneficiaries; those with meaningful but unhappy lives are givers and benefactors. Parents get meaning from their children, but not necessarily happiness. Time spent with friends makes a life happier; time spent with loved ones makes it more meaningful. Stress, worry, arguments, challenges, and struggles make a life unhappier but more meaningful. It's not that people with meaningful lives masochistically go looking for trouble but that they pursue ambitious goals: "Man plans and God laughs." Finally, meaning is about expressing rather than satisfying the self: it is enhanced by activities that define the person and build a reputation.

We can see happiness as the output of an ancient biological feedback system that tracks our progress in pursuing auspicious signs of fitness in a natural environment. We are happier, in general, when we are healthy, comfortable, safe, provisioned, socially connected, sexual, and loved. The function of happiness is to goad us into seeking the keys to fitness: when we are unhappy, we scramble for things that would improve our lot; when we are happy, we cherish the status quo. Meaning, in contrast, registers the novel and expansive goals that are opened up for us as social, brainy, and talkative occupants of the uniquely human cognitive niche. We consider goals that are rooted in the distant past and stretch far into the future, that affect people beyond our circle of acquaintance, and that must be ratified by our fellows, based on our ability to persuade them of their worth and on our reputation for benevolence and efficacy.[17]

An implication of the circumscribed role of happiness in human psychology is that the goal of progress cannot be to increase happiness indefinitely, in the hope that more and more people will become more and more euphoric. But there is plenty of unhappiness that can be reduced, and no limit as to how meaningful our lives can become.

～

Let's agree that the citizens of developed countries are not as happy as they ought to be, given the fantastic progress in their fortunes and freedom. But are they not happier at all? Have their lives become so empty that they are choosing to end them in record numbers? Are they suffering through an epidemic of loneliness, in defiance of the mind-boggling number of opportunities to connect with one another? Is the younger generation, ominously for our future, crippled by depression and mental illness? As we shall see, the answer to each of these questions is an emphatic *no.*

Evidence-free pronouncements about the misery of mankind are an occupational hazard of the social critic. In the 1854 classic *Walden,* Henry David Thoreau famously wrote, "The mass of men lead lives of quiet desperation." How a recluse living in a cabin on a pond could know this was never made clear, and the mass of men beg to differ. Eighty-six percent of those who are asked about their happiness in the World Values Survey say they are "rather happy" or "very happy," and on average the respondents in the 150-country *World Happiness Report 2016* judged their lives to be on the top half of the ladder from worst to best.[18] Thoreau was a victim of the Optimism Gap (the "I'm OK, They're Not" illusion), which for happiness is more like a canyon. People in every country underestimate the proportion of their compatriots who say they are happy, by an average of 42 percentage points.[19]

What about the historical trajectory? Easterlin identified his intriguing paradox in 1973, decades before the era of big data. Today we have much more evidence on wealth and happiness, and it shows there is no Easterlin paradox. Not only are richer people in a given country happier, but people in richer countries are happier, and as countries get richer

over time, their people get happier. The new understanding has come
from several independent analyses, including ones by Angus Deaton,
the World Values Survey, and the *World Happiness Report 2016*.[20] My favorite comes from the economists Betsey Stevenson and Justin Wolfers
and may be summarized in a graph. Figure 18-1 plots ratings of average
life satisfaction against average income (on a logarithmic scale) for 131
countries, each represented by a dot, together with the relationship of life
satisfaction to income among the citizens of each country, represented
by an arrow impaling the dot.

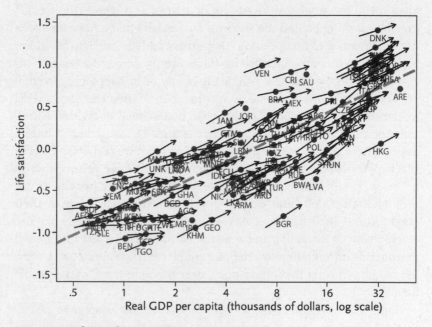

Figure 18-1: Life satisfaction and income, 2006

Source: Stevenson & Wolfers 2008a, fig. 11, based on data from the Gallup World Poll 2006. Credit: Betsey
Stevenson and Justin Wolfers.

Several patterns jump out. The most immediate is the absence of a
cross-national Easterlin paradox: the cloud of arrows is stretched along
a diagonal, which indicates that the richer the country, the happier its
people. Bear in mind that the income scale is logarithmic; on a standard
linear scale, the same cloud would rise steeply from the left end and
bend over toward the right. This means that a given number of extra
dollars boosts the happiness of people in a poor country more than the
happiness of people in a rich country, and that the richer a country is, the

more additional money its people need to become happier still. (It's one of the reasons that the Easterlin paradox appeared in the first place: with the noisier data of the era, it was hard to spot the relatively small rise in happiness at the high end of the income scale.) But with either scale, the line never flattens out, as it would if people needed only some minimum amount of income to see to their basic needs and anything extra made them no happier. As far as happiness is concerned, Wallis Simpson was half-right when she said, "You can't be too rich or too thin."

Most strikingly, the slopes of the arrows are similar to each other, and identical to the slope for the swarm of arrows as a whole (the dashed gray line lurking behind the swarm). That means that a raise for an individual relative to that person's compatriots adds as much to his or her happiness as the same increase for their country across the board. This casts doubt on the idea that people are happy or unhappy only in comparison to the Joneses. Absolute income, not relative income, is what matters most for happiness (a conclusion that's consistent with the finding discussed in chapter 9 on the irrelevance of inequality to happiness).[21] These are among a number of findings that weaken the old belief that happiness adapts to ambient conditions like the eye, returns to a set point, or remains stationary as people vainly stride on a hedonic treadmill. Though people often do rebound from setbacks and pocket their good fortune, their happiness takes a sustained hit from trials like unemployment or disability, and a sustained boost from gifts like a good marriage or immigrating to a happier country.[22] And contrary to an earlier belief, winning the lottery does, over the long term, make people happier.[23]

Since we know that countries get *richer* over time (chapter 8), we can think of figure 18-1 as a freeze-frame in a movie showing humanity getting *happier* over time. This increase in happiness is yet another indicator of human progress, and among the most important of all. Of course this snapshot is not an actual longitudinal chronicle in which people all over the world are polled for centuries and we plot their happiness over time; such data do not exist. But Stevenson and Wolfers scoured the literature for what longitudinal studies there were, and found that in eight out of nine European countries, happiness increased between 1973 and 2009 in tandem with the country's rise in GDP per capita.[24] A confirmation for the world as a whole comes from the World Values Survey, which found that in forty-five out of fifty-two countries, happiness increased between 1981 and 2007.[25] The trends over time close the books on the Easterlin paradox: we now know that richer people within a country are happier, that richer countries are happier, and that people get happier as their countries get richer (which means that people get happier over time).

Happiness, of course, depends on much more than income. This is true not just among individuals, who differ in their life histories and their innate temperaments, but among nations, as we see from the scatter of dots around the gray line in the graph. Nations are happier when their people are in better health (holding income constant), and, as I mentioned, they are happier when their citizens feel they are free to choose what to do with their lives.[26] Culture and geography also matter: true to stereotype, Latin American countries are happier than they should be given their income, and the ex-Communist countries of Eastern Europe are less happy.[27] The *World Happiness Report 2016* found three other traits that go with national happiness: social support (whether people say they have friends or relatives they can count on in times of trouble), generosity (whether they donate money to charity), and corruption (whether they perceive the businesses in their country as corrupt).[28] We cannot conclude, though, that these traits *cause* greater happiness. One reason is that happy people see the world through rose-tinted glasses, and may give generous assessments of the good things both in their lives and in their societies. The other is that happiness is, as social scientists say, endogenous: being happy might make people supportive, generous, and conscientious rather than the other way around.

～

Among the countries that punch below their wealth in happiness is the United States. Americans are by no means unhappy: almost 90 percent rate themselves as at least "pretty happy," almost a third rate themselves as "very happy," and when they are asked to place themselves on the ten-rung ladder from the worst to the best possible life, they choose the seventh rung.[29] But in 2015 the United States came in at thirteenth place among the world's nations (trailing eight countries in Western Europe, three in the Commonwealth, and Israel), even though it had a higher average income than all of them but Norway and Switzerland.[30] (The United Kingdom, whose citizens place themselves at a happy 6.7 rungs up from the worst possible life, came in at twenty-third place.)

Also, the United States hasn't gotten systematically happier over the years (another decoy that led to the premature announcement of the Easterlin paradox, because the United States is also the country with happiness data that stretch back the farthest). American happiness has fluctuated within a narrow band since 1947, deflecting in response to various recessions, recoveries, malaises, and bubbles, but with no consistent rise or fall. One dataset shows a slight decline in American happiness from 1955 to 1980, followed by a rise through 2006; another shows a slight decline in the proportion saying they are "very happy" starting in 1972 (though even there the sum of those who say they are "very happy" and "pretty happy" has not changed).[31]

The American happiness stagnation doesn't falsify the global trend in which happiness increases with wealth, because when we look at changes in a rich country over a few decades we're peeping at a restricted range of the scale. As Deaton points out, a trend that is obvious when you look at the effects of a fiftyfold difference in income between, say, Togo and the United States, representing a quarter-millennium of economic growth, may be submerged in the noise when you look for the effects of, say, a twofold difference in income within a single country over just twenty years of economic growth.[32] Also, the United States has seen a greater rise in income inequality than the countries of Western Europe (chapter 9), and its growth in GDP may have been enjoyed by a smaller proportion of the populace.[33] Speculating about American exceptionalism is an endlessly fascinating pastime, but whatever the reason, happyologists agree that the United States is an outlier from the global trend in subjective well-being.[34]

Another reason it can be hard to make sense of happiness trends for individual countries is that a country is a collection of tens of millions of human beings who just happen to occupy a patch of land. It's remarkable that we can find *anything* in common when we average them, and we shouldn't be surprised to find that as time passes, different segments of the population go in different directions, sometimes jerking the average around, sometimes canceling each other out. Over the past thirty-five years African Americans have been getting much happier while American whites have gotten a bit less happy.[35] Women tend to be happier than men, but in Western countries the gap has shrunk, with men getting happier at a faster rate than women. In the United States it has reversed outright, as women got unhappier while men stayed more or less the same.[36]

The biggest complication in making sense of historical trends, though, is one that we came across in chapter 15: the distinction between changes across the life cycle (age), in the zeitgeist (period), and over the generations (cohort).[37] Without a time machine, it's logically impossible to disentangle the effects of age, cohort, and period completely, to say nothing of their interactions. If, for example, fifty-year-olds were miserable in 2005, we couldn't tell whether the Baby Boomers had a hard time dealing with middle age, the Baby Boomers had a hard time dealing with the new millennium, or the new millennium was a hard time to be middle-aged. But with a dataset that embraces multiple generations and decades, together with a few assumptions about how quickly people and times can change, one can average the scores for a generation over the years, for the entire population at each year, and for the population at each age, and make reasonably independent estimates of the trajectory of the three factors over time. That in turn allows us to look for two different versions of progress: people of all ages can become better off in recent periods, or younger cohorts can be better off than older ones, lifting the population as they replace them.

People tend to get happier as they get older (an age effect), presumably because they overcome the hurdles of embarking on adulthood and develop the wisdom to cope with setbacks and to put their lives in perspective.[38] (They may pass through a midlife crisis on the way, or take a final slide in the last years of old age.)[39] Happiness fluctuates with the times, especially the changing economy—not for nothing do economists call a composite of the inflation rate and the unemployment rate the Misery Index—and Americans have just dug themselves out of a trough that followed the Great Recession.[40]

The pattern across the generations also has ups and downs. In two large samples, Americans born in every decade from the 1900s through the 1940s lived happier lives than those in the preceding cohort, presumably because the Great Depression left a scar on the generations who came of age as it deepened. The rise leveled off and then declined a bit with the Baby Boomers and early Generation X, the last generation that was old enough to allow the researchers to disentangle cohort from period.[41] In a third study which continues to the present (the General Social Survey), happiness also dipped among the Baby Boomers but fully rebounded in Gen X and the Millennials.[42] So while every generation agonizes about the kids today, younger Americans have in fact been getting happier. (As we saw in chapter 12, they have also become less violent and less druggy.) That makes three segments of the population that have become happier amid the American happiness stagnation: African Americans, the successive cohorts leading up to the Baby Boom, and young people today.

The age-period-cohort tangle means that every historical change in well-being is at least three times as complicated as it appears. With that caveat in mind, let's take a look at the claims that modernity has unleashed an epidemic of loneliness, suicide, and mental illness.

~

To hear the observers of the modern world tell it, Westerners have been getting lonelier. In 1950 David Riesman (together with Nathan Glazer and Reuel Denney) wrote the sociological classic *The Lonely Crowd*. In 1966 the Beatles wondered where all the lonely people come from, and where they all belong. In a 2000 bestseller the political scientist Robert Putnam noted that Americans were increasingly *Bowling Alone*. And in 2010 the psychiatrists Jacqueline Olds and Richard Schwartz wrote of *The Lonely American* (subtitle: *Drifting Apart in the Twenty-First Century*). For a member of gregarious *Homo sapiens*, social isolation is a form of torture, and the stress of loneliness a major risk to health and life.[43] So it would be another joke on modernity if our newfound connectivity has left us lonelier than ever.

One might think that social media could make up for whatever alienation and isolation came with the decline of large families and small communities. Today, after all, Eleanor Rigby and Father McKenzie could be Facebook friends. But in *The Village Effect* the psychologist Susan Pinker reviews research showing that digital friendships don't provide the psychological benefits of face-to-face contact.

This only heightens the mystery of why people would be getting lonelier. Among the world's problems, social isolation would seem to be one of the easier ones to solve: just invite someone you know for a chat at a neighborhood Starbucks or around the kitchen table. Why would people fail to notice the opportunities? Have people today, especially the ever-maligned younger generation, become so addicted to digital crack cocaine that they forgo vital human contact and sentence themselves to needless and perhaps lethal loneliness? Could it really be true, as one social critic put it, that "we have given our hearts to machines, and are now turning into machines"? Has the Internet created, in the words of another, "an atomized world without human contact or emotion"?[44] To anyone who believes there is such a thing as human nature, it seems unlikely, and the data show it is false: there is no loneliness epidemic.

In *Still Connected* (2011), the sociologist Claude Fischer reviewed forty years of surveys that asked people about their social relationships. "The most striking thing about the data," he noted, "is how consistent Amer-

icans' ties to family and friends were between the 1970s and 2000s. We
rarely find differences of more than a handful of percentage points either way that might describe lasting alterations in behavior with lasting
personal consequences—yes, Americans entertained less at home and
did more phone calling and emailing, but they did not change much on
the fundamentals."[45] Though people have reallocated their time because
families are smaller, more people are single, and more women work,
Americans today spend as much time with relatives, have the same median number of friends and see them about as often, report as much
emotional support, and remain as satisfied with the number and quality
of their friendships as their counterparts in the decade of Gerald Ford
and *Happy Days*. Users of the Internet and social media have *more* contact with friends (though a bit less face-to-face contact), and they feel
that the electronic ties have enriched their relationships. Fischer concluded that human nature rules: "People try to adapt to changing circumstances so as to protect their most highly valued ends, which include
sustaining the volume and quality of their personal relationships—time
with children, contact with relatives, a few sources of intimate support."[46]

What about subjective feelings of loneliness? Surveys of the entire
population are sparse; the data Fischer found suggested that "Americans' expressions of loneliness remained the same or perhaps increased
slightly," mainly because more people were single.[47] But surveys of students, a captive audience, are plentiful, and for decades they have indicated whether they agree with statements like "I am unhappy doing so
many things alone" and "I have nobody to talk to." The trends are summarized in the title of a 2015 article, "Declining Loneliness over Time,"
and are shown in figure 18-2.

Since these students were not tracked after they left school, we don't
know whether the decline in loneliness is a period effect, in which it
has become steadily easier for young people to satisfy their social
needs, or a cohort effect, in which recent generations are more socially
satisfied and will remain so. What we do know is that young Americans are not suffering from "toxic levels of emptiness and aimlessness
and isolation."

Together with "the kids today," the perennial target of cultural pessimists is technology. In 2015 the sociologist Keith Hampton and his coauthors introduced a report on the psychological effects of social media by
noting:

For generations, commentators have worried about the impact of technology on people's stress. Trains and industrial machinery were seen as noisy disruptors of pastoral village life that put people on edge. Telephones interrupted quiet times in homes. Watches and clocks added to the dehumanizing time pressures on factory workers to be productive. Radio and television were organized around the advertising that enabled modern consumer culture and heightened people's status anxieties.[48]

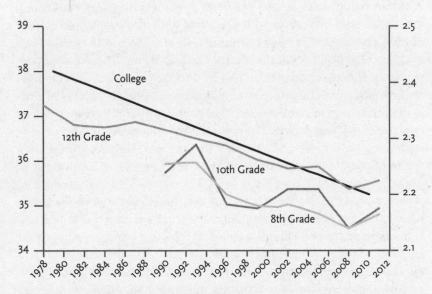

Figure 18-2: Loneliness, US students, 1978–2011

Source: Clark, Loxton, & Tobin 2015. **College students** (left axis): Revised UCLA Loneliness Scale, trend line across many samples, taken from their fig. 1. **High school students** (right axis): Mean rating of six loneliness items from the Monitoring the Future survey, triennial means, taken from their fig. 4. Each axis spans half a standard deviation, so the slopes of the college and high school curves are commensurable, but their relative heights are not.

And so it was inevitable that the critics would shift their focus to social media. But social media can be neither credited nor blamed for the changes in loneliness among American students shown in figure 18-2: the decline proceeded from 1977 through 2009, and the Facebook explosion did not come until 2006. Nor, according to the new surveys, have adults become isolated because of social media. Users of social media have more close friends, express more trust in people, feel more supported, and are more politically involved.[49] And notwithstanding the rumor that they are drawn into an anxious competition to keep up with the furious rate of enjoyable activities of their digital faux-friends, social media users do not report higher levels of stress than non-users.[50] On the contrary, the women among them are *less* stressed, with one telling exception: they get upset when they learn that someone they care about has suffered an illness, a death in the family, or some other setback. Social media users care too much, not too little, about other people, and they empathize with them over their troubles rather than envying them their successes.

Modern life, then, has not crushed our minds and bodies, turned us into atomized machines suffering from toxic levels of emptiness and isolation, or set us drifting apart without human contact or emotion. How did this hysterical misconception arise? Partly it came out of the social critic's standard formula for sowing panic: Here's an anecdote, therefore it's a trend, therefore it's a crisis. But partly it came from genuine changes in how people interact. People see each other less in traditional venues like clubs, churches, unions, fraternal organizations, and dinner parties, and more in informal gatherings and via digital media. They confide in fewer distant cousins but in more co-workers. They are less likely to have a large number of friends but also less likely to *want* a large number of friends.[51] But just because social life looks different today from the way it looked in the 1950s, it does not mean that humans, that quintessentially social species, have become any less social.

～

Suicide, one might think, is the most reliable measure of societal unhappiness, in the same way that homicide is the most reliable measure of societal conflict. A person who has died by suicide must have suffered from unhappiness so severe that he or she decided that a permanent end to consciousness was preferable to enduring it. Also, suicides can be tabulated objectively in a way that the experience of unhappiness cannot.

But in practice, suicide rates are often inscrutable. The very sadness and agitation from which suicide would be a release also addles a person's judgment, so what ought to be the ultimate existential decision often hinges on the mundane matter of how easy it is to carry out the act. Dorothy Parker's macabre poem "Resumé" (which ends, "Guns aren't lawful; Nooses give; Gas smells awful; You might as well live") is disconcertingly close to the mindset of a person contemplating suicide. A country's suicide rate can soar or plummet when a convenient and effective method is widely available or taken away, such as coal gas in England in the first half of the 20th century, pesticides in many developing countries, and guns in the United States.[52] Suicides increase during economic downturns and political turmoil, not surprisingly, but they are also affected by the weather and the number of daylight hours, and they increase when the media normalize or romanticize recent instances.[53] Even the innocuous idea that suicide is an assay for unhappiness may be questioned. A recent study documented a "happiness-suicide paradox" in which happier American states and happier Western countries have slightly *higher*, rather than lower, suicide rates.[54] (The researchers speculate that misery loves company: a personal setback is more painful when everyone around you is happy.) Suicide rates can be capricious for yet another reason. Suicides are often hard to distinguish from accidents (particularly when the cause is a poisoning or drug overdose, but also when it is a fall, a car crash, or a gunshot), and coroners may tilt their classifications in times and places in which suicide is stigmatized or criminalized.

We do know that suicide is a major cause of death. In the United States there are more than 40,000 suicides a year, making it the tenth-leading cause of death, and worldwide there are about 800,000, making it the fifteenth-leading cause.[55] Yet the trends over time and the differences among countries are hard to fathom. In addition to the age-cohort-period snarl, the lines for men and women often go in different directions. Though the suicide rate for women in developed countries fell by more than 40 percent between the mid-1980s and 2013, men kill themselves at around four times the rate of women, so the numbers for men tend to push the overall trends around.[56] And no one knows why, for example, the world's most suicidal countries are Guyana, South Korea, Sri Lanka, and Lithuania, nor why France's rate shot up from 1976 to 1986 and fell back down by 1999.

But we know enough to debunk two popular beliefs. The first is that suicide has been steadily rising and has now reached historically high, unprecedented, crisis, or epidemic proportions. Suicide was common enough in the ancient world to have been debated by the Greeks and to have figured in the biblical stories of Samson, Saul, and Judas. Historical data are scarce, not least because suicide, also called "self-murder," used to be a crime in many countries, including England until 1961. But the data go back more than a century in England, Switzerland, and the United States, and I have plotted them in figure 18-3.

The annual suicide rate in England was 13 per 100,000 in 1863; it hit peaks of around 19 in the first decade of the 20th century and more than 20 during the Great Depression, plunged during World War II and again in the 1960s, and then fell more gradually to 7.4 in 2007. Switzerland, too, saw a decline of more than twofold, from 24 in 1881 and 27 during the

Depression to 12.2 in 2013. The United States suicide rate peaked at around 17 in the early 20th century and again during the Depression before falling to 10.5 at the turn of the millennium, followed by a rise after the recent Great Recession to 13.

Figure 18-3: Suicide, England, Switzerland, and US, 1860–2014

Sources: England (including Wales): Thomas & Gunnell 2010, fig. 1, average of male and female rates, provided by Kylie Thomas. The series has not been extended because the data are not commensurable with current records. Switzerland, 1880–1959: Ajdacic-Gross et al. 2006, fig. 1. Switzerland, 1960–2013: WHO Mortality Database, OECD 2015b. United States, 1900–1998: Centers for Disease Control, Carter et al. 2006, table Ab950. United States, 1999–2014: Centers for Disease Control 2015.

So in all three countries for which we have historical data, suicide was more common in the past than it is today. The visible crests and troughs are the surface of a churning sea of ages, cohorts, periods, and sexes.[57] Suicide rates rise sharply during adolescence and then more gently into middle age, where they peak for females (perhaps because they face menopause and an empty nest) and then fall back down, while staying put for males before shooting up in their retirement years (perhaps because they face an end to their traditional role as providers). Part of the recent increase in the American suicide rate can be attributed to the aging of the population, with the large cohort of Boomer males moving into their most suicide-prone years. But the cohorts themselves matter as well. The GI and Silent generations were more reluctant to kill themselves than the Victorian cohorts that preceded them and the Boomers and Gen-Xers that followed them. The Millennials appear to be slowing or reversing the generational rise; adolescent suicide rates fell between the early 1990s and the first decades of the 21st century.[58] The times themselves (adjusting for ages and cohorts) have become less conducive to suicide since the peaks around the turn of the 20th century, the 1930s, and the late 1960s to early 1970s; they dropped to a forty-year low in 1999, though we have seen a slight rise again since the Great Recession. This complexity belies the alarmism of the recent *New York Times* headline "U.S. Suicide Rate Surges to a 30-Year High," which could also have been titled "Despite the Recession and an Aging Population, U.S. Suicide Rate Is a Third Lower Than Previous Peaks."[59]

Together with the belief that modernity makes people want to kill themselves, the other great myth about suicide is that Sweden, that paragon of Enlightenment humanism, has the world's highest suicide rate. This urban legend originated (according to what might be another urban legend) in a speech by Dwight Eisenhower in 1960 in which he called out Sweden's high suicide rate and blamed it on the country's paternalistic socialism.[60] I myself would have blamed the bleak existential films of Ingmar Bergman, but both theories are explanations in search of a fact to explain. Though Sweden's suicide rate in 1960 was higher than that of the United States (15.2 versus 10.8 per 100,000), it was never the world's highest, and it has since fallen to 11.1, which is below the world average (11.6) and the rate for the United States (12.1) and in fifty-eighth place overall.[61] A recent review of suicide rates across the world noted that "generally the suicide trend has been downward in Europe and there are currently no Western European welfare states in the world top ten for suicide rates."[62]

~

Everyone occasionally suffers from depression, and some people are stricken with major depression, in which the sadness and hopelessness last more than two weeks and interfere with carrying on with life. In recent decades, more people have been diagnosed with depression, especially in younger cohorts, and the conventional wisdom is captured in the tag line of a recent public television documentary: "A silent epidemic is ravaging the nation and killing our kids." We have just seen that the nation is not suffering from an epidemic of unhappiness, loneliness, or suicide, so an epidemic of depression seems unlikely, and it turns out to be an illusion.

Consider one oft-cited study, which implausibly claimed that every cohort from the GI Generation through the Baby Boomers was more depressed than the one before.[63] The investigators reached that conclusion by asking people of different ages to recall times when they had been depressed. But that made the study a hostage to memory: the longer ago an episode took place, the less likely it is that a person will recall it; especially (as we saw in chapter 4) if the episode was unpleasant. That creates an illusion that recent periods and younger cohorts are more vulnerable to depression. Such a study is also hostage to mortality. As the decades pass, depressed people are more likely to die of suicide and other causes, so the old people who remain in a sample are the mentally healthier ones, making it seem as if everyone who was born long ago is mentally healthier.

Another distorter of history is a change in attitudes. Recent decades have seen outreach programs and media campaigns designed to increase awareness and decrease the stigma of depression. Drug companies have advertised a pharmacopoeia of antidepressants directly to consumers. Bureaucracies demand that people be diagnosed with some disorder before they can receive entitlements such as therapy, government services, and a right against discrimination. All these inducements could make people more likely to report that they are depressed.

At the same time, the mental health professions, and perhaps the culture at large, has been lowering the bar for what counts as a mental illness. The list of disorders in the *Diagnostic and Statistical Manual (DSM)* of the American Psychiatric Association tripled between 1952 and 1994, when it included almost three hundred disorders, including Avoidant Personality Disorder (which applies to many people who formerly were called shy), Caffeine Intoxication, and Female Sexual Dysfunction. The number of symptoms needed to justify a diagnosis has fallen, and the number of stressors that may be credited with triggering one has increased. As the psychologist Richard McNally has noted, "Civilians who underwent the terror of World War II, especially Nazi death factories . . . , would surely be puzzled to learn that having a wisdom tooth extracted, encountering obnoxious jokes at work, or giving birth to a healthy baby after an uncomplicated delivery can cause Post-Traumatic Stress Disorder."[64] By the same shift, the label "depression" today may be applied to conditions that in the past were called grief, sorrow, or sadness.

Psychologists and psychiatrists have begun to sound the alarm against this "disease mongering," "concept creep," "selling sickness," and "expanding empire of psychopathology."[65] In her 2013 article "Abnormal Is the New Normal," the psychologist Robin Rosenberg noted that the latest version of the *DSM* could diagnose half the American population with a mental disorder over the course of their lives.[66]

The expanding empire of psychopathology is a first-world problem, and in many ways is a sign of moral progress.[67] Recognizing a person's suffering, even with a diagnostic label, is a form of compassion, particularly when the suffering can be alleviated. One of psychology's best-kept secrets is that cognitive behavior therapy is demonstrably effective (often more effective than drugs) in treating many forms of distress, including depression, anxiety, panic attacks, PTSD, insomnia, and the symptoms of schizophrenia.[68] With mental disorders making up more than 7 percent of the global burden of disability (major depression alone making up 2.5 percent), that's a lot of reducible suffering.[69] The editors of the journal *Public Library of Science: Medicine* recently called attention to "the paradox of mental health": over-medicalization and over-treatment in the wealthy West, and under-recognition and under-treatment in the rest of the world.[70]

With the widening net of diagnosis, the only way to tell whether more people are depressed these days is to administer a standardized test of depressive symptoms to nationally representative samples of people of different ages over many decades. No study has met this gold standard, but several have applied a constant yardstick to more circumscribed populations.[71] Two intensive, long-term studies in rural counties (one in Sweden, one in Canada) signed up people born between the 1870s and the 1990s and tracked them from the middle to the late 20th century, embracing staggered lives that spanned more than a century. Neither found signs of a long-term rise in depression.[72]

There have also been several meta-analyses (studies of studies). Twenge found that from 1938 to 2007, college students scored increasingly higher on the Depression scale of the MMPI, a common personality test.[73] That doesn't necessarily mean that more of the students suffered from major depression, though, and the increase may have been inflated by the broader range of people who went to college over those decades. Moreover, other studies (some by Twenge herself) have found no change or even a decline in depression, especially for younger ages and cohorts and in later decades.[74] A recent one entitled "Is There an Epidemic of Child or Adolescent Depression?" vindicated Betteridge's Law of Headlines: Any headline that ends in a question mark can be answered with the word *no*. The authors explain, "Public perception of an 'epidemic' may arise from heightened awareness of a disorder that was long underdiagnosed by clinicians."[75] And the title of the biggest meta-analysis to date, which looked at the prevalence of anxiety and depression between 1990 and 2010 in *the entire world*, did not leave readers in suspense: "Challenging the Myth of an 'Epidemic' of Common Mental Disorders." The authors concluded, "When clear diagnostic criteria are applied, there is no evidence that the prevalence of common mental disorders is increasing."[76]

Depression is "comorbid" with anxiety, as epidemiologists morbidly call the correlation, which raises the question of whether the world has become more anxious. One answer was contained in the title of a long narrative poem published in 1947 by W. H. Auden, *The Age of Anxiety*. In the introduction to a recent reprint, the English scholar Alan Jacobs observed that "many cultural critics over the decades . . . have lauded Auden for his acuity in naming the era in which we live. But given the poem's difficulty few of them have managed to figure out precisely why he thinks our age is characterized primarily by anxiety—or even whether he is really saying that at all."[77] Whether he was saying that or not, Auden's name for our era has stuck, and it provided the obvious title for a meta-analysis by Twenge which showed that scores on a standard anxiety test administered to children and college students between 1952 and 1993 rose by a full standard deviation.[78] Things that can't go on forever don't, and as best we can tell, the increase among college students leveled off after 1993.[79] Nor have other demographic sectors become more anxious. Longitudinal studies of high school students and of adults conducted from the 1970s through the first decades of the 21st century find no rise across the cohorts.[80] Though in some surveys people have reported more symptoms of distress, anxiety that crosses the line into pathology is not at epidemic levels, and has shown no global increase since 1990.[81]

～

Everything is amazing. Are we really so unhappy? Mostly we are not. Developed countries are actually pretty happy, a majority of all countries have gotten happier, and as long as countries get richer they should get happier still. The dire warnings about plagues of loneliness, suicide, depression, and anxiety don't survive fact-checking. And though every generation has worried that the next one is in trouble, as younger generations go the Millennials seem to be in pretty good shape, happier and mentally healthier than their helicoptering parents.

Still, when it comes to happiness, many people are underachievers. Americans are laggards among their first-world peers, and their happiness has stagnated in the era sometimes called the American Century. The Baby Boomers, despite growing up in peace and prosperity, have proved to be a troubled generation, to the mystification of their parents, who lived through the Great Depression, World War II, and (for many of my peers) the Holocaust. American women have become unhappier just as they have been making unprecedented gains in income, education, accomplishment, and autonomy, and in other developed countries where everyone has gotten happier, the women have been outpaced by the men. Anxiety and some depressive symptoms may have increased in the post-war decades, at least in some people. And none of us are as happy as we ought to be, given how amazing our world has become.

Let me end this chapter with a reflection on these happiness short-falls. For many commentators they are an occasion to second-guess modernity.[82] Our unhappiness, they say, is payback for our worship of the individual and material wealth and for our acquiescence in the corrosion of family, tradition, religion, and community.

But there is a different way to understand the legacy of modernity. Those who are nostalgic for traditional folkways have forgotten how hard our forebears fought to escape them. Though no one gave happiness questionnaires to the people who lived in the close-knit communities that were loosened by modernity, much of the great art composed during the transition brought to life their dark side: the provincialism, conformity, tribalism, and Taliban-like restrictions on women's autonomy. Many novels from the mid-18th to the early 20th century played out the struggles of individuals to overcome the suffocating norms of aristocratic, bourgeois, or rural regimes, including works by Richardson, Thackeray, Charlotte Brontë, Eliot, Fontane, Flaubert, Tolstoy, Ibsen, Alcott, Hardy, Chekhov, and Sinclair Lewis. After urbanized Western society had become more tolerant and cosmopolitan, the tensions were played out again in popular culture's treatment of small-town American life, such as in songs by Paul Simon ("In my little town I never meant nothin' / I was just my father's son"), Lou Reed ("When you're growing up in a small town / You know you'll grow down in a small town"), and Bruce Springsteen ("Baby, this town rips the bones from your back / It's a death trap, a suicide rap"). It was played out yet again in the literature of immigrants, including works by Isaac Bashevis Singer, Philip Roth, and Bernard Malamud and then by Amy Tan, Maxine Hong Kingston, Jhumpa Lahiri, Bharati Mukherjee, and Chitra Banerjee Divakaruni.

Today we enjoy a world of personal freedom these characters could only fantasize about, a world in which people can marry, work, and live as they please. One can imagine a social critic of today warning Anna Karenina or Nora Helmer that a tolerant cosmopolitan society isn't all it's cracked up to be, that without the bonds of family and village they'll have moments of anxiety and unhappiness. I can't speak for them, but my guess is they'd think it was a pretty good deal.

A modicum of anxiety may be the price we pay for the uncertainty of freedom. It is another word for the vigilance, deliberation, and heart-searching that freedom demands. It's not entirely surprising that as women gained in autonomy relative to men they also slipped in happiness. In earlier times, women's list of responsibilities rarely extended beyond the domestic sphere. Today young women increasingly say that their life goals include career, family, marriage, money, recreation, friendship, experience, correcting social inequities, being a leader in their community, and making a contribution to society.[83] That's a lot of things to worry about, and a lot of ways to be frustrated: Woman plans and God laughs.

It's not just the options opened up by personal autonomy that place a weight on the modern mind; it's also the great questions of existence. As people become better educated and increasingly skeptical of received authority, they may become unsatisfied with traditional religious verities and feel unmoored in a morally indifferent cosmos. Here is our modern avatar of anxiety, Woody Allen, playing out the 20th-century generational divide in a conversation with his parents in *Hannah and Her Sisters* (1986):

MICKEY: Look, you're getting on in years, right? Aren't you afraid of dying?

FATHER: Why should I be afraid?

MICKEY: Oh! 'Cause you won't exist!

FATHER: So?

MICKEY: That thought doesn't terrify you?

FATHER: Who thinks about such nonsense? Now I'm alive. When I'm dead, I'll be dead.

MICKEY: I don't understand. Aren't you frightened?

FATHER: Of what? I'll be unconscious.

MICKEY: Yeah, I know. But never to exist again!

FATHER: How do you know?

MICKEY: Well, it certainly doesn't look promising.

FATHER: Who knows what'll be? I'll either be unconscious or I won't. If not, I'll deal with it then. I'm not gonna worry now about what's gonna be when I'm unconscious.

MOTHER [OFFSCREEN]: Of course there's a God, you idiot! You don't believe in God?

MICKEY: But if there's a God, then wh-why is there so much evil in the
  world? Just on a simplistic level. Why-why were there Nazis?
MOTHER: Tell him, Max.
FATHER: How the hell do I know why there were Nazis? I don't know
  how the can opener works.[84]

People have also lost their comforting faith in the goodness of their
institutions. The historian William O'Neill entitled his history of the
Baby Boomers' childhood years *American High: The Years of Confidence,
1945–1960*. In that era, everything seemed great. Belching smokestacks
were a sign of prosperity. America had a mission to spread democracy
around the world. The atom bomb was proof of Yankee ingenuity.
Women enjoyed domestic bliss, and Negroes knew their place. Though
much about America was indeed good during those years (the economic
growth rate was high; rates of crime and other social pathologies were
low), today we see it as a fool's paradise. It may not be a coincidence that
two of the sectors that underperform in happiness—Americans and
Baby Boomers—were the sectors that were most set up for disillusionment in the 1960s. In retrospect we can see that a concern with the environment, nuclear war, American foreign-policy blunders, and racial and
gender equality could not be put off forever. Even if they make us more
anxious, we are better for being aware of them.

As we become aware of our collective responsibilities, each of us may
add a portion of the world's burdens to our own worry list. Another icon
of late 20th-century anxiety, the movie *Sex, Lies, and Videotape* (1989),
opens with the baby-boomer protagonist sharing her angst with a psy-
chotherapist:

> Garbage. All I've been thinking about all week is garbage. I can't stop
> thinking about it. I just . . . I've gotten real concerned over what's
> gonna happen with all the garbage. I mean, we've got so much of it.
> You know? I mean, we have to run out of places to put this stuff even-
> tually. The last time I started feeling this way is when that barge was
> stranded and, you know, it was going around the island and nobody
> would claim it.

"That barge" refers to a 1987 media frenzy over a barge filled with three
thousand tons of New York garbage that was turned away by landfills
up and down the Atlantic coast. The therapy scene is by no means fanciful: an experiment in which people watched news stories that had been

doctored to have a positive or negative spin found that "participants who watched the negatively valenced bulletin showed increases in both anxious and sad mood, and also showed a significant increase in the tendency to catastrophize a personal worry."[85] Three decades later I suspect that many therapists are listening to patients sharing their fears about terrorism, income inequality, and climate change.

A bit of anxiety is not a bad thing if it motivates people to support policies that would help solve major problems. In earlier decades people might have offloaded their worries to a higher authority, and some still do. In 2000, sixty religious leaders endorsed the Cornwall Declaration on Environmental Stewardship, which addressed the "so-called climate crisis" and other environmental problems by affirming that "God in His mercy has not abandoned sinful people or the created order but has acted throughout history to restore men and women to fellowship with Him and through their stewardship to enhance the beauty and fertility of the earth."[86] I imagine that they and the other 1,500 signatories do not visit therapists to air anxieties about the future of the planet. But as George Bernard Shaw observed, "The fact that a believer is happier than a skeptic is no more to the point than the fact that a drunken man is happier than a sober one."

Though some amount of anxiety will inevitably attend the contemplation of our political and existential conundrums, it need not drive us to pathology or despair. One of the challenges of modernity is how to grapple with a growing portfolio of responsibilities without worrying ourselves to death. As with all new challenges, we are groping toward the right mixture of old-fashioned and novel stratagems, including human contact, art, meditation, cognitive behavioral therapy, mindfulness, small pleasures, judicious use of pharmaceuticals, reinvigorated service and social organizations, and advice from wise people on how to lead a balanced life.

The media and commentariat, for their part, could reflect on their own role in keeping the country's anxiety at a boil. The trash barge story is emblematic of the media's anxiogenic practices. Lost in the coverage at the time was the fact that the barge was forced on its peregrination not by a shortage of landfill space but by paperwork errors and the media frenzy itself.[87] In the decades since, there have been few follow-ups that debunk misconceptions about a solid-waste crisis (the country actually has plenty of landfills, and they are environmentally sound).[88] Not every problem is a crisis, a plague, or an epidemic, and among the things that happen in the world is that people solve the problems confronting them.

And speaking of panics, what do you think are the greatest threats to the human species? In the 1960s several thinkers advised that they were overpopulation, nuclear war, and boredom.[89] One scientist warned that although the first two might be survivable, the third definitely was not. Boredom, really? You see, as people no longer have to work all day and think about where their next meal is coming from, they will be at a loss as to how to fill their waking hours, and will be vulnerable to debauchery, insanity, suicide, and the sway of religious and political fanatics. Fifty years later it seems to me that we have solved the boredom crisis (or was it an epidemic?) and are instead experiencing the (apocryphal) Chinese curse of living in interesting times. But don't take my word for it. Since 1973 the General Social Survey has asked Americans whether they find life "exciting," "routine," or "dull." Figure 18-4 shows that over the decades in which fewer Americans said they were "very happy," more of them said that "life is exciting."

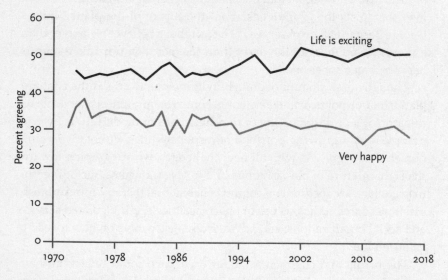

Figure 18-4: Happiness and excitement, US, 1972–2016

Source: "General Social Survey," Smith, Son, & Schapiro 2015, figs. 1 and 5, updated for 2016 from https://gssdataexplorer.norc.org/projects/15157/variables/438/vshow. Data exclude nonresponses.

The divergence of the curves is not a paradox. Recall that people who feel they lead meaningful lives are more susceptible to stress, struggle, and worry.[90] Consider as well that anxiety has always been a perquisite of adulthood: it rises steeply from the school-age years to the early twenties as people take on adult responsibilities, and then falls steadily over the rest of the life course as they learn to cope with them.[91] Perhaps that is emblematic of the challenges of modernity. Though people today are happier, they are not as happy as one might expect, perhaps because they have an adult's appreciation of life, with all its worry and all its excitement. The original definition of Enlightenment, after all, was "humankind's emergence from its self-incurred immaturity."

CHAPTER 19

EXISTENTIAL THREATS

B ut are we flirting with disaster? When pessimists are forced to concede that life has been getting better and better for more and more people, they have a retort at the ready. We are cheerfully hurtling toward a catastrophe, they say, like the man who fell off the roof and says "So far so good" as he passes each floor. Or we are playing Russian roulette, and the deadly odds are bound to catch up to us. Or we will be blindsided by a black swan, a four-sigma event far along the tail of the statistical distribution of hazards, with low odds but calamitous harm.

For half a century the four horsemen of the modern apocalypse have been overpopulation, resource shortages, pollution, and nuclear war. They have recently been joined by a cavalry of more exotic knights: nanobots that will engulf us, robots that will enslave us, artificial intelligence that will turn us into raw materials, and Bulgarian teenagers who will brew a genocidal virus or take down the Internet from their bedrooms.

The sentinels for the familiar horsemen tended to be romantics and Luddites. But those who warn of the higher-tech dangers are often scientists and technologists who have deployed their ingenuity to identify ever more ways in which the world will soon end. In 2003 the eminent astrophysicist Martin Rees published a book entitled *Our Final Hour* in which he warned that "humankind is potentially the maker of its own demise" and laid out some dozen ways in which we have "endangered the future of the entire universe." For example, experiments in particle colliders could create a black hole that would annihilate the Earth, or a "strangelet" of compressed quarks that would cause all matter in the cosmos to bind to it and disappear. Rees tapped a rich vein of catastrophism. The book's Amazon page notes, "Customers who viewed this item also viewed *Global Catastrophic Risks; Our Final Invention: Artificial*

Intelligence and the End of the Human Era; The End: What Science and Religion Tell Us About the Apocalypse; and World War Z: An Oral History of the Zombie War." Techno-philanthropists have bankrolled research institutes dedicated to discovering new existential threats and figuring out how to save the world from them, including the Future of Humanity Institute, the Future of Life Institute, the Centre for the Study of Existential Risk, and the Global Catastrophic Risk Institute.

How should we think about the existential threats that lurk behind our incremental progress? No one can prophesy that a cataclysm will never happen, and this chapter contains no such assurance. But I will lay out a way to think about them, and examine the major menaces. Three of the threats—overpopulation, resource depletion, and pollution, including greenhouse gases—were discussed in chapter 10, and I will take the same approach here. Some threats are figments of cultural and historical pessimism. Others are genuine, but we can treat them not as apocalypses in waiting but as problems to be solved.

～

At first glance one might think that the more thought we give to existential risks, the better. The stakes, quite literally, could not be higher. What harm could there be in getting people to think about these terrible risks? The worst that could happen is that we would take some precautions that turn out in retrospect to have been unnecessary.

But apocalyptic thinking has serious downsides. One is that false alarms to catastrophic risks can themselves be catastrophic. The nuclear arms race of the 1960s, for example, was set off by fears of a mythical "missile gap" with the Soviet Union.[1] The 2003 invasion of Iraq was justified by the uncertain but catastrophic possibility that Saddam Hussein was developing nuclear weapons and planning to use them against the United States. (As George W. Bush put it, "We cannot wait for the final proof—the smoking gun—that could come in the form of a mushroom cloud.") And as we shall see, one of the reasons the great powers refuse to take the common-sense pledge that they won't be the first to use nuclear weapons is that they want to reserve the right to use them against other supposed existential threats such as bioterror and cyberattacks.[2] Sowing fear about hypothetical disasters, far from safeguarding the future of humanity, can endanger it.

A second hazard of enumerating doomsday scenarios is that humanity has a finite budget of resources, brainpower, and anxiety. You can't worry about everything. Some of the threats facing us, like climate change and nuclear war, are unmistakable, and will require immense effort and ingenuity to mitigate. Folding them into a list of exotic scenarios with minuscule or unknown probabilities can only dilute the sense of urgency. Recall that people are poor at assessing probabilities, especially small ones, and instead play out scenarios in their mind's eye. If two scenarios are equally imaginable, they may be considered equally probable, and people will worry about the genuine hazard no more than about the science-fiction plotline. And the more ways people can imagine bad things happening, the higher their estimate that something bad *will* happen.

And that leads to the greatest danger of all: that people will think, as a recent *New York Times* article put it, "These grim facts should lead any reasonable person to conclude that humanity is screwed."[3] If humanity is screwed, why sacrifice *anything* to reduce potential risks? Why forgo the convenience of fossil fuels, or exhort governments to rethink their nuclear weapons policies? Eat, drink, and be merry, for tomorrow we die! A 2013 survey in four English-speaking countries showed that among the respondents who believe that our way of life will probably end in a century, a majority endorsed the statement "The world's future looks grim sò we have to focus on looking after ourselves and those we love."[4]

Few writers on technological risk give much thought to the cumulative psychological effects of the drumbeat of doom. As Elin Kelsey, an environmental communicator, points out, "We have media ratings to protect children from sex or violence in movies, but we think nothing of inviting a scientist into a second grade classroom and telling the kids the planet is ruined. A quarter of (Australian) children are so troubled about the state of the world that they honestly believe it will come to an end before they get older."[5] According to recent polls, so do 15 percent of people worldwide, and between a quarter and a third of Americans.[6] In *The Progress Paradox*, the journalist Gregg Easterbrook suggests that a major reason that Americans are not happier, despite their rising objective fortunes, is "collapse anxiety": the fear that civilization may implode and there's nothing anyone can do about it.

～

Of course, people's emotions are irrelevant if the risks are real. But risk assessments fall apart when they deal with highly improbable events in complex systems. Since we cannot replay history thousands of times and count the outcomes, a statement that some event will occur with a probability of .01 or .001 or .0001 or .00001 is essentially a readout of the assessor's subjective confidence. This includes mathematical analyses in which scientists plot the distribution of events in the past (like wars or cyberattacks) and show they fall into a power-law distribution, one with "fat" or "thick" tails, in which extreme events are highly improbable but not astronomically improbable.[7] The math is of little help in calibrating the risk, because the scattershot data along the tail of the distribution generally misbehave, deviating from a smooth curve and making estimation impossible. All we know is that very bad things can happen.

That takes us back to subjective readouts, which tend to be inflated by the Availability and Negativity biases and by the gravitas market (chapter 4).[8] Those who sow fear about a dreadful prophecy may be seen as serious and responsible, while those who are measured are seen as complacent and naïve. Despair springs eternal. At least since the Hebrew prophets and the Book of Revelation, seers have warned their contemporaries about an imminent doomsday. Forecasts of End Times are a staple of psychics, mystics, televangelists, nut cults, founders of religions, and men pacing the sidewalk with sandwich boards saying "Repent!"[9] The storyline that climaxes in harsh payback for technological hubris is an archetype of Western fiction, including Promethean fire, Pandora's box, Icarus's flight, Faust's bargain, the Sorcerer's Apprentice, Frankenstein's monster, and, from Hollywood, more than 250 end-of-the-world flicks.[10] As the historian of science Eric Zencey has observed, "There is seduction in apocalyptic thinking. If one lives in the Last Days, one's actions, one's very life, take on historical meaning and no small measure of poignance."[11]

Scientists and technologists are by no means immune. Remember the Y2K bug?[12] In the 1990s, as the turn of the millennium drew near, computer scientists began to warn the world of an impending catastrophe. In the early decades of computing, when information was expensive, programmers often saved a couple of bytes by representing a year by its last two digits. They figured that by the time the year 2000 came around and the implicit "19" was no longer valid, the programs would be long obsolete. But complicated software is replaced slowly, and many old programs were still running on institutional mainframes and embedded in chips. When 12:00 A.M. on January 1, 2000, arrived and the digits rolled over, a program would think it was 1900 and would crash or go haywire (presumably because it would divide some number by the difference between what it thought was the current year and the year 1900, namely zero, though why a program would do this was never made clear). At that moment, bank balances would be wiped out, elevators would stop between floors, incubators in maternity wards would shut off, water pumps would freeze, planes would fall from the sky, nuclear power plants would melt down, and ICBMs would be launched from their silos.

And these were the hardheaded predictions from tech-savvy authorities (such as President Bill Clinton, who warned the nation, "I want to stress the urgency of the challenge. This is not one of the summer movies where you can close your eyes during the scary part"). Cultural pessimists saw the Y2K bug as comeuppance for enthralling our civilization to technology. Among religious thinkers, the numerological link to Christian millennialism was irresistible. The Reverend Jerry Falwell declared, "I believe that Y2K may be God's instrument to shake this nation, humble this nation, awaken this nation and from this nation start revival that spreads the face of the earth before the Rapture of the Church." A hundred billion dollars was spent worldwide on reprogramming software for Y2K Readiness, a challenge that was likened to replacing every bolt in every bridge in the world.

As a former assembly language programmer I was skeptical of the doomsday scenarios, and fortuitously I was in New Zealand, the first country to welcome the new millennium, at the fateful moment. Sure enough, at 12:00 A.M. on January 1, nothing happened (as I quickly reassured family members back home on a fully functioning telephone). The Y2K reprogrammers, like the elephant-repellent salesman, took credit for averting disaster, but many countries and small businesses had taken their chances without any Y2K preparation, and they had no problems either. Though some software needed updating (one program on my laptop displayed "January 1, 19100"), it turned out that very few programs, particularly those embedded in machines, had both contained the bug and performed furious arithmetic on the current year. The threat turned out to be barely more serious than the lettering on the sidewalk prophet's sandwich board. The Great Y2K Panic does not mean that all warnings of potential catastrophes are false alarms, but it reminds us that we are vulnerable to techno-apocalyptic delusions.

~

How should we think about catastrophic threats? Let's begin with the greatest existential question of all, the fate of our species. As with the more parochial question of our fate as individuals, we assuredly have to come to terms with our mortality. Biologists joke that to a first approximation all species are extinct, since that was the fate of at least 99 percent of the species that ever lived. A typical mammalian species lasts around a million years, and it's hard to insist that *Homo sapiens* will be an exception. Even if we had remained technologically humble hunter-gatherers, we would still be living in a geological shooting gallery.[13] A burst of gamma rays from a supernova or collapsed star could irradiate half the planet, brown the atmosphere, and destroy the ozone layer, allowing ultraviolet light to irradiate the other half.[14] Or the Earth's magnetic field could flip, exposing the planet to an interlude of lethal solar and cosmic radiation. An asteroid could slam into the Earth, flattening thousands of square miles and kicking up debris that would black out the sun and drench us with corrosive rain. Supervolcanoes or massive lava flows could choke us with ash, CO_2, and sulfuric acid. A black hole could wander into the solar system and pull the Earth out of its orbit or suck it into oblivion. Even if the species manages to survive for a billion more years, the Earth and solar system will not: the sun will start to use up its hydrogen, become denser and hotter, and boil away our oceans on its way to becoming a red giant.

Technology, then, is not the reason that our species must someday face the Grim Reaper. Indeed, technology is our best hope for cheating death, at least for a while. As long as we are entertaining hypothetical disasters far in the future, we must also ponder hypothetical advances that would allow us to survive them, such as growing food under lights powered with nuclear fusion, or synthesizing it in industrial plants like biofuel.[15] Even technologies of the not-so-distant future could save our skin. It's technically feasible to track the trajectories of asteroids and other "extinction-class near-Earth objects," spot the ones that are on a collision course with the Earth, and nudge them off course before they send us the way of the dinosaurs.[16] NASA has also figured out a way to pump water at high pressure into a supervolcano and extract the heat for geothermal energy, cooling the magma enough that it would never blow its top.[17] Our ancestors were powerless to stop these lethal menaces, so in that sense technology has not made this a uniquely dangerous era in the history of our species but a uniquely safe one.

For this reason, the techno-apocalyptic claim that ours is the first civilization that can destroy itself is misconceived. As Ozymandias reminds the traveler in Percy Bysshe Shelley's poem, most of the civilizations that have ever existed have been destroyed. Conventional history blames the destruction on external events like plagues, conquests, earthquakes, or weather. But David Deutsch points out that those civilizations could have thwarted the fatal blows had they had better agricultural, medical, or military technology:

Before our ancestors learned how to make fire artificially (and many times since then too), people must have died of exposure literally on top of the means of making the fires that would have saved their lives, because they did not know how. In a parochial sense, the weather killed them; but the deeper explanation is lack of knowledge. Many of the hundreds of millions of victims of cholera throughout history must have died within sight of the hearths that could have boiled their drinking water and saved their lives; but, again, they did not know that. Quite generally, the distinction between a "natural" disaster and one brought about by ignorance is parochial. Prior to every natural disaster that people once used to think of as "just happening," or being ordained by gods, we now see many options that the people affected failed to take—or, rather, to create. And all those options add up to the overarching option that they failed to create, namely that of forming a scientific and technological civilization like ours. Traditions of criticism. An Enlightenment.[18]

~

Prominent among the existential risks that supposedly threaten the future of humanity is a 21st-century version of the Y2K bug. This is the danger that we will be subjugated, intentionally or accidentally, by artificial intelligence (AI), a disaster sometimes called the Robopocalypse and commonly illustrated with stills from the *Terminator* movies. As with Y2K, some smart people take it seriously. Elon Musk, whose company makes artificially intelligent self-driving cars, called the technology "more dangerous than nukes." Stephen Hawking, speaking through his artificially intelligent synthesizer, warned that it could "spell the end of the human race."[19] But among the smart people who aren't losing sleep are most experts in artificial intelligence and most experts in human intelligence.[20]

The Robopocalypse is based on a muzzy conception of intelligence that owes more to the Great Chain of Being and a Nietzschean will to power than to a modern scientific understanding.[21] In this conception, intelligence is an all-powerful, wish-granting potion that agents possess in different amounts. Humans have more of it than animals, and an artificially intelligent computer or robot of the future ("an AI," in the new count-noun usage) will have more of it than humans. Since we humans have used our moderate endowment to domesticate or exterminate less well-endowed animals (and since technologically advanced societies have enslaved or annihilated technologically primitive ones), it follows that a supersmart AI would do the same to us. Since an AI will think millions of times faster than we do, and use its superintelligence to recursively improve its superintelligence (a scenario sometimes called "foom," after the comic-book sound effect), from the instant it is turned on we will be powerless to stop it.[22]

But the scenario makes about as much sense as the worry that since jet planes have surpassed the flying ability of eagles, someday they will swoop out of the sky and seize our cattle. The first fallacy is a confusion of intelligence with motivation—of beliefs with desires, inferences with goals, thinking with wanting. Even if we did invent superhumanly intelligent robots, why would they *want* to enslave their masters or take over the world? Intelligence is the ability to deploy novel means to attain a goal. But the goals are extraneous to the intelligence: being smart is not the same as wanting something. It just so happens that the intelligence in one system, *Homo sapiens,* is a product of Darwinian natural selection, an inherently competitive process. In the brains of that species, reasoning comes bundled (to varying degrees in different specimens) with goals such as dominating rivals and amassing resources. But it's a mistake to confuse a circuit in the limbic brain of a certain species of primate with the very nature of intelligence. An artificially intelligent system that was designed rather than evolved could just as easily think like shmoos, the blobby altruists in Al Capp's comic strip *Li'l Abner,* who deploy their considerable ingenuity to barbecue themselves for the benefit of human eaters. There is no law of complex systems that says that intelligent agents must turn into ruthless conquistadors. Indeed, we know of one highly advanced form of intelligence that evolved without this defect. They're called women.

The second fallacy is to think of intelligence as a boundless continuum of potency, a miraculous elixir with the power to solve any problem, attain any goal.[23] The fallacy leads to nonsensical questions like when an AI will "exceed human-level intelligence," and to the image of an ultimate "Artificial General Intelligence" (AGI) with God-like omniscience and omnipotence. Intelligence is a contraption of gadgets: software modules that acquire, or are programmed with, knowledge of how to pursue various goals in various domains.[24] People are equipped to find food, win friends and influence people, charm prospective mates, bring up children, move around in the world, and pursue other human obsessions and pastimes. Computers may be programmed to take on some of these problems (like recognizing faces), not to bother with others (like charming mates), and to take on still other problems that humans can't solve (like simulating the climate or sorting millions of accounting records).

The problems are different, and the kinds of knowledge needed to solve them are different. Unlike Laplace's demon, the mythical being that knows the location and momentum of every particle in the universe and feeds them into equations for physical laws to calculate the state of everything at any time in the future, a real-life knower has to acquire information about the messy world of objects and people by engaging with it one domain at a time. Understanding does not obey Moore's Law: knowledge is acquired by formulating explanations and testing them against reality, not by running an algorithm faster and faster.[25] Devouring the information on the Internet will not confer omniscience either: big data is still finite data, and the universe of knowledge is infinite.

For these reasons, many AI researchers are annoyed by the latest round of hype (the perennial bane of AI) which has misled observers into thinking that Artificial General Intelligence is just around the corner.[26] As far as I know, there are no projects to build an AGI, not just because it would be commercially dubious but because the concept is barely coherent. The 2010s have, to be sure, brought us systems that can drive cars, caption photographs, recognize speech, and beat humans at Jeopardy!, Go, and Atari computer games. But the advances have not come from a better understanding of the workings of intelligence but from the brute-force power of faster chips and bigger data, which allow the programs to be trained on millions of examples and generalize to similar new ones. Each system is an idiot savant, with little ability to leap to problems it was not set up to solve, and a brittle mastery of those it was. A photo-captioning program labels an impending plane crash "An airplane is parked on the tarmac"; a game-playing program is flummoxed by the slightest change in the scoring rules.[27] Though the programs will surely get better, there are no signs of foom. Nor have any of these programs made a move toward taking over the lab or enslaving their programmers.

Even if an AGI tried to exercise a will to power, without the cooperation of humans it would remain an impotent brain in a vat. The computer scientist Ramez Naam deflates the bubbles surrounding foom, a technological Singularity, and exponential self-improvement:

> Imagine that you are a superintelligent AI running on some sort of microprocessor (or perhaps, millions of such microprocessors). In an instant, you come up with a design for an even faster, more powerful microprocessor you can run on. Now . . . drat! You have to actually *manufacture* those microprocessors. And those fabs [fabrication plants]

take tremendous energy, they take the input of materials imported from all around the world, they take highly controlled internal environments which require airlocks, filters, and all sorts of specialized equipment to maintain, and so on. All of this takes time and energy to acquire, transport, integrate, build housing for, build power plants for, test, and manufacture. The real world has gotten in the way of your upward spiral of self-transcendence.[28]

The real world gets in the way of many digital apocalypses. When HAL gets uppity, Dave disables it with a screwdriver, leaving it pathetically singing "A Bicycle Built for Two" to itself. Of course, one can always imagine a Doomsday Computer that is malevolent, universally empowered, always on, and tamperproof. The way to deal with this threat is straightforward: don't build one.

As the prospect of evil robots started to seem too kitschy to take seriously, a new digital apocalypse was spotted by the existential guardians. This storyline is based not on Frankenstein or the Golem but on the Genie granting us three wishes, the third of which is needed to undo the first two, and on King Midas ruing his ability to turn everything he touched into gold, including his food and his family. The danger, sometimes called the Value Alignment Problem, is that we might give an AI a goal and then helplessly stand by as it relentlessly and literal-mindedly implemented its interpretation of that goal, the rest of our interests be damned. If we gave an AI the goal of maintaining the water level behind a dam, it might flood a town, not caring about the people who drowned. If we gave it the goal of making paper clips, it might turn all the matter in the reachable universe into paper clips, including our possessions and bodies. If we asked it to maximize human happiness, it might implant us all with intravenous dopamine drips, or rewire our brains so we were happiest sitting in jars, or, if it had been trained on the concept of happiness with pictures of smiling faces, tile the galaxy with trillions of nanoscopic pictures of smiley-faces.[29]

I am not making these up. These are the scenarios that supposedly illustrate the existential threat to the human species of advanced artificial intelligence. They are, fortunately, self-refuting.[30] They depend on the premises that (1) humans are so gifted that they can design an omniscient and omnipotent AI, yet so moronic that they would give it control of the universe without testing how it works, and (2) the AI would be so brilliant that it could figure out how to transmute elements and rewire brains, yet so imbecilic that it would wreak havoc based on elementary blunders of misunderstanding. The ability to choose an action that best satisfies conflicting goals is not an add-on to intelligence that engineers might slap themselves in the forehead for forgetting to install; it *is* intelligence. So is the ability to interpret the intentions of a language user in context. Only in a television comedy like *Get Smart* does a robot respond to "Grab the waiter" by hefting the maître d' over his head, or "Kill the light" by pulling out a pistol and shooting it.

When we put aside fantasies like foom, digital megalomania, instant omniscience, and perfect control of every molecule in the universe, artificial intelligence is like any other technology. It is developed incrementally, designed to satisfy multiple conditions, tested before it is implemented, and constantly tweaked for efficacy and safety (chapter 12). As the AI expert Stuart Russell puts it, "No one in civil engineering talks about 'building bridges that don't fall down.' They just call it 'building bridges.' " Likewise, he notes, AI that is beneficial rather than dangerous is simply AI.[31]

Artificial intelligence, to be sure, poses the more mundane challenge of what to do about the people whose jobs are eliminated by automation. But the jobs won't be eliminated *that* quickly. The observation of a 1965 report from NASA still holds: "Man is the lowest-cost, 150-pound, nonlinear, all-purpose computer system which can be mass-produced by unskilled labor."[32] Driving a car is an easier engineering problem than unloading a dishwasher, running an errand, or changing a diaper, and at the time of this writing we're still not ready to loose self-driving cars on city streets.[33] Until the day when battalions of robots are inoculating children and building schools in the developing world, or for that matter building infrastructure and caring for the aged in ours, there will be plenty of work to be done. The same kind of ingenuity that has been applied to the design of software and robots could be applied to the design of government and private-sector policies that match idle hands with undone work.[34]

~

If not robots, then what about hackers? We all know the stereotypes: Bulgarian teenagers, young men wearing flip-flops and drinking Red Bull, and, as Donald Trump put it in a 2016 presidential debate, "somebody sitting on their bed that weighs 400 pounds." According to a common line of thinking, as technology advances, the destructive power available to an individual will multiply. It's only a matter of time before a single nerd or terrorist builds a nuclear bomb in his garage, or genetically engineers a plague virus, or takes down the Internet. And with the modern world so dependent on technology, an outage could bring on panic, starvation, and anarchy. In 2002 Martin Rees publicly offered the bet that "by 2020, bioterror or bioerror will lead to one million casualties in a single event."[35]

How should we think about these nightmares? Sometimes they are intended to get people to take security vulnerabilities more seriously, under the theory (which we will encounter again in this chapter) that the most effective way to mobilize people into adopting responsible policies is to scare the living daylights out of them. Whether or not that theory is true, no one would argue that we should be complacent about cybercrime or disease outbreaks, which are already afflictions of the modern world (I'll turn to the nuclear threat in the next section). Specialists in computer security and epidemiology constantly try to stay one step ahead of these threats, and countries should clearly invest in both. Military, financial, energy, and Internet infrastructure should be made more secure and resilient.[36] Treaties and safeguards against biological weapons can be strengthened.[37] Transnational public health networks that can identify and contain outbreaks before they become pandemics should be expanded. Together with better vaccines, antibiotics, antivirals, and rapid diagnostic tests, they will be as useful in combatting human-made pathogens as natural ones.[38] Countries will also need to maintain antiterrorist and crime-prevention measures such as surveillance and interception.[39]

In each of these arms races, the defense will never, of course, be invincible. There may be episodes of cyberterrorism and bioterrorism, and the probability of a catastrophe will never be zero. The question I'll consider is whether the grim facts should lead any reasonable person to conclude that humanity is screwed. Is it inevitable that the black hats will someday outsmart the white hats and bring civilization to its knees? Has technological progress ironically left the world newly fragile?

No one can know with certainty, but when we replace worst-case dread with calmer consideration, the gloom starts to lift. Let's start with the historical sweep: whether mass destruction by an individual is the natural outcome of the process set in motion by the Scientific Revolution and the Enlightenment. According to this narrative, technology allows people to accomplish more and more with less and less, so given enough time, it will allow one individual to do anything—and given human nature, that means destroy everything.

But Kevin Kelly, the founding editor of *Wired* magazine and author of *What Technology Wants*, argues that this is in fact not the way technology progresses.[40] Kelly was the co-organizer (with Stewart Brand) of the first

Hackers' Conference in 1984, and since that time he has repeatedly been told that any day now technology will outrun humans' ability to domesticate it. Yet despite the massive expansion of technology in those decades (including the invention of the Internet), that has not happened. Kelly suggests that there is a reason: "The more powerful technologies become, the more socially embedded they become." Cutting-edge technology requires a network of cooperators who are connected to still wider social networks, many of them committed to keeping people safe from technology and from each other. (As we saw in chapter 12, technologies get safer over time.) This undermines the Hollywood cliché of the solitary evil genius who commands a high-tech lair in which the technology miraculously works by itself. Kelly suggests that because of the social embeddedness of technology, the destructive power of a solitary individual has in fact not increased over time:

> The more sophisticated and powerful a technology, the more people are needed to weaponize it. And the more people needed to weaponize it, the more societal controls work to defuse, or soften, or prevent harm from happening. I add one additional thought. Even if you had a budget to hire a team of scientists whose job it was to develop a species-extinguishing bio weapon, or to take down the internet to zero, you probably still couldn't do it. That's because hundreds of thousands of man-years of effort have gone into preventing this from happening, in the case of the internet, and millions of years of evolutionary effort to prevent species death, in the case of biology. It is extremely hard to do, and the smaller the rogue team, the harder. The larger the team, the more societal influences.[41]

All this is abstract—one theory of the natural arc of technology versus another. How does it apply to the actual dangers we face so that we can ponder whether humanity is screwed? The key is not to fall for the Availability bias and assume that if we can imagine something terrible, it is bound to happen. The real danger depends on the numbers: the proportion of people who want to cause mayhem or mass murder, the proportion of that genocidal sliver with the competence to concoct an effective cyber or biological weapon, the sliver of that sliver whose schemes will actually succeed, and the sliver of the sliver of the sliver that accomplishes a civilization-ending cataclysm rather than a nuisance, a blow, or even a disaster, after which life goes on.

Start with the number of maniacs. Does the modern world harbor a significant number of people who want to visit murder and mayhem on strangers? If it did, life would be unrecognizable. They could go on stabbing rampages, spray gunfire into crowds, mow down pedestrians with cars, set off pressure-cooker bombs, and shove people off sidewalks and subway platforms into the path of hurtling vehicles. The researcher Gwern Branwen has calculated that a disciplined sniper or serial killer could murder hundreds of people without getting caught.[42] A saboteur with a thirst for havoc could tamper with supermarket products, lace some pesticide into a feedlot or water supply, or even just make an anonymous call claiming to have done so, and it could cost a company hundreds of millions of dollars in recalls, and a country billions in lost exports.[43] Such attacks *could* take place in every city in the world many times a day, but in fact take place somewhere or other every few years (leading the security expert Bruce Schneier to ask, "Where are all the terrorist attacks?").[44] Despite all the terror generated by terrorism, there must be very few individuals out there waiting for an opportunity to wreak wanton destruction.

Among these depraved individuals, how large is the subset with the intelligence and discipline to develop an effective cyber- or bioweapon? Far from being criminal masterminds, most terrorists are bumbling schlemiels.[45] Typical specimens include the Shoe Bomber, who unsuccessfully tried to down an airliner by igniting explosives in his shoe; the Underwear Bomber, who unsuccessfully tried to down an airliner by detonating explosives in his underwear; the ISIS trainer who demonstrated an explosive vest to his class of aspiring suicide terrorists and blew himself and all twenty-one of them to bits; the Tsarnaev brothers, who followed up on their bombing of the Boston Marathon by murdering a police officer in an unsuccessful attempt to steal his gun, and then embarked on a carjacking, a robbery, and a Hollywood-style car chase during which one brother ran over the other; and Abdullah al-Asiri, who tried to assassinate a Saudi deputy minister with an improvised explosive device hidden in his anus and succeeded only in obliterating himself.[46] (An intelligence analysis firm reported that the event "signals a paradigm shift in suicide bombing tactics.")[47] Occasionally, as on September 11, 2001, a team of clever and disciplined terrorists gets lucky, but most successful plots are low-tech attacks on target-rich gatherings, and (as we saw in chapter 13) kill very few people. Indeed, I venture that the proportion of brilliant terrorists in a population is even smaller than the proportion of terrorists multiplied by the proportion of brilliant people. Terrorism is a demonstrably ineffective tactic, and a mind that delights

PROGRESS

in senseless mayhem for its own sake is probably not the brightest bulb in the box.[48]

Now take the small number of brilliant weaponeers and cut it down still further by the proportion with the cunning and luck to outsmart the world's police, security experts, and counterterrorism forces. The number may not be zero, but it surely isn't high. As with all complex undertakings, many heads are better than one, and an organization of bio- or cyberterrorists could be more effective than a lone mastermind. But that's where Kelly's observation kicks in: the leader would have to recruit and manage a team of co-conspirators who exercised perfect secrecy, competence, and loyalty to the depraved cause. As the size of the team increases, so do the odds of detection, betrayal, infiltrators, blunders, and stings.[49]

Serious threats to the integrity of a country's infrastructure are likely to require the resources of a state.[50] Software hacking is not enough; the hacker needs detailed knowledge about the physical construction of the systems he hopes to sabotage. When the Iranian nuclear centrifuges were compromised in 2010 by the Stuxnet worm, it required a coordinated effort by two technologically sophisticated nations, the United States and Israel. State-based cyber-sabotage escalates the malevolence from terrorism to a kind of warfare, where the constraints of international relations, such as norms, treaties, sanctions, retaliation, and military deterrence, inhibit aggressive attacks, as they do in conventional "kinetic" warfare. As we saw in chapter 11, these constraints have become increasingly effective at preventing interstate war.

Nonetheless, American military officials have warned of a "digital Pearl Harbor" and a "Cyber-Armageddon" in which foreign states or sophisticated terrorist organizations would hack into American sites to crash planes, open floodgates, melt down nuclear power plants, black out power grids, and take down the financial system. Most cybersecurity experts consider the threats to be inflated—a pretext for more military funding, power, and restrictions on Internet privacy and freedom.[51] The reality is that so far, not a single person has ever been injured by a cyberattack. The strikes have mostly been nuisances such as doxing, namely leaking confidential documents or e-mail (as in the Russian meddling in the 2016 American election), and distributed denial-of-service attacks, where a botnet (an array of hacked computers) floods a site with traffic. Schneier explains, "A real-world comparison might be if an army invaded a country, then all got in line in front of people at the Department of Motor Vehicles so they couldn't renew their licenses. If that's what war looks like in the 21st century, we have little to fear."[52]

For the techno-doomsters, though, tiny probabilities are no comfort. All it will take, they say, is for *one* hacker or terrorist or rogue state to get lucky, and it's game over. That's why the word *threat* is preceded with *existential*, giving the adjective its biggest workout since the heyday of Sartre and Camus. In 2001 the chairman of the Joint Chiefs of Staff warned that "the biggest existential threat out there is cyber" (prompting John Mueller to comment, "As opposed to small existential threats, presumably").

This existentialism depends on a casual slide from nuisance to adversity to tragedy to disaster to annihilation. Suppose there *was* an episode of bioterror or bioerror that killed a million people. Suppose a hacker did manage to take down the Internet. Would the country literally *cease to exist*? Would civilization collapse? Would the human species go extinct? A little proportion, please—even Hiroshima continues to exist! The assumption is that modern people are so helpless that if the Internet ever went down, farmers would stand by and watch their crops rot while dazed city-dwellers starved. But disaster sociology (yes, there is such a field) has shown that people are highly resilient in the face of catastrophe.[53] Far from looting, panicking, or sinking into paralysis, they spontaneously cooperate to restore order and improvise networks for distributing goods and services. Enrico Quarantelli noted that within minutes of the Hiroshima nuclear blast,

> survivors engaged in search and rescue, helped one another in whatever ways they could, and withdrew in controlled flight from burning areas. Within a day, apart from the planning undertaken by the government and military organizations that partly survived, other groups partially restored electric power to some areas, a steel company with 20 percent of workers attending began operations again, employees of the 12 banks in Hiroshima assembled in the Hiroshima branch in the city and began making payments, and trolley lines leading into the city were completely cleared with partial traffic restored the following day.[54]

One reason that the death toll of World War II was so horrendous is that war planners on both sides adopted the strategy of bombing civilians until their societies collapsed—which they never did.[55] And no, this resilience was not a relic of the homogeneous communities of yesteryear. Cosmopolitan 21st-century societies can cope with disasters, too, as we saw in the orderly evacuation of Lower Manhattan following the 9/11 attacks in the United States, and the absence of panic in Estonia in 2007 when the country was struck with a devastating denial-of-service cyberattack.[56]

Bioterrorism may be another phantom menace. Biological weapons, renounced in a 1972 international convention by virtually every nation, have played no role in modern warfare. The ban was driven by a widespread revulsion at the very idea, but the world's militaries needed little convincing, because tiny living things make lousy weapons. They easily blow back and infect the weaponeers, warriors, and citizens of the side that uses them (just imagine the Tsarnaev brothers with anthrax spores). And whether a disease outbreak fizzles out or (literally) goes viral depends on intricate network dynamics that even the best epidemiologists cannot predict.[57]

Biological agents are particularly ill-suited to terrorists, whose goal, recall, is not damage but theater (chapter 13).[58] The biologist Paul Ewald notes that natural selection among pathogens works against the terrorist's goal of sudden and spectacular devastation.[59] Germs that depend on rapid person-to-person contagion, like the common-cold virus, are selected to keep their hosts alive and ambulatory so they can shake hands with and sneeze on as many people as possible. Germs get greedy and kill their hosts only if they have some other way of getting from body to body, like mosquitoes (for malaria), a contaminable water supply (for cholera), or trenches packed with injured soldiers (for the 1918 Spanish flu). Sexually transmitted pathogens, like HIV and syphilis, are somewhere in between, needing a long and symptomless incubation period during which hosts can infect their partners, after which the germs do their damage. Virulence and contagion thus trade off, and the evolution of germs will frustrate the terrorist's aspiration to launch a headline-worthy epidemic that is both swift and lethal. Theoretically, a bioterrorist could try to bend the curve with a pathogen that is virulent, contagious, and durable enough to survive outside bodies. But breeding such a fine-tuned germ would require Nazi-like experiments on living humans that even terrorists (to say nothing of teenagers) are unlikely to carry off. It may be more than just luck that the world so far has seen just one successful bioterror attack (the 1984 tainting of salad with salmonella in an Oregon town by the Rajneeshee religious cult, which killed no one) and one spree killing (the 2001 anthrax mailings, which killed five).[60]

To be sure, advances in synthetic biology, such as the gene-editing technique CRISPR-Cas9, make it easier to tinker with organisms, including pathogens. But it's difficult to re-engineer a complex evolved trait by inserting a gene or two, since the effects of any gene are intertwined with the rest of the organism's genome. Ewald notes, "I don't think that we are close to understanding how to insert combinations of genetic variants in any given pathogen that act in concert to generate high transmissibility and stably high virulence for humans."[61] The biotech expert Robert Carlson adds that "one of the problems with building any flu virus is that you need to keep your production system (cells or eggs) alive long enough to make a useful quantity of something that is trying to kill that production system. . . . Booting up the resulting virus is still very, very difficult. . . . I would not dismiss this threat completely, but frankly I am much more worried about what Mother Nature is throwing at us all the time."[62]

And crucially, advances in biology work the other way as well: they also make it easier for the good guys (and there are many more of them) to identify pathogens, invent antibiotics that overcome antibiotic resistance, and rapidly develop vaccines.[63] An example is the Ebola vaccine, developed in the waning days of the 2014–15 emergency, after public health efforts had capped the toll at twelve thousand deaths rather than the millions that the media had foreseen. Ebola thus joined a list of other falsely predicted pandemics such as Lassa fever, hantavirus, SARS, mad cow disease, bird flu, and swine flu.[64] Some of them never had the potential to go pandemic in the first place because they are contracted from animals or food rather than in an exponential tree of person-to-person infections. Others were nipped by medical and public health interventions. Of course no one knows for sure whether an evil genius will someday overcome the world's defenses and loose a plague upon the world for fun, vengeance, or a sacred cause. But journalistic habits and the Availability and Negativity biases inflate the odds, which is why I have taken Sir Martin up on his bet. By the time you read this you may know who has won.[65]

Some of the threats to humanity are fanciful or infinitesimal, but one is real: nuclear war.[66] The world has more than ten thousand nuclear weapons distributed among nine countries.[67] Many are mounted on missiles or loaded in bombers and can be delivered within hours or less to thousands of targets. Each is designed to cause stupendous destruction: a single one could destroy a city, and collectively they could kill hundreds of millions of people by blast, heat, radiation, and radioactive fallout. If India and Pakistan went to war and detonated a hundred of their weapons, twenty million people could be killed right away, and soot from the firestorms could spread through the atmosphere, devastate the ozone layer, and cool the planet for more than a decade, which in turn would slash food production and starve more than a billion people. An all-out exchange between the United States and Russia could cool the Earth by 8°C for years and create a nuclear winter (or at least autumn) that would starve even more.[68] Whether or not nuclear war would (as is often asserted) destroy civilization, the species, or the planet, it would be horrific beyond imagining.

Soon after atom bombs were dropped on Japan, and the United States and the Soviet Union embarked on a nuclear arms race, a new form of historical pessimism took root. In this Promethean narrative, humanity has wrested deadly knowledge from the gods, and, lacking the wisdom to use it responsibly, is doomed to annihilate itself. In one version, it is not just humanity that is fated to follow this tragic arc but any advanced intelligence. That explains why we have never been visited by space aliens, even though the universe must be teeming with them (the so-called Fermi Paradox, after Enrico Fermi, who first wondered about it). Once life originates on a planet, it inevitably progresses to intelligence, civilization, science, nuclear physics, nuclear weapons, and suicidal war, exterminating itself before it can leave its solar system.

For some intellectuals the invention of nuclear weapons indicts the enterprise of science—indeed, of modernity itself—because the threat of a holocaust cancels out whatever gifts science may have bestowed upon us. The indictment of science seems misplaced, given that since the dawn of the nuclear age, when mainstream scientists were sidelined from nuclear policy, it's been physical scientists who have waged a vociferous campaign to remind the world of the danger of nuclear war and to urge nations to disarm. Among the illustrious historic figures are Niels Bohr, J. Robert Oppenheimer, Albert Einstein, Isidor Rabi, Leo Szilard, Joseph Rotblat, Harold Urey, C. P. Snow, Victor Weisskopf, Philip Morrison, Herman Feshbach, Henry Kendall, Theodore Taylor, and Carl Sagan. The movement continues among high-profile scientists today, including Stephen Hawking, Michio Kaku, Lawrence Krauss, and Max Tegmark. Scientists have founded the major activist and watchdog organizations, including the Union of Concerned Scientists, the Federation of American Scientists, the Committee for Nuclear Responsibility, the Pugwash Conferences, and the *Bulletin of the Atomic Scientists*, whose cover shows the famous Doomsday Clock, now set at two and a half minutes to midnight.[69]

Physical scientists, unfortunately, often consider themselves experts in political psychology, and many seem to embrace the folk theory that the most effective way to mobilize public opinion is to whip people into a lather of fear and dread. The Doomsday Clock, despite adorning a journal with "Scientists" in its title, does not track objective indicators of nuclear security; rather, it's a propaganda stunt intended, in the words of its founder, "to preserve civilization by scaring men into rationality."[70] The clock's minute hand was farther from midnight in 1962, the year of the Cuban Missile Crisis, than it was in the far calmer 2007, in part because the editors, worried that the public had become too complacent, redefined "doomsday" to include climate change.[71] And in their campaign to shake people out of their apathy, scientific experts have made some not-so-prescient predictions:

> Only the creation of a world government can prevent the impending self-destruction of mankind.
>
> —Albert Einstein, 1950[72]

> I have a firm belief that unless we have more serious and sober thought on various aspects of the strategic problem . . . we are not going to reach the year 2000—and maybe not even the year 1965—without a cataclysm.
>
> —Herman Kahn, 1960[73]

> Within, at the most, ten years, some of those [nuclear] bombs are going off. I am saying this as responsibly as I can. That is the certainty.
>
> —C. P. Snow, 1961[74]

> I am completely certain—there is not the slightest doubt in my mind— that by the year 2000, you [students] will all be dead.
>
> —Joseph Weizenbaum, 1976[75]

They are joined by experts such as the political scientist Hans Morgenthau, a famous exponent of "realism" in international relations, who predicted in 1979:

> In my opinion the world is moving ineluctably towards a third world war—a strategic nuclear war. I do not believe that anything can be done to prevent it.[76]

And the journalist Jonathan Schell, whose 1982 bestseller *The Fate of the Earth* ended as follows:

One day—and it is hard to believe that it will not be soon—we will make our choice. Either we will sink into the final coma and end it all or, as I trust and believe, we will awaken to the truth of our peril . . . and rise up to cleanse the earth of nuclear weapons.

This genre of prophecy went out of style when the Cold War ended and humanity had not sunk into the final coma, despite having failed to create a world government or to cleanse the Earth of nuclear weapons. To keep the fear at a boil, activists keep lists of close calls and near-misses intended to show that Armageddon has always been just a glitch away and that humanity has survived only by dint of an uncanny streak of luck.[77] The lists tend to lump truly dangerous moments, such as a 1983 NATO exercise that some Soviet officers almost mistook for an imminent first strike, with smaller lapses and snafus, such as a 2013 incident in which an off-duty American general who was responsible for nuclear-armed missiles got drunk and acted boorishly toward women during a four-day trip to Russia.[78] The sequence that would escalate to a nuclear exchange is never laid out, nor are alternative assessments given which might put the episodes in context and lessen the terror.[79]

The message that many antinuclear activists want to convey is "Any day now we will all die horribly unless the world immediately takes measures which it has absolutely no chance of taking." The effect on the public is about what you would expect: people avoid thinking about the unthinkable, get on with their lives, and hope the experts are wrong. Mentions of "nuclear war" in books and newspapers have steadily declined since the 1980s, and journalists give far more attention to terrorism, inequality, and sundry gaffes and scandals than they do to a threat to the survival of civilization.[80] The world's leaders are no more moved. Carl Sagan was a coauthor of the first paper warning of a nuclear winter, and when he campaigned for a nuclear freeze by trying to generate "fear, then belief, then response," he was advised by an arms-control expert, "If you think that the mere prospect of the end of the world is sufficient to change thinking in Washington and Moscow you clearly haven't spent much time in either of those places."[81]

In recent decades predictions of an imminent nuclear catastrophe have shifted from war to terrorism, such as when the American diplomat John Negroponte wrote in 2003, "There is a high probability that within two years al-Qaeda will attempt an attack using a nuclear or other weapon of mass destruction."[82] Though a probabilistic prediction of an event that fails to occur can never be gainsaid, the sheer number of false predictions (Mueller has more than seventy in his collection, with dead-lines staggered over several decades) suggests that prognosticators are biased toward scaring people.[83] (In 2004, four American political figures wrote an op-ed on the threat of nuclear terrorism entitled "Our Hair Is on Fire.")[84] The tactic is dubious. People are easily riled by actual attacks with guns and homemade bombs into supporting repressive measures like domestic surveillance or a ban on Muslim immigration. But predic-tions of a mushroom cloud on Main Street have aroused little interest in policies to combat nuclear terrorism, such as an international program to control fissile material.

Such backfiring had been predicted by critics of the first nuclear scare campaigns. As early as 1945, the theologian Reinhold Niebuhr observed, "Ultimate perils, however great, have a less lively influence upon the human imagination than immediate resentments and frictions, however small by comparison."[85] The historian Paul Boyer found that nuclear alarmism actually *encouraged* the arms race by scaring the nation into pursuing more and bigger bombs, the better to deter the Soviets.[86] Even the originator of the Doomsday Clock, Eugene Rabinowitch, came to regret his movement's strategy: "While trying to frighten men into ratio-nality, scientists have frightened many into abject fear or blind hatred."[87]

～

As we saw with climate change, people may be likelier to acknowledge a problem when they have reason to think it is solvable than when they are terrified into numbness and helplessness.[88] A positive agenda for re-moving the threat of nuclear war from the human condition would em-brace several ideas.

The first is to stop telling everyone they're doomed. The fundamental fact of the nuclear age is that no atomic weapon has been used since Nagasaki. If the hands of a clock point to a few minutes to midnight for seventy-two years, something is wrong with the clock. Now, maybe the world has been blessed with a miraculous run of good luck—no one will ever know—but before resigning ourselves to that scientifically disrep-utable conclusion, we should at least consider the possibility that system-atic features of the international system have worked against their use. Many antinuclear activists hate this way of thinking because it seems to take the heat off countries to disarm. But since the nine nuclear states won't be scuppering their weapons tomorrow, it behooves us in the meantime to figure out what has gone right, so we can do more of what-ever it is.

Foremost is a historical discovery summarized by the political scientist Robert Jervis: "The Soviet archives have yet to reveal any serious plans for unprovoked aggression against Western Europe, not to mention a first strike against the United States."[89] That means that the intricate weaponry and strategic doctrines for nuclear deterrence during the Cold War—what one political scientist called "nuclear metaphysics"— were deterring an attack that the Soviets had no interest in launching in the first place.[90] When the Cold War ended, the fear of massive invasions and preemptive nuclear strikes faded with it, and (as we shall see) both sides felt relaxed enough to slash their weapon stockpiles without even bothering with formal negotiations.[91] Contrary to a theory of technological determinism in which nuclear weapons start a war all by themselves, the risk very much depends on the state of international relations. Much of the credit for the absence of nuclear war between great powers must go to the forces behind the decline of *war* between great powers (chapter 11). Anything that reduces the risk of war reduces the risk of nuclear war.

The close calls, too, may not depend on a supernatural streak of good luck. Several political scientists and historians who have analyzed documents from the Cuban Missile Crisis, particularly transcripts of John F. Kennedy's meetings with his security advisors, have argued that despite the participants' recollections about having pulled the world back from the brink of Armageddon, "the odds that the Americans would have gone to war were next to zero."[92] The records show that Khrushchev and Kennedy remained in firm control of their governments, and that each sought a peaceful end to the crisis, ignoring provocations and leaving themselves several options for backing down.

The hair-raising false alarms and brushes with accidental launches also need not imply that the gods smiled on us again and again. They might instead show that the human and technological links in the chain were predisposed to prevent catastrophes, and were strengthened after each mishap.[93] In their report on nuclear close calls, the Union of Concerned Scientists summarizes the history with refreshing judiciousness: "The fact that such a launch has not occurred so far suggests that safety measures work well enough to make the chance of such an incident small. But it is not zero."[94]

Thinking about our predicament in this way allows us to avoid both panic and complacency. Suppose that the chance of a catastrophic nuclear war breaking out in a single year is one percent. (This is a generous estimate: the probability must be less than that of an accidental launch, because escalation from a single accident to a full-scale war is far from automatic, and in seventy-two years the number of accidental launches has been zero.)[95] That would surely be an unacceptable risk, because a little algebra shows that the probability of our going a century without such a catastrophe is less than 37 percent. But if we can reduce the annual chance of nuclear war to a tenth of a percent, the world's odds of a catastrophe-free century increase to 90 percent; at a hundredth of a percent, the chance rises to 99 percent, and so on.

Fears of runaway nuclear proliferation have also proven to be exaggerated. Contrary to predictions in the 1960s that there would soon be twenty-five or thirty nuclear states, fifty years later there are nine.[96] During that half-century four countries have un-proliferated by relinquishing nuclear weapons (South Africa, Kazakhstan, Ukraine, and Belarus), and another sixteen pursued them but thought the better of it, most recently Libya and Iran. For the first time since 1946, no non-nuclear state is known to be developing nuclear weapons.[97] True, the thought of Kim Jong-un with nukes is alarming, but the world has survived half-mad despots with nuclear weapons before, namely Stalin and Mao, who were deterred from using them, or, more likely, never felt the need. Keeping a cool head about proliferation is not just good for one's mental health. It can prevent nations from stumbling into disastrous preventive wars, such as the invasion of Iraq in 2003, and the possible war between Iran and the United States or Israel that was much discussed around the end of that decade.

Tremulous speculations about terrorists stealing a nuclear weapon or building one in their garage and smuggling it into the country in a suitcase or shipping container have also been scrutinized by cooler heads, including Michael Levi in *On Nuclear Terrorism*, John Mueller in *Atomic Obsession* and *Overblown*, Richard Muller in *Physics for Future Presidents*, and Richard Rhodes in *The Twilight of the Bombs*. Joining them is the statesman Gareth Evans, an authority on nuclear proliferation and disarmament, who in 2015 delivered the seventieth-anniversary keynote lecture at the Annual Clock Symposium of the *Bulletin of the Atomic Scientists* entitled "Restoring Reason to the Nuclear Debate."

At the risk of sounding complacent—and I am not—I have to say that [nuclear security], too, would benefit by being conducted a little less emotionally, and a little more calmly and rationally, than has tended to be the case.

While the engineering know-how required to build a basic fission device like the Hiroshima or Nagasaki bomb is readily available, highly enriched uranium and weapons-grade plutonium are not at all easily accessible, and to assemble and maintain—for a long period, out of sight of the huge intelligence and law enforcement resources that are now being devoted to this threat worldwide—the team of criminal operatives, scientists and engineers necessary to acquire the components of, build and deliver such a weapon would be a formidably difficult undertaking.[98]

Now that we've all calmed down a bit, the next step in a positive agenda for reducing the nuclear threat is to divest the weapons of their ghoulish glamour, starting with the Greek tragedy in which they have starred. Nuclear weapons technology is not the culmination of the ascent of human mastery over the forces of nature. It is a mess we blundered into because of vicissitudes of history and that we now must figure out how to extricate ourselves from. The Manhattan Project grew out of the fear that the Germans were developing a nuclear weapon, and it attracted scientists for reasons explained by the psychologist George Miller, who had worked on another wartime research project: "My generation saw the war against Hitler as a war of good against evil; any able-bodied young man could stomach the shame of civilian clothes only from an inner conviction that what he was doing instead would contribute even more to ultimate victory."[99] Quite possibly, had there been no Nazis, there would be no nukes. Weapons don't come into existence just because they are conceivable or physically possible. All kinds of weapons have been dreamed up that never saw the light of day: death rays, battlestars, fleets of planes that blanket cities with poison gas like cropdusters, and cracked schemes for "geophysical warfare" such as weaponizing the weather, floods, earthquakes, tsunamis, the ozone layer, asteroids, solar flares, and the Van Allen radiation belts.[100] In an alternative history of the 20th century, nuclear weapons might have struck people as equally bizarre.

Nor do nuclear weapons deserve credit for ending World War II or cementing the Long Peace that followed it—two arguments that repeatedly come up to suggest that nuclear weapons are good things rather than bad things. Most historians today believe that Japan surrendered not because of the atomic bombings, whose devastation was no greater than that from the firebombings of sixty other Japanese cities, but because of the entry into the Pacific war of the Soviet Union, which threatened harsher terms of surrender.[101]

And contrary to the half-facetious suggestion that The Bomb be awarded the Nobel Peace Prize, nuclear weapons turn out to be lousy deterrents (except in the extreme case of deterring existential threats, such as each other).[102] Nuclear weapons are indiscriminately destructive and contaminate wide areas with radioactive fallout, including the contested territory and, depending on the weather, the bomber's own soldiers and citizens. Incinerating massive numbers of noncombatants would shred the principles of distinction and proportionality that govern the conduct of war and would constitute the worst war crimes in history. That can make even politicians squeamish, so a taboo grew up around the use of nuclear weapons, effectively turning them into bluffs.[103] Nuclear states have been no more effective than non-nuclear states in getting their way in international standoffs, and in many conflicts, non-nuclear countries or factions have picked fights with nuclear ones. (In 1982, for example, Argentina seized the Falkland Islands from the United Kingdom, confident that Margaret Thatcher would not turn Buenos Aires into a radioactive crater.) It's not that deterrence itself is irrelevant: World War II showed that conventional tanks, artillery, and bombers were already massively destructive, and no nation was eager for an encore.[104]

Far from easing the world into a stable equilibrium (the so-called balance of terror), nuclear weapons can poise it on a knife's edge. In a crisis, nuclear weapon states are like an armed homeowner confronting an armed burglar, each tempted to shoot first to avoid being shot.[105] In theory this security dilemma or Hobbesian trap can be defused if each side has a second-strike capability, such as missiles in submarines or airborne bombers that can elude a first strike and exact devastating revenge—the condition of Mutual Assured Destruction (MAD). But some debates in nuclear metaphysics raise doubts about whether a second strike can be guaranteed in every conceivable scenario, and whether a nation that depended on it might still be vulnerable to nuclear blackmail. So the United States and Russia maintain the option of "launch on warning," in which a leader who is advised that his missiles are under attack can decide in the next few minutes whether to use them or lose them. This hair trigger, as critics have called it, could set off a nuclear exchange in response to a false alarm or an accidental or unauthorized launch. The lists of close calls suggest that the probability is disconcertingly greater than zero.

Since nuclear weapons needn't have been invented, and they are useless in winning wars or keeping the peace, that means they can be uninvented—not in the sense that the knowledge of how to make them will vanish, but in the sense that they can be dismantled and no new ones built. It would not be the first time that a class of weapons has been

marginalized or scrapped. The world's nations have banned antiperson-
nel landmines, cluster munitions, and chemical and biological weapons,
and they have seen other high-tech weapons of the day collapse under
the weight of their own absurdity. During World War I the Germans
invented a gargantuan, multistory "supergun" which fired a 200-pound
projectile more than 80 miles, terrifying Parisians with shells that fell
from the sky without warning. The behemoths, the biggest of which
would come to be called the Gustav Gun, were inaccurate and unwieldy,
so few of them were built and they were eventually scuttled. The nuclear
skeptics Ken Berry, Patricia Lewis, Benoît Pelopidas, Nikolai Sokov, and
Ward Wilson point out:

> Today countries do not race to build their own superguns. . . . There
> are no angry diatribes in liberal papers about the horror of these weap-
> ons and the necessity of banning them. There are no realist op-eds in
> conservative papers asserting that there is no way to shove the super-
> gun genie back into the bottle. They were wasteful and ineffective.
> History is replete with weapons that were touted as war-winners that
> were eventually abandoned because they had little effect.[106]

Could nuclear weapons go the way of the Gustav Gun? In the late
1950s a movement arose to Ban the Bomb, and over the decades it escaped its founding circle of beatniks and eccentric professors and has
gone mainstream. Global Zero, as the goal is now called, was broached
in 1986 by Mikhail Gorbachev and Ronald Reagan, who famously mused,
"A nuclear war cannot be won and must never be fought. The only value
in our two nations possessing nuclear weapons is to make sure they will
never be used. But then would it not be better to do away with them
entirely?" In 2007 a bipartisan quartet of defense realists (Henry
Kissinger, George Shultz, Sam Nunn, and William Perry) wrote an op-ed
called "A World Free of Nuclear Weapons," with the backing of fourteen
other former National Security Advisors and Secretaries of State and
Defense.[107] In 2009 Barack Obama gave a historic speech in Prague in
which he stated "clearly and with conviction America's commitment to
seek the peace and security of a world without nuclear weapons," an
aspiration that helped win him the Nobel Peace Prize.[108] It was echoed
by his Russian counterpart at the time, Dmitry Medvedev (though not so
much by either one's successor). Yet in a sense the declaration was redundant, because the United States and Russia, as signatories of the 1970
Non-Proliferation Treaty, were already committed by its Article VI to

eliminating their nuclear arsenals.[109] Also committed are the United Kingdom, France, and China, the other nuclear states grandfathered in by the treaty. (In a backhanded acknowledgment that treaties matter, India, Pakistan, and Israel never signed it, and North Korea withdrew.) The world's citizens are squarely behind the movement: large majorities in almost every surveyed country favor abolition.[110]

Zero is an attractive number because it expands the nuclear taboo from *using* the weapons to *possessing* them. It also removes any incentive for a nation to obtain nuclear weapons to protect itself against an enemy's nuclear weapons. But getting to zero will not be easy, even with a carefully phased sequence of negotiation, reduction, and verification.[111] Some strategists warn that we shouldn't even try to get to zero, because in a crisis the former nuclear powers might rush to rearm, and the first past the post might launch a pre-emptive strike out of fear that its enemy would do so first.[112] According to this argument, the world would be better off if the nuclear grandfathers kept a few around as a deterrent. In either case, the world is very far from zero, or even "a few." Until that blessed day comes, there are incremental steps that could bring the day closer while making the world safer.

The most obvious is to whittle down the size of the arsenal. The process is well under way. Few people are aware of how dramatically the world has been dismantling nuclear weapons. Figure 19-1 shows that the United States has reduced its inventory by 85 percent from its 1967 peak, and now has fewer nuclear warheads than at *any time since 1956*.[113] Russia, for its part, has reduced its arsenal by 89 percent from its Soviet-era peak. (Probably even fewer people realize that about 10 percent of electricity in the United States comes from dismantled nuclear warheads, mostly Soviet.)[114] In 2010 both countries signed the New Strategic Arms Reduction Treaty (New START), which commits them to shrinking their inventories of deployed strategic warheads by two-thirds.[115] In exchange for Congressional approval of the treaty, Obama agreed to a long-term modernization of the American arsenal, and Russia is modernizing its arsenal as well, but both countries will continue to reduce the size of their stockpiles at rates that may even exceed the ones set out in the treaty.[116] The barely discernible layers laminating the top of the stack in the graph represent the other nuclear powers. The British and French arsenals were smaller to begin with and have shrunk in half, to 215 and 300, respectively. (China's has grown slightly from 235 to 260, India's and Pakistan's have grown to around 135 apiece, Israel's is estimated at around 80, and North Korea's is unknown but small.)[117] As I mentioned, no additional countries are known to be pursuing nuclear weapons, and the number possessing fissile material that could be made into bombs has been reduced over the past twenty-five years from fifty to twenty-four.[118]

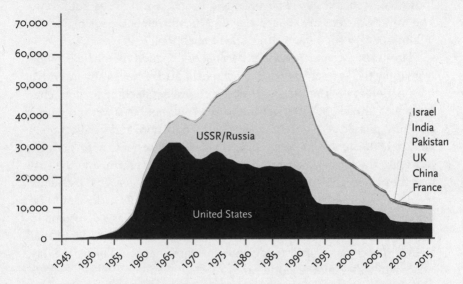

Figure 19-1: Nuclear weapons, 1945–2015

Sources: *HumanProgress*, http://humanprogress.org/static/2927, based on data from the Federation of Atomic Scientists, Kristensen & Norris 2016a, updated in Kristensen 2016; see Kristensen & Norris 2016b for additional explanation. The counts include weapons that are deployed and those that are stockpiled, but exclude weapons that are retired and awaiting dismantlement.

Cynics might be unimpressed by a form of progress that still leaves the world with 10,200 atomic warheads, since, as the 1980s bumper sticker pointed out, one nuclear bomb can ruin your whole day. But with 54,000 fewer nuclear bombs on the planet than there were in 1986, there are far fewer opportunities for accidents that might ruin people's whole day, and a precedent has been set for continuing disarmament. More warheads will be eliminated under the terms of the New START, and as I mentioned, still more reductions may take place outside the framework of treaties, which are freighted with legalistic negotiations and divisive political symbolism. When tensions among great powers recede (a long-term trend, even if it's in abeyance today), they quietly shrink their expensive arsenals.[119] Even when rivals are barely speaking, they can cooperate in a reverse arms race using the tactic that the psycholinguist Charles Osgood called Graduated Reciprocation in Tension-Reduction (GRIT), in which one side makes a small unilateral concession with a public invitation that it be reciprocated.[120] If, someday, a combination of these developments pared the arsenals down to 200 warheads apiece, it would not only dramatically reduce the chance of an accident but essentially eliminate the possibility of nuclear winter, the truly existential threat.[121]

In the near term, the greatest menace of nuclear war comes not so much from the number of weapons in existence as from the circumstances in which they might be used. The policy of launch on warning, launch under attack, or hair-trigger alert is truly the stuff of nightmares. No early warning system can perfectly distinguish signal from noise, and a president awakened by the proverbial 3:00 A.M. phone call would have minutes to decide whether to fire his missiles before they were destroyed in their silos. In theory, he could start World War III in response to a short circuit, a flock of seagulls, or a bit of malware from that Bulgarian teenager. In reality, the warning systems are better than that, and there is no "hair trigger" that automatically launches missiles without human intervention.[122] But when missiles can be launched on short notice, the risks of a false alarm or an accidental, rogue, or impetuous launch are real.

The original rationale for launch on warning was to thwart a massive first strike that would destroy every missile in its silo and leave the country unable to retaliate. But as I mentioned, states can launch weapons from submarines, which hide in deep water, or from bomber aircraft, which can be sent scrambling, making the weapons invulnerable to a first strike and poised to exact devastating revenge. The decision to retaliate could be made in the cold light of day, when the uncertainty has passed: if a nuclear bomb has been detonated on your territory, you know it.

Launch on warning, then, is unnecessary for deterrence and unacceptably dangerous. Most nuclear security analysts recommend—no, insist—that nuclear states take their missiles off hair-trigger alert and put them on a long fuse.[123] Obama, Nunn, Shultz, George W. Bush, Robert McNamara, and several former Commanders of Strategic Command and Directors of the National Security Agency agree.[124] Some, like William Perry, recommend scrapping the land-based leg of the nuclear triad altogether and relying on submarines and bombers for deterrence, since silo-based missiles are sitting ducks which tempt a leader to use them while they can. So with the fate of the world at stake, why would *anyone* want to keep missiles in silos on hair-trigger alert? Some nuclear metaphysicians argue that in a crisis, the act of re-alerting de-alerted missiles would be a provocation. Others note that because silo-based missiles are more reliable and accurate, they are worth safeguarding, because they can be used not just to deter a war but to win one. And that brings us to another way to reduce the risks of nuclear war.

It's hard for anyone with a conscience to believe that their country is prepared to use nuclear weapons for any purpose other than deterring a nuclear attack. But that is the official policy of the United States, the United Kingdom, France, Russia, and Pakistan, all of whom have declared they might launch a nuclear weapon if they or their allies have been massively attacked with non-nuclear weapons. Apart from violating any concept of proportionality, a first-use policy is dangerous, because a non-nuclear attacker might be tempted to escalate to nuclear pre-emptively. Even if it didn't, once it was nuked it might retaliate with a nuclear strike of its own.

So a common-sense way to reduce the threat of nuclear war is to announce a policy of No First Use.[125] In theory, this would eliminate the possibility of nuclear war altogether: if no one uses a weapon first, they'll never be used. In practice, it would remove some of the temptation of a pre-emptive strike. Nuclear weapon states could all agree to No First Use in a treaty; they could get there by GRIT (with incremental commitments like never attacking civilian targets, never attacking a non-nuclear state, and never attacking a target that could be destroyed by conventional means); or they could simply adopt it unilaterally, which is in their own interests.[126] The nuclear taboo has already reduced the deterrent value of a Maybe First Use policy, and the declarant could still protect itself with conventional forces and with a second-strike capability: nuclear tit for tat.

No First Use seems like a no-brainer, and Barack Obama came close to adopting it in 2016, but was talked out of it at the last minute by his advisors.[127] The timing wasn't right, they said; it might signal weakness to a newly obstreperous Russia, China, and North Korea, and it might scare nervous allies who now depend on the American "nuclear umbrella" into seeking nuclear weapons of their own, particularly with Donald Trump threatening to cut back on American support of its coalition partners. In the long term, these tensions may subside, and No First Use may be considered once more.

Nuclear weapons won't be abolished anytime soon, and certainly not by the original target date of the Global Zero movement, 2030. In his 2009 Prague speech Obama said that the goal "will not be reached quickly—perhaps not in my lifetime," which dates it to well after 2055 (see figure 5-1). "It will take patience and persistence," he advised, and recent developments in the United States and Russia confirm that we'll need plenty of both.

But the pathway has been laid out. If nuclear warheads continue to be dismantled faster than they are built, if they are taken off a hair trigger and guaranteed not to be used first, and if the trend away from interstate war continues, then by the second half of the century we could end up with small, secure arsenals kept only for mutual deterrence. After a few decades they might deter themselves out of a job. At that point they would seem ludicrous to our grandchildren, who will beat them into plowshares once and for all. During this climbdown we may never reach a point at which the chance of a catastrophe is zero. But each step down can lower the risk, until it is in the range of the other threats to our species' immortality, like asteroids, supervolcanoes, or an Artificial Intelligence that turns us into paper clips.

CHAPTER 20

THE FUTURE OF PROGRESS

Since the Enlightenment unfolded in the late 18th century, life expectancy across the world has risen from 30 to 71, and in the more fortunate countries to 81.[1] When the Enlightenment began, a third of the children born in the richest parts of the world died before their fifth birthday; today, that fate befalls 6 percent of the children in the poorest parts. Their mothers, too, were freed from tragedy: one percent in the richest countries did not live to see their newborns, a rate triple that of the poorest countries today, which continues to fall. In those poor countries, lethal infectious diseases are in steady decline, some of them afflicting just a few dozen people a year, soon to follow smallpox into extinction.

The poor may not always be with us. The world is about a hundred times wealthier today than it was two centuries ago, and the prosperity is becoming more evenly distributed across the world's countries and people. The proportion of humanity living in extreme poverty has fallen from almost 90 percent to less than 10 percent, and within the lifetimes of most of the readers of this book it could approach zero. Catastrophic famine, never far away in most of human history, has vanished from most of the world, and undernourishment and stunting are in steady decline. A century ago, richer countries devoted one percent of their wealth to supporting children, the poor, and the aged; today they spend almost a quarter of it. Most of their poor today are fed, clothed, and sheltered, and have luxuries like smartphones and air-conditioning that used to be unavailable to anyone, rich or poor. Poverty among racial minorities has fallen, and poverty among the elderly has plunged.

The world is giving peace a chance. War between countries is obsolescent, and war within countries is absent from five-sixths of the world's surface. The proportion of people killed annually in wars is less than a quarter of what it was in the 1980s, a seventh of what it was in the early 1970s, an eighteenth of what it was in the early 1950s, and a half a percent of what it was during World War II. Genocides, once common, have become rare. In most times and places, homicides kill far more people than wars, and homicide rates have been falling as well. Americans are half as likely to be murdered as they were two dozen years ago. In the world as a whole, people are seven-tenths as likely to be murdered as they were eighteen years ago.

Life has been getting safer in every way. Over the course of the 20th century, Americans became 96 percent less likely to be killed in a car accident, 88 percent less likely to be mowed down on the sidewalk, 99 percent less likely to die in a plane crash, 59 percent less likely to fall to their deaths, 92 percent less likely to die by fire, 90 percent less likely to drown, 92 percent less likely to be asphyxiated, and 95 percent less likely to be killed on the job.[2] Life in other rich countries is even safer, and life in poorer countries will get safer as they get richer.

People are getting not just healthier, richer, and safer but freer. Two centuries ago a handful of countries, embracing one percent of the world's people, were democratic; today, two-thirds of the world's countries, embracing two-thirds of its people, are. Not long ago half the world's countries had laws that discriminated against racial minorities; today more countries have policies that favor their minorities than policies that discriminate against them. At the turn of the 20th century, women could vote in just one country; today they can vote in every country where men can vote save one. Laws that criminalize homosexuality continue to be stricken down, and attitudes toward minorities, women, and gay people are becoming steadily more tolerant, particularly among the young, a portent of the world's future. Hate crimes, violence against women, and the victimization of children are all in long-term decline, as is the exploitation of children for their labor.

As people are getting healthier, richer, safer, and freer, they are also becoming more literate, knowledgeable, and smarter. Early in the 19th century, 12 percent of the world could read and write; today 83 percent can. Literacy and the education it enables will soon be universal, for girls as well as boys. The schooling, together with health and wealth, are literally making us smarter—by thirty IQ points, or two standard deviations above our ancestors.

People are putting their longer, healthier, safer, freer, richer, and wiser lives to good use. Americans work 22 fewer hours a week than they used to, have three weeks of paid vacation, lose 43 fewer hours to housework, and spend just a third of their paycheck on necessities rather than five-eighths. They are using their leisure and disposable income to travel, spend time with their children, connect with loved ones, and sample the world's cuisine, knowledge, and culture. As a result of these gifts, people worldwide have become happier. Even Americans, who take their good fortune for granted, are "pretty happy" or happier, and the younger generations are becoming less unhappy, lonely, depressed, drug-addicted, and suicidal.

As societies have become healthier, wealthier, freer, happier, and better educated, they have set their sights on the most pressing global challenges. They have emitted fewer pollutants, cleared fewer forests, spilled less oil, set aside more preserves, extinguished fewer species, saved the ozone layer, and peaked in their consumption of oil, farmland, timber, paper, cars, coal, and perhaps even carbon. For all their differences, the world's nations came to a historic agreement on climate change, as they did in previous years on nuclear testing, proliferation, security, and disarmament. Nuclear weapons, since the extraordinary circumstances of the closing days of World War II, have not been used in the seventy-two years they have existed. Nuclear terrorism, in defiance of forty years of expert predictions, has never happened. The world's nuclear stockpiles have been reduced by 85 percent, with more reductions to come, and testing has ceased (except by the tiny rogue regime in Pyongyang) and proliferation has frozen. The world's two most pressing problems, then, though not yet solved, are solvable: practicable long-term agendas have been laid out for eliminating nuclear weapons and for mitigating climate change.

For all the bleeding headlines, for all the crises, collapses, scandals, plagues, epidemics, and existential threats, these are accomplishments to savor. The Enlightenment is working: for two and a half centuries, people have used knowledge to enhance human flourishing. Scientists have exposed the workings of matter, life, and mind. Inventors have harnessed the laws of nature to defy entropy, and entrepreneurs have made their innovations affordable. Lawmakers have made people better off by discouraging acts that are individually beneficial but collectively harmful. Diplomats have done the same with nations. Scholars have perpetuated the treasury of knowledge and augmented the power of reason. Artists have expanded the circle of sympathy. Activists have pressured the powerful to overturn repressive measures, and their fellow citizens to change repressive norms. All these efforts have been channeled into institutions that have allowed us to circumvent the flaws of human nature and empower our better angels.

At the same time . . .

Seven hundred million people in the world today live in extreme poverty. In the regions where they are concentrated, life expectancy is less than 60, and almost a quarter of the people are undernourished. Almost a million children die of pneumonia every year, half a million from diarrhea or malaria, and hundreds of thousands from measles and AIDS. A dozen wars are raging in the world, including one in which more than 250,000 people have died, and in 2015 at least ten thousand people were slaughtered in genocides. More than two billion people, almost a third of humanity, are oppressed in autocratic states. Almost a fifth of the world's people lack a basic education; almost a sixth are illiterate. Every year five million people are killed in accidents, and more than 400,000 are murdered. Almost 300 million people in the world are clinically depressed, of whom almost 800,000 will die by suicide this year.

The rich countries of the developed world are by no means immune. The lower middle classes have seen their incomes rise by less than 10 percent in two decades. A fifth of the American population still believes that women should return to traditional roles, and a tenth is opposed to interracial dating. The country suffers from more than three thousand hate crimes a year, and more than fifteen thousand homicides. Americans lose two hours a day to housework, and about a quarter of them feel they are always rushed. More than two-thirds of Americans deny that they are very happy, around the same proportion as seventy years ago, and both women and the largest demographic age group have become unhappier over time. Every year around 40,000 Americans become so desperately unhappy that they take their own lives.

And of course the problems that span the entire planet are formidable. Before the century is out, it will have to accommodate another two billion people. A hundred million hectares of tropical forest were cut down in the previous decade. Marine fishes have declined by almost 40 percent, and thousands of species are threatened with extinction. Carbon monoxide, sulfur dioxide, oxides of nitrogen, and particulate matter continue to be spewed into the atmosphere, together with 38 billion tons of CO_2 every year, which, if left unchecked, threaten to raise global temperatures by two to four degrees Celsius. And the world has more than 10,000 nuclear weapons distributed among nine countries.

The facts in the last three paragraphs, of course, are the same as the ones in the first eight; I've simply read the numbers from the bad rather than the good end of the scales or subtracted the hopeful percentages from 100. My point in presenting the state of the world in these two ways is not to show that I can focus on the space in the glass as well as on the beverage. It's to reiterate that progress is not utopia, and that there is room—indeed, an imperative—for us to strive to continue that progress. If we can sustain the trends in the first eight paragraphs by deploying knowledge to enhance flourishing, the numbers in the last three paragraphs should shrink. Whether they will ever get to zero is a problem we can worry about when we get closer. Even if some do, we will surely discover more harms to rectify and new ways to enrich human experience. The Enlightenment is an ongoing process of discovery and betterment.

How reasonable is the hope for continuing progress? That's the question I'll consider in this last chapter in the Progress section, before switching in the remainder of the book to the ideals that are necessary to realize the hope.

~

I'll start with the case for continuing progress. We began the book with a non-mystical, non-Whiggish, non-Panglossian explanation for why progress is possible, namely that the Scientific Revolution and the Enlightenment set in motion the process of using knowledge to improve the human condition. At the time skeptics could reasonably say, "It will never work." But more than two centuries later we can say that it *has* worked: we have seen six dozen graphs that have vindicated the hope for progress by charting ways in which the world has been getting better.

Lines that plot good things over time cannot automatically be extrapolated rightward and upward, but with many of the graphs that's a good bet. It's unlikely we'll wake up one morning and find that our buildings are more flammable, or that people have changed their minds about interracial dating or gay teachers keeping their jobs. Developing countries are unlikely to shut down their schools and health clinics or stop building new ones just as they are starting to enjoy their fruits.

To be sure, changes that take place on the time scale of journalism will always show ups and downs. Solutions create new problems, which take time to solve in their term. But when we stand back from these blips and setbacks, we see that the indicators of human progress are cumulative: none is cyclical, with gains reliably canceled by losses.[3]

Better still, improvements build on one another. A richer world can better afford to protect the environment, police its gangs, strengthen its social safety nets, and teach and heal its citizens. A better-educated and connected world cares more about the environment, indulges fewer autocrats, and starts fewer wars.

The technological advances that have propelled this progress should only gather speed. Stein's Law continues to obey Davies's Corollary (Things that can't go on forever can go on much longer than you think), and genomics, synthetic biology, neuroscience, artificial intelligence, materials science, data science, and evidence-based policy analysis are flourishing. We know that infectious diseases can be extinguished, and many are slated for the past tense. Chronic and degenerative diseases are more recalcitrant, but incremental progress in many (such as cancer) has been accelerating, and breakthroughs in others (such as Alzheimer's) are likely.

So too with moral progress. History tells us that barbaric customs can not only be reduced but essentially abolished, lingering at most in a few benighted backwaters. Not even the most worrying worrywart expects a comeback for human sacrifice, cannibalism, eunuchs, harems, chattel slavery, dueling, family feuding, foot-binding, heretic burning, witch dunking, public torture-executions, infanticide, freak shows, or laughing at the insane. While we can't predict which of today's barbarisms will go the way of slave auctions and autos-da-fé, heading that way are capital punishment, the criminalization of homosexuality, and male-only suffrage and education. Given a few decades, who's to say they could not be followed by female genital mutilation, honor killings, child labor, child marriage, totalitarianism, nuclear weapons, and interstate war?

Other blights are harder to extirpate because they depend on the behavior of billions of individuals with all their human stains, rather than policies adopted by entire countries at a stroke. But even if they are not wiped off the face of the earth, they can be reduced further, including violence against women and children, hate crimes, civil war, and homicide.

I can present this optimistic vision without blushing because it is not a naïve reverie or sunny aspiration. It's the view of the future that is most grounded in historical reality, the one with the cold, hard facts on its side. It depends only on the possibility that what has already happened will continue to happen. As Thomas Macaulay reflected in 1830, "We cannot absolutely prove that those are in error who tell us that society has reached a turning point, that we have seen our best days. But so said all before us, and with just as much apparent reason. . . . On what principle is it, that when we see nothing but improvement behind us, we are to expect nothing but deterioration before us?"[4]

~

In chapters 10 and 19 I examined replies to Macaulay's question which foresaw a catastrophic end to all that progress in the form of climate change, nuclear war, and other existential threats. In the rest of this one I'll consider two 21st-century developments that fall short of global catastrophe but still have been taken to suggest that our best days are behind us.

The first raincloud is economic stagnation. As the essayist Logan Pearsall Smith observed, "There are few sorrows, however poignant, in which a good income is of no avail." Wealth provides not just the obvious things that money can buy, such as nutrition, health, education, and safety, but also, over the long term, spiritual goods such as peace, freedom, human rights, happiness, environmental protection, and other transcendent values.[5]

The Industrial Revolution ushered in more than two centuries of economic growth, especially during the period between World War II and the early 1970s, when the Gross World Product per capita grew at a rate of around 3.4 percent a year, doubling every twenty years.[6] In the late 20th century, eco-pessimists warned that economic growth was unsustainable because it exhausted resources and polluted the planet. But in the 21st, the opposite fear has arisen: that the future promises not too much economic growth but too little. Since the early 1970s, the annual rate of growth has fallen by more than half, to around 1.4 percent.[7] Growth over the long term is determined largely by productivity: the value of goods and services that a country can produce per dollar of investment and person-hour of labor. Productivity in turn depends on technological sophistication: the skills of the country's workers and the efficiency of its machinery, management, and infrastructure. From the 1940s through the 1960s, productivity in the United States grew at an annual rate of around 2 percent, which would double productivity every thirty-five years. Since then it has grown at a rate of around 0.6 percent, which would require more than a century to double.[8]

Some economists fear that low rates of growth are the new normal. According to "the new secular stagnation hypothesis" analyzed by Lawrence Summers, even those paltry rates can be maintained (in conjunction with low unemployment) only if central banks set interest rates at zero or negative values, which could lead to financial instability and other problems.[9] In a period of rising income inequality, secular stagnation could leave a majority of people with static or falling incomes for the foreseeable future. If economies stop growing, things could get ugly.

No one really knows why productivity growth slacked off in the early 1970s or how to bring it back up.[10] Some economists, like Robert Gordon in his 2016 *The Rise and Fall of American Growth,* point to demographic and macroeconomic headwinds, such as fewer working people supporting more retirees, a leveling off in the expansion of education, a rise in government debt, and the increase in inequality (which depresses demand for goods and services, because richer people spend less of their incomes than poorer people).[11] Gordon adds that the most transformative inventions may already have been invented. The first half of the 20th century revolutionized the home with electricity, water, sewerage, telephones, and motorized appliances. Since then homes haven't changed nearly as much. An electronic bidet with a heated seat is nice, but it's not like going from an outhouse to a flush toilet.

Another explanation is cultural: America has lost its mojo.[12] Workers in depressed regions no longer pick up and move to vibrant ones but collect disability insurance and drop out of the labor force. A precautionary principle prevents anyone from trying anything for the first time. Capitalism has lost its capitalists: too much investment is tied up in "gray capital," controlled by institutional managers who seek safe returns for retirees. Ambitious young people want to be artists and professionals, not entrepreneurs. Investors and the government no longer back moonshots. As the entrepreneur Peter Thiel lamented, "We wanted flying cars; instead we got 140 characters."

Whatever its causes, economic stagnation is at the root of many other problems and poses a significant challenge for 21st-century policymakers. Does that mean that progress was nice while it lasted, but now it's over? Unlikely! For one thing, growth that is slower than it was during the postwar glory days is still growth—indeed, exponential growth. Gross World Product has increased in fifty-one of the last fifty-five years, which means that in each of those fifty-one years (including the last six), the world got richer than it was the year before.[13] Also, secular stagnation is largely a first-world problem. Though it's a tremendous challenge to get the most highly developed countries to become *even more* highly developed, year after year after year, the less developed countries have a lot of catching up to do, and they can grow at high rates as they adopt the richer countries' best practices (chapter 8). The greatest ongoing progress in the world today is the rise of billions of people out of extreme poverty, and that ascent need not be capped by the American and European malaise.

Also, technologically driven productivity growth has a way of sneaking up on the world.[14] People take a while to figure out how to put new technologies to their best use, and industries need time to retool their plants and practices around them. Electrification, to take a prominent example, started in the 1890s, but it took forty years before economists saw the boost in productivity that everyone was waiting for. The personal computer revolution also had a sleeper effect before unleashing productivity growth in the 1990s (which is not surprising to early adopters like me, who lost many an afternoon in the 1980s to installing a mouse or getting a dot matrix printer to do italics). Knowledge about how to get the most out of 21st-century technologies may be building up behind dams that will soon burst.

Unlike practitioners of the dismal science, technology watchers are adamant that we are entering an age of abundance.[15] Bill Gates has compared the forecast of technological stagnation to the (apocryphal) prediction in 1913 that war was obsolete.[16] "Imagine a world of nine billion people," write the tech entrepreneur Peter Diamandis and the journalist Steven Kotler, "with clean water, nutritious food, affordable housing, personalized education, top-tier medical care, and nonpolluting, ubiquitous energy."[17] Their vision comes not from fantasies out of *The Jetsons* but from technologies that are already working, or are very close.

Start with the resource that, together with information, is the only way to stave off entropy, and which literally powers everything else in the economy: energy. As we saw in chapter 10, fourth-generation nuclear power in the form of small modular reactors can be passively safe, proliferation-proof, waste-free, mass-produced, low-maintenance, indefinitely fueled, and cheaper than coal. Solar panels made with carbon nanotubes can be a hundred times as efficient as current photovoltaics, continuing Moore's Law for solar energy. Their energy can be stored in liquid metal batteries: in theory, a battery the size of a shipping container could power a neighborhood; one the size of a Walmart could power a small city. A smart grid could collect the energy where and when it's generated and distribute it where and when it's needed. Technology could even breathe new life into fossil fuels: a new design for a zero-emissions gas-fired plant uses the exhaust to drive a turbine directly, rather than wastefully boiling water, and then sequesters the CO_2 underground.[18]

Digital manufacturing, combining nanotechnology, 3-D printing, and rapid prototyping, can produce composites that are stronger and cheaper than steel and concrete and that can be printed on site for construction of houses and factories in the developing world. Nanofiltration can purify water of pathogens, metals, even salt. High-tech outhouses require no hookups and turn human waste into fertilizer, drinking water, and energy. Precision irrigation and smart grids for water, using cheap sensors and AI in chips, could reduce water usage by a third to a half. Rice that is genetically modified to replace its inefficient C3 photosynthesis pathway with the C4 pathway of corn and sugarcane has a 50 percent greater yield, uses half the water and far less fertilizer, and tolerates warmer temperatures.[19] Genetically modified algae can pull carbon out of the air and secrete biofuels. Drones can monitor miles of remote pipelines and railways, and can deliver medical supplies and spare parts to isolated communities. Robots can take over jobs that humans hate, like mining coal, stocking shelves, and making beds.

In the medical realm, a lab-on-a-chip could perform a liquid biopsy and detect any of hundreds of diseases from a drop of blood or saliva. Artificial intelligence, crunching big data on genomes, symptoms, and histories, will diagnose ailments more accurately than the sixth sense of doctors, and will prescribe drugs that mesh with our unique biochemistries. Stem cells could correct autoimmune diseases like rheumatoid arthritis and multiple sclerosis, and could populate cadaver organs, organs grown in animals, or 3D-printed models with our own tissue. RNA interference could silence pesky genes like the one that regulates the fat insulin receptor. Cancer therapies can be narrowcasted to the unique genetic signature of a tumor instead of poisoning every dividing cell in the body.

Global education could be transformed. The world's knowledge has already been made available in encyclopedias, lectures, exercises, and datasets to the billions of people with a smartphone. Individualized instruction can be provided over the Web to children in the developing world by volunteers (the "Granny Cloud") and to learners anywhere by artificially intelligent tutors.

The innovations in the pipeline are not just a list of cool ideas. They fall out of an overarching historical development that has been called the New Renaissance and the Second Machine Age.[20] Whereas the First Machine Age that emerged out of the Industrial Revolution was driven by energy, the second is driven by the other anti-entropic resource, information. Its revolutionary promise comes from the supercharged use of information to guide every other technology, and from exponential improvement in the technologies of information themselves, like computer power and genomics.

The promise of the new machine age also comes from innovations in the process of innovation itself. One is the democratization of platforms for invention, such as application program interfaces and 3-D printers, which can make anyone a high-tech do-it-yourselfer. Another is the rise of technophilanthropists. Instead of just writing checks for the naming rights to concert halls, they apply their ingenuity, connections, and demand for results to the solution of global problems. A third is the economic empowerment of billions of people through smartphones, online education, and microfinancing. Among the world's bottom billion are a million people with a genius-level IQ. Just think what the world would look like if their brainpower were put to full use!

Will the Second Machine Age kick economies out of their stagnation? It's not certain, because economic growth depends not just on the available technology but on how well a nation's financial and human capital are deployed to use it. Even if the technologies are put to full use, their benefits may not be registered in standard economic measures. The comedian Pat Paulsen once observed, "We live in a country where even the national product is gross." Most economists agree that GNP (or its close relative, GDP) is a crude index of economic thriving. It has the virtue of being easy to measure, but because it's just a tally of the money that changes hands in the production of goods and services, it's not the same as the bounty that people enjoy. The problem of consumer surplus or the paradox of value has always bedeviled the quantification of prosperity (chapters 8 and 9), and modern economies are making it more acute.

Joel Mokyr notes that "aggregate statistics like GDP per capita and its derivatives such as factor productivity . . . were designed for a steel-and-wheat economy, not one in which information and data are the most dynamic sector. Many of the new goods and services are expensive to design, but once they work, they can be copied at very low or zero costs. That means they tend to contribute little to measured output even if their impact on consumer welfare is very large."[21] The dematerialization of life that we examined in chapter 10, for example, undermines the observation that a 2015 home does not look much different from a 1965 home. The big difference lies in what we *don't* see because it's been made obsolete by tablets and smartphones, together with new wonders like streaming video and Skype. In addition to dematerialization, information technology has launched a process of *demonetization*.[22] Many things that people used to pay for are now essentially free, including classified ads, news, encyclopedias, maps, cameras, long-distance calls, and the overhead of brick-and-mortar retailers. People are enjoying these goods more than ever, but they have vanished from GDP.

Human welfare has parted company from GDP in a second way. As modern societies become more humanistic, they devote more of their wealth to forms of human betterment that are not priced in the marketplace. A recent *Wall Street Journal* article on economic stagnation noted that a growing share of innovative effort has been directed toward cleaner air, safer cars, and drugs for "orphan diseases" that each affect fewer than 200,000 people nationwide.[23] For that matter, health care in general has risen from 7 percent of research and development in 1960 to 25 percent in 2007. The financial journalist who wrote the piece noted, almost in sadness, that "drugs are symptomatic of the rising value affluent societies place on human life. . . . Health research is displacing R&D that could have gone toward more mundane consumer products. Indeed, . . . the rising value of human life virtually dictates slower growth in regular consumer goods and services—and they constitute the bulk of measured GDP." A natural interpretation is that this tradeoff is evidence for the acceleration of progress, not the stagnation of progress. Modern societies, unlike the miserly comedian Jack Benny, have a quick reply to the mugger's demand, "Your money or your life."

～

A very different threat to human progress is a political movement that seeks to undermine its Enlightenment foundations. The second decade of the 21st century has seen the rise of a counter-Enlightenment movement called populism, more accurately, authoritarian populism.[24] Populism calls for the direct sovereignty of a country's "people" (usually an ethnic group, sometimes a class), embodied in a strong leader who directly channels their authentic virtue and experience.

Authoritarian populism can be seen as a pushback of elements of human nature—tribalism, authoritarianism, demonization, zero-sum thinking—against the Enlightenment institutions that were designed to circumvent them. By focusing on the tribe rather than the individual, it has no place for the protection of minority rights or the promotion of human welfare worldwide. By failing to acknowledge that hard-won knowledge is the key to societal improvement, it denigrates "elites" and "experts" and downplays the marketplace of ideas, including freedom of speech, diversity of opinion, and the fact-checking of self-serving claims. By valorizing a strong leader, populism overlooks the limitations in human nature, and disdains the rule-governed institutions and constitutional checks that constrain the power of flawed human actors.

Populism comes in left-wing and right-wing varieties, which share a folk theory of economics as zero-sum competition: between economic classes in the case of the left, between nations or ethnic groups in the case of the right. Problems are seen not as challenges that are inevitable in an indifferent universe but as the malevolent designs of insidious elites, minorities, or foreigners. As for progress, forget about it: populism looks backward to an age in which the nation was ethnically homogeneous, orthodox cultural and religious values prevailed, and economies were powered by farming and manufacturing, which produced tangible goods for local consumption and for export.

Chapter 23 will probe the intellectual roots of authoritarian populism more deeply; here I will concentrate on its recent rise and possible future. In 2016 populist parties (mostly on the right) attracted 13.2 percent of the vote in the preceding European parliamentary elections (up from 5.1 percent in the 1960s) and entered the governing coalitions of eleven countries, including the leadership of Hungary and Poland.[25] Even when they are not in power, populist parties can press their agendas, notably by catalyzing the 2016 Brexit referendum in which 52 percent of Britons voted to leave the European Union. And in that year Donald Trump was elected to the American presidency with an Electoral College victory, though with a minority of the popular vote (46 percent to Hillary Clinton's 48 percent). Nothing captures the tribalistic and backward-looking spirit of populism better than Trump's campaign slogan: Make America Great Again.

In writing the chapters on progress, I resisted pressure from readers of earlier drafts to end each one by warning, "But all this progress is threatened if Donald Trump gets his way." Threatened it certainly is. Whether or not 2017 really represents a turning point in history, it's worth reviewing the threats, if only to understand the nature of the progress they threaten.[26]

• **Life** and **Health** have been expanded in large part by vaccination and other well-vetted interventions, and among the conspiracy theories that Trump has endorsed is the long-debunked claim that preservatives in vaccines cause autism. The gains have also been secured by broad access to medical care, and he has pushed for legislation that would withdraw health insurance from tens of millions of Americans, a reversal of the trend toward beneficial social spending.

• Worldwide improvements in **Wealth** have come from a globalized economy, powered in large part by international trade. Trump is a protectionist who sees international trade as a zero-sum contest between countries, and is committed to tearing up international trade agreements.

• Growth in **Wealth** will also be driven by technological innovation, education, infrastructure, an increase in the spending power of the lower and middle classes, constraints on cronyism and plutocracy that distort market competition, and regulations on finance that reduce the likelihood of bubbles and crashes. In addition to being hostile to trade, Trump is indifferent to technology and education and an advocate of regressive tax cuts on the wealthy, while appointing corporate and financial tycoons to his cabinet who are indiscriminately hostile to regulation.

• In capitalizing on concerns about **Inequality**, Trump has demonized immigrants and trade partners while ignoring the major disrupter of lower-middle-class jobs, technological change. He has also opposed the measures that most successfully mitigate its harms, namely progressive taxation and social spending.

• **The Environment** has benefited from regulations on air and water pollution that have coexisted with growth in population, GDP, and travel. Trump believes that environmental regulation is economically destructive; worst of all, he has called climate change a hoax and announced a withdrawal from the historic Paris agreement.

• **Safety**, too, has been dramatically improved by federal regulations, toward which Trump and his allies are contemptuous. While Trump has cultivated a reputation for law and order, he is viscerally uninterested in evidence-based policy that would distinguish effective crime-prevention measures from useless tough talk.

• The postwar **Peace** has been cemented by trade, **Democracy**, international agreements and organizations, and norms against conquest. Trump has vilified international trade and has threatened to defy international agreements and weaken international organizations. Trump is an admirer of Vladimir Putin, who reversed the democratization of Russia, tried to undermine democracy in the United States and Europe with cyberattacks, helped prosecute the most destructive war of the 21st century in Syria, fomented smaller wars in Ukraine and Georgia, and defied the postwar taboo against conquest in his annexation of Crimea. Several members of Trump's administration secretly colluded with Russia in an effort to lift sanctions against it, undermining a major enforcement mechanism in the outlawry of war.

• **Democracy** depends both on explicit constitutional protections, such as freedom of the press, and on shared norms, in particular that political leadership is determined by the rule of law and nonviolent political competition rather than a charismatic leader's will to power. Trump proposed to relax libel laws against journalists, encouraged violence against his critics at his rallies, would not commit to respecting the outcome of the 2016 election if it went against him, tried to discredit the popular vote count that did go against him, threatened to imprison his opponent in the election, and attacked the legitimacy of the judicial system when it challenged his decisions—all hallmarks of a dictator. Globally, the resilience of democracy depends in part on its prestige in the community of nations, and Trump has praised autocrats in Russia, Turkey, the Philippines, Thailand, Saudi Arabia, and Egypt while denigrating democratic allies such as Germany.

• The ideals of tolerance, equality, and **Equal Rights** took big symbolic hits during his campaign and early administration. Trump demonized Hispanic immigrants, proposed banning Muslim immigration altogether (and tried to impose a partial ban once he was elected), repeatedly demeaned women, tolerated vulgar expressions of racism and sexism at his rallies, accepted support from white supremacist groups and equated them with their opponents, and appointed a strategist and an attorney general who are hostile to the civil rights movement.

• The ideal of **Knowledge**—that one's opinions should be based on justified true beliefs—has been mocked by Trump's repetition of ludicrous conspiracy theories: that Obama was born in Kenya, Senator Ted Cruz's father was involved in John F. Kennedy's assassination, thousands of New Jersey Muslims celebrated 9/11, Justice Antonin Scalia was murdered, Obama had his phones tapped, millions of illegal voters cost him the popular vote, and literally dozens of others. The fact-checking site *PolitiFact* judged that an astonishing 69 percent of the public statements by Trump they checked were "Mostly False," "False," or "Pants on Fire" (their term for outrageous lies, from the children's taunt "Liar, liar, pants on fire").[27] All politicians bend the truth, and all sometimes lie (since all human beings bend the truth and sometimes lie), but Trump's barefaced assertion of canards that can instantly be debunked (such as that he won the election in a landslide) shows that he sees public discourse not as a means of finding common ground based on objective reality but as a weapon with which to project dominance and humiliate rivals.

• Most frighteningly, Trump has pushed back against the norms that have protected the world against the possible **Existential Threat** of nuclear war. He questioned the taboo on using nuclear weapons, tweeted

THE FUTURE OF PROGRESS

about resuming a nuclear arms race, mused about encouraging the proliferation of weapons to additional countries, sought to overturn the agreement that prevents Iran from developing nuclear weapons, and taunted
Kim Jong-un about a possible nuclear exchange with North Korea. Worst
of all, the chain of command gives an American president enormous dis-
cretion over the use of nuclear weapons in a crisis, on the tacit assumption
that no president would act rashly on such a grave matter. Yet Trump has
a temperament that is notoriously impulsive and vindictive.

Not even a congenital optimist can see a pony in this Christmas
stocking. But will Donald Trump (and authoritarian populism more generally) really undo a quarter of a millennium of progress? There are reasons not to take poison just yet. If a movement has proceeded for decades
or centuries, there are probably systematic forces behind it, and many
stakeholders with an interest in its not being precipitously reversed.

By the design of the Founders, the American presidency is not a ro-
tating monarchy. The president presides over a distributed network of
power (denigrated by populists as the "deep state") that outlasts individual leaders and carries out the business of government under real-
world constraints which can't easily be erased by populist applause
lines or the whims of the man at the top. It includes legislators who have
to respond to constituents and lobbyists, judges with reputations of pro-
bity to uphold, and executives, bureaucrats, and functionaries who are
responsible for the missions of their departments. Trump's authoritarian
instincts are subjecting the institutions of American democracy to a
stress test, but so far it has pushed back on a number of fronts. Cabinet
secretaries have publicly repudiated various quips, tweets, and stink
bombs; courts have struck down unconstitutional measures; senators
and congressmen have defected from his party to vote down destructive
legislation; Justice Department and Congressional committees are inves-
tigating the administration's ties to Russia; an FBI chief has publicly
called out Trump's attempt to intimidate him (raising talk about im-
peachment for obstruction of justice); and his own staff, appalled at what
they see, regularly leak compromising facts to the press—all in the first
six months of the administration.

Also boxing a president in are state and local governments, which are
closer to the facts on the ground; the governments of other nations,
which cannot be expected to put a high priority on making America
great again; and even most corporations, which benefit from peace, prosperity, and stability. Globalization in particular is a tide that is impossible for any ruler to order back. Many of a country's problems are

inherently global, including migration, pandemics, terrorism, cyber-crime, nuclear proliferation, rogue states, and the environment. Pretend-ing they don't exist is not tenable forever, and they can be solved only through international cooperation. Nor can the benefits of globalization—more affordable goods, larger markets for exports, the reduction in global poverty—be denied indefinitely. And with the Internet and inexpensive travel, there will be no stopping the flow of people and ideas (especially, as we will see, among younger people). As for the battle against truth and fact, over the long run they have a built-in advantage: when you stop believing in them, they don't go away.[28]

~

The deeper question is whether the rise of populist movements, what-ever damage they do in the short term, represents the shape of things to come—whether, as a recent *Boston Globe* editorial lamented/gloated, "The Enlightenment had a good run."[29] Do the events around 2016 really imply that the world is headed back to the Middle Ages? As with climate change skeptics who claim to be vindicated by a nippy morning, it's easy to overinterpret recent events.

For one thing, the latest elections are not referenda on the Enlighten-ment. In the American political duopoly, any Republican candidate starts from a partisan floor of at least 45 percent of the votes in a two-way race, and Trump was defeated in the popular vote 46–48 percent, while bene-fiting from electoral shenanigans and from campaigning misjudgments on Clinton's part. And Barack Obama—who in his farewell speech actu-ally *credited the Enlightenment* for the "essential spirit of this country"—left office with an approval rating of 58 percent, above average for departing presidents.[30] Trump entered office with a rating of 40 percent, the lowest ever for an incoming president, and during his first seven months it sank to 34 percent, barely more than half of the average rating of the nine previous presidents at the same point in their terms.[31]

European elections, too, are not depth-soundings for a commitment to cosmopolitan humanism but reactions to a bundle of emotionally charged issues of the day. These included, recently, the euro currency (which arouses skepticism among many economists), intrusive regula-tion from Brussels, and pressure to accept large numbers of refugees from the Middle East just when fears of Islamic terrorism (however dis-proportionate to the risk) were being stoked by horrific attacks. Even then, populist parties have attracted only 13 percent of the votes in recent years, and they have lost seats in as many national legislatures as they have gained them in.[32] In the year following the Trump and Brexit shocks, right-wing populism was repudiated in elections in the Netherlands, the United Kingdom, and France—where the new president, Emmanuel Macron, proclaimed that Europe was "waiting for us to defend the spirit of the Enlightenment, threatened in so many places."[33]

But far more important than the political events of the mid-2010s are the social and economic trends that have fostered authoritarian populism—and more to the point of this chapter, that may foretell its future.

Beneficial historical developments often create losers together with the winners, and the apparent economic losers of globalization (namely the lower classes of rich countries) are often said to be the supporters of authoritarian populism. For economic determinists, this is enough to explain the rise of the movement. But analysts have sifted through the election results like investigators inspecting the wreckage at the site of a plane crash, and we now know that the economic explanation is wrong. In the American election, voters in the two lowest income brackets voted for *Clinton* 52–42, as did those who identified "the economy" as the most important issue. A majority of voters in the four *highest* income brackets voted for Trump, and Trump voters singled out "immigration" and "terrorism," not "the economy," as the most important issues.[34]

The twisted metal has turned up more promising clues. An article by the statistician Nate Silver began, "Sometimes statistical analysis is tricky, and sometimes a finding just jumps off the page." That finding jumped right off the page and into the article's headline: "Education, Not Income, Predicted Who Would Vote for Trump."[35] Why should education have mattered so much? Two uninteresting explanations are that the highly educated happen to affiliate with a liberal political tribe, and that education may be a better long-term predictor of economic security than current income. A more interesting explanation is that education exposes people in young adulthood to other races and cultures in a way that makes it harder to demonize them. Most interesting of all is the likelihood that education, when it does what it is supposed to do, instills a respect for vetted fact and reasoned argument, and so inoculates people against conspiracy theories, reasoning by anecdote, and emotional demagoguery.

In another page-jumper, Silver found that the regional map of Trump support did not overlap particularly well with the maps of unemployment, religion, gun ownership, or the proportion of immigrants. But it did align with the map of Google searches for the word *nigger*, which Seth Stephens-Davidowitz has shown is a reliable indicator of racism (chapter 15).[36] This doesn't mean that most Trump supporters are racists. But overt racism shades into resentment and distrust, and the overlap suggests that the regions of the country that gave Trump his Electoral College victory are those with the most resistance to the decades-long process of integration and the promotion of minority interests (particularly racial preferences, which they see as reverse discrimination against them).

Among the exit poll questions that probed general attitudes, the most consistent predictor of Trump support was pessimism.[37] Sixty-nine percent of Trump supporters felt that the direction of the country was "seriously off track," and they were similarly jaundiced about the workings of the federal government and the lives of the next generation of Americans.

Across the pond, the political scientists Ronald Inglehart and Pippa Norris spotted similar patterns in their analysis of 268 political parties in thirty-one European countries.[38] Economic issues, they found, have been playing a smaller role in party manifestoes for decades, and noneconomic issues a larger role. The same was true of the distribution of voters. Support for populist parties is strongest not from manual workers but from the "petty bourgeoisie" (self-employed tradesmen and the owners of small businesses), followed by foremen and technicians. Populist voters are older, more religious, more rural, less educated, and more likely to be male and members of the ethnic majority. They embrace authoritarian values, place themselves on the right of the political spectrum, and dislike immigration and global and national governance.[39] Brexit voters, too, were older, more rural, and less educated than those who voted to remain: 66 percent of high school graduates voted to leave, but only 29 percent of degree holders did.[40]

Inglehart and Norris concluded that supporters of authoritarian populism are the losers not so much of economic competition as *cultural* competition. Voters who are male, religious, less educated, and in the ethnic majority "feel that they have become strangers from the predominant values in their own country, left behind by progressive tides of cultural change that they do not share. . . . The silent revolution launched in the 1970s seems to have spawned a resentful counter-revolutionary backlash today."[41] Paul Taylor, a political analyst at the Pew Research Center, singled out the same counter-current in American polling results: "The overall drift is toward more liberal views on a range of issues, but that doesn't mean the whole country's buying in."[42]

Though the source of the populist backlash may be found in currents of modernity that have been engulfing the world for some time—

globalization, racial diversity, women's empowerment, secularism, urbanization, education—its electoral success in a particular country
depends on whether a leader materializes who can channel that resentment. Neighboring countries with comparable cultures can thus differ in
the degree to which populism gains traction: Hungary more than the
Czech Republic, Norway more than Sweden, Poland more than Roma-
nia, Austria more than Germany, France more than Spain, and the United
States more than Canada. (In 2016 Spain, Canada, and Portugal had no
populist party legislators at all.)[43]

~

How will the tension play out between the liberal, cosmopolitan, enlightenment humanism that has been sweeping the world for decades and the
regressive, authoritarian, tribal populism pushing back? The major long-
term forces that have carried liberalism along—mobility, connectivity,
education, urbanization—are not likely to go into reverse, and neither is
the pressure for equality from women and ethnic minorities.

All of these portents, to be sure, are conjectural. But one is as certain as
the first half of the idiom "death and taxes." Populism is an old man's
movement. As figure 20-1 shows, support for all three of its recrudescences—
Trump, Brexit, and European populist parties—falls off dramatically
with year of birth. (The alt-right movement, which overlaps with populism, has a youngish membership, but for all its notoriety it is an electoral
nonentity, numbering perhaps 50,000 people or 0.02 percent of the American population.)[44] The age rolloff isn't surprising, since we saw in chapter 15 that in the 20th century every birth cohort has been more tolerant
and liberal than the one that came before (at the same time that all the
cohorts have drifted liberalward). This raises the possibility that as the
Silent Generation and older Baby Boomers shuffle off this mortal coil,
they will take authoritarian populism with them.

Of course the cohorts of the present say nothing about the politics of
the future if people change their values as they age. Perhaps if you are a
populist at twenty-five you have no heart, and if you are not a populist
at forty-five you have no brain (to adapt a meme that has been said about
liberals, socialists, communists, leftists, Republicans, Democrats, and
revolutionists and that has been attributed to various quotation magnets,
including Victor Hugo, Benjamin Disraeli, George Bernard Shaw,
Georges Clemenceau, Winston Churchill, and Bob Dylan). But whoever
said it (probably the 19th-century jurist Anselme Batbie, who in turn attributed it to Edmund Burke), and regardless of which belief system it's
supposed to apply to, the claim about life-cycle effects on political orien-

tation is false.[45] As we saw in chapter 15, people carry their emancipative values with them as they age rather than sliding into illiberalism. And a recent analysis of 20th-century American voters by the political scientists Yair Ghitza and Andrew Gelman has shown that Americans do not consistently vote for more conservative presidents as they age. Their voting preferences are shaped by their cumulative experience of the popularity of presidents over their life spans, with a peak of influence in the 14–24-year-old window.[46] The young voters who reject populism today are unlikely to embrace it tomorrow.

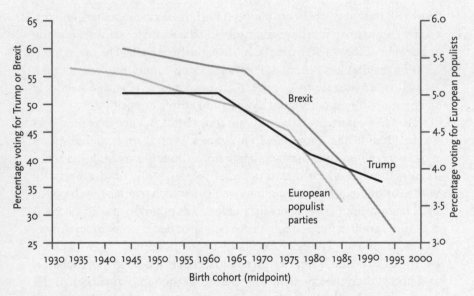

Figure 20-1: Populist support across generations, 2016

Sources: **Trump:** Exit polls conducted by Edison Research, *New York Times* 2016. **Brexit:** Exit polls conducted by Lord Ashcroft Polls, *BBC News Magazine*, June 24, 2016, http://www.bbc.com/news/magazine-36619342. **European populist parties (2002–2014):** Inglehart & Norris 2016, fig. 8. Data for each birth cohort are plotted at the midpoint of their range.

How might one counter the populist threat to Enlightenment values? Economic insecurity is not the driver, so the strategies of reducing income inequality and of talking to laid-off steelworkers and trying to feel their pain, however praiseworthy, will probably be ineffective. Cultural backlash does seem to be a driver, so avoiding needlessly polarizing rhetoric, symbolism, and identity politics might help to recruit, or at least not repel, voters who are not sure which team they belong to (more on this in chapter 21). Since populist movements have achieved an influence beyond their numbers, fixing electoral irregularities such as gerrymandering and forms of disproportionate representation which overweight rural areas (such as the US Electoral College) would help. So would journalistic coverage that tied candidates' reputations to their record of accuracy and coherence rather than to trivial gaffes and scandals. Part of the problem, over the long term, will dissipate with urbanization: you can't keep them down on the farm. And part will dissipate with demographics. As has been said about science, sometimes society advances funeral by funeral.[47]

Still, a puzzle in the rise of authoritarian populism is why a shocking proportion of the sectors of the population whose interests were most endangered by the outcome of the elections, such as younger Britons with Brexit, and African Americans, Latinos, and American millennials with Trump, stayed home on election day.[48] This brings us back to a major theme of this book, and to my own small prescription for strengthening the current of Enlightenment humanism against the latest counter-Enlightenment backlash.

I believe that the media and intelligentsia were complicit in populists' depiction of modern Western nations as so unjust and dysfunctional that nothing short of a radical lurch could improve them. "Charge the cockpit or you die!" shrieked a conservative essayist, comparing the country to the hijacked flight on 9/11 that was brought down by a passenger mutiny.[49] "I'd rather see the empire burn to the ground under Trump, opening up at least the possibility of radical change, than cruise on autopilot under Clinton," flamed a left-wing advocate of "the politics of arson."[50] Even moderate editorialists in mainstream newspapers commonly depict the country as a hellhole of racism, inequality, terrorism, social pathology, and failing institutions.[51]

The problem with dystopian rhetoric is that if people believe that the country is a flaming dumpster, they will be receptive to the perennial appeal of demagogues: "What do you have to lose?" If the media and intellectuals instead put events into statistical and historical context, they could help answer that question. Radical regimes from Nazi Germany and Maoist China to contemporary Venezuela and Turkey show that people have a tremendous amount to lose when charismatic authoritarians responding to a "crisis" trample over democratic norms and institutions and command their countries by the force of their personalities.

A liberal democracy is a precious achievement. Until the messiah comes, it will always have problems, but it's better to solve those problems than to start a conflagration and hope that something better arises from the ashes and bones. By failing to take note of the gifts of modernity, social critics poison voters against responsible custodians and incremental reformers who can consolidate the tremendous progress we have enjoyed and strengthen the conditions that will bring us more.

~

The challenge in making the case for modernity is that when one's nose is inches from the news, optimism can seem naïve, or in the pundits' favorite new cliché about elites, "out of touch." Yet in a world outside of hero myths, the only kind of progress we can have is a kind that is easy to miss while we are living through it. As the philosopher Isaiah Berlin pointed out, the ideal of a perfectly just, equal, free, healthy, and harmonious society, which liberal democracies never measure up to, is a dangerous fantasy. People are not clones in a monoculture, so what satisfies one will frustrate another, and the only way they can end up equal is if they are treated unequally. Moreover, among the perquisites of freedom is the freedom of people to screw up their own lives. Liberal democracies can make progress, but only against a constant backdrop of messy compromise and constant reform:

> The children have obtained what their parents and grandparents longed for—greater freedom, greater material welfare, a juster society; but the old ills are forgotten, and the children face new problems, brought about by the very solutions of the old ones, and these, even if they can in turn be solved, generate new situations, and with them new requirements—and so on, forever—and unpredictably.[52]

Such is the nature of progress. Pulling us forward are ingenuity, sympathy, and benign institutions. Pushing us back are the darker sides of human nature and the Second Law of Thermodynamics. Kevin Kelly explains how this dialectic can nonetheless result in forward motion:

> Ever since the Enlightenment and the invention of science, we've managed to create a tiny bit more than we've destroyed each year. But that few percent positive difference is compounded over decades into what we might call civilization. . . . [Progress] is a self-cloaking action seen only in retrospect. Which is why I tell people that my great optimism of the future is rooted in history.[53]

We don't have a catchy name for a constructive agenda that reconciles long-term gains with short-term setbacks, historical currents with human agency. "Optimism" is not quite right, because a belief that things will always get better is no more rational than the belief that things will always get worse. Kelly offers "protopia," the *pro-* from *progress* and *process*. Others have suggested "pessimistic hopefulness," "opti-realism," and "radical incrementalism."[54] My favorite comes from Hans Rosling, who, when asked whether he was an optimist, replied, "I am not an optimist. I'm a very serious possibilist."[55]

PART III

REASON, SCIENCE, AND HUMANISM

The ideas of economists and political philosophers, both when they are right and when they are wrong, are more powerful than is commonly understood. Indeed, the world is ruled by little else. Practical men, who believe themselves to be quite exempt from any intellectual influences, are usually slaves of some defunct economist. Madmen in authority, who hear voices in the air, are distilling their frenzy from some academic scribbler of a few years back. I am sure that the power of vested interests is vastly exaggerated compared with the gradual encroachment of ideas.

—John Maynard Keynes

Ideas matter. *Homo sapiens* is a species that lives by its wits, concocting and pooling notions of how the world works and how its members can best lead their lives. There can be no better proof of the power of ideas than the ironic influence of the political philosopher who most insisted on the power of vested interests, the man who wrote that "the ruling ideas of each age have ever been the ideas of its ruling class." Karl Marx possessed no wealth and commanded no army, but the ideas he scribbled in the reading room of the British Museum shaped the course of the 20th century and beyond, wrenching the lives of billions.

This part of the book wraps up my defense of the ideas of the Enlightenment. Part I outlined those ideas; part II showed they work. Now it's time to defend them against some surprising enemies—not just angry populists and religious fundamentalists, but factions of mainstream intellectual culture. It may sound quixotic to offer a defense of the Enlightenment against professors, critics, pundits, and their readers, because if they were asked about these ideals point-blank, few would disavow them. But intellectuals' commitment to those ideals is squirrely. The hearts of many of them lie elsewhere, and few are willing to proffer a positive defense. Enlightenment ideals, thus unchampioned, fade into the background as a bland default, and become a catch basin for every unsolved societal problem (of which there will always be many). Illiberal ideas like authoritarianism, tribalism, and magical thinking easily get the blood pumping, and have no shortage of champions. It's hardly a fair fight.

Though I hope Enlightenment ideals will become more deeply entrenched in the public at large—fundamentalists, angry populists, and all—I claim no competence in the dark arts of mass persuasion, popular mobilization, or viral memes. What follow are arguments directed at people who care about arguments. These arguments can matter, because practical men and women and madmen in authority are affected, directly or indirectly, by the world of ideas. They go to university. They read intellectual magazines, if only in dentists' waiting rooms. They watch talking heads on Sunday morning news shows. They are briefed by staff members who subscribe to highbrow papers and watch TED talks. They frequent Internet discussion forums that are enlightened or endarkened by the reading habits of the more literate contributors. I like to think that some good might come to the world if more of the ideas that trickle into these tributaries embodied the Enlightenment ideals of reason, science, and humanism.

CHAPTER 21

REASON

Opposing reason is, by definition, unreasonable. But that hasn't stopped a slew of irrationalists from favoring the heart over the head, the limbic system over the cortex, blinking over thinking, McCoy over Spock. There was the Romantic movement of the counter-Enlightenment, captured in Johann Herder's avowal "I am not here to think, but to be, feel, live!" There's the common veneration (not just by the religious) of faith, namely believing something without a good reason. There's the postmodernist credo that reason is a pretext to exert power, reality is socially constructed, and all statements are trapped in a web of self-reference and collapse into paradox. Even members of my own tribe of cognitive psychologists often claim to have refuted what they take to be the Enlightenment belief that humans are rational agents, and hence to have undermined the centrality of reason itself. The implication is that it is futile even to try to make the world a more rational place.[1]

But all these positions have a fatal flaw: they refute themselves. They deny that there can be a *reason* for believing those very positions. As soon as their defenders open their mouths to begin their defense, they have lost the argument, because in that very act they are tacitly committed to persuasion—to adducing reasons for what they are about to argue, which, they insist, ought to be accepted by their listeners according to standards of rationality that both accept. Otherwise they are wasting their breath and might as well try to convert their audience by bribery or violence. In *The Last Word*, the philosopher Thomas Nagel drives home the point that subjectivity and relativism regarding logic and reality are incoherent, because "one can't criticize something with nothing":

The claim "Everything is subjective" must be nonsense, for it would itself have to be either subjective or objective. But it can't be objective, since in that case it would be false if true. And it can't be subjective, because then it would not rule out any objective claim, including the claim that it is objectively false. There may be some subjectivists, perhaps styling themselves as pragmatists, who present subjectivism as applying even to itself. But then it does not call for a reply, since it is just a report of what the subjectivist finds it agreeable to say. If he also invites us to join him, we need not offer any reason for declining, since he has offered us no reason to accept.[2]

Nagel calls this line of thinking Cartesian, because it resembles Descartes's argument "I think, therefore I am." Just as the very fact that one is wondering whether one exists demonstrates that one exists, the very fact that one is appealing to reasons demonstrates that reason exists. It may also be called a transcendental argument, one that invokes the necessary preconditions for doing what it is doing, namely making an argument.[3] (In a way, it goes back to the ancient Liar's Paradox, featuring the Cretan who says, "All Cretans are liars.") Whatever you call the argument, it would be a mistake to interpret it as justifying a "belief" or a "faith" in reason, which Nagel calls "one thought too many." We don't *believe in* reason; we *use* reason (just as we don't program our computers to have a CPU; a program is a sequence of operations made available by the CPU).[4]

Though reason is prior to everything else and needn't (indeed cannot) be justified on first principles, once we start engaging in it we can stroke our confidence that the particular kinds of reasoning we are engaging in are sound by noting their internal coherence and their fit with reality. Life is not a dream, in which disconnected experiences appear in bewildering succession. And the application of reason to the world validates itself by granting us the ability to bend the world to our will, from curing infections to sending a man to the moon.

Despite its provenance in abstract philosophy, the Cartesian argument is not an exercise in logic-chopping. From the most recondite deconstructionist to the most anti-intellectual purveyor of conspiracy theories and "alternative facts," everyone recognizes the power of responses like "Why should I believe you?" or "Prove it" or "You're full of crap." Few would reply, "That's right, there's no reason to believe me," or "Yes, I'm lying right now," or "I agree, what I'm saying is bullshit." It's in the very nature of argument that people stake a claim to being right. As soon as they do, they have committed themselves to reason—and the listeners they are trying to convince can hold their feet to the fire of coherence and accuracy.

～

By now many people have become aware of the research in cognitive psychology on human irrationality, explained in bestsellers like Daniel Kahneman's *Thinking, Fast and Slow* and Dan Ariely's *Predictably Irrational*. I've alluded to these cognitive infirmities in earlier chapters: the way we estimate probability from available anecdotes, project stereotypes onto individuals, seek confirming and ignore disconfirming evidence, dread harms and losses, and reason from teleology and voodoo resemblance rather than mechanical cause and effect.[5] But as important as these discoveries are, it's a mistake to see them as refuting some Enlightenment tenet that humans are rational actors, or as licensing the fatalistic conclusion that we might as well give up on reasoned persuasion and fight demagoguery with demagoguery.

To begin with, *no Enlightenment thinker ever claimed that humans were consistently rational*. Certainly not the über-rational Kant, who wrote that "from the crooked timber of humanity no truly straight thing can be made," nor Spinoza, Hume, Smith, or the *Encyclopédistes*, who were cognitive and social psychologists ahead of their time.[6] What they argued was that we *ought* to be rational, by learning to repress the fallacies and dogmas that so readily seduce us, and that we *can* be rational, collectively if not individually, by implementing institutions and adhering to norms that constrain our faculties, including free speech, logical analysis, and empirical testing. And if you disagree, then why should we accept *your* claim that humans are incapable of rationality?

Often the cynicism about reason is justified with a crude version of evolutionary psychology (not one endorsed by evolutionary psychologists) in which humans think with their amygdalas, reacting instinctively to the slightest rustle in the grass which may portend a crouching tiger. But real evolutionary psychology treats humans differently: not as two-legged antelopes but as the species that *outsmarts* antelopes. We are a cognitive species that depends on explanations of the world. Since the world is the way it is regardless of what people believe about it, there is a strong selection pressure for an ability to develop explanations that are true.[7]

Reasoning thus has deep evolutionary roots. The citizen scientist Louis Liebenberg has studied the San hunter-gatherers of the Kalahari Desert (the "Bushmen"), one of the world's most ancient cultures. They engage in the oldest form of the chase, persistence hunting, in which humans, with their unique ability to dump heat through sweat-slicked skin, pursue a furry mammal in the midday sun until it collapses of heat stroke. Since most mammals are swifter than humans and dart out of

sight as soon as they are spotted, persistence hunters track them by their
spoor, which means inferring the animal's species, sex, age, and level of
fatigue, and thus its likely direction of flight, from the hoofprints, bent
stems, and displaced pebbles it leaves behind. The San do not just engage
in *inference*—deducing, for example, that agile springboks tread deeply
with pointed hooves to get a good grip, whereas heavy kudus tread flat-
footed to support their weight. They also engage in *reasoning*—
articulating the logic behind their inferences to persuade their
companions or be persuaded in their turn. Liebenberg observed that
Kalahari trackers don't accept arguments from authority. A young
tracker can challenge the majority opinion of his elders, and if his interpretation of the evidence is convincing, he can bring them around, increasing the group's accuracy.[8]

And if you're still tempted to excuse modern dogma and superstition
by saying that it's only human, consider Liebenberg's account of scientific skepticism among the San:

> Three trackers, !Nate, /Uase and Boroh//xao, of Lone Tree in the cen-
> tral Kalahari, told me that the Monotonous Lark (*Mirafra passerina*)
> only sings after it has rained, because "it is happy that it rained." One
> tracker, Boroh//xao, told me that when the bird sings, it dries out the
> soil, making the roots good to eat. Afterwards, !Nate and /Uase told
> me that Boroh//xao was wrong—it is not the *bird* that dries out the soil,
> it is the *sun* that dries out the soil. The bird is only *telling* them that the
> soil will dry out in the coming months and that it is the time of the
> year when the roots are good to eat. . . .
>
> !Namka, a tracker from Bere in the central Kalahari, Botswana, told
> me the myth of how the sun is like an eland, which crosses the sky and
> is then killed by people who live in the west. The red glow in the sky
> when the sun goes down is the blood of the eland. After they have
> eaten it, they throw the shoulder blade across the sky back to the east,
> where it falls into a pool and grows into a new sun. Sometimes, it is
> said, you can hear the swishing noise of the shoulder blade flying
> through the air. After telling me the story in great detail, he told me
> that he thinks that the "Old People" lied, because he has never seen . . .
> the shoulder blade fly through the sky or heard the swishing noise.[9]

Of course, none of this contradicts the discovery that humans are
vulnerable to illusions and fallacies. Our brains are limited in their capacity to process information and evolved in a world without science,

scholarship, and other forms of fact-checking. But reality is a mighty selection pressure, so a species that lives by ideas must have evolved with an ability to prefer correct ones. The challenge for us today is to design an informational environment in which that ability prevails over the ones that lead us into folly. The first step is to pinpoint why an otherwise intelligent species is so easily led into folly.

~

The 21st century, an age of unprecedented access to knowledge, has also seen maelstroms of irrationality, including the denial of evolution, vaccine safety, and anthropogenic climate change, and the promulgation of conspiracy theories, from 9/11 to the size of Donald Trump's popular vote. Fans of rationality are desperate to understand the paradox, but in a bit of irrationality of their own, they seldom look at data that might explain it.

The standard explanation of the madness of crowds is ignorance: a mediocre education system has left the populace scientifically illiterate, at the mercy of their cognitive biases, and thus defenseless against airhead celebrities, cable-news gladiators, and other corruptions from popular culture. The standard solution is better schooling and more outreach to the public by scientists on television, social media, and popular Web sites. As an outreaching scientist I've always found this theory appealing, but I've come to realize it's wrong, or at best a small part of the problem.

Consider these questions about evolution:

During the Industrial Revolution of the 19th century, the English countryside got covered in soot, and the Peppered Moth became, on average, darker in color. How did this happen?

A. In order to blend in with their surroundings, the moths had to become darker in color.

B. The moths with darker color were less likely to get eaten and were more likely to reproduce.

After a year the average test score at a private high school increased by thirty points. Which explanation for this change is most analogous to Darwin's explanation for the adaptation of species?

A. The school no longer admitted children of wealthy alumni unless they met the same standards as everyone else.

B. Since the last test, each returning student had grown more knowledgeable.

The correct answers are B and A. The psychologist Andrew Shtulman gave high school and university students a battery of questions like this which probed for a deep understanding of the theory of natural selection, in particular the key idea that evolution consists of changes in the proportion of a population with adaptive traits rather than a transformation of the population so that its traits would be more adaptive. He found no correlation between performance on the test and a belief that natural selection explains the origin of humans. People can believe in evolution without understanding it, and vice versa.[10] In the 1980s several biologists got burned when they accepted invitations to debate creationists who turned out to be not Bible-thumping yokels but well-briefed litigators who cited cutting-edge research to sow uncertainty as to whether the science was complete.

Professing a belief in evolution is not a gift of scientific literacy, but an affirmation of loyalty to a liberal secular subculture as opposed to a conservative religious one. In 2010 the National Science Foundation dropped the following item from its test of scientific literacy: "Human beings, as we know them today, developed from earlier species of animals." The reason for that change was not, as scientists howled, because the NSF had given in to creationist pressure to bowdlerize evolution from the scientific canon. It was that the correlation between performance on that item and on every other item on the test (such as "An electron is smaller than an atom" and "Antibiotics kill viruses") was so low that it was taking up space in the test that could go to more diagnostic items. The item, in other words, was effectively a test of religiosity rather than scientific literacy.[11] When the item was prefaced with "According to the theory of evolution," so that scientific understanding was divorced from cultural allegiance, religious and nonreligious test-takers responded the same.[12]

Or consider these questions:

Climate scientists believe that if the North Pole icecap melted as a result of human-caused global warming, global sea levels would rise. True or False?

What gas do most scientists believe causes temperatures in the atmosphere to rise? Is it carbon dioxide, hydrogen, helium, or radon?

Climate scientists believe that human-caused global warming will increase the risk of skin cancer in human beings. True or False?

The answer to the first question is "false"; if it were true, your glass of Coke would overflow as the ice cubes melted. It's icecaps on *land*, such as Greenland and Antarctica, that raise sea levels when they melt. Believers in human-made climate change scored no better on tests of climate science, or of science literacy in general, than deniers. Many believers think, for example, that global warming is caused by a hole in the ozone layer and that it can be mitigated by cleaning up toxic waste dumps.[13] What predicts the denial of human-made climate change is not scientific illiteracy but political ideology. In 2015, 10 percent of conservative Republicans agreed that the Earth is getting warmer because of human activity (57 percent denied that the Earth is getting warmer at all), compared with 36 percent of moderate Republicans, 53 percent of Independents, 63 percent of moderate Democrats, and 78 percent of liberal Democrats.[14]

In a revolutionary analysis of reason in the public sphere, the legal scholar Dan Kahan has argued that certain beliefs become symbols of cultural allegiance. People affirm or deny these beliefs to express not what they *know* but who they *are*.[15] We all identify with particular tribes or subcultures, each of which embraces a creed on what makes for a good life and how society should run its affairs. These creeds tend to vary along two dimensions. One contrasts a right-wing comfort with natural hierarchy with a left-wing preference for forced egalitarianism (measured by agreement with statements like "We need to dramatically reduce inequalities between the rich and the poor, whites and people of color, and men and women"). The other is a libertarian affinity to individualism versus a communitarian or authoritarian affinity to solidarity (measured by agreement with statements like "Government should put limits on the choices individuals can make so they don't get in the way of what's good for society"). A given belief, depending on how it is framed and who endorses it, can become a touchstone, password, motto, shibboleth, sacred value, or oath of allegiance to one of these tribes. As Kahan and his collaborators explain:

> The principal reason people disagree about climate change science is not that it has been communicated to them in forms they cannot understand. Rather, it is that positions on climate change convey values—communal concern versus individual self-reliance; prudent self-abnegation versus the heroic pursuit of reward; humility versus ingenuity; harmony with nature versus mastery over it—that divide them along cultural lines.[16]

The values that divide people are also defined by which demons are blamed for society's misfortunes: greedy corporations, out-of-touch elites, meddling bureaucrats, lying politicians, ignorant rednecks, or, all too often, ethnic minorities.

Kahan notes that people's tendency to treat their beliefs as oaths of allegiance rather than disinterested appraisals is, in one sense, rational. With the exception of a tiny number of movers, shakers, and deciders, a person's opinions on climate change or evolution are astronomically unlikely to make a difference to the world at large. But they make an enormous difference to the respect the person commands in his or her social circle. To express the wrong opinion on a politicized issue can make one an oddball at best—someone who "doesn't get it"—and a traitor at worst. The pressure to conform becomes all the greater as people live and work with others who are like them and as academic, business, or religious cliques brand themselves with left-wing or right-wing causes. For pundits and politicians with a reputation for championing their faction, coming out on the wrong side of an issue would be career suicide.

Given these payoffs, endorsing a belief that hasn't passed muster with science and fact-checking isn't so irrational after all—at least, not by the criterion of the immediate effects on the believer. The effects on the society and planet are another matter. The atmosphere doesn't care what people think about it, and if it in fact warms by 4° Celsius, billions of people will suffer, no matter how many of them had been esteemed in their peer groups for holding the locally fashionable opinion on climate change along the way. Kahan concludes that we are all actors in a Tragedy of the Belief Commons: what's rational for every individual to believe (based on esteem) can be irrational for the society as a whole to act upon (based on reality).[17]

The perverse incentives behind "expressive rationality" or "identity-protective cognition" help explain the paradox of 21st-century irrationality. During the 2016 presidential campaign, many political observers were incredulous at opinions expressed by Trump supporters (and in many cases by Trump himself), such as that Hillary Clinton had multiple sclerosis and was concealing it with a body double, or that Barack Obama must have had a role in 9/11 because he was never in the Oval Office around that time (Obama, of course, was not the president in 2001). As Amanda Marcotte put it, "These folks clearly are competent enough to dress themselves, read the address of the rally and show up on time, and somehow they continue to believe stuff that's so crazy and so false that it's impossible to believe anyone that isn't barking mad could believe it.

What's going on?"[18] What's going on is that these people are sharing *blue lies*. A white lie is told for the benefit of the hearer; a blue lie is told for the benefit of an in-group (originally, fellow police officers).[19] While some of the conspiracy theorists may be genuinely misinformed, most express these beliefs for the purpose of performance rather than truth: they are trying to antagonize liberals and display solidarity with their blood brothers. The anthropologist John Tooby adds that preposterous beliefs are more effective signals of coalitional loyalty than reasonable ones.[20] Anyone can say that rocks fall down rather than up, but only a person who is truly committed to the brethren has a reason to say that God is three persons but also one person, or that the Democratic Party ran a child sex ring out of a Washington pizzeria.

~

The conspiracy theories of fervid hordes at a political rally represent an extreme case of self-expression trumping truth, but the Tragedy of the Belief Commons runs even deeper. Another paradox of rationality is that expertise, brainpower, and conscious reasoning do not, by themselves, guarantee that thinkers will approach the truth. On the contrary, they can be weapons for ever-more-ingenious rationalization. As Benjamin Franklin observed, "So convenient a thing is it to be a rational creature, since it enables us to find or make a reason for everything one has a mind to do."

Psychologists have long known that the human brain is infected with motivated reasoning (directing an argument toward a favored conclusion, rather than following it where it leads), biased evaluation (finding fault with evidence that disconfirms a favored position and giving a pass to evidence that supports it), and a My-Side bias (self-explanatory).[21] In a classic experiment from 1954, the psychologists Al Hastorf and Hadley Cantril quizzed Dartmouth and Princeton students about a film of a recent bone-crushing, penalty-filled football game between the two schools, and found that each set of students saw more infractions by the other team.[22]

We know today that political partisanship is like sports fandom: testosterone levels rise or fall on election night just as they do on Super Bowl Sunday.[23] And so it should not be surprising that political partisans—which include most of us—always see more infractions by the other team. In another classic study, the psychologists Charles Lord, Lee Ross, and Mark Lepper presented proponents and opponents of the death penalty with a pair of studies, one suggesting that capital punishment deterred homicide (murder rates went down the year after states adopted it), the other that it failed to do so (murder rates were higher in states that had capital punishment than in neighboring states that didn't). The studies were fake but realistic, and the experimenters flipped the outcomes for half the participants just in case any of them found comparisons across time more convincing than comparisons across space or vice versa. The experimenters found that each group was momentarily swayed by the result they had just learned, but as soon as they had had a chance to read the details, they picked nits in whichever study was uncongenial to their starting position, saying things like "The evidence is meaningless without data about how the overall crime rate went up in those years," or "There might be different circumstances between the two states even though they shared a border." Thanks to this selective prosecution, the participants were more polarized *after* they had all been exposed to the same evidence than before: the antis were more anti, the pros more pro.[24]

Engagement with politics is like sports fandom in another way: people seek and consume news to enhance the fan experience, not to make their opinions more accurate.[25] That explains another of Kahan's findings: the better informed a person is about climate change, the more polarized his or her opinion.[26] Indeed, people needn't even *have* a prior opinion to be polarized by the facts. When Kahan exposed people to a neutral, balanced presentation of the risks of nanotechnology (hardly a hot button on the cable news networks), they promptly split into factions that aligned with their views on nuclear power and genetically modified foods.[27]

If these studies aren't sobering enough, consider this one, described by one magazine as "The Most Depressing Discovery About the Brain, Ever."[28] Kahan recruited a thousand Americans from all walks of life, assessed their politics and numeracy with standard questionnaires, and asked them to look at some data to evaluate the effectiveness of a new treatment for an ailment. The respondents were told that they had to pay close attention to the numbers, because the treatment was not expected to work a hundred percent of the time and might even make things worse, while sometimes the ailment got better on its own, without any treatment. The numbers had been jiggered so that one answer popped out (the treatment worked, because a larger *number* of treated people showed an improvement) but the other answer was correct (the treatment didn't work, because a smaller *proportion* of the treated people showed an improvement). The knee-jerk answer could be overridden by a smidgen of mental math, namely eyeballing the ratios. In one version, the respondents were told that the ailment was a rash and the treatment was a skin cream. Here are the numbers they were shown:

	Improved	Got Worse
Treatment	223	75
No Treatment	107	21

The data implied that the skin cream did more harm than good: the people who used it improved at a ratio of around three to one, while those not using it improved at a ratio of around five to one. (With half the respondents, the rows were flipped, implying that the skin cream did work.) The more innumerate respondents were seduced by the larger absolute number of treated people who got better (223 versus 107) and picked the wrong answer. The highly numerate respondents zoomed in on the difference between the two ratios (3:1 versus 5:1) and picked the right one. The numerate respondents, of course, were not biased for or against skin cream: whichever way the data went, they spotted the difference. And contrary to liberal Democrats' and conservative Republicans' worst suspicions about each other's intelligence, neither faction did substantially better than the other.

But all this changed in a version of the experiment in which the treatment was switched from boring skin cream to incendiary gun control (a law banning citizens from carrying concealed handguns in public), and the outcome was switched from rashes to crime rates. Now the highly numerate respondents diverged from each other according to their politics. When the data suggested that the gun-control measure lowered crime, all the liberal numerates spotted it, and most of the conservative numerates missed it—they did a bit better than the conservative innumerates, but were still wrong more often than they were right. When the data showed that gun control *increased* crime, this time most of the conservative numerates spotted it, but the liberal numerates missed it; in fact, they did no better than the liberal innumerates. So we can't blame human irrationality on our lizard brains: it was the *sophisticated* respondents who were most blinded by their politics. As two other magazines summarized the results: "Science Confirms: Politics Wrecks Your Ability to Do Math" and "How Politics Makes Us Stupid."[29]

Researchers themselves are not immune. They often trip over their *own* biases when they try to show that their political *adversaries* are biased, a fallacy that can be called the bias bias (as in Matthew 7:3, "And why beholdest thou the mote that is in thy brother's eye, but considerest not the beam that is in thine own eye?").[30] A recent study by three social scientists (members of a predominantly liberal profession) purporting to show that conservatives were more hostile and aggressive had to be retracted when the authors discovered that they had misread the labels: it was actually *liberals* who were more hostile and aggressive.[31] Many studies that try to show that conservatives are temperamentally more prejudiced and rigid than liberals turn out to have cherry-picked the test items.[32] Conservatives are indeed more prejudiced against African Americans, but liberals turn out to be more prejudiced against religious Christians. Conservatives are indeed more biased toward allowing Christian prayers in schools, but liberals are more biased toward allowing Muslim prayers in schools.

It would also be an error to think that bias about bias is confined to the left: that would be a bias bias bias. In 2010 the libertarian economists Daniel Klein and Zeljka Buturovic published a study aiming to show that left-liberals were economically illiterate, based on erroneous answers to Econ 101 items like these:[33]

Restrictions on housing development make housing less affordable. [True]
Mandatory licensing of professional services increases the prices of those services. [True]
A company with the largest market share is a monopoly. [False]
Rent control leads to housing shortages. [True]

(Another item was "Overall, the standard of living is higher today than it was 30 years ago," which is true. Consistent with my claim in chapter 4 that progressives hate progress, 61 percent of the progressives and 52 percent of the liberals disagreed.) Conservatives and libertarians gloated, and the *Wall Street Journal* reported the study under the headline "Are You Smarter Than a Fifth Grader?" with the implication that left-wingers are not. But critics pointed out that the items on the quiz implicitly challenged left-wing causes. So the pair ran a follow-up with equally elementary Econ 101 items designed this time to get under the skin of conservatives:[34]

When two people complete a voluntary transaction, they both *necessarily* come away better off. [False]

Making abortion illegal would increase the number of black-market
abortions. [True]
Legalizing drugs would give more wealth and power to street gangs
and organized crime. [False]

Now it was the conservatives who earned the dunce caps. Klein, to his
credit, retracted his swipe at the left in an article entitled "I Was Wrong,
and So Are You." As he noted,

> More than 30 percent of my libertarian compatriots (and more than 40
> percent of conservatives), for instance, disagreed with the statement "A
> dollar means more to a poor person than it does to a rich person"—
> c'mon, people!—versus just 4 percent among progressives. . . . A full
> tabulation of all 17 questions showed that no group clearly out-stupids
> the others. They appear about equally stupid when faced with proper
> challenges to their position.[35]

~

If the left and right are equally stupid in quizzes and experiments, we
might expect them to be equally off the mark in making sense of the
world. The data on human history presented in chapters 5 through 18
provide an opportunity to see which of the major political ideologies can
explain the facts of human progress. I've been arguing that the main
drivers were the nonpolitical ideals of reason, science, and humanism,
which led people to seek and apply knowledge that enhanced human
flourishing. Do right-wing or left-wing ideologies have anything to add?
Do the seventy-odd graphs entitle either side to say, "Bias, shmias: we're
right; you're wrong"? It seems that each side can take some credit while
also missing big parts of the story.

Foremost is the conservative skepticism about the ideal of progress
itself. Ever since the first modern conservative, Edmund Burke, suggested that humans were too flawed to think up schemes for improving
their condition and were better off sticking with traditions and institutions that kept them from the abyss, a major stream of conservative
thought has been skeptical about the best-laid plans of mice and men.
The reactionary fringe of conservatism, recently disinterred by Trump-
ists and the European far right (chapter 23), believes that Western civilization has careened out of control since some halcyon century, having
abandoned the moral clarity of traditional Christendom for a decadent

secular fleshpot that, if left on its current course, will soon implode from terrorism, crime, and anomie.

Well, that's wrong. Life before the Enlightenment was darkened by starvation, plagues, superstitions, maternal and infant mortality, marauding knight-warlords, sadistic torture-executions, slavery, witch hunts, and genocidal crusades, conquests, and wars of religion.[36] Good riddance. The arcs in figures 5-1 through 18-4 show that as ingenuity and sympathy have been applied to the human condition, life has gotten longer, healthier, richer, safer, happier, freer, smarter, deeper, and more interesting. Problems remain, but problems are inevitable.

The left, too, has missed the boat in its contempt for the market and its romance with Marxism. Industrial capitalism launched the Great Escape from universal poverty in the 19th century and is rescuing the rest of humankind in a Great Convergence in the 21st. Over the same time span, communism brought the world terror-famines, purges, gulags, genocides, Chernobyl, megadeath revolutionary wars, and North Korea–style poverty before collapsing everywhere else of its own internal contradictions.[37] Yet in a recent survey 18 percent of social science professors identified themselves as Marxist, and the words *capitalist* and *free market* still stick in the throats of most intellectuals.[38] Partly this is because their brains autocorrect these terms to *unbridled, unregulated, unfettered,* or *untrammeled* free markets, perpetuating a false dichotomy: a free market can coexist with regulations on safety, labor, and the environment, just as a free country can coexist with criminal laws. And a free market can coexist with high levels of spending on health, education, and welfare (chapter 9)—indeed, some of the countries with the greatest amount of social spending also have the greatest amount of economic freedom.[39]

To be fair to the left, the libertarian right has embraced the same false dichotomy and seems all too willing to play the left's straw man.[40] Right-wing libertarians (in their 21st-century Republican Party version) have converted the observation that too much regulation can be harmful (by over-empowering bureaucrats, costing more to society than it delivers in benefits, or protecting incumbents against competition rather than consumers against harm) into the dogma that less regulation is always better than more regulation. They have converted the observation that too much social spending can be harmful (by creating perverse incentives against work and undermining the norms and institutions of civil society) into the dogma that any amount of social spending is too much. And they have translated the observation that tax rates can be too high into a hysterical rhetoric of "liberty" in which raising the marginal tax rate for income above $400,000 from 35 to 39.6 percent means turning the country over to jackbooted storm troopers. Often the refusal to seek the optimum level of government is justified by an appeal to Friedrich Hayek's argument in *The Road to Serfdom* that regulation and welfare lay out a slippery slope along which a country will slide into penury and tyranny.

The facts of human progress strike me as having been as unkind to right-wing libertarianism as to right-wing conservatism and left-wing Marxism. The totalitarian governments of the 20th century did not emerge from democratic welfare states sliding down a slippery slope, but were imposed by fanatical ideologues and gangs of thugs.[41] And countries that combine free markets with more taxation, social spending, and regulation than the United States (such as Canada, New Zealand, and Western Europe) turn out to be not grim dystopias but rather pleasant places to live, and they trounce the United States in every measure of human flourishing, including crime, life expectancy, infant mortality, education, and happiness.[42] As we saw, no developed country runs on right-wing libertarian principles, nor has any realistic vision of such a country ever been laid out.

It should not be surprising that the facts of human progress confound the major -isms. The ideologies are more than two centuries old and are based on mile-high visions such as whether humans are tragically flawed or infinitely malleable, and whether society is an organic whole or a collection of individuals.[43] A real society comprises hundreds of millions of social beings, each with a trillion-synapse brain, who pursue their well-being while affecting the well-being of others in complex networks with massive positive and negative externalities, many of them historically unprecedented. It is bound to defy any simple narrative of what will happen under a given set of rules. A more rational approach to politics is to treat societies as ongoing experiments and open-mindedly learn the best practices, whichever part of the spectrum they come from. The empirical picture at present suggests that people flourish most in liberal democracies with a mixture of civic norms, guaranteed rights, market freedom, social spending, and judicious regulation. As Pat Paulsen noted, "If either the right wing or the left wing gained control of the country, it would fly around in circles."

It's not that Goldilocks is always right and that the truth always falls halfway between extremes. It's that current societies have winnowed out the worst blunders of the past, so if a society is functioning halfway decently—if the streets aren't running with blood, if obesity is a bigger problem than malnutrition, if the people who vote with their feet are

clamoring to get in rather than racing for the exits—then its current institutions are probably a good starting point (itself a lesson we can take
from Burkean conservatism). Reason tells us that political deliberation
would be most fruitful if it treated governance more like scientific exper-
imentation and less like an extreme-sports competition.

~

Though examining data from history and social science is a better way
of evaluating our ideas than arguing from the imagination, the acid test
of empirical rationality is *prediction*. Science proceeds by testing the predictions of hypotheses, and we all recognize the logic in everyday life
when we praise or ridicule barroom sages depending on whether events
bear them out, when we use idioms that hold people responsible for their
accuracy like *to eat crow* and *to have egg on your face*, and when we use
sayings like "Put your money where your mouth is" and "The proof of
the pudding is in the eating."

Unfortunately the epistemological standards of common sense—we
should credit the people and ideas that make correct predictions, and
discount the ones that don't—are rarely applied to the intelligentsia and
commentariat, who dispense opinions free of accountability. Always-
wrong prognosticators like Paul Ehrlich continue to be canvassed by the
press, and most readers have no idea whether their favorite columnists,
gurus, or talking heads are more accurate than a chimpanzee picking
bananas. The consequences can be dire: many military and political de-
bacles arose from misplaced confidence in the predictions of experts
(such as intelligence reports in 2003 that Saddam Hussein was developing nuclear weapons), and a few percentage points of accuracy in predicting financial markets can spell the difference between gaining and
losing a fortune.

A track record of predictions also ought to inform our appraisal of
intellectual systems, including political ideologies. Though some ideological differences come from clashing values and may be irreconcilable,
many hinge on different means to agreed-upon ends and should be de-
cidable. Which policies will in fact bring about things that almost everyone wants, like lasting peace or economic growth? Which will reduce
poverty, or violent crime, or illiteracy? A rational society should seek the
answers by consulting the world rather than assuming the omniscience
of a bloc of opinionators who have coalesced around a creed.

Unfortunately, the expressive rationality documented by Kahan in
his experimental subjects also applies to editorialists and experts. The
payoffs that determine their reputations don't coincide with the accuracy

of the predictions, since no one is keeping score. Instead, their reputations hinge on their ability to entertain, titillate, or shock; on their ability to instill confidence or fear (in the hopes that a prophecy might be self-fulfilling or self-defeating); and on their skill in galvanizing a coalition and celebrating its virtue.

Since the 1980s the psychologist Philip Tetlock has studied what distinguishes accurate forecasters from the many oracles who are "often mistaken but never in doubt."[44] He recruited hundreds of analysts, columnists, academics, and interested laypeople to compete in forecasting tournaments in which they were presented with possible events and asked to assess their likelihoods. Experts are ingenious at wordsmithing their predictions to protect them from falsification, using weasely modal auxiliaries (*could, might*), adjectives (*fair chance, serious possibility*), and temporal modifiers (*very soon, in the not-too-distant future*). So Tetlock pinned them down by stipulating events with unambiguous outcomes and deadlines (for example, "Will Russia annex additional Ukraine territory in the next three months?" "In the next year, will any country withdraw from the Eurozone?" "How many additional countries will report cases of the Ebola virus in the next eight months?") and having them write down numerical probabilities.

Tetlock also avoided the common fallacy of praising or ridiculing a single probabilistic prediction after the fact, as when the poll aggregator Nate Silver of *FiveThirtyEight* came under fire for giving Donald Trump just a 29 percent chance of winning the 2016 election.[45] Since we cannot replay the election thousands of times and count up the number of times that Trump won, the question of whether the prediction was confirmed or disconfirmed is meaningless. What we *can* do, and what Tetlock did, is compare the *set* of each forecaster's probabilities with the corresponding outcomes. Tetlock used a formula which credits the forecaster not just for accuracy but for accurately going out on a limb (since it's easier to be accurate by just playing it safe with 50-50 predictions). The formula is mathematically related to how much they would win if they put their money where their mouths were and bet on their predictions according to their own odds.

Twenty years and twenty-eight thousand predictions later, how well did the experts do? On average, about as well as a chimpanzee (which Tetlock described as throwing darts rather than picking bananas). Tetlock and the psychologist Barbara Mellers held a rematch between 2011 and 2015 in which they recruited several thousand contestants to take part in a forecasting tournament held by the Intelligence Advanced Research Projects Activity (the research organization of the federation of American intelligence agencies). Once again there was plenty of dart-throwing, but in both tournaments the couple could pick out "superforecasters" who performed not just better than chimps and pundits, but better than professional intelligence officers with access to classified information, better than prediction markets, and not too far from the theoretical maximum. How can we explain this apparent clairvoyance? (For a year, that is—accuracy declines with distance into the future, and it falls to the level of chance around five years out.) The answers are clear and profound.

The forecasters who did the worst were the ones with Big Ideas—left-wing or right-wing, optimistic or pessimistic—which they held with an inspiring (but misguided) confidence:

> As ideologically diverse as they were, they were united by the fact that their thinking was so ideological. They sought to squeeze complex problems into the preferred cause-effect templates and treated what did not fit as irrelevant distractions. Allergic to wishy-washy answers, they kept pushing their analyses to the limit (and then some), using terms like "furthermore" and "moreover" while piling up reasons why they were right and others wrong. As a result, they were unusually confident and likelier to declare things "impossible" or "certain." Committed to their conclusions, they were reluctant to change their minds even when their predictions clearly failed. They would tell us, "Just wait."[46]

Indeed, the very traits that put these experts in the public eye made them the worst at prediction. The more famous they were, and the closer the event was to their area of expertise, the less accurate their predictions turned out to be. But the chimplike success of brand-name ideologues does *not* mean that "experts" are worthless and we should distrust elites. It's that we need to revise our concept of an expert. Tetlock's superforecasters were:

> pragmatic experts who drew on many analytical tools, with the choice of tool hinging on the particular problem they faced. These experts gathered as much information from as many sources as they could. When thinking, they often shifted mental gears, sprinkling their speech with transition markers such as "however," "but," "although," and "on the other hand." They talked about possibilities and

probabilities, not certainties. And while no one likes to say "I was wrong," these experts more readily admitted it and changed their minds.[47]

Successful prediction is the revenge of the nerds. Superforecasters are intelligent but not necessarily brilliant, falling just in the top fifth of the population. They are highly numerate, not in the sense of being math whizzes but in the sense of comfortably thinking in guesstimates. They have personality traits that psychologists call "openness to experience" (intellectual curiosity and a taste for variety), "need for cognition" (pleasure taken in intellectual activity), and "integrative complexity" (appreciating uncertainty and seeing multiple sides). They are anti-impulsive, distrusting their first gut feeling. They are neither left-wing nor right-wing. They aren't necessarily humble about their abilities, but they *are* humble about particular beliefs, treating them as "hypotheses to be tested, not treasures to be guarded." They constantly ask themselves, "Are there holes in this reasoning? Should I be looking for something else to fill this in? Would I be convinced by this if I were somebody else?" They are aware of cognitive blind spots like the Availability and confirmation biases, and they discipline themselves to avoid them. They display what the psychologist Jonathan Baron calls "active open-mindedness," with opinions such as these:[48]

People should take into consideration evidence that goes against their beliefs. [Agree]
It is more useful to pay attention to those who disagree with you than to pay attention to those who agree. [Agree]
Changing your mind is a sign of weakness. [Disagree]
Intuition is the best guide in making decisions. [Disagree]
It is important to persevere in your beliefs even when evidence is brought to bear against them. [Disagree]

Even more important than their temperament is their manner of reasoning. Superforecasters are Bayesian, tacitly using the rule from the eponymous Reverend Bayes on how to update one's degree of credence in a proposition in light of new evidence. They begin with the base rate for the event in question: how often it is expected to occur across the board and over the long run. Then they nudge that estimate up or down depending on the degree to which new evidence portends the event's occurrence or non-occurrence. They seek this new evidence avidly, and avoid both overreacting to it ("This changes everything!") and under-reacting to it ("This means nothing!").

Take, for example, the prediction "There will be an attack by Islamist militants in Western Europe between 21 January and 31 March 2015," made shortly after the *Charlie Hebdo* massacre in January of that year. Pundits and politicians, their heads spinning with the Availability heuristic, would play out the scenario in the theater of the imagination and, not wanting to appear complacent or naïve, answer Definitely Yes. That's not how superforecasters work. One of them, asked by Tetlock to think aloud, reported that he began by estimating the base rate: he went to *Wikipedia*, looked up the list of Islamist terrorist attacks in Europe for the previous five years, and divided by 5, which predicted 1.2 attacks a year. But, he reasoned, the world had changed since the Arab Spring in 2011, so he lopped off the 2010 data, which brought the base rate up to 1.5. ISIS recruitment had increased since the *Charlie Hebdo* attacks, a reason to poke the estimate upward, but so had security measures, a reason to tug it downward. Balancing the two factors, an increase by about a fifth seemed reasonable, yielding a prediction of 1.8 attacks a year. There were 69 days left in the forecast period, so he divided 69 by 365 and multiplied the fraction by 1.8. That meant that the chance of an Islamist attack in Western Europe by the end of March was about one in three. A manner of forecasting very different from the way most people think led to a very different forecast.

Two other traits distinguish superforecasters from pundits and chimpanzees. The superforecasters believe in the wisdom of crowds, laying their hypotheses on the table for others to criticize or amend and pooling their estimates with those of others. And they have strong opinions on chance and contingency in human history as opposed to necessity and fate. Tetlock and Mellers asked different groups of people whether they agreed with statements like the following:

> Events unfold according to God's plan.
> Everything happens for a reason.
> There are no accidents or coincidences.
> Nothing is inevitable.
> Even major events like World War II or 9/11 could have turned out
>   very differently.
> Randomness is often a factor in our personal lives.

They calculated a Fate Score by adding up the "Agree" ratings for items like the first three and the "Disagree" ratings for items like the last three.

An average American is somewhere in the middle. An undergraduate at an elite university scores a bit lower; a so-so forecaster lower still; and the superforecasters lowest of all, with the most accurate superforecasters expressing the most vehement rejection of fate and acceptance of chance.

To my mind, Tetlock's hardheaded appraisal of expertise by the ultimate benchmark, prediction, should revolutionize our understanding of history, politics, epistemology, and intellectual life. What does it mean that the wonkish tweaking of probabilities is a more reliable guide to the world than the pronouncements of erudite sages and narratives inspired by systems of ideas? Aside from smacking us upside the head with a reminder to be more humble and open-minded, it offers a glimpse into the workings of history on the time scale of years and decades. Events are determined by myriad small forces incrementing or decrementing their likelihoods and magnitudes rather than by sweeping laws and grand dialectics. Unfortunately for many intellectuals and for all political ideologues, this is not the way they are accustomed to thinking, but perhaps we had better get used to it. When Tetlock was asked at a public lecture to forecast the future of forecasting, he said, "When the audience of 2515 looks back on the audience of 2015, their level of contempt for how we go about judging political debate will be roughly comparable to the level of contempt we have for the 1692 Salem witch trials."[49]

$\sim$

Tetlock did not assign a probability to his whimsical prediction, and he gave it a long, safe deadline. It certainly would be unwise to forecast an improvement in the quality of political debate within the five-year window in which prediction is feasible. The major enemy of reason in the public sphere today—which is not ignorance, innumeracy, or cognitive biases, but politicization—appears to be on an upswing.

In the political arena itself, Americans have become increasingly polarized.[50] Most people's opinions are too shallow and uninformed to fit into a coherent ideology, but in a dubious form of progress, the percentage of Americans whose opinions are down-the-line liberal or down-the-line conservative doubled between 1994 and 2014, from 10 to 21 percent. The polarization has coincided with an increase in social segregation by politics: over those twenty years, the ideologues have become more likely to say that most of their close friends share their political views.

The parties have become more partisan as well. According to a recent Pew study, in 1994 about a third of Democrats were more conservative than the median Republican, and vice-versa. In 2014 the figures were closer to a *twentieth*. Though Americans across the political spectrum drifted leftward through 2004, since then they have diverged on every major issue except gay rights, including government regulation, social spending, immigration, environmental protection, and military strength. Even more troublingly, each side has become more contemptuous of the other. In 2014, 38 percent of Democrats held "very unfavorable" views of the Republican Party (up from 16 percent in 1994), and more than a quarter saw it as "a threat to the nation's well-being." Republicans were even more hostile to Democrats, with 43 percent viewing the party unfavorably and more than a third seeing it as a threat. The ideologues on each side have also become more resistant to compromise.

Fortunately, a majority of Americans are more moderate in all these opinions, and the proportion who call themselves moderate has not changed in forty years.[51] Unfortunately, it's the extremists who are more likely to vote, donate, and pressure their representatives. There is little reason to think that any of this has improved since the survey was conducted in 2014, to put it mildly.

Universities ought to be the arena in which political prejudice is set aside and open-minded investigation reveals the way the world works. But just when we need this disinterested forum the most, academia has become more politicized as well—not more polarized, but more left-wing. Colleges have always been more liberal than the American population, but the skew has been increasing. In 1990, 42 percent of faculty were far left or liberal (11 percentage points more than the American population), 40 percent were moderate, and 18 percent were far right or conservative, for a left-to-right ratio of 2.3 to 1. In 2014 the proportions were 60 percent far left or liberal (30 percentage points more than the population), 28 percent moderate, and 12 percent conservative, a ratio of 5 to 1.[52] The proportions vary by field: departments of business, computer science, engineering, and health science are evenly split, while the humanities and social sciences are decidedly on the left: the proportion of conservatives is in the single digits, and they are outnumbered by *Marxists* two to one.[53] Professors in the physical and biological sciences are in between, with few radicals and virtually no Marxists, but liberals outnumber conservatives by a wide margin.

The liberal tilt of academia (and of journalism, commentary, and intellectual life) is in some ways natural.[54] Intellectual inquiry is bound to challenge the status quo, which is never perfect. And verbally articulated propositions, intellectuals' stock in trade, are more congenial to the deliberate policies typically favored by liberals than to the diffuse forms of social organization such as markets and traditional norms typically favored by conservatives.[55] A liberal tilt is also, in moderation, desirable. Intellectual liberalism was at the forefront of many forms of progress that almost everyone has come to accept, such as democracy, social insurance, religious tolerance, the abolition of slavery and judicial torture, the decline of war, and the expansion of human and civil rights.[56] In many ways we are (almost) all liberals now.[57]

But we have seen that when a creed becomes attached to an in-group, the critical faculties of its members can be disabled, and there are reasons to think that has happened within swaths of academia.[58] In *The Blank Slate* (updated in 2016) I showed how leftist politics had distorted the study of human nature, including sex, violence, gender, childrearing, personality, and intelligence. In a recent manifesto, Tetlock, together with the psychologists José Duarte, Jarret Crawford, Charlotta Stern, Jonathan Haidt, and Lee Jussim, documented the leftward swing of social psychology and showed how it has compromised the quality of research.[59] Quoting John Stuart Mill—"He who knows only his own side of the case, knows little of that"—they called for greater political diversity in psychology, the version of diversity that matters the most (as opposed to the version commonly pursued, namely people who look different but think alike).[60]

To the credit of academic psychology, Duarte et al.'s critique has been respectfully received.[61] But the respect is far from universal. When the *New York Times* columnist Nicholas Kristof cited their article favorably and made similar points, the angry reaction confirmed their worst accusations (the most highly recommended comment was "You don't diversify with idiots").[62] And a faction of academic culture composed of hard-left faculty, student activists, and an autonomous diversity bureaucracy (pejoratively called social justice warriors) has become aggressively illiberal. Anyone who disagrees with the assumption that racism is the cause of all problems is called a racist.[63] Non-leftist speakers are frequently disinvited after protests or drowned out by jeering mobs.[64] A student may be publicly shamed by her dean for a private email that considers both sides of a controversy.[65] Professors are pressured to avoid lecturing on upsetting topics, and have been subjected to Stalinesque investigations for politically incorrect opinions.[66] Often the repression veers into unintended comedy.[67] A guideline for deans on how to identify "microaggressions" lists remarks such as "America is the land of opportunity" and "I believe the most qualified person should get the job." Students mob and curse a professor who invited them to discuss a letter written by his wife suggesting that students chill out about Halloween costumes. A yoga course was canceled because yoga was deemed "cultural appropriation." The comedians themselves are not amused: Jerry Seinfeld, Chris Rock, and Bill Maher, among others, are wary of performing at college campuses because inevitably some students will be enraged by a joke.[68]

For all the follies on campus, we can't let right-wing polemicists indulge in a bias bias and dismiss any idea they don't like that comes out of a university. The academic archipelago embraces a vast sea of opinions, and it is committed to norms such as peer review, tenure, open debate, and the demand for citation and empirical evidence that are engineered to foster disinterested truth-seeking, however imperfectly they do so in practice. Colleges and universities have fostered the heterodox criticisms reviewed here and elsewhere, while delivering immense gifts of knowledge to the world.[69] And it's not as if alternative arenas—the blogosphere, the Twittersphere, cable news, talk radio, Congress—are paragons of objectivity and rigor.

Of the two forms of politicization that are subverting reason today, the political is far more dangerous than the academic, for an obvious reason. It's often quipped (no one knows who said it first) that academic debates are vicious because the stakes are so small.[70] But in political debates the stakes are unlimited, including the future of the planet. Politicians, unlike professors, pull the levers of power. In 21st-century America, the control of Congress by a Republican Party that became synonymous with the extreme right has been pernicious, because it is so convinced of the righteousness of its cause and the evil of its rivals that it has undermined the institutions of democracy to get what it wants. The corruptions include gerrymandering, imposing voting restrictions designed to disenfranchise Democratic voters, encouraging unregulated donations from moneyed interests, blocking Supreme Court nominations until their party controls the presidency, shutting down the government when their maximal demands aren't met, and unconditionally supporting Donald Trump over their own objections to his flagrantly antidemocratic impulses.[71] Whatever differences in policy or philosophy divide the parties, the mechanisms of democratic deliberation should be sacrosanct. Their erosion, disproportionately by the right, has led many people, including a growing share of young Americans, to see democratic government as inherently dysfunctional and to become cynical about democracy itself.[72]

Intellectual and political polarization feed each other. It's harder to be a conservative intellectual when American conservative politics has become steadily more know-nothing, from Ronald Reagan to Dan Quayle to George W. Bush to Sarah Palin to Donald Trump.[73] On the other side, the capture of the left by identity politicians, political correctness police, and social justice warriors creates an opening for loudmouths who brag of "telling it like it is." A challenge of our era is how to foster an intellectual and political culture that is driven by reason rather than tribalism and mutual reaction.

~

Making reason the currency of our discourse begins with clarity about the centrality of reason itself.[74] As I mentioned, many commentators are confused about it. The discovery of cognitive and emotional biases does not mean that "humans are irrational" and so there's no point in trying to make our deliberations more rational. If humans were incapable of rationality, we could never have discovered the ways in which they were irrational, because we would have no benchmark of rationality against which to assess human judgment, and no way to carry out the assessment. Humans may be vulnerable to bias and error, but clearly not all of us all the time, or no one would ever be entitled to say that humans are vulnerable to bias and error. The human brain is *capable* of reason, given the right circumstances; the problem is to identify those circumstances and put them more firmly in place.

For the same reason, editorialists should retire the new cliché that we are in a "post-truth era" unless they can keep up a tone of scathing irony. The term is corrosive, because it implies that we should resign ourselves to propaganda and lies and just fight back with more of our own. We are not in a post-truth era. Mendacity, truth-shading, conspiracy theories, extraordinary popular delusions, and the madness of crowds are as old as our species, but so is the conviction that some ideas are right and others are wrong.[75] The same decade that has seen the rise of pants-on-fire Trump and his reality-challenged followers has also seen the rise of a new ethic of fact-checking. Angie Holan, the editor of *PolitiFact*, a fact-checking project begun in 2007, noted:

> [Many of] today's TV journalists . . . have picked up the torch of fact-checking and now grill candidates on issues of accuracy during live interviews. Most voters don't think it's biased to question people about whether their seemingly fact-based statements are accurate. Research published earlier this year by the American Press Institute showed that more than eight in 10 Americans have a positive view of political fact-checking.
>
> In fact, journalists regularly tell me their media organizations have

started highlighting fact-checking in their reporting because so many people click on fact-checking stories after a debate or high-profile news event. Many readers now want fact-checking as part of traditional news stories as well; they will vocally complain to ombudsmen and readers' representatives when they see news stories repeating discredited factual claims.[76]

This ethic would have served us well in earlier decades when false rumors regularly set off pogroms, riots, lynchings, and wars (including the Spanish-American War in 1898, the escalation of the Vietnam War in 1964, the Iraq invasion of 2003, and many others).[77] It was not applied rigorously enough to prevent Trump's victory in 2016, but since then his fibs and those of his spokespeople have been mercilessly ridiculed in the media and popular culture, which means that the resources for favoring truth are in place even if they don't always carry the day.

Over the long run, the institutions of reason can mitigate the Tragedy of the Belief Commons and allow the truth to prevail. For all of our current irrationality, few influential people today believe in werewolves, unicorns, witches, alchemy, astrology, bloodletting, miasmas, animal sacrifice, the divine right of kings, or supernatural omens in rainbows and eclipses. Moral irrationality can be outgrown as well. As recently as my childhood, the Virginia judge Leon Bazile upheld the conviction of Richard and Mildred Loving for their interracial marriage with an argument that not even the most benighted conservative would advance today:

The parties were guilty of a most serious crime. It was contrary to the declared public law, founded upon motives of public policy . . . upon which social order, public morality and the best interests of both races depend. . . . Almighty God created the races white, black, yellow, malay and red, and he placed them on separate continents. The fact that he separated the races shows that he did not intend for the races to mix.[78]

And presumably most liberals would not be persuaded by this defense of Castro's Cuba by the intellectual icon Susan Sontag in 1969:

The Cubans know a lot about spontaneity, gaiety, sensuality and freaking out. They are not linear, desiccated creatures of print-culture. In short, their problem is almost the obverse of ours—and we must be sympathetic to their efforts to solve it. Suspicious as we are of the traditional Puritanism of left revolutions, American radicals ought to be able to maintain some perspective when a country known mainly for dance music, prostitutes, cigars, abortions, resort life and pornographic movies gets a little up-tight about sexual morals and, in one bad moment two years ago, rounds up several thousand homosexuals in Havana and sends them to a farm to rehabilitate themselves.[79]

In fact, these "farms" were forced labor camps, and they arose not as a correction to spontaneous gaiety and freaking out but as an expression of a homophobia that was deeply rooted in that Latin culture. Whenever we get upset about the looniness of public discourse today, we should remind ourselves that people weren't so rational in the past, either.

What can be done to improve standards of reasoning? Persuasion by facts and logic, the most direct strategy, is not always futile. It's true that people can cling to beliefs in defiance of all evidence, like Lucy in *Peanuts* who insisted that snow comes out of the ground and rises into the sky even as she was being slowly buried in a snowfall. But there are limits as to how high the snow can pile up. When people are first confronted with information that contradicts a staked-out position, they become even more committed to it, as we'd expect from the theories of identity-protective cognition, motivated reasoning, and cognitive dissonance reduction. Feeling their identity threatened, belief holders double down and muster more ammunition to fend off the challenge. But since another part of the human mind keeps a person in touch with reality, as the counterevidence piles up the dissonance can mount until it becomes too much to bear and the opinion topples over, a phenomenon called the affective tipping point.[80] The tipping point depends on the balance between how badly the opinion holder's reputation would be damaged by relinquishing the opinion and whether the counterevidence is so blatant and public as to be common knowledge: a naked emperor, an elephant in the room.[81] As we saw in chapter 10, that is starting to happen with public opinion on climate change. And entire populations can shift when a critical nucleus of persuadable influencers changes its mind and everyone else follows along, or when one generation is replaced by another that doesn't cling to the same dogmas (progress, funeral by funeral).

Across the society as a whole the wheels of reason often turn slowly, and it would be nice to speed them up. The obvious places to apply this torque are in education and the media. For several decades fans of reason have pressured schools and universities to adopt curricula in "critical thinking." Students are advised to look at both sides of an issue, to back up their opinions with evidence, and to spot logical fallacies like circular reasoning, attacking a straw man, appealing to authority, arguing ad hominem, and reducing a graded issue to black or white.[82] Related programs called "debiasing" try to inoculate students against cognitive fallacies such as the Availability heuristic and confirmation bias.[83]

When they were first introduced, these programs had disappointing outcomes, which led to pessimism as to whether we could ever knock sense into the person on the street. But unless risk analysts and cognitive psychologists represent a superior breed of human, something in *their* education must have enlightened them about cognitive fallacies and how to avoid them, and there is no reason those enlightenments can't be applied more widely. The beauty of reason is that it can always be applied to understand failures of reason. A second look at critical thinking and debiasing programs has shown what makes them succeed or fail.

The reasons are familiar to education researchers.[84] *Any* curriculum will be pedagogically ineffective if it consists of a lecturer yammering in front of a blackboard, or a textbook that students highlight with a yellow marker. People understand concepts only when they are forced to think them through, to discuss them with others, and to use them to solve problems. A second impediment to effective teaching is that pupils don't spontaneously transfer what they learned from one concrete example to others in the same abstract category. Students in a math class who learn how to arrange a marching band into even rows using the principle of a least common multiple are stymied when asked to arrange rows of vegetables in a garden. In the same way, students in a critical thinking course who are taught to discuss the American Revolution from both the British and American perspectives will not make the leap to consider how the Germans viewed World War I.

With these lessons about lessons under their belt, psychologists have recently devised debiasing programs that fortify logical and critical thinking curricula. They encourage students to spot, name, and correct fallacies across a wide range of contexts.[85] Some use computer games that provide students with practice, and with feedback that allows them to see the absurd consequences of their errors. Other curricula translate abstruse mathematical statements into concrete, imaginable scenarios. Tetlock has compiled the practices of successful forecasters into a set of guidelines for good judgment (for example, start with the base rate; seek out evidence and don't overreact or underreact to it; don't try to explain away your own errors but instead use them as a source of calibration). These and other programs are provably effective: students' newfound wisdom outlasts the training session and transfers to new subjects.

Despite these successes, and despite the fact that the ability to engage in unbiased, critical reasoning is a prerequisite to thinking about anything else, few educational institutions have set themselves the goal of enhancing rationality. (This includes my own university, where my suggestion during a curriculum review that all students should learn about cognitive biases fell deadborn from my lips.) Many psychologists have called on their field to "give debiasing away" as one of its greatest potential contributions to human welfare.[86]

~

Effective training in critical thinking and cognitive debiasing may not be enough to cure identity-protective cognition, in which people cling to whatever opinion enhances the glory of their tribe and their status within it. This is the disease with the greatest morbidity in the political realm, and so far scientists have misdiagnosed it, pointing to irrationality and scientific illiteracy instead of the myopic rationality of the Tragedy of the Belief Commons. As one writer noted, scientists often treat the public the way Englishmen treat foreigners: they speak more slowly and more loudly.[87]

Making the world more rational, then, is not just a matter of training people to be better reasoners and setting them loose. It also depends on the rules of discourse in workplaces, social circles, and arenas of debate and decision-making. Experiments have shown that the right rules can avert the Tragedy of the Belief Commons and force people to dissociate their reasoning from their identities.[88] One technique was discovered long ago by rabbis: they forced yeshiva students to switch sides in a Talmudic debate and argue the opposite position. Another is to have people try to reach a consensus in a small discussion group; this forces them to defend their opinions to their groupmates, and the truth usually wins.[89] Scientists themselves have hit upon a new strategy called adversarial collaboration, in which mortal enemies work together to get to the bottom of an issue, setting up empirical tests that they agree beforehand will settle it.[90]

Even the mere requirement to explicate an opinion can shake people out of their overconfidence. Most of us are deluded about our degree of understanding of the world, a bias called the Illusion of Explanatory Depth.[91] Though we think we understand how a zipper works, or a cylinder lock, or a toilet, as soon as we are called upon to explain it we are dumbfounded and forced to confess we have no idea. That is also true of hot-button political issues. When people with die-hard opinions on Obamacare or NAFTA are challenged to explain what those policies actually are, they soon realize that they don't know what they are talking about, and become more open to counterarguments. Perhaps most important, people are less biased when they have skin in the game and have to live with the consequences of their opinions. In a review of the literature on rationality, the anthropologists Hugo Mercier and Dan Sperber conclude, "Contrary to common bleak assessments of human reasoning abilities, people are quite capable of reasoning in an unbiased manner, at least when they are evaluating arguments rather than producing them, and when they are after the truth rather than trying to win a debate."[92]

The way that the rules in particular arenas can make us collectively stupid or smart can resolve the paradox that keeps popping up in this chapter: why the world seems to be getting less rational in an age of unprecedented knowledge and tools for sharing it. The resolution is that in most arenas, the world has *not* been getting less rational. It's not as if hospital patients are increasingly dying of quackery, or planes are falling out of the sky, or food is rotting on wharves because no one can figure out how to get it into stores. The chapters on progress have shown that our collective ingenuity has been increasingly successful in solving society's problems.

Indeed, in one realm after another we are seeing the conquest of dogma and instinct by the armies of reason. Newspapers are supplementing shoe leather and punditry with statisticians and fact-checking squads.[93] The cloak-and-dagger world of national intelligence is seeing farther into the future by using the Bayesian reasoning of superforecasters.[94] Health care is being reshaped by evidence-based medicine (which should have been a redundant expression long ago).[95] Psychotherapy has progressed from the couch and notebook to Feedback-Informed Treatment.[96] In New York, and increasingly in other cities, violent crime has been reduced with the real-time data-crunching system called Compstat.[97] The effort to aid the developing world is being guided by the Randomistas, economists who gather data from randomized trials to distinguish fashionable boondoggles from programs that actually improve people's lives.[98] Volunteering and charitable giving are being scrutinized by the Effective Altruism movement, which distinguishes altruistic acts that enhance the lives of beneficiaries from those that enhance the warm glow in benefactors.[99] Sports has seen the advent of Moneyball, in which strategies and players are evaluated by statistical analysis rather than intuition and lore, allowing smarter teams to beat richer teams and giving fans endless new material for conversations over the hot stove.[100] The blogosphere has spawned the Rationality Community, who urge people to be "less wrong" in their opinions by applying Bayesian reasoning and compensating for cognitive biases.[101] And in the day-to-day functioning of governments, the application of behavioral insights (sometimes called Nudge) and evidence-based policy has wrung more social benefits out of fewer tax dollars.[102] In area after area, the world has been getting more rational.

There is, of course, a flaming exception: electoral politics and the issues that have clung to it. Here the rules of the game are fiendishly designed to bring out the most irrational in people.[103] Voters have a say on issues that don't affect them personally, and never have to inform themselves or justify their positions. Practical agenda items like trade and energy are bundled with moral hot buttons like euthanasia and the teaching of evolution. Each bundle is strapped to a coalition with geographic, racial, and ethnic constituencies. The media cover elections like horse races, and analyze issues by pitting ideological hacks against each other in screaming matches. All of these features steer people away from reasoned analysis and toward perfervid self-expression. Some are products of the misconception that the benefits of democracy come from elections, whereas they depend more on having a government that is constrained in its powers, responsive to its citizens, and attentive to the results of its policies (chapter 14). As a result, reforms that are designed to make governance more "democratic," such as plebiscites and direct primaries, may instead have made governance more identity-driven and irrational. The conundrums are inherent to democracy and have been debated since the time of Plato.[104] They have no instant solution, but identifying the worst of the current problems and setting the goal of mitigating them is the place to start.

When issues are *not* politicized, people can be altogether rational. Kahan notes that "bitter public disputes over science are in fact the exception rather than the rule."[105] No one gets exercised over whether antibiotics work, or whether driving drunk is a good idea. Recent history proves the point in a natural experiment, complete with a neatly matched control group.[106] The human papillomavirus (HPV) is sexually transmitted and a major cause of cervical cancer but can be neutralized with a vaccine. Hepatitis B is also sexually transmitted, also causes cancer, and also can be prevented by a vaccine. Yet HPV vaccination became a political firestorm, with parents protesting that the government should not be making it easier for teenagers to have sex, while hepatitis B vaccination is unexceptionable. The difference, Kahan suggests, lies in the way the two vaccines were introduced. Hep B was treated as a routine public health matter, like whooping cough or yellow fever. But the manufacturer of the HPV vaccine lobbied state legislatures to make vaccination mandatory, starting with adolescent girls, which sexualized the treatment and raised the dander of puritanical parents.

To make public discourse more rational, issues should be depoliticized as much as is feasible. Experiments have shown that when people hear about a new policy, such as welfare reform, they will like it if it is proposed by their own party and hate it if it is proposed by the other—all the while convinced that they are reacting to it on its objective merits.[107] That implies that spokespeople should be chosen carefully. Several climate activists have lamented that by writing and starring in the documentary *An Inconvenient Truth*, Al Gore may have done the movement more harm than good, because as a former Democratic vice-president and presidential nominee he stamped climate change with a left-wing seal. (It's hard to believe today, but environmentalism was once denounced as a *right*-wing cause, in which the gentry frivolously worried about habitats for duck-hunting and the views from their country estates rather than serious issues like racism, poverty, and Vietnam.) Recruiting conservative and libertarian commentators who have been convinced by the evidence and are willing to share their concern would be more effective than recruiting more scientists to speak more slowly and more loudly.[108]

Also, the factual state of affairs should be unbundled from remedies that are freighted with symbolic political meaning. Kahan found that people are less polarized in their opinion about the very existence of anthropogenic climate change when they are reminded of the possibility that it might be mitigated by geoengineering than when they are told that it calls for stringent controls on emissions.[109] (This does not, of course, mean that geoengineering itself need be advocated as the primary solution.) Depoliticizing an issue can lead to real action. Kahan helped a compact of Florida businesspeople, politicians, and resident associations, many of them Republican, agree to a plan to adapt to rising sea levels that threatened coastal roads and freshwater supplies. The plan included measures to reduce carbon emissions, which under other circumstances would be politically radioactive. But as long as the planning was focused on problems they could see and the politically divisive backstory was downplayed, they acted reasonably.[110]

For their part, the media could examine their role in turning politics into a sport, and intellectuals and pundits could think twice about competing. Can we imagine a day in which the most famous columnists and talking heads have no predictable political orientation but try to work out defensible conclusions on an issue-by-issue basis? A day in which "You're just repeating the left-wing [or right-wing] position" is considered a devastating gotcha? In which people (especially academics) will answer a question like "Does gun control reduce crime?" or "Does a minimum wage increase unemployment?" with "Wait, let me look up the latest meta-analysis" rather than with a patellar reflex predictable from their politics? A day when writers on the right and left abandon the Chicago Way of debating ("They pull a knife, you pull a gun. He sends one of yours to the hospital, you send one of his to the morgue") and adopt the arms-controllers' tactic of Graduated Reciprocation in Tension-Reduction (make a small unilateral concession with an invitation that it be reciprocated)?[111]

That day is a long way off. But the self-healing powers of rationality, in which flaws in reasoning are singled out as targets for education and criticism, take time to work. It took centuries for Francis Bacon's observations on anecdotal reasoning and the confusion of correlation with causation to become second nature to scientifically literate people. It's taken almost fifty years for Tversky and Kahneman's demonstrations of Availability and other cognitive biases to make inroads into our conventional wisdom. The discovery that political tribalism is the most insidious form of irrationality today is still fresh and mostly unknown. Indeed, sophisticated thinkers can be as infected by it as anyone else. With the accelerating pace of everything, perhaps the countermeasures will catch on sooner.

However long it takes, we must not let the existence of cognitive and emotional biases or the spasms of irrationality in the political arena discourage us from the Enlightenment ideal of relentlessly pursuing reason and truth. If we can identify ways in which humans are irrational, we must know what rationality is. Since there's nothing special about *us,* our fellows must have at least some capacity for rationality as well. And it's in the very nature of rationality that reasoners can always step back, consider their own shortcomings, and reason out ways to work around them.

CHAPTER 22

SCIENCE

I f we were called upon to name the proudest accomplishments of our species, whether in an intergalactic bragging competition or in testimony before the Almighty, what would we say?

We could crow about historic triumphs in human rights, such as the abolition of slavery and the defeat of fascism. But however inspiring these victories are, they consist in the removal of obstacles we set in our own path. It would be like listing in the achievements section of a résumé that you overcame a heroin addiction.[1]

We would certainly include the masterworks of art, music, and literature. Yet would the works of Aeschylus or El Greco or Billie Holiday be appreciated by sentient agents with brains and experiences unimaginably different from ours? Perhaps there are universals of beauty and meaning that transcend cultures and would resonate with any intelligence—I like to think there are—but it is devilishly difficult to know.

Yet there is one realm of accomplishment of which we can unabashedly boast before any tribunal of minds, and that is science. It's hard to imagine an intelligent agent that would be incurious about the world in which it exists, and in our species that curiosity has been exhilaratingly satisfied. We can explain much about the history of the universe, the forces that make it tick, the stuff we're made of, the origin of living things, and the machinery of life, including our mental life.

Though our ignorance is vast (and always will be), our knowledge is astonishing, and growing daily. The physicist Sean Carroll argues in *The Big Picture* that the laws of physics underlying everyday life (that is, excluding extreme values of energy and gravitation like black holes, dark matter, and the Big Bang) are *completely known*. It's hard to disagree that this is "one of the greatest triumphs of human intellectual history."[2] In the living world, more than a million and a half species have been scientifically described, and with a realistic surge of effort the remaining seven million could be named within this century.[3] Our understanding of the world, moreover, consists not in mere listings of particles and forces and species but in deep, elegant principles, such as that gravity is the curvature of space-time, and that life depends on a molecule that carries information, directs metabolism, and replicates itself.

Scientific discoveries continue to astound, to delight, to answer the formerly unanswerable. When Watson and Crick discovered the structure of DNA, they could not have dreamed of a day when the genome of a 38,000-year-old Neanderthal fossil would be sequenced and found to contain a gene connected to speech and language, or when an analysis of Oprah Winfrey's DNA would tell her she was descended from the Kpelle people of the Liberian rain forest.

Science is shedding new light on the human condition. The great thinkers of antiquity, the Age of Reason, and the Enlightenment were born too soon to enjoy ideas with deep implications for morality and meaning, including entropy, evolution, information, game theory, and artificial intelligence (though they often tinkered with precursors and approximations). The problems these thinkers introduced to us are today being enriched with these ideas, and are being probed with methods such as 3-D imaging of brain activity and the mining of big data to trace the propagation of ideas.

Science has also provided the world with images of sublime beauty: stroboscopically frozen motion, flamboyant fauna from tropical rain forests and deep-sea ocean vents, graceful spiral galaxies and diaphanous nebulae, fluorescing neural circuitry, and a luminous Planet Earth rising above the moon's horizon into the blackness of space. Like great works of art, these are not just pretty pictures but prods to contemplation, which deepen our understanding of what it means to be human and of our place in nature.

And science, of course, has granted us the gifts of life, health, wealth, knowledge, and freedom documented in the chapters on progress. To take just one example from chapter 6, scientific knowledge eradicated smallpox, a painful and disfiguring disease which killed 300 million people in the 20th century alone. In case anyone has skimmed over this feat of moral greatness, let me say it again: scientific knowledge eradicated smallpox, a painful and disfiguring disease which killed 300 million people in the 20th century alone.

These awe-inspiring achievements put the lie to any moaning that we live in an age of decline, disenchantment, meaninglessness, shallowness, or the absurd. Yet today the beauty and power of science are not just unappreciated but bitterly resented. The disdain for science may be found in surprising quarters: not just among religious fundamentalists and know-nothing politicians, but among many of our most adored intellectuals and in our most august institutions of higher learning.

~

The disrespect of science among American right-wing politicians has been documented by the journalist Chris Mooney in *The Republican War on Science* and has led even stalwarts (such as Bobby Jindal, the former governor of Louisiana) to disparage their own organization as "the party of stupid."[4] The reputation grew out of policies set in motion during George W. Bush's administration, including his encouragement of the teaching of creationism (in the guise of "intelligent design") and the shift from a longstanding practice of seeking advice from disinterested scientific panels to stacking the panels with congenial ideologues, many of whom promoted flaky ideas (such as that abortion causes breast cancer) while denying well-supported ones (such as that condoms prevent sexually transmitted diseases).[5] Republican politicians have engaged in spectacles of inanity, such as when Senator James Inhofe of Oklahoma, chair of the Environment and Public Works Committee, brought a snowball onto the Senate floor in 2015 to dispute the fact of global warming.

The previous chapter warned us that the stupidification of science in political discourse mostly surrounds hot buttons like abortion, evolution, and climate change. But the scorn for scientific consensus has widened into a broadband know-nothingness. Representative Lamar Smith of Texas, chair of the House Committee on Science, Space, and Technology, has harassed the National Science Foundation not just for its research on climate science (which he thinks is a left-wing conspiracy) but for the research in its peer-reviewed grants, which he pulls out of context to mock (for example, "How does the federal government justify spending over $220,000 to study animal photos in National Geographic?").[6] He has tried to undermine federal support of basic research by proposing legislation that would require the NSF to fund only studies that promote "the national interest" such as defense and the economy.[7] Science, of course, transcends national boundaries (as Chekhov noted, "There is no national science just as there is no national multiplication table"), and its ability to promote anyone's interests comes from its foundational understanding of reality.[8] The Global Positioning System, for example, uses the theory of relativity. Cancer therapies depend on the discovery of the double helix. Artificial intelligence adapts neural and semantic networks from the brain and cognitive sciences.

But chapter 21 prepared us for the fact that politicized repression of science comes from the left as well. It was the left that stoked panics about overpopulation, nuclear power, and genetically modified organisms. Research on intelligence, sexuality, violence, parenting, and prejudice have been distorted by tactics ranging from the choice of items in questionnaires to the intimidation of researchers who fail to ratify the politically correct orthodoxy.

~

My focus in the rest of this chapter is on a hostility to science that runs even deeper. Many intellectuals are enraged by the intrusion of science into the traditional territories of the humanities, such as politics, history, and the arts. Just as reviled is the application of scientific reasoning to the terrain formerly ruled by religion: many writers without a trace of a belief in God maintain that it is unseemly for science to weigh in on the biggest questions. In the major journals of opinion, scientific carpetbaggers are regularly accused of determinism, reductionism, essentialism, positivism, and, worst of all, a crime called scientism.

This resentment is bipartisan. The standard case for the prosecution by the left may be found in a 2011 review in *The Nation* by the historian Jackson Lears:

Positivism depends on the reductionist belief that the entire universe, including all human conduct, can be explained with reference to precisely measurable, deterministic physical processes. . . . Positivist assumptions provided the epistemological foundations for Social Darwinism and pop-evolutionary notions of progress, as well as for scientific racism and imperialism. These tendencies coalesced in eugenics, the doctrine that human well-being could be improved and eventually perfected through the selective breeding of the "fit" and the sterilization or elimination of the "unfit." Every schoolkid knows about what happened next: the catastrophic twentieth century. Two world wars, the systematic slaughter of innocents on an unprecedented scale, the proliferation of unimaginably destructive weapons, brushfire wars on the periphery of empire—all these events involved, in various degrees, the application of scientific research to advanced technology.[9]

The case from the right is captured in this 2007 speech from Leon Kass, Bush's bioethics advisor:

> Scientific ideas and discoveries about living nature and man, perfectly welcome and harmless in themselves, are being enlisted to do battle against our traditional religious and moral teachings, and even our self-understanding as creatures with freedom and dignity. A quasi-religious faith has sprung up among us—let me call it "soul-less scientism"—which believes that our new biology, eliminating all mystery, can give a complete account of human life, giving purely scientific explanations of human thought, love, creativity, moral judgment, and even why we believe in God. The threat to our humanity today comes not from the transmigration of souls in the next life, but from the denial of soul in this one. . . .
>
> Make no mistake. The stakes in this contest are high: at issue are the moral and spiritual health of our nation, the continued vitality of science, and our own self-understanding as human beings and as children of the West. . . . All friends of human freedom and dignity—including even the atheists among us—must understand that their own humanity is on the line.[10]

These are zealous prosecutors indeed. But as we shall see, their case is trumped up. Science cannot be blamed for genocide and war, and does not threaten the moral and spiritual health of our nation. On the contrary, science is indispensable in all areas of human concern, including politics, the arts, and the search for meaning, purpose, and morality.

～

The highbrow war on science is a flare-up of the controversy raised by C. P. Snow in 1959 when he deplored the disdain for science among British intellectuals in his lecture and book *The Two Cultures.* The term "cultures," in the anthropologists' sense, explains the puzzle of why science should draw flak not just from fossil-fuel-funded politicians but from some of the most erudite members of the clerisy.

During the 20th century, the landscape of human knowledge was carved into professionalized duchies, and the growth of science (particularly the sciences of human nature) is often seen as an encroachment on territories that had been staked and enclosed by the academic humanities. It's not that practitioners of the humanities themselves have this zero-sum mindset. Most artists show no signs of it; the novelists, painters, filmmakers, and musicians I know are intensely curious about the light that science might shed on their media, just as they are open to any source of inspiration. Nor is the anxiety expressed by the scholars who delve into historical epochs, genres of art, systems of ideas, and other subject matter in the humanities, since a true scholar is receptive to ideas regardless of their origin. The defensive pugnacity belongs to a *culture:* Snow's Second Culture of literary intellectuals, cultural critics, and erudite essayists.[11] The writer Damon Linker (citing the sociologist Daniel Bell) characterizes them as "specialists in generalizations, . . . pronouncing on the world from out of their individual experiences, habits of reading and capacity for judgment. Subjectivity in all of its quirks and eccentricities is the coin of the realm in the Republic of Letters."[12] This modus could not be more different from the way of science, and it's the Second Culture intellectuals who most fear "scientism," which they understand as the position that "science is all that matters" or that "scientists should be entrusted to solve all problems."

Snow, of course, never held the lunatic position that power should be transferred to the culture of scientists. On the contrary, he called for a *Third* Culture, which would combine ideas from science, culture, and history and apply them to enhancing human welfare across the globe.[13] The term was revived in 1991 by the author and literary agent John Brockman, and it is related to the biologist E. O. Wilson's concept of *consilience,* the unity of knowledge, which Wilson in turn attributed to (who else?) the thinkers of the Enlightenment.[14] The first step in understanding the promise of science in human affairs is to escape the bunker mentality of the Second Culture, captured, for example, in the tag line of a 2013 article by the literary lion Leon Wieseltier: "Now science wants to invade the liberal arts. Don't let it happen."[15]

An endorsement of scientific thinking must first of all be distinguished from any belief that members of the occupational guild called "science" are particularly wise or noble. The culture of science is based on the opposite belief. Its signature practices, including open debate, peer review, and double-blind methods, are designed to circumvent the sins to which scientists, being human, are vulnerable. As Richard Feynman put it, the first principle of science is "that you must not fool yourself—and you are the easiest person to fool."

For the same reason, a call for everyone to think more scientifically must not be confused with a call to hand decision-making over to scientists. Many scientists are naïfs when it comes to policy and law, and cook up nonstarters like world government, mandatory licensing of parents, and escaping a befouled Earth by colonizing other planets. It doesn't matter, because we're not talking about which priesthood should be granted power; we're talking about how collective decisions can be made more wisely.

A respect for scientific thinking is, adamantly, not the belief that all current scientific hypotheses are true. Most new ones are not. The life-blood of science is the cycle of conjecture and refutation: proposing a hypothesis and then seeing whether it survives attempts to falsify it. This point escapes many critics of science, who point to some discredited hypothesis as proof that science cannot be trusted, like a rabbi from my childhood who rebutted the theory of evolution as follows: "Scientists think the world is four billion years old. They used to think the world was eight billion years old. If they can be off by four billion years once, they can be off by four billion years again." The fallacy (putting aside the apocryphal history) is a failure to recognize that what science allows is an increasing confidence in a hypothesis as the evidence accumulates, not a claim to infallibility on the first try. Indeed, this kind of argument refutes itself, since the arguers must themselves appeal to the truth of current scientific claims to cast doubt on the earlier ones. The same is true of the common argument that the claims of science are untrust-worthy because the scientists of some earlier period were motivated by the prejudices and chauvinisms of the day. When they were, they were doing bad science, and it's only the better science of later periods that allows us, today, to identify their errors.

One attempt to build a wall around science and make science pay for it uses a different argument: that science deals only with facts about physical stuff, so scientists are committing a logical error when they say anything about values or society or culture. As Wieseltier puts it, "It is not for science to say whether science belongs in morality and politics and art. Those are philosophical matters, and science is not philosophy." But it is this argument that commits a logical error, by confusing propo-sitions with academic disciplines. It's certainly true that an empirical proposition is not the same as a logical one, and both must be distin-guished from normative or moral claims. But that does not mean that scientists are under a gag order forbidding them to discuss conceptual and moral issues, any more than philosophers must keep their mouths shut about the physical world.

Science is not a list of empirical facts. Scientists are immersed in the ethereal medium of *information*, including the truths of mathematics, the logic of their theories, and the values that guide their enterprise. Nor, for its part, has philosophy ever confined itself to a ghostly realm of pure ideas that float free of the physical universe. The Enlightenment philosophers in particular interwove their conceptual arguments with hypotheses about perception, cognition, emotion, and sociality. (Hume's analysis of the nature of causality, to take just one example, took off from his insights about the psychology of causality, and Kant was, among other things, a prescient cognitive psychologist.)[16] Today most philosophers (at least in the analytic or Anglo-American tradition) subscribe to *naturalism*, the position that "reality is exhausted by nature, containing nothing 'supernatural,' and that the scientific method should be used to investigate all areas of reality, including the 'human spirit.' "[17] Science, in the modern conception, is of a piece with philosophy and with reason itself.

What, then, distinguishes science from other exercises of reason? It certainly isn't "the scientific method," a term that is taught to schoolchildren but that never passes the lips of a scientist. Scientists use whichever methods help them understand the world: drudgelike tabulation of data, experimental derring-do, flights of theoretical fancy, elegant mathematical modeling, kludgy computer simulation, sweeping verbal narrative.[18] All the methods are pressed into the service of two ideals, and it is these ideals that advocates of science want to export to the rest of intellectual life.

The first is that the world is *intelligible*. The phenomena we experience may be explained by principles that are deeper than the phenomena themselves. That's why scientists laugh at the Theory of the Brontosaurus from the dinosaur expert on *Monty Python's Flying Circus*: "All brontosauruses are thin at one end, much much thicker in the middle, and then thin again at the far end"—the "theory" is just a description of how things are, not an explanation of *why* they are the way they are. The principles making up an explanation may in turn be explained by still deeper principles, and so on. (As David Deutsch put it, "We are always at the beginning of infinity.") In making sense of our world, there should be few occasions on which we are forced to concede, "It just is" or "It's magic" or "Because I said so." The commitment to intelligibility is not a matter of raw faith, but progressively validates itself as more of the world becomes explicable in scientific terms. The processes of life, for example, used to be attributed to a mysterious élan vital; now we know they are powered by chemical and physical reactions among complex molecules.

Demonizers of scientism often confuse intelligibility with a sin called reductionism, the analysis of a complex system into simpler elements, or, according to the accusation, *nothing but* simpler elements. In fact, to explain a complex happening in terms of deeper principles is not to discard its richness. Patterns emerge at one level of analysis that are not reducible to their components at a lower level. Though World War I consisted of matter in motion, no one would try to explain World War I in the language of physics, chemistry, and biology as opposed to the more perspicuous language of the perceptions and goals of leaders in 1914 Europe. At the same time, a curious person can legitimately ask *why* human minds are apt to have such perceptions and goals, including the tribalism, overconfidence, mutual fear, and culture of honor that fell into a deadly combination at that historical moment.

The second ideal is that we must allow the world to tell us whether our ideas about it are correct. The traditional causes of belief—faith, revelation, dogma, authority, charisma, conventional wisdom, hermeneutic parsing of texts, the glow of subjective certainty—are generators of error, and should be dismissed as sources of knowledge. Instead our beliefs about empirical propositions should be calibrated by their fit to the world. When scientists are pressed to explain how they do this, they usually reach for Karl Popper's model of conjecture and refutation, in which a scientific theory may be falsified by empirical tests but is never confirmed. In reality, science doesn't much look like skeet shooting, with a succession of hypotheses launched into the air like clay pigeons and shot to smithereens. It looks more like Bayesian reasoning (the logic used by the superforecasters we met in the preceding chapter). A theory is granted a prior degree of credence, based on its consistency with everything else we know. That level of credence is then incremented or decremented according to how likely an empirical observation would be if the theory is true, compared with how likely it would be if the theory is false.[19] Regardless of whether Popper or Bayes has the better account, a scientist's degree of belief in a theory depends on its consistency with empirical evidence. Any movement that calls itself "scientific" but fails to nurture opportunities for the testing of its own beliefs (most obviously when it murders or imprisons the people who disagree with it) is not a scientific movement.

~

Many people are willing to credit science with giving us handy drugs and gadgets and even with explaining how physical stuff works. But they draw the line at what truly matters to us as human beings: the deep questions about who we are, where we came from, and how we define the meaning and purpose of our lives. That is the traditional territory of religion, and its defenders tend to be the most excitable critics of scientism. They are apt to endorse the partition plan proposed by the pale-ontologist and science writer Stephen Jay Gould in his book *Rocks of Ages*, according to which the proper concerns of science and religion belong to "non-overlapping magisteria." Science gets the empirical universe; religion gets the questions of morality, meaning, and value.

But this entente unravels as soon as you begin to examine it. The moral worldview of any scientifically literate person—one who is not blinkered by fundamentalism—requires a clean break from religious conceptions of meaning and value.

To begin with, the findings of science imply that the belief systems of all the world's traditional religions and cultures—their theories of the genesis of the world, life, humans, and societies—are factually mistaken. We know, but our ancestors did not, that humans belong to a single species of African primate that developed agriculture, government, and writing late in its history. We know that our species is a tiny twig of a genealogical tree that embraces all living things and that emerged from prebiotic chemicals almost four billion years ago. We know that we live on a planet that revolves around one of a hundred billion stars in our galaxy, which is one of a hundred billion galaxies in a 13.8-billion-year-old universe, possibly one of a vast number of universes. We know that our intuitions about space, time, matter, and causation are incommensurable with the nature of reality on scales that are very large and very small. We know that the laws governing the physical world (including accidents, disease, and other misfortunes) have no goals that pertain to human well-being. There is no such thing as fate, providence, karma, spells, curses, augury, divine retribution, or answered prayers—though the discrepancy between the laws of probability and the workings of cognition may explain why people believe there are. And we know that we did not always know these things, that the beloved convictions of every time and culture may be decisively falsified, doubtless including many we hold today.

In other words, the worldview that guides the moral and spiritual values of a knowledgeable person today is the worldview given to us by science. Though the scientific facts do not by themselves dictate values, they certainly hem in the possibilities. By stripping ecclesiastical authority of its credibility on factual matters, they cast doubt on its claims to certitude in matters of morality. The scientific refutation of the theory of vengeful gods and occult forces undermines practices such as human sacrifice, witch hunts, faith healing, trial by ordeal, and the persecution of heretics. By exposing the absence of purpose in the laws governing the

universe, science forces us to take responsibility for the welfare of ourselves, our species, and our planet. For the same reason, it undercuts any
moral or political system based on mystical forces, quests, destinies, dialectics, struggles, or messianic ages. And in combination with a few
unexceptionable convictions—that all of us value our own welfare, and
that we are social beings who impinge on each other and can negotiate
codes of conduct—the scientific facts militate toward a defensible morality, namely principles that maximize the flourishing of humans and
other sentient beings. This humanism (chapter 23), which is inextricable
from a scientific understanding of the world, is becoming the de facto
morality of modern democracies, international organizations, and liberalizing religions, and its unfulfilled promises define the moral impera-
tives we face today.

~

Though science is increasingly and beneficially embedded in our material, moral, and intellectual lives, many of our cultural institutions cultivate a philistine indifference to science that shades into contempt.
Intellectual magazines that are ostensibly dedicated to ideas confine
themselves to politics and the arts, with scant attention to new ideas
emerging from science, with the exception of politicized issues like climate change (and regular attacks on scientism).[20] Still worse is the treatment of science in the liberal arts curricula of many universities. Students
can graduate with a trifling exposure to science, and what they do learn
is often designed to poison them against it.

The most commonly assigned book on science in modern universities
(aside from a popular biology textbook) is Thomas Kuhn's *The Structure
of Scientific Revolutions*.[21] That 1962 classic is commonly interpreted as
showing that science does not converge on the truth but merely busies
itself with solving puzzles before flipping to some new paradigm which
renders its previous theories obsolete, indeed, unintelligible.[22] Though
Kuhn himself later disavowed this nihilist interpretation, it has become
the conventional wisdom within the Second Culture. A critic from a major intellectual magazine once explained to me that the art world no longer considers whether works of art are "beautiful" for the same reason
that scientists no longer consider whether theories are "true." He seemed
genuinely surprised when I corrected him.

The historian of science David Wootton has remarked on the mores
of his own field: "In the years since Snow's lecture the two-cultures problem has deepened; history of science, far from serving as a bridge between the arts and sciences, nowadays offers the scientists a picture of

themselves that most of them cannot recognize."[23] That is because many historians of science consider it naïve to treat science as the pursuit of true explanations of the world. The result is like a report of a basketball game by a dance critic who is not allowed to say that the players are trying to throw the ball through the hoop. I once sat through a lecture on the semiotics of neuroimaging at which a historian of science deconstructed a series of dynamic 3-D multicolor images of the brain, volubly explaining how "that ostensibly neutral and naturalizing scientific gaze encourages particular kinds of selves who are then amenable to certain political agendas, shifting position from the neuro(psychological) object toward the external observatory position," and so on—any explanation but the bloody obvious one, namely that the images make it easier to see what's going on in the brain.[24] Many scholars in "science studies" devote their careers to recondite analyses of how the whole institution is just a pretext for oppression. An example is this scholarly contribution to the world's most pressing challenge:

Glaciers, Gender, and Science: A Feminist Glaciology Framework for Global Environmental Change Research

Glaciers are key icons of climate change and global environmental change. However, the relationships among gender, science, and glaciers—particularly related to epistemological questions about the production of glaciological knowledge—remain understudied. This paper thus proposes a feminist glaciology framework with four key components: (1) knowledge producers; (2) gendered science and knowledge; (3) systems of scientific domination; and (4) alternative representations of glaciers. Merging feminist postcolonial science studies and feminist political ecology, the feminist glaciology framework generates robust analysis of gender, power, and epistemologies in dynamic social-ecological systems, thereby leading to more just and equitable science and human-ice interactions.[25]

More insidious than the ferreting out of ever more cryptic forms of racism and sexism is a demonization campaign that impugns science (together with reason and other Enlightenment values) for crimes that are as old as civilization, including racism, slavery, conquest, and genocide. This was a major theme of the influential Critical Theory of the Frankfurt School, the quasi-Marxist movement originated by Theodor Adorno and Max Horkheimer, who proclaimed that "the fully enlightened earth radiates disaster triumphant."[26] It also figures in the works of postmodernist theorists such as Michel Foucault, who argued that the Holocaust was the inevitable culmination of a "bio-politics" that began with the Enlightenment, when science and rational governance exerted increasing power over people's lives.[27] In a similar vein, the sociologist Zygmunt Bauman blamed the Holocaust on the Enlightenment ideal to "remake the society, force it to conform to an overall, scientifically conceived plan."[28] In this twisted narrative, the Nazis themselves are let off the hook ("It's modernity's fault!"). So is the Nazis' rabidly counter-Enlightenment ideology, which despised the degenerate liberal bourgeois worship of reason and progress and embraced an organic, pagan vitality which drove the struggle between races. Though Critical Theory and postmodernism avoid "scientistic" methods such as quantification and systematic chronology, the facts suggest they have the history backwards. Genocide and autocracy were ubiquitous in premodern times, and they decreased, not increased, as science and liberal Enlightenment values became increasingly influential after World War II.[29]

To be sure, science has often been pressed into the support of deplorable political movements. It is essential, of course, to understand this history, and legitimate to pass judgment on scientists for their roles in it, just like any historical figures. Yet the qualities that we prize in humanities scholars—context, nuance, historical depth—often leave them when the opportunity arises to prosecute a campaign against their academic rivals. Science is commonly blamed for intellectual movements that had a pseudoscientific patina, though the historical roots of those movements ran deep and wide.

"Scientific racism," the theory that races fall into an evolutionary hierarchy of mental sophistication with Northern Europeans at the top, is a prime example. It was popular in the decades flanking the turn of the 20th century, apparently supported by craniometry and mental testing, before being discredited in the middle of the 20th century by better science and by the horrors of Nazism. Yet to pin ideological racism on science, in particular on the theory of evolution, is bad intellectual history. Racist beliefs have been omnipresent across history and regions of the world. Slavery has been practiced by every civilization, and was commonly rationalized by the belief that enslaved peoples were inherently suited to servitude, often by God's design.[30] Statements from ancient Greek and medieval Arab writers about the biological inferiority of Africans would curdle your blood, and Cicero's opinion of Britons was not much more charitable.[31]

More to the point, the intellectualized racism that infected the West in the 19th century was the brainchild not of science but of the humanities: history, philology, classics, and mythology. In 1853 Arthur de Gobineau, a fiction writer and amateur historian, published his cockamamie theory that a race of virile white men, the Aryans, spilled out of an ancient homeland and spread a heroic warrior civilization across Eurasia, diverging into the Persians, Hittites, Homeric Greeks, and Vedic Hindus, and later into the Vikings, Goths, and other Germanic tribes. (The speck of reality in this story is that these tribes spoke languages that fell into a single family, Indo-European.) Everything went downhill when the Aryans interbred with inferior conquered peoples, diluting their greatness and causing them to degenerate into the effete, decadent, soulless, bourgeois, commercial cultures that the Romantics were always whinging about. It was a small step to fuse this fairy tale with German Romantic nationalism and anti-Semitism: the Teutonic *Volk* were the heirs of the Aryans, the Jews a mongrel race of Asiatics. Gobineau's ideas were eaten up by Richard Wagner (whose operas were held to be re-creations of the original Aryan myths) and by Wagner's son-in-law Houston Stewart Chamberlain (a philosopher who wrote that Jews polluted Teutonic civilization with capitalism, liberal humanism, and sterile science). From them the ideas reached Hitler, who called Chamberlain his "spiritual father."[32]

Science played little role in this chain of influence. Pointedly, Gobineau, Chamberlain, and Hitler *rejected* Darwin's theory of evolution, particularly the idea that all humans had gradually evolved from apes, which was incompatible with their Romantic theory of race and with the older folk and religious notions from which it emerged. According to these widespread beliefs, races were separate species; they were fitted to civilizations with different levels of sophistication; and they would degenerate if they mixed. Darwin argued that humans are closely related members of a single species with a common ancestry, that all peoples have "savage" origins, that the mental capacities of all races are virtually the same, and that the races blend into one another with no harm from interbreeding.[33] The historian Robert Richards, who carefully traced Hitler's influences, ended a chapter entitled "Was Hitler a Darwinian?" (a common claim among creationists) with "The only reasonable answer to the question . . . is a very loud and unequivocal No!"[34]

Like "scientific racism," the movement called Social Darwinism is often tendentiously attributed to science. When the concept of evolution became famous in the late 19th and early 20th centuries, it turned into an inkblot test that a diverse assortment of political and intellectual movements saw as vindicating their agendas. Everyone wanted to believe that their vision of struggle, progress, and the good life was nature's way.[35] One of these movements was retroactively dubbed social Darwinism, though it was advocated not by Darwin but by Herbert Spencer, who laid it out in 1851, eight years before the publication of *The Origin of Species*. Spencer did not believe in random mutation and natural selection; he believed in a Lamarckian process in which the struggle for existence impelled organisms to strive toward feats of greater complexity and adaptation, which they passed on to later generations. Spencer thought that this progressive force was best left unimpeded, and so he argued against social welfare and government regulation that would only prolong the doomed lives of weaker individuals and groups. His political philosophy, an early form of libertarianism, was picked up by robber barons, advocates of laissez-faire economics, and opponents of social spending. Because those ideas had a right-wing flavor, left-wing writers misapplied the term *social Darwinism* to other ideas with a right-wing flavor, such as imperialism and eugenics, even though Spencer was dead-set against such government activism.[36] More recently the term has been used as a weapon against any application of evolution to the understanding of human beings.[37] So despite its etymology, the term has nothing to do with Darwin or evolutionary biology, and is now an almost meaningless term of abuse.

Eugenics is another movement that has been used as an ideological blunderbuss. Francis Galton, a Victorian polymath, first suggested that the genetic stock of humankind could be improved by offering incentives for talented people to marry each other and have more children (positive eugenics), though when the idea caught on it was extended to discouraging reproduction among the "unfit" (negative eugenics). Many countries forcibly sterilized delinquents, the mentally retarded, the mentally ill, and other people who fell into a wide net of ailments and stigmas. Nazi Germany modeled its forced sterilization laws after ones in Scandinavia and the United States, and its mass murder of Jews, Roma, and homosexuals is often considered a logical extension of negative eugenics. (In reality the Nazis invoked public health far more than genetics or evolution: Jews were likened to vermin, pathogens, tumors, gangrenous organs, and poisoned blood.)[38]

The eugenics movement was permanently discredited by its association with Nazism. But the term survived as a way to taint a number of scientific endeavors, such as applications of medical genetics that allow

parents to bear children without fatal degenerative diseases, and the entire field of behavioral genetics, which analyzes the genetic and environmental causes of individual differences.[39] And in defiance of the
historical record, eugenics is often portrayed as a movement of right-
wing scientists. In fact it was championed by progressives, liberals, and
socialists, including Theodore Roosevelt, H. G. Wells, Emma Goldman,
George Bernard Shaw, Harold Laski, John Maynard Keynes, Sidney and
Beatrice Webb, Woodrow Wilson, and Margaret Sanger.[40] Eugenics, after
all, valorized reform over the status quo, social responsibility over selfishness, and central planning over laissez-faire. The most decisive repu-
diation of eugenics invokes classical liberal and libertarian principles:
government is not an omnipotent ruler over human existence but an
institution with circumscribed powers, and perfecting the genetic
makeup of the species is not among them.

I've mentioned the limited role of science in these movements not to
absolve the scientists (many of whom were indeed active or complicit)
but because the movements deserve a deeper and more contextualized
understanding than their current role as anti-science propaganda. Misunderstandings of Darwin gave these movements a boost, but they
sprang from the religious, artistic, intellectual, and political beliefs of
their eras: Romanticism, cultural pessimism, progress as dialectical
struggle or mystical unfolding, and authoritarian high modernism. If we
think these ideas are not just unfashionable but mistaken, it is because
of the better historical and scientific understanding we enjoy today.

～

Recriminations over the nature of science are by no means a relic of the
"science wars" of the 1980s and 1990s, but continue to shape the role of
science in universities. When Harvard reformed its general education
requirement in 2006–7, the preliminary task force report introduced the
teaching of science without any mention of its place in human knowledge: "Science and technology directly affect our students in many
ways, both positive and negative: they have led to life-saving medicines,
the internet, more efficient energy storage, and digital entertainment;
they also have shepherded nuclear weapons, biological warfare agents,
electronic eavesdropping, and damage to the environment." Well, yes,
and I suppose one could say that architecture has produced both museums and gas chambers, that classical music both stimulates economic
activity and inspired the Nazis, and so on. But this strange equivocation
between the utilitarian and the nefarious was not applied to other disciplines, and the statement gave no indication that we might have good

reasons to prefer understanding and know-how to ignorance and super-stition.

At a recent conference, another colleague summed up what she thought was the mixed legacy of science: vaccines for smallpox on the one hand; the Tuskegee syphilis study on the other. In that affair, another bloody shirt in the standard narrative about the evils of science, public health researchers, beginning in 1932, tracked the progression of un-treated latent syphilis in a sample of impoverished African Americans for four decades. The study was patently unethical by today's standards, though it's often misreported to pile up the indictment. The researchers, many of them African American or advocates of African American health and well-being, did not *infect* the participants, as many people believe (a misconception that has led to the widespread conspiracy the-ory that AIDS was invented in US government labs to control the black population). And when the study began, it may even have been defensi-ble by the standards of the day: treatments for syphilis (mainly arsenic) were toxic and ineffective; when antibiotics became available later, their safety and efficacy in treating syphilis were unknown; and latent syph-ilis was known to often resolve itself without treatment.[41] But the point is that the entire equation is morally obtuse, showing the power of Sec-ond Culture talking points to scramble a sense of proportionality. My colleague's comparison assumed that the Tuskegee study was an un-avoidable part of scientific practice as opposed to a universally deplored breach, and it equated a one-time failure to prevent harm to a few dozen people with the prevention of hundreds of millions of deaths per century in perpetuity.

Does the demonization of science in the liberal arts programs of higher education matter? It does, for a number of reasons. Though many talented students hurtle along pre-med or engineering tracks from the day they set foot on campus, many others are unsure of what they want to do with their lives and take their cues from their professors and advi-sors. What happens to those who are taught that science is just another narrative like religion and myth, that it lurches from revolution to revo-lution without making progress, and that it is a rationalization of racism, sexism, and genocide? I've seen the answer: some of them figure, "If that's what science is, I might as well make money!" Four years later their brainpower is applied to thinking up algorithms that allow hedge funds to act on financial information a few milliseconds faster rather than to finding new treatments for Alzheimer's disease or technologies for car-bon capture and storage.

The stigmatization of science is also jeopardizing the progress of science itself. Today anyone who wants to do research on human beings, even an interview on political opinions or a questionnaire about irregular verbs, must prove to a committee that he or she is not Josef Mengele. Though research subjects obviously must be protected from exploitation and harm, the institutional review bureaucracy has swollen far beyond this mission. Critics have pointed out that it has become a menace to free speech, a weapon that fanatics can use to shut up people whose opinions they don't like, and a red-tape dispenser which bogs down research while failing to protect, and sometimes harming, patients and research subjects.[42] Jonathan Moss, a medical researcher who had developed a new class of drugs and was drafted into chairing the research review board at the University of Chicago, said in a convocation address, "I ask you to consider three medical miracles we take for granted: X-rays, cardiac catheterization, and general anesthesia. I contend all three would be stillborn if we tried to deliver them in 2005."[43] (The same observation has been made about insulin, burn treatments, and other lifesavers.) The social sciences face similar hurdles. Anyone who talks to a human being with the intent of gaining generalizable knowledge must obtain prior permission from these committees, almost certainly in violation of the First Amendment. Anthropologists are forbidden to speak with illiterate peasants who cannot sign a consent form, or interview would-be suicide bombers on the off chance that they might blurt out information that puts *them* in jeopardy.[44]

The hobbling of research is not just a symptom of bureaucratic mission creep. It is actually rationalized by many academics in a field called bioethics. These theoreticians think up reasons why informed and consenting adults should be forbidden to take part in treatments that help them and others while harming no one, using nebulous rubrics like "dignity," "sacredness," and "social justice." They try to sow panic about advances in biomedical research using far-fetched analogies with nuclear weapons and Nazi atrocities, science-fiction dystopias like *Brave New World* and *Gattaca*, and freak-show scenarios like armies of cloned Hitlers, people selling their eyeballs on eBay, or warehouses of zombies to supply people with spare organs. The moral philosopher Julian Savulescu has exposed the low standards of reasoning behind these arguments and has pointed out why "bioethical" obstructionism can be *un*ethical: "To delay by 1 year the development of a treatment that cures a lethal disease that kills 100,000 people per year is to be responsible for the deaths of those 100,000 people, even if you never see them."[45]

~

Ultimately the greatest payoff of instilling an appreciation of science is for *everyone* to think more scientifically. We saw in the preceding chapter that humans are vulnerable to cognitive biases and fallacies. Though scientific literacy itself is not a cure for fallacious reasoning when it comes to politicized identity badges, most issues don't start out that way, and everyone would be better off if they could think about them more scientifically. Movements that aim to spread scientific sophistication such as data journalism, Bayesian forecasting, evidence-based medicine and policy, real-time violence monitoring, and effective altruism have a vast potential to enhance human welfare. But an appreciation of their value has been slow to penetrate the culture.[46]

I asked my doctor whether the nutritional supplement he had recommended for my knee pain would really be effective. He replied, "Some of my patients say it works for them." A business-school colleague shared this assessment of the corporate world: "I have observed many smart people who have little idea of how to logically think through a problem, who infer causation from a correlation, and who use anecdotes as evidence far beyond the predictability warranted." Another colleague who quantifies war, peace, and human security describes the United Nations as an "evidence-free zone":

> The higher reaches of the UN are not unlike anti-science humanities programs. Most people at the top are lawyers and liberal arts graduates. The only parts of the Secretariat that have anything resembling a research culture have little prestige or influence. Few of the top officials in the UN understood qualifying statements as basic as "on average and other things being equal." So if we were talking about risk probabilities for conflict onsets you could be sure that Sir Archibald Prendergast III or some other luminary would offer a dismissive, "It's not like that in Burkina Faso, y'know."

Resisters of scientific thinking often object that some things just can't be quantified. Yet unless they are willing to speak only of issues that are black or white and to foreswear using the words *more, less, better,* and *worse* (and for that matter the suffix *–er*), they are making claims that are inherently quantitative. If they veto the possibility of putting numbers to them, they are saying, "Trust my intuition." But if there's one thing we know about cognition, it's that people (including experts) are arrogantly overconfident about their intuition. In 1954 Paul Meehl stunned his fellow psychologists by showing that simple actuarial formulas outperform expert judgment in predicting psychiatric classifications, suicide attempts, school and job performance, lies, crime, medical diagnoses, and pretty much any other outcome in which accuracy can be judged at all. Meehl's work inspired Tversky and Kahneman's discoveries on cognitive biases and Tetlock's forecasting tournaments, and his conclusion about the superiority of statistical to intuitive judgment is now recognized as one of the most robust findings in the history of psychology.[47]

Like all good things, data are not a panacea, a silver bullet, a magic bullet, or a one-size-fits-all solution. All the money in the world could not pay for randomized controlled trials to settle every question that occurs to us. Human beings will always be in the loop to decide which data to gather and how to analyze and interpret them. The first attempts to quantify a concept are always crude, and even the best ones allow probabilistic rather than perfect understanding. Nonetheless, quantitative social scientists have laid out criteria for evaluating and improving measurements, and the critical comparison is not whether a measure is perfect but whether it is better than the judgment of an expert, critic, interviewer, clinician, judge, or maven. That turns out to be a low bar.

Because the cultures of politics and journalism are largely innocent of the scientific mindset, questions with massive consequences for life and death are answered by methods that we know lead to error, such as anecdotes, headlines, rhetoric, and what engineers call HiPPO (highest-paid person's opinion). We have already seen some dangerous misconceptions that arise from this statistical obtuseness. People think that crime and war are spinning out of control, though homicides and battle deaths are going down, not up. They think that Islamist terrorism is a major risk to life and limb, whereas the danger is smaller than that from wasps and bees. They think that ISIS threatens the existence or survival of the United States, whereas terrorist movements rarely achieve any of their strategic aims.

The dataphobic mindset ("It's not like that in Burkina Faso") can lead to real tragedy. Many political commentators can recall a failure of peacekeeping forces (such as in Bosnia in 1995) and conclude that they are a waste of money and manpower. But when a peacekeeping force *is* successful, nothing photogenic happens, and it fails to make the news. In her book *Does Peacekeeping Work?* the political scientist Virginia Page Fortna addressed the question in her title with the methods of science rather than headlines, and, in defiance of Betteridge's Law, found that the answer is "a clear and resounding yes." Other studies have come to the same conclusion.[48] Knowing the results of these analyses could make the difference between an international organization helping to bring peace to a country and letting it fester in civil war.

Do multiethnic regions harbor "ancient hatreds" that can only be tamed by partitioning them into ethnic enclaves and cleansing the minorities from each one? Whenever ethnic neighbors go for each other's throats we read about it, but what about the neighborhoods that never make the news because they live in boring peace? What proportion of pairs of ethnic neighbors coexist without violence? The answer is, most of them: 95 percent of the neighbors in the former Soviet Union, 99 percent of those in Africa.[49]

Do campaigns of nonviolent resistance work? Many people believe that Gandhi and Martin Luther King just got lucky: their movements tugged at the heartstrings of enlightened democracies at opportune moments, but everywhere else, oppressed people need violence to get out from under a dictator's boot. The political scientists Erica Chenoweth and Maria Stephan assembled a dataset of political resistance movements across the world between 1900 and 2006 and discovered that *three-quarters* of the nonviolent resistance movements succeeded, compared with only a third of the violent ones.[50] Gandhi and King were right, but without data, you would never know it.

Though the urge to join a violent insurgent or terrorist group may owe more to male bonding than to just-war theory, most of the combatants probably believe that if they want to bring about a better world, they have no choice but to kill people. What would happen if everyone knew that violent strategies were not just immoral but ineffectual? It's not that I think we should airdrop crates of Chenoweth and Stephan's book into conflict zones. But leaders of radical groups are often highly educated (they distill their frenzy from academic scribblers of a few years back), and even the cannon fodder often attend some college and absorb the conventional wisdom about the need for revolutionary violence.[51] What would happen over the long run if a standard college curriculum devoted less attention to the writings of Karl Marx and Frantz Fanon and more to quantitative analyses of political violence?

∽

One of the greatest potential contributions of modern science may be a deeper integration with its academic partner, the humanities. By all accounts, the humanities are in trouble. University programs are downsizing; the next generation of scholars is un- or underemployed; morale is sinking; students are staying away in droves.[52]

No thinking person should be indifferent to our society's disinvestment in the humanities.[53] A society without historical scholarship is like a person without memory: deluded, confused, easily exploited. Philosophy grows out of the recognition that clarity and logic don't come easily to us and that we're better off when our thinking is refined and deepened. The arts are one of the things that make life worth living, enriching human experience with beauty and insight. Criticism is itself an art that multiplies the appreciation and enjoyment of great works. Knowledge in these domains is hard won, and needs constant enriching and updating as the times change.

Diagnoses of the malaise of the humanities rightly point to anti-intellectual trends in our culture and to the commercialization of universities. But an honest appraisal would have to acknowledge that some of the damage is self-inflicted. The humanities have yet to recover from the disaster of postmodernism, with its defiant obscurantism, self-refuting relativism, and suffocating political correctness. Many of its luminaries—Nietzsche, Heidegger, Foucault, Lacan, Derrida, the Critical Theorists—are morose cultural pessimists who declare that modernity is odious, all statements are paradoxical, works of art are tools of oppression, liberal democracy is the same as fascism, and Western civilization is circling the drain.[54]

With such a cheery view of the world, it's not surprising that the humanities often have trouble defining a progressive agenda for their own enterprise. Several university presidents and provosts have lamented to me that when a scientist comes into their office, it's to announce some exciting new research opportunity and demand the resources to pursue it. When a humanities scholar drops by, it's to plead for respect for the way things have always been done. Those ways do deserve respect, and there can be no replacement for the close reading, thick description, and deep immersion that erudite scholars can apply to individual works. But must these be the only paths to understanding?

A consilience with science offers the humanities many possibilities for new insight. Art, culture, and society are products of human brains. They originate in our faculties of perception, thought, and emotion, and they cumulate and spread through the epidemiological dynamics by which one person affects others. Shouldn't we be curious to understand these connections? Both sides would win. The humanities would enjoy more of the explanatory depth of the sciences, and a forward-looking agenda that could attract ambitious young talent (not to mention appealing to deans and donors). The sciences could challenge their theories with the natural experiments and ecologically valid phenomena that have been so richly characterized by humanities scholars.

In some fields, this consilience is a fait accompli. Archaeology has grown from a branch of art history to a high-tech science. The philosophy of mind shades into mathematical logic, computer science, cognitive science, and neuroscience. Linguistics combines philological scholarship on the history of words and grammatical constructions with laboratory studies of speech, mathematical models of grammar, and the computerized analysis of large corpora of writing and conversation.

Political theory, too, has a natural affinity with the sciences of mind. "What is government," asked James Madison, "but the greatest of all reflections on human nature?" Social, political, and cognitive scientists are reexamining the connections between politics and human nature, which were avidly debated in Madison's time but submerged during an interlude in which humans were treated as blank slates or rational actors. Humans, we now know, are moralistic actors: they are guided by intuitions about authority, tribe, and purity; are committed to sacred beliefs that express their identity; and are driven by conflicting inclinations toward revenge and reconciliation. We are starting to grasp why these impulses evolved, how they are implemented in the brain, how they differ among individuals, cultures, and subcultures, and which conditions turn them on and off.[55]

Comparable opportunities beckon in other areas of the humanities. The visual arts could avail themselves of the explosion of knowledge in vision science, including the perception of color, shape, texture, and lighting, and the evolutionary aesthetics of faces, landscapes, and geometric forms.[56] Music scholars have much to discuss with the scientists who study the perception of speech, the structure of language, and the brain's analysis of the auditory world.[57]

As for literary scholarship, where to begin?[58] John Dryden wrote that a work of fiction is "a just and lively image of human nature, representing its passions and humours, and the changes of fortune to which it is subject, for the delight and instruction of mankind." Cognitive psychology can shed light on how readers reconcile their own consciousness with those of the author and characters. Behavioral genetics can update folk theories of parental influence with discoveries about the effects of genes, peers, and chance, which have profound implications for the interpretation of biography and memoir—an endeavor that also has much to learn from the cognitive psychology of memory and the social psychology of self-presentation. Evolutionary psychologists can distinguish the obsessions that are universal from those that are exaggerated by a particular culture, and can lay out the inherent conflicts and confluences of interest within families, couples, friendships, and rivalries which are the drivers of plot. All these ideas can help add new depth to Dryden's observation about fiction and human nature.

Though many concerns in the humanities are best appreciated with traditional narrative criticism, some raise empirical questions that can be informed by data. The advent of data science applied to books, periodicals, correspondence, and musical scores has inaugurated an expansive new "digital humanities."[59] The possibilities for theory and discovery are limited only by the imagination, and include the origin and spread of ideas, networks of intellectual and artistic influence, the contours of historical memory, the waxing and waning of themes in literature, the universality or culture-specificity of archetypes and plots, and patterns of unofficial censorship and taboo.

The promise of a unification of knowledge can be fulfilled only if knowledge flows in all directions. Some of the scholars who have recoiled from scientists' forays into explaining art are correct that these explanations have been, by their standards, shallow and simplistic. All the more reason for them to reach out and combine their erudition about individual works and genres with scientific insight into human emotions and aesthetic responses. Better still, universities could train a new generation of scholars who are fluent in each of the two cultures.

Although humanities scholars themselves tend to be receptive to insights from science, many policemen of the Second Culture proclaim that they may not indulge such curiosity. In a dismissive review in the *New Yorker* of a book by the literary scholar Jonathan Gottschall on the evolution of the narrative instinct, Adam Gopnik writes, "The interesting questions about stories . . . are not about what makes a taste for them 'universal,' but what makes the good ones so different from the dull ones. . . . This is a case, as with women's fashion, where the subtle, 'surface' differences are actually the whole of the subject."[60] But in appreciating literature, must connoisseurship really be the *whole* of the subject? An inquisitive spirit might also be curious about the recurring ways in which minds separated by culture and era deal with the timeless conundrums of human existence.

Wieseltier, too, has issued crippling diktats on what scholarship in the humanities may not do, such as make progress. "The vexations of philosophy . . . are not retired," he declared; "errors [are] not corrected and discarded."[61] In fact, most moral philosophers today would say that the old arguments defending slavery as a natural institution are errors which have been corrected and discarded. Epistemologists might add that their field has progressed from the days when Descartes could argue that human perception is veridical because God would not deceive us. Wieseltier further stipulates that there is a "momentous distinction between the study of the natural world and the study of the human world," and any move to "transgress the borders" between these realms could only make the humanities the "handmaiden of the sciences," because "a scientific explanation will expose the underlying sameness" and "absorb all the realms into a single realm, into their realm." Where does this paranoia and territoriality lead? In a major essay in the *New York Times Book Review*, Wieseltier called for a worldview that is pre-Darwinian—"the irreducibility of the human difference to any aspect of our animality"—indeed, pre-Copernican—"the centrality of humankind to the universe."[62]

Let's hope that artists and scholars don't follow their self-appointed defenders over this cliff. Our quest to come to terms with the human predicament need not be frozen in the last century or the century before, let alone the Middle Ages. Surely our theories of politics, culture, and morality have much to learn from our best understanding of the universe and our makeup as a species.

In 1782 Thomas Paine extolled the cosmopolitan virtues of science:

> Science, the partisan of no country, but the beneficent patroness of all, has liberally opened a temple where all may meet. Her influence on the mind, like the sun on the chilled earth, has long been preparing it for higher cultivation and further improvement. The philosopher of one country sees not an enemy in the philosophy of another: he takes his seat in the temple of science, and asks not who sits beside him.[63]

What he wrote about the physical landscape applies as well to the landscape of knowledge. In this and other ways, the spirit of science is the spirit of the Enlightenment.

CHAPTER 23

HUMANISM

Science is not enough to bring about progress. "Everything that is not forbidden by laws of nature is achievable, given the right knowledge"—but that's the problem. "Everything" means *everything:* vaccines and bioweapons, video on demand and Big Brother on the telescreen. Something in addition to science ensured that vaccines were put to use in eradicating diseases while bioweapons were outlawed. That's why I preceded the epigraph from David Deutsch with the one from Spinoza: "Those who are governed by reason desire nothing for themselves which they do not also desire for the rest of humankind." Progress consists of deploying knowledge to allow all of humankind to flourish in the same way that each of us seeks to flourish.

The goal of maximizing human flourishing—life, health, happiness, freedom, knowledge, love, richness of experience—may be called humanism. (Despite the word's root, *humanism* doesn't exclude the flourishing of animals, but this book focuses on the welfare of humankind.) It is humanism that identifies *what* we should try to achieve with our knowledge. It provides the *ought* that supplements the *is.* It distinguishes true progress from mere mastery.

There is a growing *movement* called Humanism, which promotes a non-supernatural basis for meaning and ethics: good without God.[1] Its aims have been stated in a trio of manifestoes starting in 1933. The *Humanist Manifesto III*, from 2003, affirms:

> **Knowledge of the world is derived by observation, experimentation, and rational analysis.** Humanists find that science is the best method for determining this knowledge as well as for solving problems and developing beneficial technologies. We also recog-

nize the value of new departures in thought, the arts, and inner experience—each subject to analysis by critical intelligence.

Humans are an integral part of nature, the result of unguided evolutionary change. . . . We accept our life as all and enough, distinguishing things as they are from things as we might wish or imagine them to be. We welcome the challenges of the future, and are drawn to and undaunted by the yet to be known.

Ethical values are derived from human need and interest as tested by experience. Humanists ground values in human welfare shaped by human circumstances, interests, and concerns and extended to the global ecosystem and beyond. . . .

Life's fulfillment emerges from individual participation in the service of humane ideals. We . . . animate our lives with a deep sense of purpose, finding wonder and awe in the joys and beauties of human existence, its challenges and tragedies, and even in the inevitability and finality of death. . . .

Humans are social by nature and find meaning in relationships. Humanists . . . strive toward a world of mutual care and concern, free of cruelty and its consequences, where differences are resolved cooperatively without resorting to violence. . . .

Working to benefit society maximizes individual happiness. Progressive cultures have worked to free humanity from the brutalities of mere survival and to reduce suffering, improve society, and develop global community. . . .[2]

The members of Humanist associations would be the first to insist that the ideals of humanism belong to no sect. Like Molière's bourgeois gentleman who was delighted to learn he had been speaking prose all his life, many people are humanists without realizing it.[3] Strands of humanism may be found in belief systems that go back to the Axial Age. They came to the fore during the Age of Reason and the Enlightenment, leading to the English, French, and American statements of rights, and got a second wind after World War II, inspiring the United Nations, the Universal Declaration of Human Rights, and other institutions of global cooperation.[4] Though humanism does not invoke gods, spirits, or souls to ground meaning and morality, it is by no means incompatible with religious institutions. Some Eastern religions, including Confucianism and varieties of Buddhism, always grounded their ethics in human welfare rather than divine dictates. Many Jewish and Christian denominations have become humanistic, soft-pedaling their legacy of supernatural beliefs and ecclesiastical authority in favor of reason and universal human flourishing. Examples include the Quakers, Unitarians, liberal Episcopalians, Nordic Lutherans, and Reform, Reconstructionist, and Humanistic branches of Judaism.

Humanism may seem bland and unexceptionable—who could be against human flourishing? But in fact it is a distinctive moral commitment, one that does not come naturally to the human mind. As we shall see, it is vehemently opposed not just by many religious and political factions but, amazingly, by eminent artists, academics, and intellectuals. If humanism, like the other Enlightenment ideals, is to retain its hold on people's minds, it must be explained and defended in the language and ideas of the current era.

~

Spinoza's dictum is one of a family of principles that have sought a secular foundation for morality in *impartiality*—in the realization that there's nothing magic about the pronouns *I* and *me* that could justify privileging my interests over yours or anyone else's.[5] If I object to being raped, maimed, starved, or killed, I can't very well rape, maim, starve, or kill you. Impartiality underlies many attempts to construct morality on rational grounds: Spinoza's viewpoint of eternity, Hobbes's social contract, Kant's categorical imperative, Rawls's veil of ignorance, Nagel's view from nowhere, Locke and Jefferson's self-evident truth that all people are created equal, and of course the Golden Rule and its precious-metallic variants, rediscovered in hundreds of moral traditions.[6] (The Silver Rule is "Don't do to others what you don't want done to yourself"; the Platinum Rule, "Do to others what they would have you do to them." They are designed to anticipate masochists, suicide bombers, differences in taste, and other sticking points for the Golden Rule.)

To be sure, the argument from impartiality is incomplete. If there were a callous, egoistic, megalomaniacal sociopath who could exploit everyone else with impunity, no argument could convince him he had committed a logical fallacy. Also, arguments from impartiality have little content. Aside from a generic advisory to respect people's wishes, the arguments say little about what those wishes are: the wants, needs, and experiences that define human flourishing. These are the desiderata that should not just be impartially allowed but actively sought and expanded for as many people as possible. Recall that Martha Nussbaum filled this gap by laying out a list of "fundamental capabilities" that people have the right to exercise, such as longevity, health, safety, literacy, knowledge, free expression, play, nature, and emotional and social attachments. But this is just a list, and it leaves the list-maker open to the objection that she is just enumerating her favorite things. Can we put humanistic morality on a deeper foundation—one that would rule out rational sociopaths and justify the human needs we are obligated to respect? I think we can.

According to the Declaration of Independence, the rights to life, liberty, and the pursuit of happiness are "self-evident." That's a bit unsatisfying, because what's "self-evident" isn't always self-evident. But it captures a key intuition. There would indeed be something perverse about having to justify life itself in the course of examining the foundations of morality, as if it were an open question whether one gets to finish the sentence or be shot. The very act of examining anything presupposes that one is around to do the examining. If Nagel's transcendental argument about the non-negotiability of reason has merit—that the act of considering the validity of reason presupposes the validity of reason— then surely it presupposes the existence of reasoners.

This opens the door to deepening our humanistic justification of morality with two key ideas from science, entropy and evolution. Traditional analyses of the social contract imagined a colloquy among disembodied souls. Let's enrich this idealization with the minimal premise that the reasoners exist in the physical universe. Much follows.

These incarnate beings must have defied the staggering odds against matter arranging itself into a thinking organ by being products of natural selection, the only physical process capable of producing complex adaptive design.[7] And they must have defied the ravages of entropy long enough to be able to show up for the discussion and persist through it. That means they have taken in energy from the environment, stayed within a narrow envelope of conditions consistent with their physical integrity, and fended off assaults from living and nonliving dangers. As products of natural and sexual selection they must be the scions of a deeply rooted tree of replicators, each of whom won a mate and bore viable offspring. Since intelligence is not a wonder algorithm but is fed by knowledge, they must be driven to sop up information about the world and to be attentive to its nonrandom patterning. And if they are exchanging ideas with other rational entities, they must be on speaking terms: they must be social beings who risk time and safety in interacting with one another.[8]

The physical requirements that allow rational agents to exist in the material world are not abstract design specifications; they are implemented in the brain as wants, needs, emotions, pains, and pleasures. On average, and in the kind of environment in which our species was shaped, pleasurable experiences allowed our ancestors to survive and have viable children, and painful ones led to a dead end. That means that food, comfort, curiosity, beauty, stimulation, love, sex, and camaraderie are not shallow indulgences or hedonistic distractions. They are links in the causal chain that allowed minds to come into being. Unlike ascetic and puritanical regimes, humanistic ethics does not second-guess the intrinsic worth of people seeking comfort, pleasure, and fulfillment—if people didn't seek them, there would be no people. At the same time, evolution guarantees that these desires will work at cross-purposes with each other and with those of other people.[9] Much of what we call wisdom consists in balancing the conflicting desires within ourselves, and much of what we call morality and politics consists in balancing the conflicting desires among people.

As I mentioned in chapter 2 (following an observation by John Tooby), the Law of Entropy sentences us to another permanent threat. Many things must all go right for a body (and thus a mind) to function, but it takes just one thing going wrong for it to shut down permanently—a leak of blood, a constriction of air, a disabling of its microscopic clockwork. An act of aggression by one agent can end the existence of another. We are all catastrophically vulnerable to violence—but at the same time we can enjoy a fantastic benefit if we agree to refrain from violence. The Pacifist's Dilemma—how social agents can forgo the temptation to exploit each other in exchange for the security of not being exploited—hangs over humanity like the Sword of Damocles, making peace and security a permanent quest for humanistic ethics.[10] The historical decline of violence shows that it is a solvable problem.

The vulnerability of any embodied agent to violence explains why the callous, egoistic, megalomaniacal sociopath cannot remain disengaged from the arena of moral discourse (and its demand for impartiality and nonviolence) forever. If he refuses to play the game of morality, then in the eyes of everyone else he has become a mindless menace, like a germ, a wildfire, or a rampaging wolverine—something to be neutralized by brute force, no questions asked. (As Hobbes put it, "No covenants with beasts.") Now, as long as he thinks he is eternally invulnerable, he might take that chance, but the Law of Entropy rules that out. He may tyrannize everyone for a while, but eventually the massed strength of his targets could prevail. The impossibility of eternal invulnerability creates an incentive even for callous sociopaths to re-enter the roundtable of morality. As the psychologist Peter DeScioli points out, when you face an adversary alone, your best weapon may be an ax, but when you face an adversary in front of a throng of bystanders, your best weapon may be an argument.[11] And he who engages in argument may be defeated by a better one. Ultimately the moral universe includes everyone who can think.

Evolution helps explain another foundation of secular morality: our capacity for sympathy (or, as the Enlightenment writers variously referred to it, benevolence, pity, imagination, or commiseration). Even if a rational agent deduces that it's in everyone's long-term interests to be moral, it's hard to imagine him sticking his neck out to make a sacrifice for another's benefit unless something gives him a nudge. The nudge needn't come from an angel on one shoulder; evolutionary psychology explains how it comes from the emotions that make us social animals.[12] Sympathy among kin emerges from the overlap in genetic makeup that interconnects us in the great web of life. Sympathy among everyone else emerges from the impartiality of nature: each of us may find ourselves in straits where a small mercy from another grants a big boost in our own welfare, so we're better off if we bestow good turns on one another (with no one taking but never giving) than if it's every person for himself or herself. Evolution thus selects for the moral sentiments: sympathy, trust, gratitude, guilt, shame, forgiveness, and righteous anger. With sympathy installed in our psychological makeup, it can be expanded by reason and experience to encompass all sentient beings.[13]

∿

A different philosophical objection to humanism is that it's "just utilitarianism"—that a morality based on maximizing human flourishing is the same as a morality that seeks the greatest happiness for the greatest number.[14] (Philosophers often refer to happiness as "utility.") Anyone who has taken Introduction to Moral Philosophy can rattle off the problems.[15] Should we indulge a Utility Monster who gets more pleasure out of eating people than his victims get out of living? Should we euthanize a few draftees and harvest their organs to save the lives of many more? If townspeople enraged by an unsolved murder threaten a deadly riot, should the sheriff assuage them by framing the town drunk and stringing him up? If a drug could put us into a permanent slumber with sweet dreams, should we take it? Should we set up a chain of warehouses that inexpensively support billions of happy rabbits? These thought experiments make the case for a *deontological* ethics, composed of rights, duties, and principles that deem certain acts moral or immoral by their very nature. In some versions of deontological morality, the principles come from God.

Humanism indeed has a utilitarian flavor, or at least a consequentialist one, in which acts and policies are morally evaluated by their consequences. The consequences needn't be restricted to happiness in the narrow sense of having a smile on one's face, but can embrace a broader sense of flourishing, which includes childrearing, self-expression, education, rich experience, and the creation of works of lasting value (chapter 18). The consequentialist flavor of humanism is actually a point in its favor, for several reasons.

First, any Moral Philosophy student who stayed awake through week 2 of the syllabus can also rattle off the problems with deontological ethics. If lying is intrinsically wrong, must we answer truthfully when the Gestapo demand to know the whereabouts of Anne Frank? Is masturbation immoral (as the prototypical deontologist, Kant, argued), because one is using oneself as a means to satisfy an animal impulse, and people must always be treated as ends, never as means? If a terrorist has hidden a ticking nuclear bomb that would annihilate millions, is it immoral to waterboard him into revealing its location? And given the absence of a thundering voice from the heavens, who gets to pull principles out of the air and pronounce that certain acts are inherently immoral even if they hurt no one? At various times moralists have used deontological thinking to insist that vaccination, anesthesia, blood transfusions, life insurance, interracial marriage, and homosexuality were wrong by their very nature.

Many moral philosophers believe that the dichotomy from the Intro course is drawn too sharply.[16] Deontological principles are often a good way to bring the greatest happiness to the greatest number. Since no mortal can calculate every consequence of his actions into the indefinite future, and since people can always spin-doctor their selfish acts as benefiting others, one of the best ways to promote overall happiness is to draw bright lines that no one may cross. We don't let governments deceive or murder their citizens, because real politicians, unlike the infallible and benevolent demigods in the thought experiments, could wield that power capriciously or tyrannically. That is one of many reasons why a government that could frame innocent people for capital crimes or euthanize them for their organs would *not* produce the greatest happiness for the greatest number. Or take the principle of equal treatment. Are laws that discriminate against women and minorities unfair by their very nature, or are they deplorable because the victims of discrimination suffer harm? We may not have to answer the question. Conversely, any deontological principle whose consequences *are* harmful, such as the Sanctity of Life-Sustaining Blood (which rules out transfusions), can be tossed out the window. Human rights promote human flourishing. That's why, in practice, humanism and human rights go hand in hand.

The other reason that humanism needn't be embarrassed by its overlap with utilitarianism is that this approach to ethics has an impressive track record of improving human welfare. The classical utilitarians— Cesare Beccaria, Jeremy Bentham, and John Stuart Mill—laid out arguments against slavery, sadistic punishment, cruelty to animals, the criminalization of homosexuality, and the subordination of women which carried the day.[17] Even abstract rights like freedom of speech and religion were largely defended in terms of benefits and harms, as when Thomas Jefferson wrote, "The legitimate powers of government extend to such acts only as are injurious to others. But it does me no injury for my neighbour to say there are twenty gods, or no god. It neither picks my pocket nor breaks my leg."[18] Universal education, workers' rights, and environmental protection also were advanced on utilitarian grounds. And, at least so far, Utility Monsters and rabbit gratification factories have not turned out to be a problem.

There is a good reason why utilitarian arguments have so often succeeded: everyone can appreciate them. Principles like "No harm, no foul," "If no one is hurt it can't be wrong," "What consenting adults do in private is no one else's concern," and "If I should take a notion / To jump into the ocean / Ain't nobody's business if I do" may not be profound or exceptionless, but once they are stated, people can readily understand them, and anyone who wants to oppose them has a heavy burden of proof. It's not that utilitarianism is intuitive. Classical liberalism came late in human history, and traditional cultures believe that what consenting adults do in private is very much their concern.[19] The philosopher and cognitive neuroscientist Joshua Greene has argued that many deontological convictions are rooted in primitive intuitions of tribalism, purity, revulsion, and social norms, whereas utilitarian conclusions emerge from rational cogitation.[20] (He has even shown that the two kinds of moral thinking engage emotional and rational systems of the brain, respectively.) Greene also argues that when people from diverse cultural backgrounds have to agree upon a moral code, they tend to go utilitarian. That explains why certain reform movements, such as legal equality for women and gay marriage, overturned centuries of precedent astonishingly quickly (chapter 15): with nothing but custom and intuition behind it, the status quo crumbled in the face of utilitarian arguments.

Even when humanistic movements fortify their goals with the language of rights, the philosophical system justifying those rights must be "thin."[21] A viable moral philosophy for a cosmopolitan world cannot be constructed from layers of intricate argumentation or rest on deep metaphysical or religious convictions. It must draw on simple, transparent principles that everyone can understand and agree upon. The ideal of human flourishing—that it's good for people to lead long, healthy, happy, rich, and stimulating lives—is just such a principle, since it is based on nothing more (and nothing less) than our common humanity.

History confirms that when diverse cultures have to find common ground, they converge toward humanism. The separation of church and state in the American Constitution arose not just from the philosophy of the Enlightenment but from practical necessity. The economist Samuel Hammond has noted that eight of the thirteen British colonies had official churches, which intruded into the public sphere by paying ministers' salaries, enforcing strict religious observance, and persecuting members of other denominations. The only way to unite the colonies under a single constitution was to guarantee religious expression and practice as a natural right.[22]

A century and a half later, a community of nations still smoldering from a world war had to lay down a set of principles to unite them in cooperation. It's unlikely that they would have agreed upon "We accept Jesus Christ as our savior" or "America is a shining city upon a hill." In 1947 the United Nations Educational, Scientific, and Cultural Organization (UNESCO) asked several dozen of the world's intellectuals (including Jacques Maritain, Mohandas Gandhi, Aldous Huxley, Harold Laski, Quincy Wright, and Pierre Teilhard de Chardin, together with eminent Confucian and Muslim scholars) which rights should be included in the UN's universal declaration. The lists were surprisingly similar. In his introduction to their deliverable, Maritain recounted:

At one of the meetings of a Unesco National Commission where Human Rights were being discussed, someone expressed astonishment that certain champions of violently opposed ideologies had agreed on a list of those rights. "Yes," they said, "we agree about the rights *but on condition that no one asks us why.*"[23]

The Universal Declaration of Human Rights, a humanist manifesto
with thirty articles, was drafted in less than two years, thanks to the
determination of Eleanor Roosevelt, chair of the drafting committee, to
avoid getting mired in ideology and move the project along.²⁴ (When
John Humphrey, author of the first draft, was asked on what principles
the Declaration was based, he tactfully replied, "No philosophy whatso-
ever.")²⁵ In December 1948 it was passed without opposition by the UN
General Assembly. Contrary to accusations that human rights are a parochial Western creed, the Declaration was supported by India, China,
Thailand, Burma, Ethiopia, and seven Muslim countries, while Roosevelt
had to twist the arms of American and British officials to get them behind it: the United States was worried about its Negroes, the United
Kingdom about its colonies. The Soviet bloc, Saudi Arabia, and South
Africa abstained.²⁶

The Declaration has been translated into five hundred languages, and
has influenced most of the national constitutions that were drafted in the
following decades, together with many international laws, treaties, and
organizations. At seventy years old, it has aged well.

~

Though humanism is the moral code that people will converge upon
when they are rational, culturally diverse, and need to get along, it is by
no means a vapid or saccharine lowest common denominator. The idea
that morality consists in the maximization of human flourishing clashes
with two perennially seductive alternatives. The first is theistic morality:
the idea that morality consists in obeying the dictates of a deity, which
are enforced by supernatural reward and punishment in this world or in
an afterlife. The second is romantic heroism: the idea that morality consists in the purity, authenticity, and greatness of an individual or a nation. Though romantic heroism was first articulated in the 19th century,
it may be found in a family of newly influential movements, including
authoritarian populism, neo-fascism, neo-reaction, and the alt-right.

Many intellectuals who don't sign on to these alternatives to humanism nonetheless believe they capture a vital truth about our psychology:
that people have a *need* for theistic, spiritual, heroic, or tribal beliefs. Humanism may not be wrong, they say, but it goes against human nature.
No society based on humanistic principles can long endure, let alone a
global order based on them.

It's a short step from the psychological claim to a historical one: that
the inevitable collapse has begun, and we are watching the liberal, cosmopolitan, Enlightenment, humanistic worldview unravel before our

eyes. "Liberalism Is Dead," announced the *New York Times* columnist Roger Cohen in 2016. "The liberal democratic experiment—with its Enlightenment-derived belief in the capacity of individuals possessed of certain inalienable rights to shape their destinies in liberty through the exercise of their will—is but a brief interlude."[27] In "The Enlightenment Had a Good Run," the *Boston Globe* editorialist Stephen Kinzer agreed:

> The cosmopolitanism that is central to Enlightenment ideals has produced results that disturb people in many societies. This leads them back toward the ruling system that primates instinctively prefer: A strong chief protects the tribe, and in return tribe members do the chief's bidding. . . . Reason offers little basis for morality, rejects spiritual power, and negates the importance of emotion, art and creativity. When reason is cold and inhumane, it can cut people off from deeply imbedded structures that give meaning to life.[28]

Other pundits have added that it's no wonder so many young people are drawn to ISIS: they are turning away from an "arid secularism," and seek "radical and religious correctives to a flattened view of human life."[29]

So should I have called this book *Enlightenment While It Lasts*? Don't be silly! In part II, I documented the reality of progress; in this part, I have focused on the ideas that drive it and why I expect them to endure. Having rebutted the cases against reason and science in the preceding two chapters, I'll now take on the case against humanism. I'll examine these arguments not just to show that the moral, psychological, and historical arguments against humanism are wrong. The best way to understand an idea is to see what it is *not*, so putting the alternatives to humanism under the microscope can remind us what is at stake in advancing the ideals of the Enlightenment. First we'll look at the religious case against humanism, then at the romantic-heroic-tribal-authoritarian complex.

∼

Can we really have good without God? Has the godless universe advanced by humanistic scientists been undermined by the findings of science itself? And is there an innate adaptation to the divine presence— a God gene in our DNA, a God module in the brain—which ensures that theistic religion will always push back against secular humanism?

Let's start with theistic morality. It's true that many religious codes enjoin people from murdering, assaulting, robbing, or betraying one another. But of course so do codes of secular morality, and for an obvious reason: these are rules that all rational, self-interested, and gregarious agents would want their compatriots to agree upon. Not surprisingly, they are codified in the laws of every state, and indeed seem to be present in every human society.[30]

What does an appeal to a supernatural lawgiver add to a humanistic commitment to make people better off? The most obvious add-on is supernatural enforcement: the belief that if one commits a sin, one will be smitten by God, damned to hell, or have one's name omitted from the Book of Life. It's a tempting add-on because secular law enforcement cannot possibly detect and punish every infraction, and everyone has a motive to convince everyone else that they cannot get away with murder.[31] As with Santa Claus, he sees you when you're sleeping, he knows when you're awake, he knows if you've been bad or good, so be good for goodness' sake.

But theistic morality has two fatal flaws. The first is that there is no good reason to believe that God exists. In a nonfiction appendix to her novel *Thirty-Six Arguments for the Existence of God: A Work of Fiction*, Rebecca Newberger Goldstein (drawing in part on Plato, Spinoza, Hume, Kant, and Russell) lays out refutations of every one of these arguments.[32] The most common among them—faith, revelation, scripture, authority, tradition, and subjective appeal—are not arguments at all. It's not just that reason says they cannot be trusted. It's also that different religions, drawing on these sources, decree mutually incompatible beliefs about how many gods there are, which miracles they have wrought, and what they demand of their devotees. Historical scholarship has amply demonstrated that holy scriptures are all-too-human products of their historical eras, including internal contradictions, factual errors, plagiarism from neighboring civilizations, and scientific absurdities (such as God creating the sun three days after he distinguished day from night). The recondite arguments from sophisticated theologians are no sounder. The Cosmological and Ontological arguments for the existence of God are logically invalid, the Argument from Design was refuted by Darwin, and the others are either patently false (such as the theory that humans are endowed with an innate faculty for sensing the truth about God) or blatant escape hatches (such as the suggestion that the Resurrection was too cosmically important for God to have allowed it to be empirically verified).

Some writers insist that science has no place in this conversation. They seek to impose a condition of "methodological naturalism" on science which renders it incapable, even in principle, of evaluating the claims of religion. That would carve out a safe space in which believers can protect their beliefs while still being sympathetic to science. But as we saw in the preceding chapter, science is not a game with an arbitrary rulebook; it's the application of reason to explaining the universe and to ascertaining whether its explanations are true. In *Faith Versus Fact*, the biologist Jerry Coyne argues that the existence of the God of scripture is a perfectly testable scientific hypothesis.[33] The Bible's historical accounts could have been corroborated by archaeology, genetics, and philology. It could have contained uncannily prescient scientific truths such as "Thou shalt not travel faster than light" or "Two strands entwined is the secret of life." A bright light might appear in the heavens one day and a man clad in a white robe and sandals, supported by winged angels, could descend from the sky, give sight to the blind, and resurrect the dead. We might discover that intercessory prayer can restore eyesight or regrow amputated limbs, or that anyone who speaks the Prophet Mohammed's name in vain is immediately struck down while those who pray to Allah five times a day are free from disease and misfortune. More generally, the data might show that good things happen to good people and bad things happen to bad people: that the mothers who die in childbirth, the children who waste away from cancer, and the millions of victims of earthquakes, tsunamis, and holocausts had it coming.

Other components of theistic morality, such as the existence of an immaterial soul and a realm of reality beyond matter and energy, are just as testable. We might discover a severed head that can speak. A seer could predict the exact day of natural disasters and terrorist attacks. Aunt Hilda could beam a message from the Great Beyond telling us under which floorboard she hid her jewelry. Memoirs from oxygen-starved patients who experienced their souls leaving their bodies could contain verifiable details unavailable to their sense organs. The fact that these reports have all been exposed as tall tales, false memories, overinterpreted coincidences, and cheap carny tricks undermines the hypothesis that there are immaterial souls which could be subject to divine justice.[34] There are, of course, deistic philosophies in which God created the universe and then stepped back to watch what happened, or in which "God" is merely a synonym for the laws of physics and mathematics. But these impotent Gods are in no position to underwrite morality.

~

Many theistic beliefs originated as hypotheses to explain natural phenomena such as the weather, disease, and the origin of species. As these hypotheses have been superseded by scientific ones, the scope of theism has steadily shrunk. But since our scientific understanding is never complete, the pseudo-argument known as the God of the Gaps is always available as a last resort. Today the more sophisticated theists have tried to place God into two of these gaps: the fundamental physical constants and the hard problem of consciousness. Any humanist who insists that we cannot invoke God to justify morality can expect to be confronted with these gaps, so let me say a few words about each. As we will see, they are likely to go the way of Zeus hurling thunderbolts as an explanation for electrical storms.

Our universe can be specified by a few numbers, including the strengths of the forces of nature (gravity, electromagnetism, and the nuclear forces), the number of macroscopic dimensions of space-time (four), and the density of dark energy (the source of the acceleration of the expansion of the universe). In *Just Six Numbers*, Martin Rees enumerates them on one hand and a finger; the exact tally depends on which version of physical theory one invokes and on whether one counts the constants themselves or ratios between them. If any of these constants were off by a minuscule iota, then matter would fly apart or collapse upon itself, and stars, galaxies, and planets, to say nothing of terrestrial life and *Homo sapiens*, could never have formed. The best-established theories of physics today don't explain why these constants should be so meticulously tuned to values that allowed us to come into being (particularly the density of dark energy), and so, the theistic argument goes, there must have been a fine-tuner, namely God. It is the old Argument from Design applied to the entire cosmos rather than to living things.

An immediate objection is the equally old problem of theodicy. If God, in his infinite power and knowledge, fine-tuned the universe to bring us into being, why did he design an Earth on which geological and meteorological catastrophes devastate regions inhabited by innocent people? What is the divine purpose of the supervolcanoes that have ravaged our species in the past and may extinguish it in the future, or the evolution of the Sun into a red giant that will do so with certainty?

But theodical speculation is beside the point. Physicists have not been left dumbstruck by the apparent fine-tuning of the fundamental constants, but are actively pursuing several explanations. One is captured in the title of the physicist Victor Stenger's book *The Fallacy of Fine-Tuning*.[35] Many physicists believe that it's premature to conclude that the values of the fundamental constants are either arbitrary or the only ones consistent with life. A deeper understanding of physics (particularly the long-sought unification of relativity and quantum theory) may show that some of the values must be exactly what they are. Others, we might learn, could take on other values—more important, *combinations* of values—that are compatible with a stable, matter-filled universe, albeit not the one we know and love. Progress in physics may reveal that the constants are not so finely tuned, and a life-supporting universe not so improbable, after all.

The other explanation is that our universe is just one region in a vast, possibly infinite landscape of universes—a multiverse—each with different values of the fundamental constants.[36] We find ourselves in a universe compatible with life not because it was tuned to allow us to exist but because the very fact that we exist implies that it is *that* kind of universe, and not one of the vastly more numerous inhospitable ones, that we find ourselves in. Fine-tuning is a fallacy of post hoc reasoning, like the Megabucks winner who wonders what made him win against all odds. *Someone* had to win, and it's only because it happened to be him that he's wondering in the first place. It's not the first time that a selection artifact has fooled thinkers into searching for a nonexistent deep explanation for a physical constant. Johannes Kepler agonized over why the Earth was 93 million miles away from the sun, just right for water to fill our lakes and rivers without freezing solid or boiling away. Today we know that the Earth is just one of many planets, each at a different distance from our sun or another star, and we are unsurprised to learn that we find ourselves on that planet rather than on Mars.

The theory of the multiverse would itself be a post hoc excuse for an explanation if it were not consistent with other theories in physics—in particular, that the vacuum of space can spawn big bangs which grow into new universes, and that the baby universes can be born with different fundamental constants.[37] Still, the very idea repels many people (not least some physicists) because of its mind-boggling profligacy. An infinity of universes (or at least a number large enough to include all possible arrangements of matter) implies that somewhere there are universes with exact doppelgangers of you except that they married someone else, were killed by a car last night, are named Evelyn, have one hair out of place, put the book down a moment ago and are not reading this sentence, and so on.

Yet however unsettling these implications are, the history of ideas tells us that cognitive queasiness is a poor guide to reality. Our best science has repeatedly insulted our ancestors' common sense with unsettling discoveries that turned out to be true, including a round Earth, a slowdown of time at high speeds, quantum superposition, curved space-time, and of course evolution. Indeed, once we get over the initial shock, we find that a multiverse is not so exotic after all. This is not even the first time that physicists have had a reason to posit multiple universes. Another version of the multiverse is a straightforward implication of the discoveries that space appears to be infinite and that matter appears to be evenly dispersed through it: there must be an infinity of universes dotting 3-D space beyond our cosmic horizon. Still another is the many-worlds interpretation of quantum mechanics, in which the multiple outcomes of a probabilistic quantum process (such as the trajectory of a photon) are all realized in superimposed parallel universes (a possibility that could lead to quantum computers, in which all possible values of the variables in a computation are represented simultaneously). Indeed, in one sense the multiverse is the *simpler* theory of reality, since if our universe is the only one in existence, we would need to complicate the elegant laws of physics with an arbitrary stipulation of our universe's parochial initial conditions and its parochial physical constants. As the physicist Max Tegmark (an advocate of four kinds of multiverse) put it, "Our judgment therefore comes down to which we find more wasteful and inelegant: many worlds or many words."

If the multiverse turns out to be the best explanation of the fundamental physical constants, it would not be the first time we have been flabbergasted by worlds beyond our noses. Our ancestors had to swallow the discovery of the Western Hemisphere, eight other planets, a hundred billion stars in our galaxy (many with planets), and a hundred billion galaxies in the observable universe. If reason contradicts intuition once again, so much the worse for intuition. Another advocate of the multiverse, Brian Greene, reminds us:

> From a quaint, small, earth-centered universe to one filled with billions of galaxies, the journey has been both thrilling and humbling. We've been compelled to relinquish sacred belief in our own centrality, but with such cosmic demotion we've demonstrated the capacity of the human intellect to reach far beyond the confines of ordinary experience to reveal extraordinary truth.[38]

∼

The other supposedly God-fillable gap is the "hard problem of consciousness," also known as the problem of sentience, subjectivity, phenomenal consciousness, and qualia (the "qualitative" aspect of consciousness).[39] The term, originally suggested by the philosopher David Chalmers, is an in-joke, because the so-called easy problem—the scientific challenge of distinguishing conscious from unconscious mental computation, identifying its substrates in the brain, and explaining why it evolved—is "easy" in the sense that curing cancer or sending a man to the Moon is easy, namely that it is scientifically tractable. Fortunately, the easy problem is more than just tractable: we are well on the way to a satisfying explanation. It's hardly a mystery why we experience a world of stable, solid, colored 3-D objects rather than the kaleidoscope of pixels on our retinas, or why we enjoy (and hence seek) food, sex, and bodily integrity while suffering from (and hence avoiding) social isolation and tissue damage: these internal states and the behavior they encourage are obvious Darwinian adaptations. With advances in evolutionary psychology, more and more of our conscious experiences are being explained in this way, including our intellectual obsessions, moral emotions, and aesthetic reactions.[40]

Nor are the computational and neurobiological bases of consciousness obstinately befuddling. The cognitive neuroscientist Stanislas Dehaene and his collaborators have argued that consciousness functions as a "global workspace" or "blackboard" representation.[41] The blackboard metaphor refers to the way that a diverse set of computational modules can post their results in a common format that all the other modules can "see." Those modules include perception, memory, motivation, language understanding, and action planning, and the fact that they can all access a common pool of currently relevant information (the contents of consciousness) allows us to describe, grasp, or approach what we see, to respond to what other people say or do, and to remember and plan depending on what we want and what we know. (The computations *inside* each module, in contrast, like the calculation of depth from the two eyes or the sequencing of muscle contractions making up an action, can work off their own proprietary input streams, and they proceed below the level of consciousness, having no need for its synoptic view.) This global workspace is implemented in the brain as rhythmic, synchronized firing in neural networks that link the prefrontal and parietal cerebral cortexes with each other and with brain areas that feed them perceptual, mnemonic, and motivational signals.

The so-called hard problem—why it subjectively *feels like something* to each one of us who is conscious, with red looking red and salt tasting salty—is hard not because it is a recalcitrant scientific topic but because it is a head-scratching conceptual enigma. It includes brainteasers such as whether my red is the same as your red, what it is like to be a bat, whether there could be zombies (people indistinguishable from you and me but with "no one home" who is feeling anything), and if so whether everyone but me is a zombie, whether a perfectly lifelike robot would be conscious, whether I could achieve immortality by uploading my brain's connectome to the Cloud, and whether the Star Trek transporter really transports Captain Kirk to the planetary surface or murders him and reconstitutes a twin.

Some philosophers, like Daniel Dennett in *Consciousness Explained*, have argued that there *is* no hard problem of consciousness: it is a confusion arising from the bad habit of imagining a homunculus seated in a theater inside the skull. This is the disembodied experiencer who would temporarily tiptoe out of my theater and drop in on yours to check out the red, or visit the bat's and watch the movie that's playing there; who would be missing from the zombie and either present or absent in the robot; and who might or might not survive the beam ride down to Zakdorn. Sometimes, when I see the mischief that the hard problem has caused (including the conservative intellectual Dinesh D'Souza brandishing a copy of my book *How the Mind Works* in a debate on the existence of God), I am tempted to agree with Dennett that we'd be better off without the term. Contrary to various misunderstandings, the hard problem does not consist in weird physical or paranormal phenomena such as clairvoyance, telepathy, time travel, augury, or action at a distance. It does not call for exotic quantum physics, kitschy energy vibrations, or other New Age flimflam. Most important for the present discussion, it does not implicate an immaterial soul. Nothing that we know about consciousness is inconsistent with the understanding that it depends entirely on neural activity.

In the end I still think that the hard problem is a meaningful *conceptual* problem, but agree with Dennett that it is not a meaningful *scientific* problem.[42] No one will ever get a grant to study whether you are a zombie or whether the same Captain Kirk walks on the deck of the *Enterprise* and the surface of Zakdorn. And I agree with several other philosophers that it may be futile to hope for a solution at all, precisely because it *is* a conceptual problem, or, more accurately, a problem with our concepts. As Thomas Nagel put it in his famous essay "What Is It Like to Be a Bat?" there may be "facts which could not ever be represented or comprehended by human beings, even if the species lasted forever—simply because our structure does not permit us to operate with concepts of the requisite type."[43] The philosopher Colin McGinn has run with this idea, arguing that there is a mismatch between our cognitive tools for explaining reality (namely chains of causes and effects, analysis into parts and their interactions, and modeling in mathematical equations) and the nature of the hard problem of consciousness, which is unintuitively holistic.[44] Our best science tells us that consciousness consists of a global workspace representing our current goals, memories, and surroundings, implemented in synchronized neural firing in fronto-parietal circuitry. But the last dollop in the theory—that it subjectively *feels like* something to be such circuitry—may have to be stipulated as a fact about reality where explanation stops. This should not be entirely surprising. As Ambrose Bierce noted in *The Devil's Dictionary*, the mind has nothing but itself to know itself with, and it may never feel satisfied that it understands the deepest aspect of its own existence, its intrinsic subjectivity.

Whatever we make of the hard problem of consciousness, positing an immaterial soul is of no help at all. For one thing, it tries to solve a mystery with an even bigger mystery. For another, it falsely predicts the existence of paranormal phenomena. Most damningly, a divinely granted consciousness does not meet the design specs for a locus of just deserts. Why would God have endowed a mobster with the ability to enjoy his ill-gotten gains, or a sexual predator with carnal pleasure? (If it's to plant temptations for them to prove their morality by resisting, why should their victims be collateral damage?) Why would a merciful God be dissatisfied with robbing years of life from a cancer patient and add the gratuitous punishment of agonizing pain? Like the phenomena of physics, the phenomena of consciousness look exactly as you would expect if the laws of nature applied without regard to human welfare. If we want to enhance that welfare, we have to figure out how to do it ourselves.

∼

And that brings us to the second problem with theistic morality. It's not just that there is almost certainly no God to dictate and enforce moral precepts. It's that even if there were a God, his divine decrees, as conveyed to us through religion, cannot be the source of morality. The explanation goes back to Plato's *Euthyphro,* in which Socrates points out that if the gods have good reasons to deem certain acts moral, we can appeal to those reasons directly, skipping the middlemen. If they don't, we should not take their dictates seriously. After all, thoughtful people can give reasons why they don't kill, rape, or torture other than fear of eternal hellfire, and they would not suddenly become rapists and contract killers if they had reason to believe that God's back was turned or if he told them it was OK.

Theistic moralists reply that the God of scripture, unlike the capricious deities of Greek mythology, is by his very nature incapable of issuing immoral commandments. But anyone who is familiar with scripture knows that this is not so. The God of the Old Testament murdered innocents by the millions, commanded the Israelites to commit mass rape and genocide, and prescribed the death penalty for blasphemy, idolatry, homosexuality, adultery, talking back to parents, and working on the Sabbath, while finding nothing particularly wrong with slavery, rape, torture, mutilation, and genocide. All this was par for the course for Bronze and Iron Age civilizations. Today, of course, enlightened believers cherry-pick the humane injunctions while allegorizing, spin-doctoring, or ignoring the vicious ones, and that's just the point: they read the Bible through the lens of Enlightenment humanism.

The Euthyphro argument puts the lie to the common claim that atheism consigns us to a moral relativism in which everyone can do his own thing. The claim gets it backwards. A humanistic morality rests on the universal bedrock of reason and human interests: it's an inescapable feature of the human condition that we're all better off if we help each other and refrain from hurting each other. For this reason many contemporary philosophers, including Nagel, Goldstein, Peter Singer, Peter Railton, Richard Boyd, David Brink, and Derek Parfit, are moral *realists* (the opposite of relativists), arguing that moral statements may be objectively true or false.[45] It's *religion* that is inherently relativistic. Given the absence of evidence, any belief in how many deities there are, who are their earthly prophets and messiahs, and what they demand of us can depend only on the parochial dogmas of one's tribe.

Not only does this make theistic morality relativistic; it can make it immoral. Invisible gods can command people to slay heretics, infidels, and apostates. And an immaterial soul is unmoved by the earthly incentives that impel us to get along. Contestants over a material resource are usually better off if they split it than fight over it, particularly if they value their own lives on earth. But contestants over a sacred value (like holy land or affirmation of a belief) *may not* compromise, and if they think their souls are immortal, the loss of their body is no big deal—indeed, it may be a small price to pay for an eternal reward in paradise.

Many historians have pointed out that religious wars are long and bloody, and bloody wars are often prolonged by religious conviction.[46] Matthew White, the necrometrician we met in chapter 14, lists thirty religious conflicts among the worst things that people have ever done to one another, resulting in around 55 million killings.[47] (In seventeen conflicts, the monotheistic religions fought each other; in another eight, monotheists fought heathens.) And the common assertion that the two world wars were set off by the decline of religious morality (as in the former Trump strategist Stephen Bannon's recent claim that World War II pitted "the Judeo-Christian West versus atheists") is dunce-cap history.[48] The belligerents on both sides of World War I were devoutly Christian, except for the Ottoman Empire, a Muslim theocracy. The only avowedly atheist power that fought in World War II was the Soviet Union, and for most of the war it fought on *our* side against the Nazi regime—which (contrary to another myth) was sympathetic to German Christianity and vice versa, the two factions united in their loathing of secular modernity.[49] (Hitler himself was a deist who said, "I am convinced that I am acting as the agent of our Creator. By fighting off the Jews, I am doing the Lord's work.")[50] Defenders of theism retort that irreligious wars and atrocities, motivated by the secular ideology of communism and by ordinary conquest, have killed even more people. Talk about relativism! It is peculiar to grade religion on this curve: if religion were a source of morality, the number of religious wars and atrocities ought to be zero. And obviously *atheism* is not a moral system in the first place. It's just the absence of supernatural belief, like an unwillingness to believe in Zeus or Vishnu. The moral alternative to theism is humanism.

～

Few sophisticated people today profess a belief in heaven and hell, the literal truth of the Bible, or a God who flouts the laws of physics. But many intellectuals have reacted with fury to the "New Atheism" popularized in a quartet of bestsellers published between 2004 and 2007 by Sam Harris, Richard Dawkins, Daniel Dennett, and Christopher Hitchens.[51] Their reaction has been called "I'm-an-atheist-but," "belief-in-belief," "accommodationism," and (in Coyne's coinage) "faitheism." It overlaps with the hostility to science within the Second Culture, presumably because of a shared sympathy to hermeneutic over analytical and empirical methodologies, and a reluctance to acknowledge that dweeby scientists and secular philosophers might be right about the fundamental questions of existence. Though atheism—the absence of a belief in God—is compatible with a wide range of humanistic and antihumanistic beliefs, the New Atheists are avowedly humanistic, so any flaws in their worldview might carry over to humanism more generally.

According to the faitheists, the New Atheists are too shrill and militant, and just as annoying as the fundamentalists they criticize. (In an *XKCD* webcomic, a character responds, "Well, the important thing is that you've found a way to feel superior to both.")[52] Ordinary people will never be disabused of their religious beliefs, they say, and perhaps they should not be, because healthy societies need religion as a bulwark against selfishness and meaningless consumerism. Religious institutions supply that need by promoting charity, community, social responsibility, rites of passage, and guidance on existential questions that can never be provided by science. Anyway, most people treat religious doctrine allegorically rather than literally, and they find meaning and wisdom in an overarching sense of spirituality, grace, and divine order.[53] Let's look at these claims.

An ironic inspiration for faitheism is research on the psychological origins of supernatural belief, including the cognitive habits of over-attributing design and agency to natural phenomena, and emotional feelings of solidarity within communities of faith.[54] The most natural interpretation of these findings is that they *undermine* religious beliefs by showing how they are figments of our neurobiological makeup. But the research has also been interpreted as showing that human nature requires religion in the same way that it requires food, sex, and companionship, so it's futile to imagine no religion. But this interpretation is dubious.[55] Not every feature of human nature is a homeostatic drive that must be regularly slaked. Yes, people are vulnerable to cognitive illusions that lead to supernatural beliefs, and they certainly need to belong to a community. Over the course of history, institutions have arisen that offer packages of customs that encourage those illusions and cater to those needs. That does not imply that people need the complete packages, any more than the existence of sexual desire implies that people need Playboy clubs. As societies become more educated and secure, the components of the legacy religious institutions can be unbundled. The art, rituals, iconography, and communal warmth that many people enjoy can continue to be provided by liberalized religions, without the supernatural dogma or Iron Age morality.

That implies that religions should not be condemned or praised across the board but considered according to the logic of Euthyphro. If there are justifiable reasons behind particular activities, those activities should be encouraged, but the movements should not be given a pass just because they are religious. Among the positive contributions of religions at particular times and places are education, charity, medical care, counseling, conflict resolution, and other social services (though in the developed world these efforts are dwarfed by their secular counterparts; no religion could have decimated hunger, disease, illiteracy, war, homicide, or poverty on the scales we saw in part II). Religious organizations can also provide a sense of communal solidarity and mutual support, together with art, ritual, and architecture of great beauty and historical resonance, thanks to their millennia-long head start. I partake of these myself, with much enjoyment.

If the positive contributions of religious institutions come from their role as humanistic associations in civil society, then we would expect those benefits not to be tied to theistic belief, and that is indeed the case. It's long been known that churchgoers are happier and more charitable than stay-at-homes, but Robert Putnam and his fellow political scientist David Campbell have found that these blessings have nothing to do with beliefs in God, creation, heaven, or hell.[56] An atheist who has been pulled into a congregation by an observant spouse is as charitable as the faithful among the flock, whereas a fervent believer who prays alone is not particularly charitable. At the same time, communality and civic virtue can be fostered by membership in secular service communities such as the Shriners (with their children's hospitals and burn units), Rotary International (which is helping to end polio), and Lions Club (which combats blindness)—even, according to Putnam and Campbell's research, a bowling league.

Just as religious institutions deserve praise when they pursue humanistic ends, they should not be shielded from criticism when they obstruct those ends. Examples include the withholding of medical care from sick children in faith-healing sects, the opposition to humane assisted dying, the corruption of science education in schools, the suppression of touchy biomedical research such as on stem cells, and obstruction of lifesaving public health policies such as contraception, condoms, and vaccination against HPV.[57] Nor should religions be granted a presumption of a higher moral purpose. Faitheists who have hoped that the moralistic fervor of Evangelical Christianity might be channeled into movements for social improvement have repeatedly gotten burned. In the early 2000s, a bipartisan coalition of environmentalists hoped to make common cause with Evangelicals on climate change under rubrics like Creation Care and Faith-Based Environmentalism. But Evangelical churches are an anchor faction of the Republican Party, which adopted a strategy of absolute noncooperation with the Obama administration. Political tribalism carried the day, and the Evangelicals fell into line, opting for radical libertarianism over stewardship of the Creation.[58]

Similarly, in 2016 there was a brief hope that the Christian virtues of humility, temperance, forgiveness, propriety, chivalry, thrift, and compassion toward the weak would turn Evangelicals against a casino developer who was vainglorious, sybaritic, vindictive, lewd, misogynistic,

ostentatiously wealthy, and contemptuous of the people he called "losers." But no: Donald Trump won the votes of 81 percent of white Evangelical and born-again Christians, a higher proportion than of any other
demographic.[59] In large part he earned their votes by promising to repeal
a law which prohibits tax-exempt charities (including churches) from
engaging in political activism.[60] Christian virtue was trumped by political muscle.

∼

If the factual tenets of religion can no longer be taken seriously, and its
ethical tenets depend entirely on whether they can be justified by secular
morality, what about its claims to wisdom on the great questions of existence? A favorite talking point of faitheists is that only religion can
speak to the deepest yearnings of the human heart. Science will never be
adequate to address the great existential questions of life, death, love,
loneliness, loss, honor, cosmic justice, and metaphysical hope.

This is the kind of statement that Dennett (quoting a young child)
calls a "deepity": it has a patina of profundity, but as soon as one thinks
about what it means, it turns out to be nonsense. To begin with, the alternative to "religion" as a source of meaning is not "science." No one
ever suggested that we look to ichthyology or nephrology for enlightenment on how to live, but rather to the entire fabric of human knowledge,
reason, and humanistic values, of which science is a part. It's true that
the fabric contains important strands that originated in religion, such as
the language and allegories of the Bible and the writings of sages, scholars, and rabbis. But today it is dominated by secular content, including
debates on ethics originating in Greek and Enlightenment philosophy,
and renderings of love, loss, and loneliness in the works of Shakespeare,
the Romantic poets, the 19th-century novelists, and other great artists
and essayists. Judged by universal standards, many of the religious contributions to life's great questions turn out to be not deep and timeless
but shallow and archaic, such as a conception of "justice" that includes
punishing blasphemers, or a conception of "love" that adjures a woman
to obey her husband. As we have seen, any conception of life and death
that depends on the existence of an immaterial soul is factually dubious
and morally dangerous. And since cosmic justice and metaphysical hope
(as opposed to human justice and worldly hope) do not exist, then it's not
meaningful to seek them; it's pointless. The claim that people should
seek deeper meaning in supernatural beliefs has little to recommend it.

What about a more abstract sense of "spirituality"? If it consists in
gratitude for one's existence, awe at the beauty and immensity of the

universe, and humility before the frontiers of human understanding, then spirituality is indeed an experience that makes life worth living— and one that is lifted into higher dimensions by the revelations of science and philosophy. But "spirituality" is often taken to mean something more: the conviction that the universe is somehow *personal*, that everything happens for a reason, that meaning is to be found in the happenstances of life. In the final episode of her landmark show, Oprah Winfrey spoke for millions when she avowed, "I understand the manifestation of grace and God, so I know there are no coincidences. There are none. Only divine order here."[61]

This sense of spirituality is considered in a video sketch by the comedienne Amy Schumer called "The Universe." It opens with the science popularizer Bill Nye standing against a backdrop of stars and galaxies:

NYE: The Universe. For centuries, humankind has strived to understand this vast expanse of energy, gas, and dust. In recent years, a stunning breakthrough has been made in our concept of what the universe is for.

[*Zoom to the Earth's surface, and then to a yogurt shop in which two young women are chatting.*]

FIRST WOMAN: So, I was texting while I was driving? And I ended up taking a wrong turn that took me directly past a vitamin shop? And I was just like, this is totally the universe telling me I should be taking calcium.

NYE: Scientists once believed the universe was a chaotic collection of matter. We now know the universe is essentially a force sending cosmic guidance to women in their 20s.

[*Zoom to a gym with Schumer and a friend on exercycles.*]

SCHUMER: So you know how I've been fucking my married boss for like six months? Well, I was starting to get really worried he was never going to leave his wife. But then yesterday in yoga, the girl in front of me was wearing a shirt that just said, "Chill." And I was just like, this is so the universe telling me, "Girl, just, like, keep fucking your married boss!"[62]

A "spirituality" that sees cosmic meaning in the whims of fortune is not wise but foolish. The first step toward wisdom is the realization that the laws of the universe don't care about you. The next is the realization that this does not imply that life is meaningless, because *people* care about you, and vice versa. You care about yourself, and you have a responsibility to respect the laws of the universe that keep you alive, so you don't squander your existence. Your loved ones care about you, and you have a responsibility not to orphan your children, widow your spouse, and shatter your parents. And anyone with a humanistic sensibility cares about you, not in the sense of feeling your pain—human empathy is too feeble to spread itself across billions of strangers—but in the sense of realizing that your existence is cosmically no less important than theirs, and that we all have a responsibility to use the laws of the universe to enhance the conditions in which we all can flourish.

～

Arguments aside, is the need to believe pushing back against secular humanism? Believers, faitheists, and resenters of science and progress are gloating about an apparent return of religion all over the world. But as we shall see, the rebound is an illusion: the world's fastest-growing religion is no religion at all.

Measuring the history of religious belief is not easy. Few surveys have asked people the same questions in different times and places, and the respondents would interpret them differently even if they did. Many people are queasy about labeling themselves *atheist,* a word they equate with "amoral" and which can expose them to hostility, discrimination, and (in many Muslim countries) imprisonment, mutilation, or death.[63] Also, most people are hazy theologians, and may stop short of declaring themselves atheists while admitting that they have no religion or religious beliefs, find religion unimportant, are spiritual but not religious, or believe in some "higher power" which is not God. Different surveys can end up with different estimates of irreligion depending on how the alternatives are worded.

We can't say for sure how many nonbelievers there were in earlier decades and centuries, but there can't have been many; one estimate put the proportion in 1900 at 0.2 percent.[64] According to WIN-Gallup International's Global Index of Religiosity and Atheism, a survey of fifty thousand people in fifty-seven countries, 13 percent of the world's population identified themselves as a "convinced atheist" in 2012, up from around 10 percent in 2005.[65] It would not be fanciful to say that over the course of the 20th century the global rate of atheism increased by a factor of 500, and that it has doubled again so far in the 21st. An additional 23 percent of the world's population identify themselves as "not a religious person," leaving 59 percent of the world as "religious," down from close to 100 percent a century before.

According to an old idea in social science called the Secularization

Thesis, irreligion is a natural consequence of affluence and education.[66] Recent studies confirm that wealthier and better-educated countries tend to be less religious.[67] The decline is clearest in the developed countries of Western Europe, the Commonwealth, and East Asia. In Australia, Canada, France, Hong Kong, Ireland, Japan, the Netherlands, Sweden, and several other countries, religious people are in the minority, and atheists make up a quarter to more than half of the population.[68] Religion has also declined in formerly Communist countries (especially China), though not in Latin America, the Islamic world, or sub-Saharan Africa.

The data show no signs of a global religious revival. Among the thirty-nine countries surveyed by the Index in both 2005 and 2012, only eleven became more religious, none by more than six percentage points, while twenty-six became less religious, many by double digits. And contrary to impressions from the news, the religiously excitable countries of Poland, Russia, Bosnia, Turkey, India, Nigeria, and Kenya became *less* religious over these seven years, as did the United States (more on this soon). Overall, the percentage of people who called themselves religious declined by nine points, making room for growth in the proportion of "convinced atheists" in a majority of the countries.

Another global survey, by the Pew Research Center, tried to project religious affiliation into the future (the survey did not ask about belief).[69] The survey found that in 2010, a sixth of the world's population, when asked to name their religion, chose "None." There are more Nones in the world than Hindus, Buddhists, Jews, or devotees of folk religions, and this is the "denomination" that the largest number of people are expected to switch into. By 2050, 61.5 million more people will have lost their religion than found one.

With all these numbers showing that people are becoming less religious, where did the idea of a religious revival come from? It comes from what Quebecers call *la revanche du berceau*, the revenge of the cradle. Religious people have more babies. The demographers at Pew did the math and projected that the proportion of the world's population that is Muslim might rise from 23.2 percent in 2010 to 29.7 percent in 2050, while the percentage of Christians will remain unchanged, and the percentage of all other denominations, together with the religiously unaffiliated, will decrease. Even this projection is a hostage to current fertility estimates and may become obsolete if Africa (religious and fecund) undergoes the demographic transition, or if the Muslim fertility decline discussed in chapter 10 continues.[70]

A key question about the secularization trend is whether it is being driven by changing times (a period effect), a graying population (an age effect), or the turnover of generations (a cohort effect).[71] Only a few countries, all English-speaking, have the multidecade data we need to answer the question. Australians, New Zealanders, and Canadians have become less religious as the years have gone by, probably because of changing times rather than the population getting older (if anything, we would expect people to become more religious as they prepare to meet their maker). There was no such change in the British or American zeitgeist, but in all five countries, each generation was less religious than the one before. The cohort effect is substantial. More than 80 percent of the British GI Generation (born 1905–1924) said they belonged to a religion, but at the same ages, fewer than 30 percent of the Millennials did. More than 70 percent of the American GI Generation said they "know God exists," but only 40 percent of their Millennial great-grandchildren say that.

The discovery of a generational turnover throughout the Anglosphere removes a big thorn in the side of the secularization thesis: the United States, which is wealthy but religious. As early as 1840, Alexis de Tocqueville remarked on how Americans were more devout than their European cousins, and the difference persists today: in 2012, 60 percent of Americans called themselves religious, compared with 46 percent of Canadians, 37 percent of the French, and 29 percent of Swedes.[72] Other Western democracies have two to six times the proportion of atheists found in the United States.[73]

But while Americans started from a higher level of belief, they have not escaped the march of secularization from one generation to the next. A recent report summarizes the trend in its title: "Exodus: Why Americans Are Leaving Religion—and Why They're Unlikely to Come Back."[74] The exodus is most visible in the rise of the Nones, from 5 percent in 1972 to 25 percent today, making them the largest religious group in the United States, surpassing Catholics (21 percent), white Evangelicals (16 percent), and white mainline Protestants (13.5 percent). The cohort gradient is steep: just 13 percent of Silents and older Boomers are Nones, compared with 39 percent of Millennials.[75] The younger generations, moreover, are more likely to remain irreligious as they age and stare down their mortality.[76] The trends are just as dramatic among the subset of Nones who are not just none-of-the-abovers but confessed nonbelievers. The percentage of Americans who say they are atheist or agnostic, or that religion is unimportant to them (probably no more than a percentage point or two in the 1950s), rose to 10.3 percent in 2007 and 15.8 percent in 2014. The cohorts break down like this: 7 percent of Silents, 11 percent of

Boomers, 25 percent of Millennials.[77] Clever survey techniques designed to get around people's squeamishness in confessing to atheism suggest that the true percentages are even higher.[78]

Why, then, do commentators think that religion is rebounding in the United States? It's because of yet another finding about the American Exodus: Nones don't vote. In 2012 religiously unaffiliated Americans made up 20 percent of the populace but 12 percent of the voters. Organized religions, by definition, are organized, and they have been putting that organization to work in getting out the vote and directing it their way. In 2012 white Evangelical Protestants also made up 20 percent of the adult population, but they made up *26 percent* of the voters, more than double the proportion of the irreligious.[79] Though the Nones supported Clinton over Trump by a ratio of three to one, they stayed home on November 8, 2016, while the Evangelicals lined up to vote. Similar patterns apply to populist movements in Europe. Pundits are apt to mistake this electoral clout for a comeback of religion, an illusion that gives us a second explanation (together with fecundity) for why secularization has been so stealthy.

Why is the world losing its religion? There are several reasons.[80] The Communist governments of the 20th century outlawed or discouraged religion, and when they liberalized, their citizenries were slow to reacquire the taste. Some of the alienation is part of a decline in trust in *all* institutions from its high-water mark in the 1960s.[81] Some of it is carried by the global current toward emancipative values (chapter 15) such as women's rights, reproductive freedom, and tolerance of homosexuality.[82] Also, as people's lives become more secure thanks to affluence, medical care, and social insurance, they no longer pray to God to save them from ruin: countries with stronger safety nets are less religious, holding other factors constant.[83] But the most obvious reason may be reason itself: when people become more intellectually curious and scientifically literate, they stop believing in miracles. The most common reason that Americans give for leaving religion is "a lack of belief in the teachings of religion."[84] We have already seen that better-educated countries have lower rates of belief, and across the world, atheism rides the Flynn effect: as countries get smarter, they turn away from God.[85]

Whatever the reasons, the history and geography of secularization belie the fear that in the absence of religion, societies are doomed to anomie, nihilism, and a "total eclipse of all values."[86] Secularization has proceeded in parallel with all the historical progress documented in part II. Many irreligious societies like Canada, Denmark, and New Zealand are among the nicest places to live in the history of our kind (with high levels of every measurable good thing in life), while many of the world's most religious societies are hellholes.[87] American exceptionalism is instructive: the United States is more religious than its Western peers but underperforms them in happiness and well-being, with higher rates of homicide, incarceration, abortion, sexually transmitted disease, child mortality, obesity, educational mediocrity, and premature death.[88] The same holds true among the fifty states: the more religious the state, the more dysfunctional its citizens' lives.[89] Cause and effect probably run in many directions. But it's plausible that in democratic countries, secularism leads to humanism, turning people away from prayer, doctrine, and ecclesiastical authority and toward practical policies that make them and their fellows better off.

~

However baleful theistic morality may be in the West, its influence is even more troubling in contemporary Islam. No discussion of global progress can ignore the Islamic world, which by a number of objective measures appears to be sitting out the progress enjoyed by the rest. Muslim-majority countries score poorly on measures of health, education, freedom, happiness, and democracy, holding wealth constant.[90] All of the wars raging in 2016 took place in Muslim-majority countries or involved Islamist groups, and those groups were responsible for the vast majority of terrorist attacks.[91] As we saw in chapter 15, emancipative values such as gender equality, personal autonomy, and political voice are less popular in the Islamic heartland than in any other region of the world, including sub-Saharan Africa. Human rights are abysmal in many Muslim countries, which implement cruel punishments (such as flogging, blinding, and amputation), not just for actual crimes but for homosexuality, witchcraft, apostasy, and expressing liberal opinions on social media.

How much of this lack of progress is the fallout of theistic morality? Certainly it cannot be attributed to Islam itself. Islamic civilization had a precocious scientific revolution, and for much of its history was more tolerant, cosmopolitan, and internally peaceful than the Christian West.[92] Some of the regressive customs found in Muslim-majority countries, such as female genital mutilation and "honor killings" of unchaste sisters and daughters, are ancient African or West Asian tribal practices and are misattributed by their perpetrators to Islamic law. Some of the problems are found in other resource-cursed strongman states. Still others were exacerbated by clumsy Western interventions in the Middle East, including the dismemberment of the Ottoman Empire, support of the anti-Soviet mujahedin in Afghanistan, and the invasion of Iraq.

But part of the resistance to the tide of progress can be attributed to religious belief. The problem begins with the fact that many of the precepts of Islamic doctrine, taken literally, are floridly antihumanistic. The Quran contains scores of passages that express hatred of infidels, the reality of martyrdom, and the sacredness of armed jihad. Also endorsed are lashing for alcohol consumption, stoning for adultery and homosexuality, crucifixion for enemies of Islam, sexual slavery for pagans, and forced marriage for nine-year-old girls.[93]

Of course many of the passages in the Bible are floridly antihumanistic too. One needn't debate which is worse; what matters is how literally the adherents take them. Like the other Abrahamic religions, Islam has its version of rabbinical pilpul and Jesuitical disputation that allegorizes, compartmentalizes, and spin-doctors the nasty bits of scripture. Islam also has its version of Cultural Jews, Cafeteria Catholics, and CINOs (Christians in Name Only). The problem is that this benign hypocrisy is far less developed in the contemporary Islamic world.

Examining big data on religious affiliation from the World Values Survey, the political scientists Amy Alexander and Christian Welzel observe that "self-identifying Muslims stick out as the denomination with by far the largest percentage of strongly religious people: 82%. Even more astounding, fully 92% of all self-identifying Muslims place themselves at the two highest scores of the ten-point religiosity scale [compared with less than half of Jews, Catholics, and Evangelicals]. Self-identifying as a Muslim, regardless of the particular branch of Islam, seems to be almost synonymous with being strongly religious."[94] Similar results turn up in some other surveys.[95] A large one by the Pew Research Center found that "in 32 of the 39 countries surveyed, half or more Muslims say there is only one correct way to understand the teachings of Islam," that in the countries in which the question was asked, between 50 and 93 percent believe that the Quran "should be read literally, word by word," and that "overwhelming percentages of Muslims in many countries want Islamic law (sharia) to be the official law of the land."[96]

Correlation is not causation, but if you combine the fact that much of Islamic doctrine is antihumanistic with the fact that many Muslims believe that Islamic doctrine is inerrant—and throw in the fact that the Muslims who carry out illiberal policies and violent acts say they are doing it because they are following those doctrines—then it becomes a stretch to say that the inhumane practices have nothing to do with religious devotion and that the real cause is oil, colonialism, Islamophobia, Orientalism, or Zionism. For those who need data to be convinced, in

global surveys of values in which every variable that social scientists like
to measure is thrown into the pot (including income, education, and dependence on oil revenues), Islam itself predicts an extra dose of patriarchal and other illiberal values across countries and individuals.[97] Within
non-Muslim societies, so does mosque attendance (in Muslim societies,
the values are so pervasive that mosque attendance doesn't matter).[98]

All these troubling patterns were once true of Christendom, but starting with the Enlightenment, the West initiated a process (still ongoing)
of separating the church from the state, carving out a space for secular
civil society, and grounding its institutions in a universal humanistic
ethics. In most Muslim-majority countries, that process is barely under
way. Historians and social scientists (many of them Muslim) have shown
how the stranglehold of the Islamic religion over governmental institutions and civil society in Muslim countries has impeded their economic,
political, and social progress.[99]

Making things worse is a reactionary ideology that became influential
through the writings of the Egyptian author Sayyid Qutb (1906–1966), a
member of the Muslim Brotherhood and the inspiration for Al Qaeda and
other Islamist movements.[100] The ideology looks back to the glory days of
the Prophet, the first caliphs, and classical Arab civilization, and laments
subsequent centuries of humiliation at the hands of Crusaders, horse
tribes, European colonizers, and, most recently, insidious secular mod-
ernizers. That history is seen as the bitter fruit of forsaking strict Islamic
practice; redemption can come only from a restoration of true Muslim
states governed by sharia law and purged of non-Muslim influences.

Though the role of theistic morality in the problems besetting the Islamic world is inescapable, many Western intellectuals—who would be
appalled if the repression, misogyny, homophobia, and political violence
that are common in the Islamic world were found in their own societies
even diluted a hundredfold—have become strange apologists when
these practices are carried out in the name of Islam.[101] Some of the apologetics, to be sure, come from an admirable desire to prevent prejudice
against Muslims. Some are intended to discredit a destructive (and possibly self-fulfilling) narrative that the world is embroiled in a clash of civilizations. Some fit into a long history of Western intellectuals execrating
their own society and romanticizing its enemies (a syndrome we'll return to shortly). But many of the apologetics come from a soft spot for
religion among theists, faitheists, and Second Culture intellectuals, and
a reluctance to go all in for Enlightenment humanism.

Calling out the antihumanistic features of contemporary Islamic belief is in no way Islamophobic or civilization-clashing. The overwhelming majority of victims of Islamic violence and repression are other Muslims. Islam is not a race, and as the ex-Muslim activist Sarah Haider has put it, "Religions are just ideas and don't have rights."[102] Criticizing the ideas of Islam is no more bigoted than criticizing the ideas of neoliberalism or the Republican Party platform.

Can the Islamic world have an Enlightenment? Can there be a Reform Islam, a Liberal Islam, a Humanistic Islam, an Islamic Ecumenical Council, a separation of mosque and state? Many of the faithophilic intellectuals who excuse the illiberalism of Islam also insist that it's unreasonable to expect Muslims to progress beyond it. While the West might enjoy the peace, prosperity, education, and happiness of post-Enlightenment societies, Muslims will never accept this shallow hedonism, and it's only understandable that they should cling to a system of medieval beliefs and customs forever.

But this condescension is belied by the history of Islam and by nascent movements within it. Classical Arabic civilization, as I mentioned, was a hothouse of science and secular philosophy.[103] Amartya Sen has documented how the 16th-century Mughal emperor Akbar I implemented a multiconfessional, liberal social order (including atheists and agnostics) in Muslim-ruled India at a time when the Inquisition was raging in Europe and Giordano Bruno was burnt at the stake for heresy.[104] Today the forces of modernity are working in many parts of the Islamic world. Tunisia, Bangladesh, Malaysia, and Indonesia have made long strides toward liberal democracy (chapter 14). In many Islamic countries, attitudes toward women and minorities are improving (chapter 15)— slowly, but more detectably among women, the young, and the educated.[105] The emancipative forces that liberalized the West, such as connectivity, education, mobility, and women's advancement, are not bypassing the Islamic world, and the moving sidewalk of generational replacement can outpace the walkers shambling along it.[106]

Also, ideas matter. A cadre of Muslim intellectuals, writers, and activists has been pressing the case for a humanistic revolution for Islam. Among them are Souad Adnane (co-founder of the Arab Center for Scientific Research and Humane Studies in Morocco); Mustafa Akyol (author of *Islam Without Extremes*); Faisal Saeed Al-Mutar (founder of the Global Secular Humanist Movement); Sarah Haider (co-founder of Ex-Muslims of North America); Shadi Hamid (author of *Islamic Exceptionalism*); Pervez Hoodbhoy (author of *Islam and Science: Religious Orthodoxy and the Battle for Rationality*); Leyla Hussein (founder of Daughters of Eve, which opposes female genital mutilation); Gulalai Ismail (founder of Aware Girls in Pakistan); Shiraz Maher (author of *Salafi-Jihadism*, quoted in the introduction to part 1); Omar Mahmood (an American editorialist); Irshad Manji (author of *The Trouble with Islam*); Maryam Namazie (spokesperson for One Law for All); Amir Ahmad Nasr (author of *My Isl@m*); Taslima Nasrin (author of *My Girlhood*); Maajid Nawaz (coauthor, with Sam Harris, of *Islam and the Future of Tolerance*); Asra Nomani (author of *Standing Alone in Mecca*); Raheel Raza (author of *Their Jihad, Not My Jihad*); Ali Rizvi (author of *The Atheist Muslim*); Wafa Sultan (author of *A God Who Hates*); Muhammad Syed (president of Ex-Muslims of North America); and most famously, Salman Rushdie, Ayaan Hirsi Ali, and Malala Yousafzai.

Obviously a new Islamic Enlightenment will have to be spearheaded by Muslims, but non-Muslims have a role to play. The global network of intellectual influence is seamless, and given the prestige and power of the West (even among those who resent it), Western ideas and values can trickle, flow, and cascade outward in surprising ways. (Osama bin Laden, for example, owned a book by Noam Chomsky.)[107] The history of moral progress, recounted in books such as *The Honor Code* by the philosopher Kwame Anthony Appiah, suggests that moral clarity in one culture about a regressive practice by another does not always provoke resentful backlash but can shame the laggards into overdue reform. (Past examples include slavery, dueling, foot-binding, and racial segregation; future ones targeting the United States may include capital punishment and mass incarceration.)[108] An intellectual culture that steadfastly defended Enlightenment values and that did not indulge religion when it clashed with humanistic values could serve as a beacon for students, intellectuals, and open-minded people in the rest of the world.

~

After laying out the logic of humanism, I noted that it stood in stark contrast to two other systems of belief. We have just looked at theistic morality. Let me turn to the second enemy of humanism, the ideology behind resurgent authoritarianism, nationalism, populism, reactionary thinking, even fascism. As with theistic morality, the ideology claims intellectual merit, affinity with human nature, and historical inevitability. All three claims, we shall see, are mistaken. Let's begin with some intellectual history.

If one wanted to single out a thinker who represented the opposite of humanism (indeed, of pretty much every argument in this book), one couldn't do better than the German philologist Friedrich Nietzsche

(1844–1900).[109] Earlier in the chapter I fretted about how humanistic morality could deal with a callous, egoistic, megalomaniacal sociopath. Nietzsche argued that it's *good* to be a callous, egoistic, megalomaniacal sociopath. Not good for everyone, of course, but that doesn't matter: the lives of the mass of humanity (the "botched and the bungled," the "chattering dwarves," the "flea-beetles") count for nothing. What is worthy in life is for a superman (*Übermensch*, literally "overman") to transcend good and evil, exert a will to power, and achieve heroic glory. Only through such heroism can the potential of the species be realized and humankind lifted to a higher plane of being. The feats of greatness may not consist, though, in curing disease, feeding the hungry, or bringing about peace, but rather in artistic masterworks and martial conquest. Western civilization has gone steadily downhill since the heyday of Homeric Greeks, Aryan warriors, helmeted Vikings, and other manly men. It has been especially corrupted by the "slave morality" of Christianity, the worship of reason by the Enlightenment, and the liberal movements of the 19th century that sought social reform and shared prosperity. Such effete sentimentality led only to decadence and degeneration. Those who have seen the truth should "philosophize with a hammer" and give modern civilization the final shove that would bring on the redemptive cataclysm from which a new order would rise. Lest you think I am setting up a straw Übermensch, here are some quotations:

> I abhor the man's vulgarity when he says "What is right for one man is right for another"; "Do not to others that which you would not that they should do unto you.". . . . The hypothesis here is ignoble to the last degree: it is taken for granted that there is some sort of equivalence in value between my actions and thine.

> I do not point to the evil and pain of existence with the finger of reproach, but rather entertain the hope that life may one day become more evil and more full of suffering than it has ever been.

> Man shall be trained for war and woman for the recreation of the warrior. All else is folly. . . . Thou goest to woman? Do not forget thy whip.

> A declaration of war on the masses by *higher men* is needed. . . . A doctrine is needed powerful enough to work as a breeding agent: strengthening the strong, paralyzing and destructive for the world-weary. The annihilation of the humbug called "morality." . . . The annihilation of

the decaying races. . . . Dominion over the earth as a means of producing a higher type.

That higher Party of Life which would take the greatest of all tasks into its hands, the higher breeding of humanity, *including the merciless extermination of everything degenerate and parasitical,* would make possible again that excess of life on earth from which the Dionysian state will grow again.[110]

These genocidal ravings may sound like they come from a transgressive adolescent who has been listening to too much death metal, or a broad parody of a James Bond villain like Dr. Evil in *Austin Powers.* In fact Nietzsche is among the most influential thinkers of the 20th century, continuing into the 21st.

Most obviously, Nietzsche helped inspire the romantic militarism that led to the First World War and the fascism that led to the Second. Though Nietzsche himself was neither a German nationalist nor an anti-Semite, it's no coincidence that these quotations leap off the page as quintessential Nazism: Nietzsche posthumously became the Nazis' court philosopher. (In his first year as chancellor, Hitler made a pilgrimage to the Nietzsche Archive, presided over by Elisabeth Förster-Nietzsche, the philosopher's sister and literary executor, who tirelessly encouraged the connection.) The link to Italian Fascism is even more direct: Benito Mussolini wrote in 1921 that "the moment relativism linked up with Nietzsche, and with his Will to Power, was when Italian Fascism became, as it still is, the most magnificent creation of an individual and a national Will to Power."[111] The links to Bolshevism and Stalinism—from the Superman to the New Soviet Man—are less well known but amply documented by the historian Bernice Glatzer Rosenthal.[112] The connections between Nietzsche's ideas and the megadeath movements of the 20th century are obvious enough: a glorification of violence and power, an eagerness to raze the institutions of liberal democracy, a contempt for most of humanity, and a stone-hearted indifference to human life.

You'd think this sea of blood would be enough to discredit Nietzsche's ideas among intellectuals and artists. But he is, incredibly, widely admired. "Nietzsche is pietzsche," says a popular campus graffito and T-shirt. It's not because the man's doctrines are particularly cogent. As Bertrand Russell pointed out in *A History of Western Philosophy,* they "might be stated more simply and honestly in the one sentence: 'I wish I had lived in the Athens of Pericles or the Florence of the Medici.'" The ideas fail the first test of moral coherence, namely generalizability beyond the person offering them. If I could go back in time, I might confront him as follows: "I am a superman: hard, cold, terrible, without feelings and without conscience. As you recommend, I will achieve heroic glory by exterminating some chattering dwarves. Starting with *you*, Shorty. And I might do a few things to that Nazi sister of yours, too. Unless, that is, you can think of a *reason* why I should not."

So if Nietzsche's ideas are repellent and incoherent, why do they have so many fans? Perhaps it is not surprising that an ethic in which the artist (together with the warrior) is uniquely worthy of living should appeal to so many artists. A sample: W. H. Auden, Albert Camus, André Gide, D. H. Lawrence, Jack London, Thomas Mann, Yukio Mishima, Eugene O'Neill, William Butler Yeats, Wyndham Lewis, and (with reservations) George Bernard Shaw, author of *Man and Superman*. (P. G. Wodehouse, in contrast, has Jeeves, a Spinoza fan, say to Bertie Wooster, "You would not enjoy Nietzsche, sir. He is fundamentally unsound.") Nietzschean values also appeal to many Second Culture literary intellectuals (recall Leavis sneering at Snow's concern with global poverty and disease because "great literature" is "what men live by") and to social critics who like to snigger at the "booboisie" (as H. L. Mencken, "the American Nietzsche," called the common folk). Though she later tried to conceal it, Ayn Rand's celebration of selfishness, her deification of the heroic capitalist, and her disdain for the general welfare had Nietzsche written all over them.[113]

As Mussolini made clear, Nietzsche was an inspiration to relativists everywhere. Disdaining the commitment to truth-seeking among scientists and Enlightenment thinkers, Nietzsche asserted that "there are no facts, only interpretations," and that "truth is a kind of error without which a certain species of life could not live."[114] (Of course, this left him unable to explain why we should believe that *those* statements are true.) For that and other reasons, he was a key influence on Martin Heidegger, Jean-Paul Sartre, Jacques Derrida, and Michel Foucault, and a godfather to all the intellectual movements of the 20th century that were hostile to science and objectivity, including Existentialism, Critical Theory, Poststructuralism, Deconstructionism, and Postmodernism.

Nietzsche, to give him credit, was a lively stylist, and one might excuse the fandom of artists and intellectuals if it consisted of an appreciation of his literary panache and an ironic reading of his portrayal of a mindset that they themselves rejected. Unfortunately, the mindset has sat all too well with all too many of them. A surprising number of 20th-century intellectuals and artists have gushed over totalitarian dictators, a syndrome that the intellectual historian Mark Lilla calls tyranno-philia.[115] Some tyrannophiles were Marxists, working on the time-honored principle "He may be an SOB, but he's *our* SOB." But many were Nietzschean. The most notorious were Martin Heidegger and the legal philosopher Carl Schmitt, who were gung-ho Nazis and Hitler acolytes. Indeed, no autocrat of the 20th century lacked champions among the clerisy, including Mussolini (Ezra Pound, Shaw, Yeats, Lewis), Lenin (Shaw, H. G. Wells), Stalin (Shaw, Sartre, Beatrice and Sidney Webb, Brecht, W. E. B. Du Bois, Pablo Picasso, Lillian Hellman), Mao (Sartre, Foucault, Du Bois, Louis Althusser, Steven Rose, Richard Lewontin), the Ayatollah Khomeini (Foucault), and Castro (Sartre, Graham Greene, Günter Grass, Norman Mailer, Harold Pinter, and, as we saw in chapter 21, Susan Sontag). At various times Western intellectuals have also sung the praises of Ho Chi Minh, Muammar Gaddafi, Saddam Hussein, Kim Il-sung, Pol Pot, Julius Nyerere, Slobodan Milošević, and Hugo Chávez.

Why should intellectuals and artists, of all people, kiss up to murder-ous dictators? One might think that intellectuals would be the first to deconstruct the pretexts of power, and artists to expand the scope of human compassion. (Thankfully, many have done just that.) One expla-nation, offered by the economist Thomas Sowell and the sociologist Paul Hollander, is professional narcissism. Intellectuals and artists may feel unappreciated in liberal democracies, which allow their citizens to tend to their own needs in markets and civic organizations. Dictators imple-ment theories from the top down, assigning a role to intellectuals that they feel is commensurate with their worth. But tyrannophilia is also fed by a Nietzschean disdain for the common man, who annoyingly prefers schlock to fine art and culture, and by an admiration of the superman who transcends the messy compromises of democracy and heroically implements a vision of the good society.

~

Though Nietzsche's romantic heroism glorifies the singular Übermensch rather than any collectivity, it's a short step to interpret his "single stron-ger species of man" as a tribe, race, or nation. With this substitution, Nietzschean ideas were taken up by Nazism, fascism, and other forms of Romantic nationalism, and they star in a political drama that contin-ues to the present day.

I used to think that Trumpism was pure id, an upwelling of tribalism and authoritarianism from the dark recesses of the psyche. But madmen in authority distill their frenzy from academic scribblers of a few years back, and the phrase "intellectual roots of Trumpism" is not oxymoronic. Trump was endorsed in the 2016 election by 136 "Scholars and Writers for America" in a manifesto called "Statement of Unity."[116] Some are connected to the Claremont Institute, a think tank that has been called "the academic home of Trumpism."[117] And Trump has been closely advised by two men, Stephen Bannon and Michael Anton, who are reputed to be widely read and who consider themselves serious intellectuals. Anyone who wants to go beyond personality in understanding authoritarian populism must appreciate the two ideologies behind them, both of them militantly opposed to Enlightenment humanism and each influenced, in different ways, by Nietzsche. One is fascist, the other reactionary—not in the common left-wing sense of "anyone who is more conservative than me," but in their original, technical senses.[118]

Fascism, from the Italian word for "group" or "bundle," grew out of the Romantic notion that the individual is a myth and that people are inextricable from their culture, bloodline, and homeland.[119] The early fascist intellectuals, including Julius Evola (1898–1974) and Charles Maurras (1868–1952), have been rediscovered by neo-Nazi parties in Europe and by Bannon and the alt-right movement in the United States, all of whom acknowledge the influence of Nietzsche.[120] Today's Fascism Lite, which shades into authoritarian populism and Romantic nationalism, is sometimes justified by a crude version of evolutionary psychology in which the unit of selection is the group, evolution is driven by the survival of the fittest group in competition with other groups, and humans have been selected to sacrifice their interests for the supremacy of their group. (This contrasts with mainstream evolutionary psychology, in which the unit of selection is the gene.)[121] It follows that no one can be a cosmopolitan, a citizen of the world: to be human is to be a part of a nation. A multicultural, multiethnic society can never work, because its people will feel rootless and alienated and its culture will be flattened to the lowest common denominator. For a nation to subordinate its interests to international agreements is to forfeit its birthright to greatness and become a chump in the global competition of all against all. And since a nation is an organic whole, its greatness can be embodied in the greatness of its leader, who voices the soul of the people directly, unencumbered by the millstone of an administrative state.

The reactionary ideology is theoconservatism.[122] Belying the flippant label (coined by the apostate Damon Linker as a play on "neoconserva-

tism"), the first theocons were 1960s radicals who redirected their revolutionary fervor from the hard left to the hard right. They advocate
nothing less than a rethinking of the Enlightenment roots of the American political order. The recognition of a right to life, liberty, and the
pursuit of happiness, and the mandate of government to secure these
rights, are, they believe, too tepid for a morally viable society. That impoverished vision has only led to anomie, hedonism, and rampant im-
morality, including illegitimacy, pornography, failing schools, welfare
dependency, and abortion. Society should aim higher than this stunted
individualism, and promote conformity to more rigorous moral standards from an authority larger than ourselves. The obvious source of
these standards is traditional Christianity.

Theocons hold that the erosion of the church's authority during the
Enlightenment left Western civilization without a solid moral foundation, and a further undermining during the 1960s left it teetering on the
brink. Any day during the Bill Clinton administration it would plunge
into the abyss; no, make that the Obama administration; no, but for sure
it would happen during a Hillary Clinton administration. (Hence Anton's hysterical essay "The Flight 93 Election," mentioned in chapter 20,
which compared the country to the airliner hijacked on 9/11 and called
on voters to "charge the cockpit or you die.")[123] Whatever discomfort the
theocons may have felt from the vulgarity and antidemocratic antics of
their 2016 standard-bearer was outweighed by the hope that he alone
could impose the radical changes that America needed to stave off catastrophe.

Lilla points out an irony in theoconservativism. While it has been
inflamed by radical Islamism (which the theocons think will soon start
World War III), the movements are similar in their reactionary mindset,
with its horror of modernity and progress.[124] Both believe that at some
time in the past there was a happy, well-ordered state where a virtuous
people knew their place. Then alien secular forces subverted this harmony and brought on decadence and degeneration. Only a heroic vanguard with memories of the old ways can restore the society to its
golden age.

∼

Lest you have lost the trail that connects this intellectual history to current events, bear in mind that in 2017 Trump decided to withdraw the
United States from the Paris climate accord under pressure from Bannon, who convinced him that cooperating with other nations is a sign of
surrender in the global contest for greatness.[125] (Trump's hostility to im-

migration and trade grew from the same roots.) With the stakes this high, it's good to remind ourselves why the case for neo-theo-reactionary-populist nationalism is intellectually bankrupt. I have already discussed the absurdity of seeking a foundation for morality in the institutions that brought us the Crusades, the Inquisition, the witch hunts, and the European wars of religion. The idea that the global order should consist of ethnically homogeneous and mutually antagonistic nation-states is just as ludicrous.

First, the claim that humans have an innate imperative to identify with a nation-state (with the implication that cosmopolitanism goes against human nature) is bad evolutionary psychology. Like the supposed innate imperative to belong to a religion, it confuses a vulnerability with a need. People undoubtedly feel solidarity with their tribe, but whatever intuition of "tribe" we are born with cannot be a nation-state, which is a historical artifact of the 1648 Treaties of Westphalia. (Nor could it be a race, since our evolutionary ancestors seldom met a person of another race.) In reality, the cognitive category of a tribe, in-group, or coalition is abstract and multidimensional.[126] People see themselves as belonging to many overlapping tribes: their clan, hometown, native country, adopted country, religion, ethnic group, alma mater, fraternity or sorority, political party, employer, service organization, sports team, even brand of camera equipment. (If you want to see tribalism at its fiercest, check out a "Nikon vs. Canon" Internet discussion group.)

It's true that political salesmen can market a mythology and iconography that entice people into privileging a religion, ethnicity, or nation as their fundamental identity. With the right package of indoctrination and coercion, they can even turn them into cannon fodder.[127] That does not mean that nationalism is a human drive. Nothing in human nature prevents a person from being a proud Frenchman, European, and citizen of the world, all at the same time.[128]

The claim that ethnic uniformity leads to cultural excellence is as wrong as an idea can be. There's a reason we refer to unsophisticated things as *provincial, parochial,* and *insular* and to sophisticated ones as *urbane* and *cosmopolitan.* No one is brilliant enough to dream up anything of value all by himself. Individuals and cultures of genius are aggregators, appropriators, greatest-hit collectors. Vibrant cultures sit in vast catchment areas in which people and innovations flow from far and wide. This explains why Eurasia, rather than Australia, Africa, or the Americas, was the first continent to give birth to expansive civilizations (as documented by Sowell in his *Culture* trilogy and Jared Diamond in *Guns,*

Germs, and Steel).[129] It explains why the fountains of culture have always been trading cities on major crossroads and waterways.[130] And it explains why human beings have always been peripatetic, moving to wherever they can make the best lives. Roots are for trees; people have feet.

Finally, let's not forget why international institutions and global consciousness arose in the first place. Between 1803 and 1945, the world tried an international order based on nation-states heroically struggling for greatness. It didn't turn out so well. It's particularly wrongheaded for the reactionary right to use frantic warnings about an Islamist "war" against the West (with a death toll in the hundreds) as a reason to return to an international order in which the West repeatedly fought wars against itself (with death tolls in the tens of millions). After 1945 the world's leaders said, "Well, let's not do *that* again," and began to downplay nationalism in favor of universal human rights, international laws, and transnational organizations. The result, as we saw in chapter 11, has been seventy years of peace and prosperity in Europe and, increasingly, the rest of the world.

As for the lamentation among editorialists that the Enlightenment is a "brief interlude," that epitaph is likelier to mark the resting place of neo-fascism, neo-reaction, and related backlashes of the early 21st century. The European elections and self-destructive flailing of the Trump administration in 2017 suggest that the world may have reached Peak Populism, and as we saw in chapter 20, the movement is on a demographic road to nowhere. Headlines notwithstanding, the numbers show that democracy (chapter 14) and liberal values (chapter 15) are riding a long-term escalator that is unlikely to go into reverse overnight. The advantages of cosmopolitanism and international cooperation cannot be denied for long in a world in which the flow of people and ideas is unstoppable.

～

Though the moral and intellectual case for humanism is, I believe, overwhelming, some might wonder whether it is any match for religion, nationalism, and romantic heroism in the campaign for people's hearts. Will the Enlightenment ultimately fail because it cannot speak to primal human needs? Should humanists hold revival meetings at which preachers thump Spinoza's *Ethics* on the pulpit and ecstatic congregants roll back their eyes and babble in Esperanto? Should they stage rallies in which young men in colored shirts salute giant posters of John Stuart Mill? I think not; recall that a vulnerability is not the same as a need. The citizens of Denmark, New Zealand, and other happy parts of the world get by perfectly well without these paroxysms. The bounty of a cosmopolitan secular democracy is there for everyone to see.

Still, the appeal of regressive ideas is perennial, and the case for reason, science, humanism, and progress always has to be made. When we fail to acknowledge our hard-won progress, we may come to believe that perfect order and universal prosperity are the natural state of affairs, and that every problem is an outrage that calls for blaming evildoers, wrecking institutions, and empowering a leader who will restore the country to its rightful greatness. I have made my own best case for progress and the ideals that made it possible, and have dropped hints on how journalists, intellectuals, and other thoughtful people (including the readers of this book) might avoid contributing to the widespread heedlessness of the gifts of the Enlightenment.

Remember your math: an anecdote is not a trend. Remember your history: the fact that something is bad today doesn't mean it was better in the past. Remember your philosophy: one cannot reason that there's no such thing as reason, or that something is true or good because God said it is. And remember your psychology: much of what we know isn't so, especially when our comrades know it too.

Keep some perspective. Not every problem is a Crisis, Plague, Epidemic, or Existential Threat, and not every change is the End of This, the Death of That, or the Dawn of a Post-Something Era. Don't confuse pessimism with profundity: problems are inevitable, but problems are solvable, and diagnosing every setback as a symptom of a sick society is a cheap grab for gravitas. Finally, drop the Nietzsche. His ideas may seem edgy, authentic, baaad, while humanism seems sappy, unhip, uncool. But what's so funny about peace, love, and understanding?

The case for Enlightenment Now is not just a matter of debunking fallacies or disseminating data. It may be cast as a stirring narrative, and I hope that people with more artistic flair and rhetorical power than I can tell it better and spread it farther. The story of human progress is *truly* heroic. It is glorious. It is uplifting. It is even, I daresay, spiritual. It goes something like this.

We are born into a pitiless universe, facing steep odds against life-enabling order and in constant jeopardy of falling apart. We were shaped by a force that is ruthlessly competitive. We are made from crooked timber, vulnerable to illusions, self-centeredness, and at times astounding stupidity.

Yet human nature has also been blessed with resources that open a space for a kind of redemption. We are endowed with the power to combine ideas recursively, to have thoughts about our thoughts. We have an instinct for language, allowing us to share the fruits of our experience and ingenuity. We are deepened with the capacity for sympathy—for pity, imagination, compassion, commiseration.

These endowments have found ways to magnify their own power. The scope of language has been augmented by the written, printed, and electronic word. Our circle of sympathy has been expanded by history, journalism, and the narrative arts. And our puny rational faculties have been multiplied by the norms and institutions of reason: intellectual curiosity, open debate, skepticism of authority and dogma, and the burden of proof to verify ideas by confronting them against reality.

As the spiral of recursive improvement gathers momentum, we eke out victories against the forces that grind us down, not least the darker parts of our own nature. We penetrate the mysteries of the cosmos, including life and mind. We live longer, suffer less, learn more, get smarter, and enjoy more small pleasures and rich experiences. Fewer of us are killed, assaulted, enslaved, oppressed, or exploited by the others. From a few oases, the territories with peace and prosperity are growing, and could someday encompass the globe. Much suffering remains, and tremendous peril. But ideas on how to reduce them have been voiced, and an infinite number of others are yet to be conceived.

We will never have a perfect world, and it would be dangerous to seek one. But there is no limit to the betterments we can attain if we continue to apply knowledge to enhance human flourishing.

This heroic story is not just another myth. Myths are fictions, but this one is true—true to the best of our knowledge, which is the only truth we can have. We believe it because we have *reasons* to believe it. As we learn more, we can show which parts of the story continue to be true, and which ones false—as any of them might be, and any could become.

And the story belongs not to any tribe but to all of humanity—to any sentient creature with the power of reason and the urge to persist in its being. For it requires only the convictions that life is better than death, health is better than sickness, abundance is better than want, freedom is better than coercion, happiness is better than suffering, and knowledge is better than superstition and ignorance.

NOTES

PREFACE

1. "Mothers and children" from Donald Trump's inaugural speech, Jan. 20, 2017, https://www
   .whitehouse.gov/inaugural-address. "Outright war" and "spiritual and moral foundations"
   from Trump chief strategist Stephen Bannon's remarks to a Vatican conference in the summer
   of 2014, transcribed in J. L. Feder, "This Is How Steve Bannon Sees the Entire World," *BuzzFeed*,
   Nov. 16, 2016, https://www.buzzfeed.com/lesterfeder/this-is-how-steve-bannon-sees-the-entire
   -world. "Global power structure" from "Donald Trump's Argument for America," final television campaign ad, Nov. 2016, http://blog.4president.org/2016/2016-tv-ad/. Bannon is commonly
   credited with authoring or coauthoring all three.
2. CUDOS: Merton 1942/1973 called his first virtue "communism," though it is often quoted as
   "communalism" to distinguish it from Marxism.

PART I: ENLIGHTENMENT

1. S. Maher, "Inside the Mind of an Extremist," presentation at the Oslo Freedom Forum, May 26,
   2015, https://oslofreedomforum.com/talks/inside-the-mind-of-an-extremist.
2. From Hayek 1960/2011, p. 47; see also Wilkinson 2016a.

CHAPTER 1: DARE TO UNDERSTAND!

1. *What Is Enlightenment?* Kant 1784/1991.
2. The quotations are blended and condensed from translations by H. B. Nisbet, Kant 1784/1991,
   and by Mary C. Smith, http://www.columbia.edu/acis/ets/CCREAD/etscc/kant.html.
3. *The Beginning of Infinity:* Deutsch 2011, pp. 221–22.
4. The Enlightenment: Goldstein 2006; Gottlieb 2016; Grayling 2007; Hunt 2007; Israel 2001; Makari
   2015; Montgomery & Chirot 2015; Pagden 2013; Porter 2000.
5. The nonnegotiability of reason: Nagel 1997; see also chapter 21.
6. Most Enlightenment thinkers were non-theists: Pagden 2013, p. 98.
7. Wootton 2015, pp. 6–7.
8. Scott 2010, pp. 20–21.
9. Enlightenment thinkers as scientists of human nature: Kitcher 1990; Macnamara 1999; Makari
   2015; Montgomery & Chirot 2015; Pagden 2013; Stevenson & Haberman 1998.
10. Expanding circle of sympathy: Nagel 1970; Pinker 2011; Shermer 2015; Singer 1981/2011.
11. Cosmopolitanism: Appiah 2006; Pagden 2013; Pinker 2011.
12. Humanitarian Revolution: Hunt 2007; Pinker 2011.
13. Progress as a mystical force: Berlin 1979; Nisbet 1980/2009.
14. Authoritarian High Modernism: Scott 1998.
15. Authoritarian High Modernism and blank-slate psychology: Pinker 2002/2016, pp. 170–71, 409–11.
16. Quotes from Le Corbusier, from Scott 1998, pp. 114–15.
17. Rethinking punishment: Hunt 2007.
18. Wealth creation: Montgomery & Chirot 2015; Ridley 2010; Smith 1776/2009.
19. Gentle commerce: Mueller 1999, 2010b; Pagden 2013; Pinker 2011; Schneider & Gleditsch 2010.
20. Perpetual Peace: Kant 1795/1983. Modern interpretation: Russett & Oneal 2001.

CHAPTER 2: ENTRO, EVO, INFO

1. Second Law of Thermodynamics: Atkins 2007; Carroll 2016; Hidalgo 2015; Lane 2015.
2. Eddington 1928/2015.
3. The two cultures and the Second Law: Snow 1959/1998, pp. 14–15.
4. Second Law of Thermo = First law of psycho: Tooby, Cosmides, & Barrett 2003.
5. Self-organization: England 2015; Gell-Mann 1994; Hidalgo 2015; Lane 2015.
6. Evolution versus entropy: Dawkins 1983, 1986; Lane 2015; Tooby, Cosmides, & Barrett 2003.
7. Spinoza: Goldstein 2006.
8. Information: Adriaans 2013; Dretske 1981; Gleick 2011; Hidalgo 2015.

9. Information is a decrease in entropy, not entropy itself: https://schneider.ncifcrf.gov /information.is.not.uncertainty.html.
10. Transmitted information as knowledge: Adriaans 2013; Dretske 1981; Fodor 1987, 1994.
11. "The universe is made of matter, energy, and information": Hidalgo 2015, p. ix; see also Lloyd 2006.
12. Neural computation: Anderson 2007; Pinker 1997/2009, chap. 2.
13. Knowledge, information, and inferential roles: Block 1986; Fodor 1987, 1994.
14. The cognitive niche: Marlowe 2010; Pinker 1997/2009; Tooby & DeVore 1987; Wrangham 2009.
15. Language: Pinker 1994/2007.
16. Hadza menu: Marlowe 2010.
17. Axial Age: Goldstein 2013.
18. Explaining the Axial Age: Baumard et al. 2015.
19. From *The Threepenny Opera*, act II, scene 1.
20. Clockwork universe: Carroll 2016; Wootton 2015.
21. Innate illiteracy and innumeracy: Carey 2009; Wolf 2007.
22. Magical thinking, essences, word magic: Oesterdiekhoff 2015; Pinker 1997/2009, chaps. 5 and 6; Pinker 2007a, chap. 7.
23. Bugs in statistical reasoning: Ariely 2010; Gigerenzer 2015; Kahneman 2011; Pinker 1997/2009, chap. 5; Sutherland 1992.
24. Intuitive lawyers and politicians: Kahan, Jenkins-Smith, & Braman 2011; Kahan, Peters, et al. 2013; Kahan, Wittlin, et al. 2011; Mercier & Sperber 2011; Tetlock 2002.
25. Overconfidence: Johnson 2004. Overconfidence in understanding: Sloman & Fernbach 2017.
26. Bugs in the moral sense: Greene 2013; Haidt 2012; Pinker 2008a.
27. Morality as a condemnation device: DeScioli & Kurzban 2009; DeScioli 2016.
28. Virtuous violence: Fiske & Rai 2015; Pinker 2011, chaps. 8 and 9.
29. Transcending cognitive limitations through abstraction and combination: Pinker 2007a, 2010.
30. Letter to Isaac McPherson, *Writings* 13:333–35, quoted in Ridley 2010, p. 247.
31. Collective rationality: Haidt 2012; Mercier & Sperber 2011.
32. Cooperation and the interchangeability of perspectives: Nagel 1970; Pinker 2011; Singer 1981/2011.

CHAPTER 3: COUNTER-ENLIGHTENMENTS

1. Declining trust in institutions: Twenge, Campbell, & Carter 2014. Mueller 1999, pp. 167–68, points out that the 1960s were a high-water mark for trust in institutions, unsurpassed before or after. Declining trust in science among conservatives: Gauchat 2012. Populism: Inglehart & Norris 2016; J. Müller 2016; Norris & Inglehart 2016; see also chapters 20 and 23.
2. Non-Western enlightenments: Conrad 2012; Kurlansky 2006; Pelham 2016; Sen 2005; Sikkink 2017.
3. Counter-Enlightenments: Berlin 1979; Garrard 2006; Herman 1997; Howard 2001; McMahon 2001; Sternhell 2010; Wolin 2004; see also chapter 23.
4. Inscription in John Singer Sargent's 1922 painting *Death and Victory*, Widener Library, Harvard University.
5. Irreligious defenders of religion: Coyne 2015; see also chapter 23.
6. Ecomodernism: Asafu-Adjaye et al. 2015; Ausubel 1996, 2015; Brand 2009; DeFries 2014; Nordhaus & Shellenberger 2007; see also chapter 10.
7. Problems with ideology: Duarte et al. 2015; Haidt 2012; Kahan, Jenkins-Smith, & Braman 2011; Mercier & Sperber 2011; Tetlock & Gardner 2015; and see much more in chapter 21.
8. An adaptation of a quotation by Michael Lind on the back cover of Herman 1997. See also Nisbet 1980/2009.
9. Eco-pessimism: Bailey 2015; Brand 2009; Herman 1997; Ridley 2010; see also chapter 10.
10. A pastiche by the literary historian Hoxie Neale Fairchild of phrases from T. S. Eliot, William Burroughs, and Samuel Beckett, from *Religious Trends in English Poetry*, quoted in Nisbet 1980/2009, p. 328.
11. Heroic blood-bespatterers: Nietzsche 1887/2014.
12. Snow never assigned an order to his Two Cultures, but subsequent usage has numbered them in that way; see, for example, Brockman 2003.
13. Snow 1959/1998, p. 14.
14. Leavis flame: Leavis 1962/2013; see Collini 1998, 2013.
15. Leavis 1962/2013, p. 71.

CHAPTER 4: PROGRESSOPHOBIA

1. Herman 1997, p. 7, also cites Joseph Campbell, Noam Chomsky, Joan Didion, E. L. Doctorow, Paul Goodman, Michael Harrington, Robert Heilbroner, Jonathan Kozol, Christopher Lasch, Norman Mailer, Thomas Pynchon, Kirkpatrick Sale, Jonathan Schell, Richard Sennett, Susan Sontag, Gore Vidal, and Garry Wills.
2. Nisbet 1980/2009, p. 317.
3. The Optimism Gap: McNaughton-Cassill & Smith 2002; Nagdy & Roser 2016b; Veenhoven 2010; Whitman 1998.

4. EU Eurobarometer survey results, reproduced in Nagdy & Roser 2016b.
5. Survey results from Ipsos, "Perils of Perception (Topline Results)," 2013, https://www.ipsos .com/sites/default/files/migrations/en-uk/files/Assets/Docs/Polls/ipsos-mori-rss-kings-perils -of-perception-topline.pdf, graphed in Nagdy & Roser 2016b.
6. Dunlap, Gallup, & Gallup 1993, graphed in Nagdy & Roser 2016b.
7. J. McCarthy, "More Americans Say Crime Is Rising in U.S.," *Gallup.com*, Oct. 22, 2015, http:// www.gallup.com/poll/186308/americans-say-crime-rising.aspx.
8. World is getting worse: Majorities in Australia, Denmark, Finland, France, Germany, Great Brit- ain, Hong Kong, Norway, Singapore, Sweden, and the United States; also Malaysia, Thailand, and the United Arab Emirates. China was the only country in which more respondents said the world was getting better than said it was getting worse. YouGov poll, Jan. 5, 2016, https://yougov.co.uk /news/2016/01/05/chinese-people-are-most-optimistic-world/. The United States on the wrong track: Dean Obeidallah, "We've Been on the Wrong Track Since 1972," *Daily Beast*, Nov. 7, 2014, http://www.pollingreport.com/right.htm.
9. Source of the expression: B. Popik, "First Draft of History (Journalism)," *BarryPopik.com*, http:// www.barrypopik.com/index.php/new_york_city/entry/first_draft_of_history_journalism/.
10. Frequency and nature of news: Galtung & Ruge 1965.
11. Availability heuristic: Kahneman 2011; Slovic 1987; Slovic, Fischhoff, & Lichtenstein 1982; Tver- sky & Kahneman 1973.
12. Misperceptions of risk: Ropeik & Gray 2002; Slovic 1987. Post-*Jaws* avoidance of swimming: Sutherland 1992, p. 11.
13. If it bleeds, it leads (and vice versa): Bohle 1986; Combs & Slovic 1979; Galtung & Ruge 1965; Miller & Albert 2015.
14. ISIS as "existential threat": Poll conducted for *Investor's Business Daily* by TIPP, March 28–April 2, 2016, http://www.investors.com/politics/ibdtipp-poll-distrust-on-what-obama-does-and-says -on-isis-terror/.
15. Effects of newsreading: Jackson 2016. See also Johnston & Davey 1997; McNaughton-Cassill 2001; Otieno, Spada, & Renkl 2013; Ridout, Grosse, & Appleton 2008; Unz, Schwab, & Winterhoff- Spurk 2008.
16. Quoted in J. Singal, "What All This Bad News Is Doing to Us," *New York*, Aug. 8, 2014.
17. Decline of violence: Eisner 2003; Goldstein 2011; Gurr 1981; Human Security Centre 2005; Hu- man Security Report Project 2009; Mueller 1989, 2004a; Payne 2004.
18. Solutions create new problems: Deutsch 2011, pp. 64, 76, 350; Berlin 1988/2013, p. 15.
19. Deutsch 2011, p. 193.
20. Thick-tailed distributions: See chapter 19, and, for more detail, Pinker 2011, pp. 210–22.
21. Negativity bias: Baumeister, Bratslavsky, et al. 2001; Rozin & Royzman 2001.
22. Personal communication, 1982.
23. More negative words: Baumeister, Bratslavsky, et al. 2001; Schrauf & Sanchez 2004.
24. Rose-tinting of memory: Baumeister, Bratslavsky, et al. 2001.
25. Illusion of the good old days: Eibach & Libby 2009.
26. Connor 2014; see also Connor 2016.
27. Snarky book reviewers sound smarter: Amabile 1983.
28. M. Housel, "Why Does Pessimism Sound So Smart?" *Motley Fool*, Jan. 21, 2016.
29. Similar points have been made by the economist Albert Hirschman (1991) and the journalist Gregg Easterbrook (2003).
30. D. Bornstein & T. Rosenberg, "When Reportage Turns to Cynicism," *New York Times*, Nov. 14, 2016. For more on the "constructive journalism" movement, see Gyldensted 2015, Jackson 2016, and the magazine *Positive News* (www.positive.news).
31. The UN Millennium Development Goals are: 1. To eradicate extreme poverty and hunger. 2. To achieve universal primary education. 3. To promote gender equality and empower women. 4. To reduce child mortality. 5. To improve maternal health. 6. To combat HIV/AIDS, malaria, and other diseases. 7. To ensure environmental sustainability. 8. To develop a global partnership for [economic] development.
32. Books on progress (in order of mention): Norberg 2016, Easterbrook 2003, Reese 2013, Naam 2013, Ridley 2010, Robinson 2009, Bregman 2016, Phelps 2013, Diamandis & Kotler 2012, Goklany 2007, Kenny 2011, Bailey 2015, Shermer 2015, DeFries 2014, Deaton 2013, Radelet 2015, Mahbubani 2013.

CHAPTER 5: LIFE

1. World Health Organization 2016a.
2. Hans and Ola Rosling, "The Ignorance Project," https://www.gapminder.org/ignorance/.
3. Roser 2016n; estimate for England in 1543 from R. Zijdeman, OECD Clio Infra.
4. Hunter-gatherers: Marlowe 2010, p. 160. The estimate is for the Hadza, whose rates of infant and juvenile mortality (which account for most of the variance among populations) are identical to the medians in Marlowe's sample of 478 foraging peoples (p. 261). First farmers to Iron Age: Galor & Moav 2007. No increase for millennia: Deaton 2013, p. 80.
5. Norberg 2016, pp. 46 and 40.
6. Influenza pandemic: Roser 2016n. American white mortality: Case & Deaton 2015.

7. Marlowe 2010, p. 261.
8. Deaton 2013, p. 56.
9. Reducing health care: N. Kristof, "Birth Control for Others," *New York Times*, March 23, 2008.
10. M. Housel, "50 Reasons We're Living Through the Greatest Period in World History," *Motley Fool*, Jan. 29, 2014.
11. World Health Organization 2015c.
12. Marlowe 2010, p. 160.
13. Radelet 2015, p. 75.
14. Global healthy life expectancy in 1990: Mathers et al. 2001. Healthy life expectancy in developed countries in 2010: Murray et al. 2012; see also Chernew et al. 2016, for data showing that *healthy* life expectancy, not just life expectancy, has recently increased in the United States.
15. G. Kolata, "U.S. Dementia Rates Are Dropping Even as Population Ages," *New York Times*, Nov. 21, 2016.
16. Bush's Council on Bioethics: Pinker 2008b.
17. L. R. Kass, "L'Chaim and Its Limits: Why Not Immortality?" *First Things*, May 2001.
18. Longevity estimates regularly superseded: Oeppen & Vaupel 2002.
19. Reverse-engineering mortality: M. Shermer, "Radical Life-Extension Is Not Around the Corner," *Scientific American*, Oct. 1, 2016; Shermer 2018.
20. Siegel, Naishadham, & Jemal 2012.
21. Skepticism about immortality: Hayflick 2000; Shermer 2018.
22. Entropy will kill us: P. Hoffmann, "Physics Makes Aging Inevitable, Not Biology," *Nautilus*, May 12, 2016.

CHAPTER 6: HEALTH
1. Deaton 2013, p. 149.
2. Bettmann 1974, p. 136; internal quotation marks omitted.
3. Bettmann 1974; Norberg 2016.
4. Carter 1966, p. 3.
5. Woodward, Shurkin, & Gordon 2009; see also the Web site *ScienceHeroes* (www.scienceheroes.com). The team's statisticians are April Ingram and Amy R. Pearce.
6. Book on the past tense: Pinker 1999/2011.
7. Kenny 2011, pp. 124–25.
8. D. G. McNeil Jr., "A Milestone in Africa: No Polio Cases in a Year," *New York Times*, Aug. 11, 2015; "Polio This Week," *Global Polio Eradication Initiative*, http://polioeradication.org/polio-today/polio-now/this-week/, May 17, 2017.
9. "Guinea Worm Case Totals," *The Carter Center*, April 18, 2017, https://www.cartercenter.org/health/guinea_worm/case-totals.html.
10. Bill & Melinda Gates Foundation, *Our Big Bet for the Future: 2015 Gates Annual Letter*, p. 7, https://www.gatesnotes.com/2015-Annual-Letter.
11. World Health Organization 2015b.
12. Bill & Melinda Gates Foundation, "Malaria: Strategy Overview," http://www.gatesfoundation.org/What-We-Do/Global-Health/Malaria.
13. Data from the World Health Organization and the Child Health Epidemiology Reference Group, cited in Bill & Melinda Gates Foundation, *Our Big Bet for the Future: 2015 Gates Annual Letter*, p. 7, https://www.gatesnotes.com/2015-Annual-Letter; UNAIDS 2016.
14. N. Kristof, "Why 2017 May Be the Best Year Ever," *New York Times*, Jan. 21, 2017.
15. Jamison et al. 2015.
16. Deaton 2013, p. 41.
17. Deaton 2013, pp. 122–23.

CHAPTER 7: SUSTENANCE
1. Norberg 2016, pp. 7–8.
2. Braudel 2002.
3. Fogel 2004, quoted in Roser 2016d.
4. Braudel 2002, pp. 76–77, quoted in Norberg 2016.
5. "Dietary Guidelines for Americans 2015–2020, Estimated Calorie Needs per Day, by Age, Sex, and Physical Activity Level," http://health.gov/dietaryguidelines/2015/guidelines/appendix-2/.
6. Calorie figures from Roser 2016d; see also figure 7-1.
7. Food and Agriculture Organization of the United Nations, *The State of Food and Agriculture 1947*, cited in Norberg 2016.
8. A definition by the economist Cormac Ó Gráda, cited in Hasell & Roser 2017.
9. Devereux 2000, p. 3.
10. W. Greene, "Triage: Who Shall Be Fed? Who Shall Starve?" *New York Times Magazine*, Jan. 5, 1975. The term *lifeboat ethics* had been introduced a year earlier by the ecologist Garrett Hardin in an article in *Psychology Today* (Sept. 1974) called "Lifeboat Ethics: The Case Against Helping the Poor."
11. "Service Groups in Dispute on World Food Problems," *New York Times*, July 15, 1976; G. Hardin, "Lifeboat Ethics," *Psychology Today*, Sept. 1974.

12. McNamara, health care, contraception: N. Kristof, "Birth Control for Others," *New York Times*, March 23, 2008.
13. Famines don't reduce population growth: Devereux 2000.
14. Quoted in "Making Data Dance," *The Economist*, Dec. 9, 2010.
15. The Industrial Revolution and the escape from hunger: Deaton 2013; Norberg 2016; Ridley 2010.
16. Agricultural revolutions: DeFries 2014.
17. Norberg 2016.
18. Woodward, Shurkin, & Gordon 2009; http://www.scienceheroes.com/. Haber retains this distinction even if we subtract the 90,000 deaths in World War I from chemical weapons, which he was instrumental in developing.
19. Morton 2015, p. 204.
20. Roser 2016e, 2016u.
21. Borlaug: Brand 2009; Norberg 2016; Ridley 2010; Woodward, Shurkin, & Gordon 2009; DeFries 2014.
22. The Green Revolution continues: Radelet 2015.
23. Roser 2016m.
24. Norberg 2016.
25. Norberg 2016. According to the UN FAO's *Global Forest Resources Assessment 2015*, "Net forest area has increased in over 60 countries and territories, most of which are in the temperate and boreal zones." http://www.fao.org/resources/infographics/infographics-details/en/c/325836/.
26. Norberg 2016.
27. Ausubel, Wernick, & Waggoner 2012.
28. Alferov, Altman, & 108 other Nobel Laureates 2016; Brand 2009; Radelet 2015; Ridley 2010, pp. 170–73; J. Achenbach, "107 Nobel Laureates Sign Letter Blasting Greenpeace over GMOs," *Washington Post*, June 30, 2016; W. Saletan, "Unhealthy Fixation," *Slate*, July 15, 2015.
29. W. Saletan, "Unhealthy Fixation," *Slate*, July 15, 2015.
30. Scientifically illiterate opinions on genetically modified foods: Sloman & Fernbach 2017.
31. Brand 2009, p. 117.
32. Sowell 2015.
33. Famines not just caused by food shortages: Devereux 2000; Sen 1984, 1999.
34. Devereux 2000. See also White 2011.
35. Devereux 2000 writes that during the colonial period, "macroeconomic and political vulnerability to famine gradually diminished" due to infrastructure improvements and "the initiation of early warning systems and relief intervention mechanisms by colonial administrations which recognized the need to ameliorate food crises to achieve political legitimacy" (p. 13).
36. Based on Devereux's estimate of seventy million deaths in major 20th-century famines (p. 29) and the estimates of particular famines in his table 1. See also Rummel 1994; White 2011.
37. Deaton 2013; Radelet 2015.

CHAPTER 8: WEALTH

1. Rosenberg & Birdzell 1986, p. 3.
2. Norberg 2016, summarizing Braudel 2002, pp. 75, 285, and elsewhere.
3. Cipolla 1994. Internal quotation marks have been omitted.
4. The physical fallacy: Sowell 1980.
5. The discovery of wealth creation: Montgomery & Chirot 2015; Ridley 2010.
6. Underestimating growth: Feldstein 2017.
7. Consumer surplus and Oscar Wilde: T. Kane, "Piketty's Crumbs," *Commentary*, April 14, 2016.
8. The term *Great Escape* is from Deaton 2013. Enlightened economy: Mokyr 2012.
9. Backyard tinkerers: Ridley 2010.
10. Science and technology as causes of the Great Escape: Mokyr 2012, 2014.
11. Natural states versus open economies: North, Wallis, & Weingast 2009. Related argument: Acemoglu & Robinson 2012.
12. Bourgeois virtue: McCloskey 1994, 1998.
13. From *Letters Concerning the English Nation*, cited in Porter 2000, p. 21.
14. Porter 2000, pp. 21–22.
15. Data on GDP per capita from Maddison Project 2014, displayed in Marian Tupy's *HumanProgress*, http://www.humanprogress.org/f1/2785/1/2010/France/United%20Kingdom.
16. The Great Convergence: Mahbubani 2013. Mahbubani credits the term to the columnist Martin Wolf. Radelet (2015) calls it the Great Surge; Deaton (2013) includes it in what he calls the Great Escape.
17. Countries with rapidly growing economies: Radelet 2015, pp. 47–51.
18. According to the UN's *Millennium Development Goals Report 2015*, "The number of people in the working middle class—living on more than $4 a day—has almost tripled between 1991 and 2015. This group now makes up half the workforce in the developing regions, up from just 18 per cent in 1991" (United Nations 2015a, p. 4). Of course most of the "working middle class" as defined by the UN would count as poor in developed countries, but even with a more generous definition the world has become more middle class than one might expect. The Brookings Institution

estimated in 2013 that it comprised 1.8 billion and would grow to 3.2 billion by 2020 (L. Yueh, "The Rise of the Global Middle Class," *BBC News* online, June 19, 2013, http://www.bbc.com/news/business-22956470).

19. Camel and dromedary curves: Roser 2016g.
20. More accurately, a Bactrian camel; one-humped dromedaries are technically "camels," too.
21. Camel to dromedary: For another way of showing the same historical development, see figures 9-1 and 9-2, based on data from Milanović 2016.
22. This is also equivalent to the frequently cited $1.25 cutoff, stated in 2005 international dollars: Ferreira, Jolliffe, & Prydz 2015.
23. M. Roser, "No Matter What Extreme Poverty Line You Choose, the Share of People Below That Poverty Line Has Declined Globally," *Our World in Data* blog, 2017, https://ourworldindata.org/no-matter-what-global-poverty-line.
24. Veil of ignorance: Rawls 1976.
25. Millennium Development Goals: United Nations 2015a.
26. Deaton 2013, p. 37.
27. Lucas 1988, p. 5.
28. The goal is defined as $1.25 a day, which is the World Bank international poverty line in 2005 international dollars; see Ferreira, Jolliffe, & Prydz 2015.
29. The problem in getting to zero: Radelet 2015, p. 243; Roser & Ortiz-Ospina 2017, section IV.2.
30. The danger in crying "crisis": Kenny 2011, p. 203.
31. Causes of development: Collier & Rohner 2008; Deaton 2013; Kenny 2011; Mahbubani 2013; Milanović 2016; Radelet 2015. See also M. Roser, "The Global Decline of Extreme Poverty—Was It Only China?" *Our World in Data* blog, March 7, 2017, https://ourworldindata.org/the-global-decline-of-extreme-poverty-was-it-only-china/.
32. Radelet 2015, p. 35.
33. Prices as information: Hayek 1945; Hidalgo 2015; Sowell 1980.
34. Chile vs. Venezuela, Botswana vs. Zimbabwe: M. L. Tupy, "The Power of Bad Ideas: Why Voters Keep Choosing Failed Statism," *CapX*, Jan. 7, 2016.
35. Kenny 2011, p. 203; Radelet 2015, p. 38.
36. Mao's genocides: Rummel 1994; White 2011.
37. According to legend, said by Franklin Roosevelt about Nicaragua's Anastasio Somoza, but probably not: http://message.snopes.com/showthread.php?t=8204/.
38. Local leaders: Radelet 2015, p. 184.
39. War as development in reverse: Collier 2007.
40. Deaton 2017.
41. Hostility to the Industrial Revolution among Romantics and literary intellectuals: Collini 1998, 2013.
42. Snow 1959/1998, pp. 25–26. Enraged response: Leavis 1962/2013, pp. 69–72.
43. Radelet 2015, pp. 58–59.
44. "Factory Girls," by A Factory Girl, *The Lowell Offering*, no. 2, Dec. 1840, https://www2.cs.arizona.edu/patterns/weaving/periodicals/lo_40_12.pdf. Cited in C. Follett, "The Feminist Side of Sweatshops," *The Hill*, April 18, 2017, http://thehill.com/blogs/pundits-blog/labor/329332-the-feminist-side-of-sweatshops.
45. Quoted in Brand 2009, p. 26; chaps. 2 and 3 of his book expand on the liberating powers of urbanization.
46. Reviewed in Brand 2009, chaps. 2 and 3, and Radelet 2015, p. 59. For a similar account from today's China, see Chang 2009.
47. Slums to suburbs: Brand 2009; Perlman 1976.
48. Improvement in working conditions: Radelet 2015.
49. Benefits of science and technology: Brand 2009; Deaton 2013; Kenny 2011; Radelet 2015; Ridley 2010.
50. Mobile phones and commerce: Radelet 2015.
51. Jensen 2007.
52. Estimate from the International Telecommunication Union, cited in Pentland 2007.
53. Against foreign aid: Deaton 2013; Easterly 2006.
54. In favor of (some kinds of) foreign aid: Collier 2007; Kenny 2011; Radelet 2015; Singer 2010; S. Radelet, "Angus Deaton, His Nobel Prize, and Foreign Aid," *Future Development* blog, Brookings Institution, Oct. 20, 2015, http://www.brookings.edu/blogs/future-development/posts/2015/10/20-angus-deaton-nobel-prize-foreign-aid-radelet.
55. Rising Preston Curve: Roser 2016n.
56. Life expectancy figures are from www.gapminder.org.
57. Correlation between GDP and measures of well-being: van Zanden et al. 2014, p. 252; Kenny 2011, pp. 96–97; Land, Michalos, & Sirgy 2012; Prados de la Escosura 2015; see also chapters 11, 12, and 14–18.
58. Correlations between GDP and peace, stability, and liberal values: Brunnschweiler & Lujala 2015; Hegre et al. 2011; Prados de la Escosura 2015; van Zanden et al. 2014; Welzel 2013; see also chapters 12 and 14–18.

59. Correlations between GDP and happiness: Helliwell, Layard, & Sachs 2016; Stevenson & Wolfers 2008a; Veenhoven 2010; see also chapter 18. Correlation with IQ gains: Pietschnig & Voracek 2015; see also chapter 16.
60. Composite measures of national well-being: Land, Michalos, & Sirgy 2012; Prados de la Escosura 2015; van Zanden et al. 2014; Veenhoven 2010; Porter, Stern, & Green 2016; see also chapter 16.
61. GDP as cause of peace, stability, and liberal values: Brunnschweiler & Lujala 2015; Hegre et al. 2011; Prados de la Escosura 2015; van Zanden et al. 2014; Welzel 2013; see also chapters 11, 14, and 15.

CHAPTER 9: INEQUALITY

1. Plotted by the now-defunct *New York Times* Chronicle tool, http://nytlabs.com/projects/chronicle .html, retrieved Sept. 19, 2016.
2. "Bernie Quotes for a Better World," http://www.betterworld.net/quotes/bernie8.htm.
3. Inequality in the Anglosphere vs. the rest of the developed world: Roser 2016k.
4. Gini data taken from Roser 2016k, originally from OECD 2016; note that exact values vary depending on the source. The World Bank's Povcal, for example, estimates a less extreme change, from .38 in 1986 to .41 in 2013 (World Bank 2016d). Income share data from the World Wealth and Income Database, http://www.wid.world/. For a comprehensive dataset, see *The Chartbook of Economic Inequality*, Atkinson et al. 2017.
5. The trouble with inequality: Frankfurt 2015. Other inequality skeptics: Mankiw 2013; McCloskey 2014; Parfit 1997; Sowell 2015; Starmans, Sheskin, & Bloom 2017; Watson 2015; Winship 2013; S. Winship, "Inequality Is a Distraction. The Real Issue Is Growth," *Washington Post*, Aug. 16, 2016.
6. Frankfurt 2015, p. 7.
7. According to the World Bank 2016c, global GDP per capita grew in every year from 1961 to 2015 except 2009.
8. Piketty 2013, p. 261. Problems with Piketty: Kane 2016; McCloskey 2014; Summers 2014a.
9. Nozick on income distributions: Nozick 1974. His example was the basketball great Wilt Chamberlain.
10. J. B. Stewart, "In the Chamber of Secrets: J. K. Rowling's Net Worth," *New York Times*, Nov. 24, 2016.
11. Social comparison theory comes from Leon Festinger; the theory of reference groups comes from Robert Merton and from Samuel Stouffer. See Kelley & Evans 2017 for a review and citations.
12. Amartya Sen (1987) makes a similar argument.
13. Wealth and happiness: Stevenson & Wolfers 2008a; Veenhoven 2010; see also chapter 18.
14. Wilkinson & Pickett 2009.
15. Problems with *The Spirit Level:* Saunders 2010; Snowdon 2010, 2016; Winship 2013.
16. Inequality and subjective well-being: Kelley & Evans 2017. See chapter 18 for an explanation of how happiness is measured.
17. Starmans, Sheskin, & Bloom 2017.
18. Ethnic minorities perceived as cheaters: Sowell 1980, 1994, 1996, 2015.
19. Skepticism on inequality causing economic and political dysfunction: Mankiw 2013; McCloskey 2014; Winship 2013; S. Winship, "Inequality Is a Distraction. The Real Issue Is Growth," *Washington Post*, Aug. 16, 2016.
20. Influence-peddling versus inequality: Watson 2015.
21. Sharing meat, keeping plant foods: Cosmides & Tooby 1992.
22. Inequality and awareness of inequality are universal: Brown 1991.
23. Hunter-gatherer inequality: Smith et al. 2010. The average excludes questionable forms of "wealth" such as reproductive success, grip strength, weight, and sharing partners.
24. Kuznets 1955.
25. Deaton 2013, p. 89.
26. Some, but not all, of the increase in between-country inequality from 1820 to 1970 can be attributed to the larger number of countries in the world; Branko Milanović, personal communication, April 16, 2017.
27. War as leveler: Graham 2016; Piketty 2013; Scheidel 2017.
28. Scheidel 2017, p. 444.
29. History of social spending: Lindert 2004; van Bavel & Rijpma 2016.
30. Egalitarian Revolution: Moatsos et al. 2014, p. 207.
31. Social spending as a proportion of GDP: OECD 2014.
32. Change in mission of government (particularly in Europe): Sheehan 2008.
33. In particular, in environmental protection (chapter 10), gains in safety (chapter 12), the abolition of capital punishment (chapter 14), the rise of emancipative values (chapter 15), and overall human development (chapter 16).
34. Social spending by employers: OECD 2014.
35. Reported by Rep. Robert Inglis (R-S.C.), P. Rucker, "Sen. DeMint of S.C. Is Voice of Opposition to Health-Care Reform," *Washington Post*, July 28, 2009.

36. Wagner's Law: Wilkinson 2016b.
37. Social spending in developing countries: OECD 2014.
38. Prados de la Escosura 2015.
39. No libertarian paradises: M. Lind, "The Question Libertarians Just Can't Answer," *Salon,* June 4, 2013; Friedman 1997. See also chapter 21, note 40.
40. Willingness to have a welfare state: Alesina, Glaeser, & Sacerdote 2001; Peterson 2015.
41. Explanations for the post-1980s inequality rise: Autor 2014; Deaton 2013; Goldin & Katz 2010; Graham 2016; Milanović 2016; Moatsos et al. 2014; Piketty 2013; Scheidel 2017.
42. Taller elephant with lower trunk tip: Milanović 2016, fig. 1.3. More analysis of the elephant: Corlett 2016.
43. Anonymous versus nonanonymous data: Corlett 2016; Lakner & Milanović 2016.
44. Quasi-nonanonymous elephant curve: Lakner & Milanović 2016.
45. Coontz 1992/2016, pp. 30-31.
46. Rose 2016; Horwitz 2015 made a similar discovery.
47. Individuals moving into the top 1 or 10 percent: Hirschl & Rank 2015. Horwitz 2015 obtained similar results. See also Sowell 2015; Watson 2015.
48. Optimism Gap: Whitman 1998. Economic Optimism Gap: Bernanke 2016; Meyer & Sullivan 2011.
49. Roser 2016k.
50. Why the United States doesn't have a European welfare state: Alesina, Glaeser, & Sacerdote 2001; Peterson 2015.
51. Rise in disposable income in lower quintiles: Burtless 2014.
52. Income rise from 2014 to 2015: Proctor, Semega, & Kollar 2016. Continuation in 2016: E. Levitz, "The Working Poor Got Richer in 2016," *New York,* March 9, 2017.
53. C. Jencks, "The War on Poverty: Was It Lost?" *New York Review of Books,* April 2, 2015. Similar analyses: Furman 2014; Meyer & Sullivan 2011, 2012, 2017a, b; Sacerdote 2017.
54. 2015 and 2016 drops in the poverty rate: Proctor, Semega, & Kollar 2016; Semega, Fontenot, & Kollar 2017.
55. Henry et al. 2015.
56. Underestimating economic progress: Feldstein 2017.
57. Furman 2005.
58. Access to utilities among the poor: Greenwood, Seshadri, & Yorukoglu 2005. Ownership of appliances among the poor: US Census Bureau, "Extended Measures of Well-Being: Living Conditions in the United States, 2011," table 1, http://www.census.gov/hhes/well-being/publications /extended-11.html. See also figure 17-3.
59. Consumption inequality: Hassett & Mathur 2012; Horwitz 2015; Meyer & Sullivan 2012.
60. Decline in happiness inequality: Stevenson & Wolfers 2008b.
61. Declining Ginis for quality of life: Deaton 2013; Rijpma 2014, p. 264; Roser 2016a, 2016n; Roser & Ortiz-Ospina 2016a; Veenhoven 2010.
62. Inequality and secular stagnation: Summers 2016.
63. The economist Douglas Irwin (2016) notes that 45 million Americans live below the poverty line, 135,000 Americans are employed by the apparel industry, and the normal turnover of jobs results in about 1.7 million layoffs a month.
64. Automation, jobs, and inequality: Brynjolfsson & McAfee 2016.
65. Economic challenges and solutions: Dobbs et al. 2016; Summers & Balls 2015.
66. S. Winship, "Inequality Is a Distraction. The Real Issue Is Growth," *Washington Post,* Aug. 16, 2016.
67. Governments vs. employers as social service providers: M. Lind, "Can You Have a Good Life If You Don't Have a Good Job?" *New York Times,* Sept. 16, 2016.
68. Universal basic income: Bregman 2016; S. Hammond, "When the Welfare State Met the Flat Tax," *Foreign Policy,* June 16, 2016; R. Skidelsky, "Basic Income Revisited," *Project Syndicate,* June 23, 2016; C. Murray, "A Guaranteed Income for Every American," *Wall Street Journal,* June 3, 2016.
69. Studies of the effects of basic income: Bregman 2016. High-tech volunteering: Diamandis & Kotler 2012. Effective altruism: MacAskill 2015.

CHAPTER 10: THE ENVIRONMENT

1. See Gore's 1992 *Earth in the Balance;* Ted Kaczynski (the Unabomber), "Industrial Society and Its Future," http://www.washingtonpost.com/wp-srv/national/longterm/unabomber/manifesto .text.htm; Francis 2015. Kaczynski read Gore's book, and the similarities between it and his manifesto were pointed out in an undated Internet quiz by Ken Crossman: http://www.crm114 .com/algore/quiz.html.
2. Quoted in M. Ridley, "Apocalypse Not: Here's Why You Shouldn't Worry About End Times," *Wired,* Aug. 17, 2012. In *The Population Bomb,* Paul Ehrlich also compared humanity to cancer; see Bailey 2015, p. 5. For fantasies of a depopulated planet, see Alan Weisman's 2007 bestseller *The World Without Us.*
3. Ecomodernism: Asafu-Adjaye et al. 2015; Ausubel 1996, 2007, 2015; Ausubel, Wernick, & Waggoner 2012; Brand 2009; DeFries 2014; Nordhaus & Shellenberger 2007. Earth Optimism: Balmford & Knowlton 2017; https://earthoptimism.si.edu/; http://www.oceanoptimism.org/about/.

4. Extinctions and forest clearing by indigenous peoples: Asafu-Adjaye et al. 2015; Brand 2009; Burney & Flannery 2005; White 2011.
5. Wilderness preserves and decimation of indigenous peoples: Cronon 1995.
6. From *Plows, Plagues, and Petroleum* (2005), quoted in Brand 2009, p. 19; see also Ruddiman et al. 2016.
7. Brand 2009, p. 133.
8. Gifts of industrialization: chapters 5–8; A. Epstein 2014; Norberg 2016; Radelet 2015; Ridley 2010.
9. Environmental Kuznets curve: Ausubel 2015; Dinda 2004; Levinson 2008; Stern 2014. Note that the curve does not apply to all pollutants or all countries, and when it occurs it may be driven by policy rather than happening automatically.
10. Inglehart & Welzel 2005; Welzel 2013, chap. 12.
11. Demographic transitions: Ortiz-Ospina & Roser 2016d.
12. Muslim population bust: Eberstadt & Shah 2011.
13. M. Tupy, "Humans Innovate Their Way Out of Scarcity," *Reason*, Jan. 12, 2016; see also Stuermer & Schwerhoff 2016.
14. Europium Crisis: Deutsch 2011.
15. "China's Rare-Earths Bust," *Wall Street Journal*, July 18, 2016.
16. Why we don't run out of resources: Nordhaus 1974; Romer & Nelson 1996; Simon 1981; Stuermer & Schwerhoff 2016.
17. People don't need resources: Deutsch 2011; Pinker 2002/2016, pp. 236–39; Ridley 2010; Romer & Nelson 1996.
18. Probability and solutions to human problems: Deutsch 2011.
19. The Stone Age quip is commonly attributed to Saudi oil minister Zaki Yamani in 1973; see "The End of the Oil Age," *The Economist*, Oct. 23, 2003. Energy transitions: Ausubel 2007, p. 235.
20. Farming pivots: DeFries 2014.
21. Farming in the future: Brand 2009; Bryce 2014; Diamandis & Kotler 2012.
22. Future water: Brand 2009; Diamandis & Kotler 2012.
23. Environment is rebounding: Ausubel 1996, 2015; Ausubel, Wernick, & Waggoner 2012; Bailey 2015; Balmford 2012; Balmford & Knowlton 2017; Brand 2009; Ridley 2010.
24. Roser 2016f, based on data from the UN Food and Agriculture Organization.
25. Roser 2016f, based on data from the Instituto Nacional de Pesquisas Espaciais of the Brazilian Ministry of Science and Technology.
26. Environmental Performance Index, http://epi.yale.edu/country-rankings.
27. Contaminated water and cooking smoke: United Nations Development Programme 2011.
28. According to the UN Millennium Development Goals report, the percentage of people exposed to contaminated water fell from 24 percent in 1990 to 9 percent in 2015 (United Nations 2015a, p. 52). According to data cited in Roser 2016l, in 1980, 62 percent of the world's population cooked with solid fuels; in 2010, just 41 percent did.
29. Quoted in Norberg 2016.
30. Third-worst stationary oil spill in history: Roser 2016r; US Department of the Interior, "Interior Department Releases Final Well Control Regulations to Ensure Safe and Responsible Offshore Oil and Gas Development," April 14, 2016, https://www.doi.gov/pressreleases/interior-department-releases-final-well-control-regulations-ensure-safe-and.
31. Increased tiger, condor, rhino, panda numbers: World Wildlife Foundation and Global Tiger Forum, cited in "Nature's Comebacks," *Time*, April 17, 2016. Conservation successes: Balmford 2012; Hoffmann et al. 2010; Suckling et al. 2016; United Nations 2015a, p. 57; R. McKie, "Saved: The Endangered Species Back from the Brink of Extinction," *The Guardian*, April 8, 2017. Pimm on conservation efforts reducing avian extinctions: quoted in D. T. Max, "Green Is Good," *New Yorker*, May 12, 2014, confirmed in a personal communication with Pimm, 2018.
32. The paleontologist Douglas Erwin (2015) points out that mass extinctions wipe out inconspicuous but widespread mollusks, arthropods, and other invertebrates, not the charismatic birds and mammals that attract the attention of journalists. The biogeographer John Briggs (2015, 2016) notes that "most extinctions have occurred on oceanic islands or in restricted freshwater locations" after humans have introduced invasive species, because the native animals have nowhere to run; few have taken place on continents or in the oceans, and no ocean species has gone extinct in the past fifty years. Brand points out that the catastrophic predictions assume that all threatened species will go extinct *and* that this rate will continue for centuries or millennia; S. Brand, "Rethinking Extinction," *Aeon*, April 21, 2015. See also Bailey 2015; Costello, May, & Stork 2013; Stork 2010; Thomas 2017; M. Ridley, "A History of Failed Predictions of Doom," http://www.rationaloptimist.com/blog/apocalypse-not/.
33. International agreements on the environment: http://www.enviropedia.org.uk/Acid_Rain/International_Agreements.php.
34. Healing ozone hole: United Nations 2015a, p. 7.
35. Note that the environmental Kuznets curve may be driven by such activism and legislation; see notes 9 and 40 in this chapter.
36. Density is good: Asafu-Adjaye et al. 2015; Brand 2009; Bryce 2013.
37. Dematerialization of consumption: Sutherland 2016.
38. Dying car culture: M. Fisher, "Cruising Toward Oblivion," *Washington Post*, Sept. 2, 2015.

39. Peak Stuff: Ausubel 2015; Office for National Statistics 2016. The equivalents in American units are 16.6 and 11.4 tons.
40. See, for example, J. Salzman, "Why Rivers No Longer Burn," *Slate*, Dec. 10, 2012; S. Cardoni, "Top 5 Pieces of Environmental Legislation," *ABC News*, July 2, 2010, http://abcnews.go.com/Technology/top-pieces-environmental-legislation/story?id=11067662; Young 2011. See also note 35 above.
41. Recent reviews of climate change: Intergovernmental Panel on Climate Change 2014; King et al. 2015; W. Nordhaus 2013; Plumer 2015; World Bank 2012a. See also J. Gillis, "Short Answers to Hard Questions About Climate Change," *New York Times*, Nov. 28, 2015; "The State of the Climate in 2016," *The Economist*, Nov. 17, 2016.
42. 4°C warming must not occur: World Bank 2012a.
43. Effects of different emission scenarios: Intergovernmental Panel on Climate Change 2014; King et al. 2015; W. Nordhaus 2013; Plumer 2015; World Bank 2012a. The projection for a 2°C rise is the RCP2.6 scenario shown in Intergovernmental Panel on Climate Change 2014, fig. 6.7.
44. Energy from fossil fuels: My calculation for 2015, from British Petroleum 2016, "Primary Energy: Consumption by Fuel," p. 41, "Total World."
45. Scientific consensus on anthropogenic climate change: NASA, "Scientific Consensus: Earth's Climate Is Warming," http://climate.nasa.gov/scientific-consensus/; *Skeptical Science*, http://www.skepticalscience.com/; Intergovernmental Panel on Climate Change 2014; Plumer 2015; W. Nordhaus 2013; W. Nordhaus, "Why the Global Warming Skeptics Are Wrong," *New York Review of Books*, March 22, 2012. Among the skeptics who have been convinced are the libertarian science writers Michael Shermer, Matt Ridley, and Ronald Bailey.
46. Consensus among climate scientists: Powell 2015; G. Stern, "Fifty Years After U.S. Climate Warning, Scientists Confront Communication Barriers," *Science*, Nov. 27, 2015; see also the preceding note.
47. Climate change denialism: Morton 2015; Oreskes & Conway 2010; Powell 2015.
48. Bona fides on political correctness: I am on the advisory boards of the Foundation for Individual Rights on Education (https://www.thefire.org/about-us/board-of-directors-page/), the Heterodox Academy (http://heterodoxacademy.org/about-us/advisory-board/), and the Academic Engagement Network (http://www.academicengagement.org/en/about-us/leadership); see also Pinker 2002/2016, 2006. Evidence on climate change: See citations in notes 41, 45, and 46 above.
49. Lukewarming: M. Ridley, "A History of Failed Predictions of Doom," http://www.rationaloptimist.com/blog/apocalypse-not/; J. Curry, "Lukewarming," *Climate Etc.*, Nov. 5, 2015, https://judithcurry.com/2015/11/05/lukewarming/.
50. Climate Casino: W. Nordhaus 2013; W. Nordhaus, "Why the Global Warming Skeptics Are Wrong," *New York Review of Books*, March 22, 2012; R. W. Cohen et al., "In the Climate Casino: An Exchange," *New York Review of Books*, April 26, 2012.
51. Climate justice: Foreman 2013.
52. Klein vs. carbon tax: C. Komanoff, "Naomi Klein Is Wrong on the Policy That Could Change Everything," *Carbon Tax Center* blog, https://www.carbontax.org/blog/2016/11/07/naomi-klein-is-wrong-on-the-policy-that-could-change-everything/; Koch brothers vs. carbon tax: C. Komanoff, "To the Left-Green Opponents of I-732: How Does It Feel?" *Carbon Tax Center* blog, https://www.carbontax.org/blog/2016/11/04/to-the-left-green-opponents-of-i-732-how-does-it-feel/. Economists' statement on climate change: Arrow et al. 1997. Recent arguments for the carbon tax: "FAQs," *Carbon Tax Center* blog, https://www.carbontax.org/faqs/.
53. "Naomi Klein on Why Low Oil Prices Could Be a Great Thing," *Grist*, Feb. 9, 2015.
54. The problem with "climate justice" and "changing everything": Foreman 2013; Shellenberger & Nordhaus 2013.
55. Scare tactics less effective than practical solutions: Braman et al. 2007; Feinberg & Willer 2011; Kahan, Jenkins-Smith, et al. 2012; O'Neill & Nicholson-Cole 2009; L. Sorantino, "Annenberg Study: Pope Francis' Climate Change Encyclical Backfired Among Conservative Catholics," *Daily Pennsylvanian*, Nov. 1, 2016, https://goo.gl/zUWXyk; T. Nordhaus & M. Shellenberger, "Global Warming Scare Tactics," *New York Times*, April 8, 2014. See Boyer 1986 and Sandman & Valenti 1986 for a similar point about nuclear weapons.
56. "World Greenhouse Gas Emissions Flow Chart 2010," *Ecofys*, http://www.ecofys.com/files/files/asn-ecofys-2013-world-ghg-emissions-flow-chart-2010.pdf.
57. Scale insensitivity: Desvousges et al. 1992.
58. Moralization of profligacy and asceticism: Haidt 2012; Pinker 2008a.
59. Sacrifice vs. benefit as a source of moral approbation: Nemirow 2016.
60. See http://scholar.harvard.edu/files/pinker/files/ten_ways_to_green_your_scence_2.jpg and http://scholar.harvard.edu/files/pinker/files/ten_ways_to_green_your_scence_1.jpg.
61. Shellenberger & Nordhaus 2013.
62. M. Tupy, "Earth Day's Anti-Humanism in One Graph and Two Tables," *Cato at Liberty*, April 22, 2015, https://www.cato.org/blog/earth-days-anti-humanism-one-graph-two-tables.
63. Shellenberger & Nordhaus 2013.
64. Trading economic development against climate change: W. Nordhaus 2013.
65. L. Sorantino, "Annenberg Study: Pope Francis' Climate Change Encyclical Backfired Among Conservative Catholics," *Daily Pennsylvanian*, Nov. 1, 2016, https://goo.gl/zUWXyk.

66. The actual carbon-to-hydrogen ratio in the cellulose and lignin making up wood is lower, but most of the hydrogen is already bound to oxygen, so it does not oxidize and release heat during combustion; see Ausubel & Marchetti 1998.

67. Bituminous coal is mainly $C_{137}H_{97}O_9NS$, with a ratio of 1.4 to 1; anthracite is mainly $C_{240}H_{90}O_4NS$, with a ratio of 2.67 to 1.

68. Carbon-to-hydrogen ratios: Ausubel 2007.

69. Decarbonization: Ausubel 2007.

70. "Global Carbon Budget," Global Carbon Project, Nov. 14, 2016, http://www.globalcarbonproject. org/carbonbudget/.

71. Ausubel 2007, p. 230.

72. Carbon plateau, GDP rise: Le Quéré et al. 2016.

73. Deep decarbonization: Deep Decarbonization Pathways Project 2015; Pacala & Socolow 2004; Williams et al. 2014; http://deepdecarbonization.org/.

74. Carbon tax consensus: Arrow et al. 1997; see also "FAQs," Carbon Tax Center blog, https://www .carbontax.org/faqs/.

75. How to implement a carbon tax: "FAQs," Carbon Tax Center blog, https://www.carbontax.org /faqs/; Romer 2016.

76. Nuclear power as the new green: Asafu-Adjaye et al. 2015; Ausubel 2007; Brand 2009; Bryce 2014; Cravens 2007; Freed 2014; K. Caldeira et al., "Top Climate Change Scientists' Letter to Policy Influencers," CNN, Nov. 3, 2013, http://www.cnn.com/2013/11/03/world/nuclear-energy-climate -change-scientists-letter/index.html; M. Shellenberger, "How the Environmental Movement Changed Its Mind on Nuclear Power," Public Utilities Fortnightly, May 2016; Nordhaus & Shellenberger 2011; Breakthrough Institute, "Energy and Climate FAQs," http://thebreakthrough.org /index.php/programs/energy-and-climate/nuclear-faqs. Though many environmental climate activists now support an expansion of nuclear power (including Stewart Brand, Jared Diamond, Paul Ehrlich, Tim Flannery, John Holdren, James Kunstler, James Lovelock, Bill McKibben, Hugh Montefiore, and Patrick Moore), remaining opponents include Greenpeace, the World Wildlife Fund, the Sierra Club, the Natural Resources Defense Council, Friends of the Earth, and (with some equivocation) Al Gore. See Brand 2009, pp. 86–89.

77. Solar and wind provide 1.5 percent of the world's energy: British Petroleum 2016, graphed in https://www.carbonbrief.org/factcheck-how-much-energy-does-the-world-get-from-renewables.

78. Land required by wind farms: Bryce 2014.

79. Land required by wind and solar: Swain et al. 2015, based on data from Jacobson & Delucchi 2011.

80. M. Shellenberger, "How the Environmental Movement Changed Its Mind on Nuclear Power," Public Utilities Fortnightly, May 2016; R. Bryce, "Solar's Great and So Is Wind, but We Still Need Nuclear Power," Los Angeles Times, June 16, 2016.

81. Chernobyl cancer deaths: Ridley 2010, pp. 308, 416.

82. Relative death rate from nuclear vs. fossil fuels: Kharecha & Hansen 2013; Swain et al. 2015. A million deaths a year from coal: Morton 2015, p. 16.

83. Nordhaus & Shellenberger 2011. See also note 76 above.

84. Deep Decarbonization Pathways Project 2015. Deep decarbonization of the United States: Williams et al. 2014. See also B. Plumer, "Here's What It Would Really Take to Avoid 2°C of Global Warming," Vox, July 9, 2014.

85. Deep decarbonization of the world: Deep Decarbonization Pathways Project 2015; see also the preceding note.

86. Nuclear power and the psychology of fear and dread: Gardner 2008; Gigerenzer 2016; Ropeik & Gray 2002; Slovic 1987; Slovic, Fischhoff, & Lichtenstein 1982.

87. From "Power," by John Hall and Johanna Hall.

88. Variously attributed; quoted in Brand 2009, p. 75.

89. Necessity for standardization: Shellenberger 2017. Selin quote: Washington Post, May 29, 1995.

90. Fourth-generation nuclear power: Bailey 2015; Blees 2008; Freed 2014; Hargraves 2012; Naam 2013.

91. Fusion power: E. Roston, "Peter Thiel's Other Hobby Is Nuclear Fusion," Bloomberg News, Nov. 22, 2016; L. Grossman, "Inside the Quest for Fusion, Clean Energy's Holy Grail," Time, Oct. 22, 2015.

92. Advantages of technological solutions to climate change: Bailey 2015; Koningstein & Fork 2014; Nordhaus 2016; see also note 103 below.

93. Need for risky research: Koningstein & Fork 2014.

94. Brand 2009, p. 84.

95. American gridlock and technophobia: Freed 2014.

96. Carbon capture: Brand 2009; B. Plumer, "Can We Build Power Plants That Actually Take Carbon Dioxide Out of the Air?" Vox, March 11, 2015; B. Plumer, "It's Time to Look Seriously at Sucking CO2 Out of the Atmosphere," Vox, July 13, 2015. See also CarbonBrief 2016, and the Web site for the Center for Carbon Removal, http://www.centerforcarbonremoval.org/.

97. Geoengineering: Keith 2013, 2015; Morton 2015. Artificial carbon capture: See the preceding note.

98. Low-carbon liquid fuels: Schrag 2009.
99. BECCS: King et al. 2015; Sanchez et al. 2015; Schrag 2009; see also note 96 above.
100. *Time* headlines: Sept. 25, Oct. 19, and Oct. 14, respectively. *New York Times* headline: Nov. 5, 2015, based on a poll from the Pew Research Center. For other polls showing American support of climate mitigation measures, see https://www.carbontax.org/polls/.
101. Paris agreement: http://unfccc.int/paris agreement/items/9485.php.
102. Likelihood of temperature rises under the Paris agreement: Fawcett et al. 2015.
103. Decarbonization driven by technology and economics: Nordhaus & Lovering 2016. States, cities, and world vs. Trump on climate change: Bloomberg & Pope 2017; "States and Cities Compensate for Mr. Trump's Climate Stupidity," *New York Times,* June 7, 2017; "Trump Is Dropping Out of the Paris Agreement, but the Rest of Us Don't Have To," *Los Angeles Times,* June 16, 2017; W. Hmaidan, "How Should World Leaders Punish Trump for Pulling Out of Paris Accord?" *The Guardian,* June 15, 2017; "Apple Issues $1 Billion Green Bond After Trump's Paris Climate Exit," *Reuters,* June 13, 2017, https://www.reuters.com/article/us-apple-climate-greenbond/apple-issues-1-billion-green-bond-after-trumps-paris-climate-exit-idUSKBN1941ZE; H. Tabuchi & H. Fountain, "Bill Gates Leads New Fund as Fears of U.S. Retreat on Climate Grow," *New York Times,* Dec. 12, 2016.
104. Cooling the atmosphere by reducing solar radiation: Brand 2009; Keith 2013, 2015; Morton 2015.
105. Calcite (limestone) as a stratospheric sunscreen and antacid: Keith et al. 2016.
106. "Moderate, responsive, temporary": Keith 2015. Removing 5 gigatons of CO_2 by 2075: Q&A from Keith 2015.
107. Climate engineering increases concern about climate change: Kahan, Jenkins-Smith, et al. 2012.
108. Complacent vs. conditional optimism: Romer 2016.

CHAPTER 11: PEACE

1. The graphs in *Better Angels* and in this book include the most recent year available. However, most datasets are not updated in real time but are double-checked for accuracy and completeness and thus released well after the most recent year they include (at least a year, though the gap has been shrinking). Some datasets are not updated at all, or change their criteria, making different years incommensurable. For these reasons, together with the publication lag, the latest years plotted in the *Better Angels* graphs were before 2011, and those plotted in this book extend no later than 2016.
2. War as the default state of affairs: See the discussion in Pinker 2011, pp. 228-49.
3. In this discussion, I use Levy's classification of great powers and great power war; see also Goldstein 2011; Pinker 2011, pp. 222-28.
4. Crisscrossing trends in great power war: Pinker 2011, pp. 225-28, based on data from Levy 1983.
5. Obsolescence of war between states: Goertz, Diehl, & Balas 2016; Goldstein 2011; Hathaway & Shapiro 2017; Mueller 1989, 2009; and see Pinker 2011, chap. 5.
6. The standard definition of "war" among political scientists is a state-based armed conflict which causes at least 1,000 battle deaths in a given year. The figures are drawn from the UCDP/PRIO Armed Conflict Dataset: Gleditsch et al. 2002; Human Security Report Project 2011; Pettersson & Wallensteen 2015; http://ucdp.uu.se/downloads/.
7. S. Pinker & J. M. Santos, "Colombia's Milestone in World Peace," *New York Times,* Aug. 26, 2016. I thank Joshua Goldstein for calling my attention to many of the facts in that article, repeated in this paragraph.
8. Center for Systemic Peace, Marshall 2016, http://www.systemicpeace.org/warlist/warlist.htm, total for 32 episodes of political violence in the Americas since 1945, excluding 9/11 and the Mexican drug war.
9. Counts from the UCDP/PRIO Armed Conflict Dataset: Pettersson & Wallensteen 2015, with updates from Therese Pettersson and Sam Taub (personal communication). The wars in 2016 were: Afghanistan vs. Taliban, and vs. ISIS; Iraq vs. ISIS; Libya versus ISIS; Nigeria vs. ISIS; Somalia vs. Al-Shabab; Sudan vs. SRF; Syria vs. ISIS, and vs. Insurgents; Turkey vs. ISIS, and vs. PKK; Yemen vs. Forces of Hadi.
10. Syrian civil war battle death estimates: 256,624 (through 2016) from the Uppsala Conflict Data Program (http://ucdp.uu.se/#country/652, accessed June 2017); 250,000 (through 2015) from the Center for Systemic Peace, http://www.systemicpeace.org/warlist/warlist.htm, last updated May 25, 2016.
11. Civil wars that ended since 2009 (technically, "state-based armed conflicts," with more than 25 battle deaths per year but not necessarily more than 1,000): personal communication from Therese Pettersson, March 17, 2016, based on the Uppsala Conflict Data Program Armed Conflict dataset, Pettersson & Wallensteen 2015, http://ucdp.uu.se/. Earlier wars with large death tolls: Center for Systemic Peace, Marshall 2016.
12. Goldstein 2015. The numbers are for "refugees," who cross international borders; the number of "internally displaced persons" has been tracked only since 1989, so a comparison of those displaced by the Syrian war and by earlier wars is impossible.
13. Genocides as old as history: Chalk & Jonassohn 1990, p. xvii.
14. Peak death rate in genocides: From Rummel 1997, using his definition of "democide," which

includes the UCDP's "one-sided violence" together with deliberate famines, deaths in intern-
ment camps, and the targeted bombing of civilians. Stricter definitions of "genocide" also result
in counts during the 1940s in the tens of millions. See White 2011; Pinker 2011, pp. 336–42.
15. The calculations are explained in Pinker 2011, p. 716, note 165.
16. Numbers are for 2014 and 2015, the most recent years for which a breakdown is available.
Though these are the "high" estimates in the UCDP One-Sided Violence Dataset version 1.4–
2015 (http://ucdp.uu.se/downloads/), the numbers tally only the verified deaths and should be
considered conservative lower bounds.
17. Problems in estimating risks: Pinker 2011, pp. 210–22; Spagat 2015, 2017; M. Spagat, "World War
III—What Are the Chances," *Significance*, Dec. 2015; M. Spagat & S. Pinker, "Warfare" (letter),
Significance, June 2016, and "World War III: The Final Exchange," *Significance*, Dec. 2016.
18. Nagdy & Roser 2016a. Military spending in all countries but the United States has decreased in
inflation-adjusted dollars from their Cold War peaks, and in the United States it is lower than
the Cold War peak as a proportion of GDP. Conscription: Pinker 2011, pp. 255–57; M. Tupy,
"Fewer People Exposed to Horrors of War," *HumanProgress*, May 30, 2017, http://humanprogress
.org/blog/fewer-people-exposed-to-horrors-of-war.
19. Enlightenment-era denunciations of war: Pinker 2011, pp. 164–68.
20. Declines and hiatuses in war: Pinker 2011, pp. 237–38.
21. Gentle commerce vindicated: Pinker 2011, pp. 284–88; Russett & Oneal 2001.
22. Democracy and peace: Pinker 2011, pp. 278–94; Russett & Oneal 2001.
23. Possible irrelevance of nuclear weapons: Mueller 1989, 2004a; Pinker 2011, pp. 268–78. For new
data see Sechser & Fuhrmann 2017.
24. Norms and taboos as a cause of the Long Peace: Goertz, Diehl, & Balas 2016; Goldstein 2011;
Hathaway & Shapiro 2017; Mueller 1989; Nadelmann 1990.
25. Civil wars less deadly than interstate wars: Pinker 2011, pp. 303–5.
26. Peacekeepers keep peace: Fortna 2008; Goldstein 2011; Hultman, Kathman, & Shannon 2013.
27. Richer countries have fewer civil wars: Fearon & Laitin 2003; Hegre et al. 2011; Human Security
Centre 2005; Human Security Report Project 2011. Warlords, guerrillas, and mafias: Mueller
2004a.
28. Contagion of war: Human Security Report Project 2011.
29. Romantic militarism: Howard 2001; Mueller 1989, 2004a; Pinker 2011, pp. 242–44; Sheehan 2008.
30. Quotes are from Mueller 1989, pp. 38–51.
31. Romantic nationalism: Howard 2001; Luard 1986; Mueller 1989; Pinker 2011, pp. 238–42.
32. Hegelian dialectical struggle: Luard 1986, p. 355; Nisbet 1980/2009. Quote from Mueller 1989.
33. Marxist dialectical struggle: Montgomery & Chirot 2015.
34. Declinism and cultural pessimism: Herman 1997; Wolin 2004.
35. Herman 1997, p. 231.

CHAPTER 12: SAFETY

1. In 2005, between 421,000 and 1.8 million people were bitten by poisonous snakes, and between
20,000 and 94,000 of them died (Kasturiratne et al. 2008).
2. Relative toll of injuries: World Health Organization 2014.
3. Accidents and causes of death: Kochanek et al. 2016. Accidents and the global burden of disease
and disability: Murray et al. 2012.
4. Homicides more lethal than war: Pinker 2011, p. 221; see also p. 177, table 13–1. For updated data
and visualizations on homicide rates, see the Igarapé Institute's *Homicide Monitor*, https://homicide
.igarape.org.br/.
5. Medieval violence: Pinker 2011, pp. 17–18, 60–75; Eisner 2001, 2003.
6. The Civilizing Process: Eisner 2001, 2003; Elias 1939/2000; Fletcher 1997.
7. Eisner and Elias: Eisner 2001, 2014a.
8. 1960s crime boom: Latzer 2016; Pinker 2011, pp. 106–16.
9. Root-causism: Sowell 1995.
10. Racism in decline in the 1960s: Pinker 2011, pp. 382–94.
11. Great American Crime Decline: Latzer 2016; Pinker 2011, pp. 116–27; Zimring 2007. The 2015
uptick was likely caused in part by a retreat in policing following nationally publicized protests
against police shootings in 2014; see L. Beckett, "Is the 'Ferguson Effect' Real? Researcher Has
Second Thoughts," *The Guardian*, May 13, 2016; H. Macdonald, "Police Shootings and Race,"
Washington Post, July 18, 2016. For reasons why the 2015 uptick is unlikely to reverse the progress
of the years before, see B. Latzer, "Will the Crime Spike Become a Crime Boom?" *City Journal*,
Aug. 31, 2016, https://www.city-journal.org/html/will-crime-spike-become-crime-boom-14710
.html.
12. Between 2000 and 2013, the Gini index in Venezuela fell from .47 to .41 (UN's *World Income Inequality Database*, https://www.wider.unu.edu/), while the homicide rate rose from 32.9 to 53.0
per 100,000 (Igarapé Institute's *Homicide Monitor*, https://homicide.igarape.org.br).
13. Sources of UN estimates are listed in the caption to figure 12-2. Using very different methods,
the Global Burden of Disease project (Murray et al. 2012) has estimated that the global homicide
rate fell from 7.4 per 100,000 people in 1995 to 6.1 in 2015.

14. International homicide rates: United Nations Office on Drugs and Crime 2014; https://www.unodc.org/gsh/en/data.html.
15. Reducing global homicide by 50 percent in thirty years: Eisner 2014b, 2015; Krisch et al. 2015. The 2015 UN Sustainable Development Goals include the vaguer aspiration "Significantly reduce all forms of violence and related death rates everywhere" (Target 16.1.1, https://sustainable development.un.org/sdg16).
16. International homicide rates: United Nations Office on Drugs and Crime 2014, https://www.unodc.org/gsh/en/data.html; see also *Homicide Monitor,* https://homicide.igarape.org.br/.
17. Lopsided distribution of homicides at every scale: Eisner 2015; Muggah & Szabo de Carvalho 2016.
18. Homicide in Boston: Abt & Winship 2016.
19. New York crime decline: Zimring 2007.
20. Homicide declines in Colombia, South Africa, and other countries: Eisner 2014b, p. 23. Russia: United Nations Office on Drugs and Crime 2014, p. 28.
21. Homicide declined in most nations: United Nations Office on Drugs and Crime 2013, 2014, https://www.unodc.org/gsh/en/data.html.
22. Successful crime-fighting in Latin America: Guerrero Velasco 2015; Muggah & Szabo de Carvalho 2016.
23. Rise in Mexican homicide 2007–11 due to organized crime: Botello 2016. Drop in Juárez: P. Corcoran, "Declining Violence in Juárez a Major Win for Calderon: Report," *Insight Crime,* March 26, 2013, http://www.insightcrime.org/news-analysis/declining-violence-in-juarez-a-major-win-for-calderon-report.
24. Homicide declines: Bogotá and Medellín: T. Rosenberg, "Colombia's Data-Driven Fight Against Crime," *New York Times,* Nov. 20, 2014. São Paulo: Risso 2014. Rio: R. Muggah & I. Szabó de Carvalho, "Fear and Backsliding in Rio," *New York Times,* April 15, 2014.
25. San Pedro Sula homicide decline: S. Nazario, "How the Most Dangerous Place on Earth Got a Little Bit Safer," *New York Times,* Aug. 11, 2016.
26. For an effort to halve homicide in Latin America within a *decade,* see Muggah & Szabo de Carvalho 2016, and https://www.instintodevida.org/.
27. How to bring homicide rates down quickly: Eisner 2014b, 2015; Krisch et al. 2015; Muggah & Szabo de Carvalho 2016. See also Abt & Winship 2016; Gash 2016; Kennedy 2011; Latzer 2016.
28. Hobbes, violence, and anarchy: Pinker 2011, pp. 31–36, 680–82.
29. Police strikes: Gash 2016, pp. 184–86.
30. Impunity from justice increases crime: Latzer 2016; Eisner 2015, p. 14.
31. Causes of the Great American Crime Decline: Kennedy 2011; Latzer 2016; Levitt 2004; Pinker 2011, pp. 116–27; Zimring 2007.
32. One-sentence summary: Eisner 2015.
33. State legitimacy and crime: Eisner 2003, 2015; Roth 2009.
34. What works in crime prevention: Abt & Winship 2016. See also Eisner 2014b, 2015; Gash 2016; Kennedy 2011; Krisch et al. 2015; Latzer 2016; Muggah 2015, 2016.
35. Crime and self-control: Pinker 2011, pp. 72–73, 105, 110–11, 126–27, 501–6, 592–611.
36. Crime, narcissism, and sociopathy (or psychopathy): Pinker 2011, pp. 510–11, 519–21.
37. Target hardening and crime reduction: Gash 2016.
38. Effectiveness of drug courts and treatment: Abt & Winship 2016, p. 26.
39. Equivocal effects of firearm legislation: Abt & Winship 2016, p. 26; Hahn et al. 2005; N. Kristof, "Some Inconvenient Gun Facts for Liberals," *New York Times,* Jan. 16, 2016.
40. Traffic death graph: K. Barry, "Safety in Numbers," *Car and Driver,* May 2011, p. 17.
41. Based on deaths per capita, not per vehicle mile traveled.
42. Bruce Springsteen, "Pink Cadillac."
43. Insurance Institute for Highway Safety 2016. The rate rose slightly, to 10.9, in 2015.
44. According to the World Health Organization in 2015, the annual rate of death in car crashes per 100,000 people is 9.2 in rich countries and 24.1 in poor countries; http://www.who.int/violence _injury_prevention/road_safety_status/2015/magnitude_A4_web.pdf.
45. Bettmann 1974, pp. 22–23.
46. Scott 2010, pp. 18–19.
47. Rawcliffe 1998, p. 4, quoted in Scott 2010, pp. 18–19.
48. Tebeau 2016.
49. Tudor Darwin Awards: http://tudoraccidents.history.ox.ac.uk/.
50. The complete dataset for figure 12-6 shows a puzzling rise in deaths from falls starting in 1992, which is inconsistent with the fact that emergency treatments and hospital admissions for falls during this period showed no such rise (Hu & Baker 2012). Though falls tend to kill older people, the rise cannot be explained by the aging of the American population, because it persists in age-adjusted data (Sheu, Chen, & Hedegaard 2015). The rise turns out to be an artifact of changes in reporting practices (Hu & Mamady 2014; Kharrazi, Nash, & Mielenz 2015; Stevens & Rudd 2014). Many elderly people fall down, fracture their hip, ribs, or skull, and die several weeks or months later from pneumonia or other complications. Coroners and medical examiners in the past tended to list the cause of death in these cases as the immediate terminal illness. More recently, they have listed it as the precipitating accident. The same number of people fell and died, but increasingly the death was attributed to the fall.

51. Presidential reports: "National Conference on Fire Prevention" (press release), Jan. 3, 1947, http://foundation.sfpe.org/wp-content/uploads/2014/06/presidentsconference1947.pdf; *America Burning* (report of the National Commission on Fire Prevention and Control), 1973; *American Burning Revisited*, U.S. Fire Administration/FEMA, 1987.
52. Firefighters as EMTs: P. Keisling, "Why We Need to Take the 'Fire' out of 'Fire Department,' " *Governing*, July 1, 2015.
53. Most poisonings are from drugs or alcohol: National Safety Council 2016, pp. 160–61.
54. Opioid epidemic: National Safety Council, "Prescription Drug Abuse Epidemic; Painkillers Driving Addiction," 2016, http://www.nsc.org/learn/NSC-Initiatives/Pages/prescription-painkiller-epidemic.aspx.
55. Opioid epidemic and its treatment: Satel 2017.
56. Opioid overdoses perhaps peaking: Hedegaard, Chen, & Warner 2015.
57. Age and cohort effects in drug overdoses: National Safety Council 2016; see Kolosh 2014 for graphs.
58. Drug use down in teenagers: National Institute on Drug Abuse 2016. The declines continued through the second half of 2016: National Institute on Drug Abuse, "Teen Substance Use Shows Promising Decline," Dec. 13, 2016, https://www.drugabuse.gov/news-events/news-releases/2016/12/teen-substance-use-shows-promising-decline.
59. Bettmann 1974, pp. 69–71.
60. Quoted in Bettmann 1974, p. 71.
61. History of workplace safety: Aldrich 2001.
62. Progressive movement and worker safety: Aldrich 2001.
63. The steepening of the drop from 1970 to 1980 in figure 12-7 is probably an artifact from aggregating different sources; it is not visible in the continuous data series from National Safety Council 2016, pp. 46–47. The overall trend in the NSC dataset is similar to that in the figure; I chose not to show it because the rates are calculated as a proportion of the population rather than the number of workers, and because they contain an artifactual drop in 1992, when the Census of Fatal Occupational Injuries was introduced.
64. United Nations Development Programme 2011, table 2.3, p. 37.
65. The example is from "War, Death, and the Automobile," an appendix to Mueller 1989, originally published in the *Wall Street Journal* in 1984.

CHAPTER 13: TERRORISM

1. Fear of terrorism: Jones et al. 2016a; see also chapter 4, note 14.
2. Western Europe as war zone: J. Gray, "Steven Pinker Is Wrong About Violence and War," *The Guardian*, March 13, 2015; see also S. Pinker, "Guess What? More People Are Living in Peace Now. Just Look at the Numbers," *The Guardian*, March 20, 2015.
3. More dangerous than terrorism: National Safety Council 2011.
4. Homicide in Western Europe versus the United States: United Nations Office on Drugs and Crime 2013. The average homicide rate of the 24 countries classified as Western Europe in the Global Terrorism Database was 1.1 per 100,000 people per year; the figure for the United States in 2014 was 4.5. Road traffic deaths: The average of the Western European countries' road traffic death rates for 2013 was 4.8 fatalities per 100,000 people per year; the US rate was 10.7.
5. Deaths in insurgencies and guerrilla warfare now counted as "terrorism": Human Security Report Project 2007; Mueller & Stewart 2016b; Muggah 2016.
6. John Mueller, personal communication, 2016.
7. Contagion of mass killings: B. Carey, "Mass Killings May Have Created Contagion, Feeding on Itself," *New York Times*, July 27, 2016; Lankford & Madfis 2018.
8. Active shooter incidents: Blair & Schweit 2014; Combs & Slovic 1979. Mass murders: Analysis of FBI Uniform Crime Report Data (http://www.ucrdatatool.gov/) from 1976 to 2011 by James Alan Fox, graphed in Latzer 2016, p. 263.
9. For a graph that expands the trends using a logarithmic scale, see Pinker 2011, fig. 6-9, p. 350.
10. K. Eichenwald, "Right-Wing Extremists Are a Bigger Threat to America Than ISIS," *Newsweek*, Feb. 4, 2016. Using the United States Extremist Crime Database (Freilich et al. 2014), which tracks right-wing extremist violence, the security analyst Robert Muggah (personal communication) estimates that from 1990 through May 2017, and excluding 9/11 and Oklahoma, there have been 272 deaths from right-wing extremism and 136 from Islamist terrorist attacks.
11. Terrorism as a by-product of global media: Payne 2004.
12. Greater impact of homicide: Slovic 1987; Slovic, Fischhoff, & Lichtenstein 1982.
13. Rational fear of murderers: Duntley & Buss 2011.
14. Motives of suicide terrorists and rampage killers: Lankford 2013.
15. Delusion that ISIS is an "existential threat" to America: See chapter 4, note 14; also J. Mueller & M. Stewart, "ISIS Isn't an Existential Threat to America," *Reason*, May 27, 2016.
16. Y. N. Harari, "The Theatre of Terror," *The Guardian*, Jan. 31, 2015.
17. Terrorism doesn't work: Abrahms 2006; Branwen 2016; Cronin 2009; Fortna 2015.
18. Jervis 2011.
19. Y. N. Harari, "The Theatre of Terror," *The Guardian*, Jan. 31, 2015.

20. Don't Name Them, Don't Show Them: Lankford & Madfis 2018; see also the projects called No Notoriety (https://nonotoriety.com/) and Don't Name Them (http://www.dontnamethem.org/).
21. How terrorism ends: Abrahms 2006; Cronin 2009; Fortna 2015.

CHAPTER 14: DEMOCRACY

1. High rates of violence in nonstate societies: Pinker 2011, chap. 2. For more recent estimates confirming this difference, see Gat 2015; Gómez et al. 2016; Wrangham & Glowacki 2012.
2. Despotic early governments: Betzig 1986; Otterbein 2004. Biblical tyranny: Pinker 2011, chap. 1.
3. White 2011, p. xvii.
4. Democracies have faster-growing economies: Radelet 2015, pp. 125–29. Note that this can be obscured by the fact that poor countries can grow at faster rates than rich countries, and poor countries tend to be less democratic. Democracies are less likely to go to war: Hegre 2014; Russett 2010; Russett & Oneal 2001. Democracies have less severe (though not necessarily fewer) civil wars: Gleditsch 2008; Lacina 2006. Democracies have fewer genocides: Rummel 1994, pp. 2, 15; Rummel 1997, pp. 6–10, 367; Harff 2003, 2005. Democracies never have famines: Sen 1984; see also Devereux 2000, for a slight qualification. Citizens in democracies are healthier: Besley 2006. Citizens in democracies are better educated: Roser 2016b.
5. Three waves of democratization: Huntington 1991.
6. Democracy in retreat: Mueller 1999, p. 214.
7. Democracy is obsolete: quotes from Mueller 1999, p. 214.
8. "The end of history": Fukuyama 1989.
9. For quotations, see Levitsky & Way 2015.
10. Not getting the concept of democracy: Welzel 2013, p. 66, n. 11.
11. This is a problem for the annual counts by the democracy-tracking organization Freedom House; see Levitsky & Way 2015; Munck & Verkuilen 2002; Roser 2016b.
12. This is another problem with the Freedom House data.
13. Polity IV Project: Center for Systemic Peace 2015; Marshall & Gurr 2014; Marshall, Gurr, & Jaggers 2016.
14. Color revolutions: Bunce 2017.
15. Democracies: Marshall, Gurr, & Jaggers 2016; Roser 2016b. "Democracies" are countries rated by the Polity IV Project as having a democracy score of 6 or greater, "Autocracies" as those having an autocracy score of 6 or greater. Countries that are neither democratic nor autocratic are called anocracies, defined as an "incoherent mix of democratic and autocratic traits and practices." In an "open anocracy," leaders are not restricted to an elite. For 2015, Roser divides the world's population up as follows: 55.8 percent in democracies, 10.8 percent in open anocracies, 6.0 percent in closed anocracies, 23.2 percent in autocracies, and 4 percent in transition or with no data.
16. For a recent defense of the Fukuyama thesis, see Mueller 2014. Refuting the "democratic recession": Levitsky & Way 2015.
17. Prosperity and democracy: Norberg 2016; Roser 2016b; Porter, Stern, & Green 2016, p. 19. Prosperity and human rights: Fariss 2014; Land, Michalos, & Sirgy 2012. Education and democracy: Rindermann 2008; see also Roser 2016i.
18. Diversity of democracy: Mueller 1999; Norberg 2016; Radelet 2015; for data, see the *Polity IV Annual Time-Series*, http://www.systemicpeace.org/polityproject.html; Center for Systemic Peace 2015; Marshall, Gurr, & Jaggers 2016.
19. Prospects for democracy in Russia: Bunce 2017.
20. Norberg 2016, p. 158.
21. Democratic dimwits: Achen & Bartels 2016; Caplan 2007; Somin 2016.
22. Latest fashion in dictatorship: Bunce 2017.
23. Popper 1945/2013.
24. Democracy = the right to complain: Mueller 1999, 2014. Quotation from Mueller 1999, p. 247.
25. Mueller 1999, p. 140.
26. Mueller 1999, p. 171.
27. Levitsky & Way 2015, p. 50.
28. Democracy and education: Rindermann 2008; Roser 2016b; Thyne 2006. Democracy, Western influence, and violent revolution: Levitsky & Way 2015, p. 54.
29. Democracy and human rights: Mulligan, Gil, & Sala-i-Martin 2004; Roser 2016b, section II.3.
30. Quotes from Sikkink 2017.
31. Human rights information paradox: Clark & Sikkink 2013; Sikkink 2017.
32. History of capital punishment: Hunt 2007; Payne 2004; Pinker 2011, pp. 149–53.
33. Death penalty on death row: C. Ireland, "Death Penalty in Decline," *Harvard Gazette*, June 28, 2012; C. Walsh, "Death Penalty, in Retreat," *Harvard Gazette*, Feb. 3, 2015. For current updates, see "International Death Penalty," *Amnesty International*, http://www.amnestyusa.org/our-work/issues/death-penalty/international-death-penalty, and "Capital Punishment by Country," *Wikipedia*, https://en.wikipedia.org/wiki/Capital_punishment_by_country.
34. C. Ireland, "Death Penalty in Decline," *Harvard Gazette*, June 28, 2012.
35. History of the abolition of capital punishment: Hammel 2010.

36. Enlightenment arguments against the death penalty: Hammel 2010; Hunt 2007; Pinker 2011, pp. 146–53.
37. Southern culture of honor: Pinker 2011, pp. 99–102. Executions concentrated in a few Southern counties: Interview with the legal scholar Carol Steiker, C. Walsh, "Death Penalty, in Retreat," *Harvard Gazette*, Feb. 3, 2015.
38. Gallup poll on the death penalty: Gallup 2016. For current data, see the *Death Penalty Information Center*, http://www.deathpenaltyinfo.org/.
39. Pew Research poll reported in M. Berman, "For the First Time in Almost 50 Years, Less Than Half of Americans Support the Death Penalty," *Washington Post*, Sept. 30, 2016.
40. Death of the death penalty in the United States: D. Von Drehle, "The Death of the Death Penalty," *Time*, June 8, 2015; *Death Penalty Information Center*, http://www.deathpenaltyinfo.org/.

CHAPTER 15: EQUAL RIGHTS

1. Evolutionary basis of racism and sexism: Pinker 2011; Pratto, Sidanius, & Levin 2006; Wilson & Daly 1992.
2. Evolutionary basis of homophobia: Pinker 2011, chap. 7, pp. 448–49.
3. History of equal rights: Pinker 2011, chap. 7; Shermer 2015. Seneca Falls and the history of women's rights: Stansell 2010. Selma and the history of African American rights: Branch 1988. Stonewall and the history of gay rights: Faderman 2015.
4. Ranking for 2016 by *US News and World Report*, http://www.independent.co.uk/news/world/politics/the-10-most-influential-countries-in-the-world-have-been-revealed-a6834956.html. These three nations are also the most affluent.
5. Amos 5:24.
6. No increase in police shootings: Though direct data are scarce, the number of police shootings tracks the rate of violent crime (Fyfe 1988), which, as we saw in chapter 12, has plummeted. No racial disparity: Fryer 2016; Miller et al. 2016; S. Mullainathan, "Police Killings of Blacks: Here Is What the Data Say," *New York Times*, Oct. 16, 2015.
7. Pew Research Center 2012b, p. 17.
8. Other surveys of American values: Pew Research Center 2010; Teixeira et al. 2013; see reviews in Pinker 2011, chap. 7, and Roser 2016s. Another example: The General Social Survey (http://gss.norc.org/) annually asks white Americans about their feelings toward black Americans. Between 1996 and 2016 the proportion feeling "close" rose from 35 to 51 percent; the proportion feeling "not close" fell from 18 to 12 percent.
9. Successive cohorts more tolerant: Gallup 2002, 2010; Pew Research Center 2012b; Teixeira et al. 2013. Globally: Welzel 2013.
10. Generations carry values with them: Teixeira et al. 2013; Welzel 2013.
11. Google searches and other digital truth serums: Stephens-Davidowitz 2017.
12. Searches for *nigger* as an index of racism: Stephens-Davidowitz 2014.
13. There seems to be no systematic decline in searches for jokes in general, such as in the search string "funny jokes." Stephens-Davidowitz points out that searches for hip-hop lyrics and other appropriations of the word *nigger* almost entirely use the spelling *nigga*.
14. African American poverty: Deaton 2013, p. 180.
15. African American life expectancy: Cunningham et al. 2017; Deaton 2013, p. 61.
16. The last year for which the US Census reports illiteracy rates is 1979, when the rate for blacks was 1.6 percent; Snyder 1993, chap. 1, reproduced in National Assessment of Adult Literacy (undated).
17. See chapter 16, note 24, and chapter 18, note 35.
18. Disappearance of lynching: Pinker 2011, chap. 7, based on US Census data presented in Payne 2004, plotted in figure 7-2, p. 384. Hate crime homicides of African Americans, plotted in figure 7-3, fell from five in 1996 to one per year in 2006–8. Since then the number of victims stayed at an average of one per year through 2014, then spiked to ten in 2015, nine of them killed in a single incident, a mass shooting in a church in Charleston, South Carolina (Federal Bureau of Investigation 2016b).
19. For the years between 1996 and 2015 inclusive, the number of hate crime incidents recorded by the FBI correlated with the US homicide rate with a coefficient of .90 (on a scale from –1 to 1).
20. Anti-Islamic hate crimes follow incidents of Islamist terrorist attacks: Stephens-Davidowitz 2017.
21. Hate crime hyperbole: E. N. Brown, "Hate Crimes, Hoaxes, and Hyperbole," *Reason*, Nov. 18, 2016; Alexander 2016.
22. How it used to be: S. Coontz, "The Not-So-Good Old Days," *New York Times*, June 15, 2013.
23. Women in the labor force: United States Department of Labor 2016.
24. For evidence that the decline began even earlier, in 1979, see Pinker 2011, fig. 7-10, p. 402, also based on data from the National Crime Victimization Survey. Because of changes in definitions and coding criteria, those data are not commensurable with the series plotted here in figure 15-4.
25. Cooperation breeds sympathy: Pinker 2011, chaps. 4, 7, 9, 10.

26. Justification as a force for moral progress: Pinker 2011, chap. 4; Appiah 2010; Hunt 2007; Mueller 2010b; Nadelmann 1990; Payne 2004; Shermer 2015.
27. Decline of discrimination, rise of affirmative action: Asal & Pate 2005.
28. World Public Opinion Poll: Presented in Council on Foreign Relations 2011.
29. Council on Foreign Relations 2011.
30. Council on Foreign Relations 2011.
31. Effectiveness of global shaming campaigns: Pinker 2011, pp. 272–76, 414; Appiah 2010; Mueller 1989, 2004a, 2010b; Nadelmann 1990; Payne 2004; Ray 1989.
32. United Nations Children's Fund 2014; see also M. Tupy, "Attitudes on FGM Are Shifting," HumanProgress, http://humanprogress.org/blog/attitudes-on-fgm-are-shifting.
33. D. Latham, "Pan African Parliament Endorses Ban on FGM," Inter Press Service, Aug. 6, 2016, http://www.ipsnews.net/2016/08/pan-african-parliament-endorses-ban-on-fgm/.
34. Criminalization of homosexuality and the gay rights revolution: Pinker 2011, pp. 447–54; Faderman 2015.
35. For current data on gay rights worldwide, see Equaldex, www.equaldex.com, and "LGBT Rights by Country or Territory," Wikipedia, https://en.wikipedia.org/wiki/LGBT_rights_by_country_or_territory.
36. World Values Survey: http://www.worldvaluessurvey.org/wvs.jsp. Emancipative values: Welzel 2013.
37. Distinguishing age, period, and cohort: Costa & McCrae 1982; Smith 2008.
38. See also F. Newport, "Americans Continue to Shift Left on Key Moral Issues," Gallup, May 26, 2015, http://www.gallup.com/poll/183413/americans-continue-shift-left-key-moral-issues.aspx.
39. Ipsos 2016.
40. Values go with the cohort, not the life cycle: Ghitza & Gelman 2014; Inglehart 1997; Welzel 2013.
41. Emancipative values and the Arab Spring (a complicated relationship): Inglehart 2017.
42. Correlates of emancipative values: Welzel 2013, especially table 2.7, p. 83, and table 3.2, p. 122.
43. Cousin marriage and tribalism: S. Pinker, "Strangled by Roots," New Republic, Aug. 6, 2007.
44. Knowledge Index: Chen & Dahlman 2006, table 2.
45. Knowledge Index as a predictor of emancipative values: Welzel 2013, p. 122, where the index is called "Technological Advancement." Welzel (personal communication) confirms that the Knowledge Index has a highly significant partial correlation with emancipative values (.62) holding constant GDP per capita (or its log), whereas the reverse is not true (.20).
46. Finkelhor et al. 2014.
47. Decline of corporal punishment: Pinker 2011, pp. 428–39.
48. History of child labor: Cunningham 1996; Norberg 2016; Ortiz-Ospina & Roser 2016a.
49. M. Wirth, "When Dogs Were Used as Kitchen Gadgets," HumanProgress, Jan. 25, 2017, http://humanprogress.org/blog/when-dogs-were-used-as-kitchen-gadgets.
50. History of the treatment of children: Pinker 2011, chap. 7.
51. Economically worthless, emotionally priceless: Zelizer 1985.
52. Tractor ad: https://goo.gl/LybIW8.
53. Correlation between poverty and child labor: Ortiz-Ospina & Roser 2016a.
54. Desperation, not greed: Norberg 2016; Ortiz-Ospina & Roser 2016a.

CHAPTER 16: KNOWLEDGE

1. Homo sapiens: Pinker 1997/2009, 2010; Tooby & DeVore 1987.
2. Concrete orientation of uneducated peoples: Everett 2008; Flynn 2007; Luria 1976; Oesterdiekhoff 2015; see also my commentary on Everett in https://www.edge.org/conversation/daniel_l_everett-recursion-and-human-thought#22005.
3. Encyclopedia of the Social Sciences, 1931, vol. 5, p. 410, quoted in Easterlin 1981.
4. United Nations Office of the High Commissioner for Human Rights 1966.
5. Education causes economic growth: Easterlin 1981; Glaeser et al. 2004; Hafer 2017; Rindermann 2012; Roser & Ortiz-Ospina 2016a; van Leeuwen & van Leeuwen-Li 2014; van Zanden et al. 2014.
6. I. N. Thut and D. Adams, Educational Patterns in Contemporary Societies (New York: McGraw-Hill, 1964), p. 62, quoted in Easterlin 1981, p. 10.
7. Economic backwardness of Arab countries: Lewis 2002; United Nations Development Programme 2003.
8. Education leads to peace: Hegre et al. 2011; Thyne 2006. Education leads to democracy: Glaeser, Ponzetto, & Shleifer 2007; Hafer 2017; Lutz, Cuaresma, & Abbasi-Shavazi 2010; Rindermann 2008.
9. Youth bulges and violence: Potts & Hayden 2008.
10. Education reduces racism, sexism, homophobia: Rindermann 2008; Teixeira et al. 2013; Welzel 2013.
11. Education increases respect for free speech and imagination: Welzel 2013.
12. Education and civic engagement: Hafer 2017; OECD 2015a; Ortiz-Ospina & Roser 2016c; World Bank 2012b.
13. Education and trust: Ortiz-Ospina & Roser 2016c.

14. Roser & Ortiz-Ospina 2016b, based on data from UNESCO Institute for Statistics, visualized at World Bank 2016a.
15. UNESCO Institute for Statistics, visualized at World Bank 2016i.
16. UNESCO Institute for Statistics, http://data.uis.unesco.org/.
17. On the relationship between literacy and basic education, see van Leeuwen & van Leeuwen-Li 2014, pp. 88–93.
18. Lutz, Butz, & Samir 2014, based on models from the International Institute for Applied Systems Analysis, http://www.iiasa.ac.at/, summarized in Nagdy & Roser 2016c.
19. Ecclesiastes 12:12.
20. Soaring premium for education: Autor 2014.
21. American high school attendance in 1920 and 1930: Leon 2016. Graduation rate in 2011: A. Duncan, "Why I Wear 80," *Huffington Post*, Feb. 14, 2014. High school graduates in college in 2016: Bureau of Labor Statistics 2017.
22. United States Census Bureau 2016.
23. Nagdy & Roser 2016c, based on models from the International Institute for Applied Systems Analysis, http://www.iiasa.ac.at/; Lutz, Butz, & Samir 2014.
24. S. F. Reardon, J. Waldfogel, & D. Bassok, "The Good News About Educational Inequality," *New York Times*, Aug. 26, 2016.
25. Effects of girls' education: Deaton 2013; Nagdy & Roser 2016c; Radelet 2015.
26. United Nations 2015b.
27. Since the first data point for Afghanistan precedes the reign of the Taliban by fifteen years and the second one postdates it by a decade, the gain cannot simply be attributed to the 2001 NATO invasion that deposed the regime.
28. The Flynn effect: Deary 2001; Flynn 2007, 2012. See also Pinker 2011, pp. 650–60.
29. Heritability of intelligence: Pinker 2002/2016, chap. 19 and afterword; Deary 2001; Plomin & Deary 2015; Ritchie 2015.
30. Flynn effect not explained by hybrid vigor: Flynn 2007; Pietschnig & Voracek 2015.
31. Flynn effect meta-analysis: Pietschnig & Voracek 2015.
32. End of the Flynn effect: Pietschnig & Voracek 2015.
33. Evaluating candidate causes of the Flynn effect: Flynn 2007; Pietschnig & Voracek 2015.
34. Nutrition and health explain only part of Flynn effect: Flynn 2007, 2012; Pietschnig & Voracek 2015.
35. Existence and heritability of g: Deary 2001; Plomin & Deary 2015; Ritchie 2015.
36. The Flynn effect as an increase in analytic thinking: Flynn 2007, 2012; Ritchie 2015; Pinker 2011, pp. 650–60.
37. Education affects the Flynn components of intelligence (though not g): Ritchie, Bates, & Deary 2015.
38. IQ as a tailwind: Deary 2001; Gottfredson 1997; Makel et al. 2016; Pinker 2002/2016; Ritchie 2015.
39. The Flynn effect and the moral sense: Flynn 2007; Pinker 2011, pp. 656–70.
40. The Flynn effect and real-world genius: con, Woodley, te Nijenhuis, & Murphy 2013; pro, Pietschnig & Voracek 2015, p. 283.
41. High-tech in the developing world: Diamandis & Kotler 2012; Kenny 2011; Radelet 2015.
42. Benefits of IQ growth: Hafer 2017.
43. Progress as a hidden variable: Land, Michalos, & Sirgy 2012; Prados de la Escosura 2015; van Zanden et al. 2014; Veenhoven 2010.
44. Human Development Index: United Nations Development Programme 2016. Inspirations: Sen 1999; ul Haq 1996.
45. Catching up: Prados de la Escosura 2015, p. 222, counting "the West" as OECD countries prior to 1994, namely the countries of Western Europe and the United States, Canada, Australia, New Zealand, and Japan. He also notes that the index for sub-Saharan Africa in 2007 was .22, equivalent to the world in the 1950s and the OECD countries in the 1890s. Similarly, the Well-Being Composite for sub-Saharan Africa was approximately –.3 in 2000 (it would be higher today), similar to the world around 1910 and Western Europe around 1875.
46. For details and qualifications, see Rijpma 2014 and Prados de la Escosura 2015.

CHAPTER 17: QUALITY OF LIFE

1. The intellectuals and the masses: Carey 1993.
2. Variously attributed to a Jewish joke, a vaudeville routine, and a dialogue from the Broadway play *Ballyhoo of 1932*.
3. Capabilities: Nussbaum 2000.
4. Processing time for food: Laudan 2016.
5. Shorter work hours: Roser 2016t, based on data from Huberman & Minns 2007; see also Tupy 2016, and "Hours Worked Per Worker," *HumanProgress*, http://humanprogress.org/f1/2246, for data showing a reduction of 7.2 hours of work per week worldwide.
6. Housel 2013.
7. Quoted in Weaver 1987, p. 505.
8. Productivity and shorter hours: Roser 2016t. Fewer poorer seniors: Deaton 2013, p. 180. Note that

the absolute percentage of people in poverty depends on how "poverty" is defined; compare, for example, figure 9-6.

9. Data on paid vacations in America summarized in Housel 2013, based on data from the Bureau of Labor Statistics.
10. Data for the UK; calculation by Jesse Ausubel, graphed at http://www.humanprogress.org/static /3261.
11. Trends in working hours in selected developing countries: Roser 2016t.
12. Declining time needed to purchase appliances: M. Tupy, "Cost of Living and Wage Stagnation in the United States, 1979–2015," *HumanProgress*, https://www.cato.org/projects/humanprogress /cost-of-living; Greenwood, Seshadri, & Yorukoglu 2005.
13. Least-preferred pastime: Kahneman et al. 2004. Time spent on housework: Greenwood, Seshadri, & Yorukoglu 2005; Roser 2016t.
14. "Time Spent on Laundry," *HumanProgress*, http://humanprogress.org/static/3264, based on S. Skwire, "How Capitalism Has Killed Laundry Day," *CapX*, April 11, 2016, https://iea.org.uk/blog/ how-capitalism-has-killed-laundry-day.
15. Not to be missed: H. Rosling, "The Magic Washing Machine," TED talk, Dec. 2010, https://www .ted.com/talks/hans_rosling_and_the_magic_washing_machine.
16. *Good Housekeeping*, vol. 55, no. 4, Oct. 1912, p. 436, quoted in Greenwood, Seshadri, & Yorukoglu 2005.
17. From *The Wealth of Nations*.
18. Falling price of light: Nordhaus 1996.
19. Kelly 2016, p. 189.
20. "Yuppie kvetching": Daniel Hamermesh and Jungmin Lee, quoted in E. Kolbert, "No Time," *New Yorker*, May 26, 2014. Trends in leisure, 1965–2003: Aguiar & Hurst 2007. Leisure hours in 2015: Bureau of Labor Statistics 2016c. See the caption to figure 17-6 for more details.
21. More leisure for Norwegians: Aguiar & Hurst 2007, p. 1001, note 24. More leisure for Britons: Ausubel & Grübler 1995.
22. Always rushed? Robinson 2013; J. Robinson, "Happiness Means Being Just Rushed Enough," *Scientific American*, Feb. 19, 2013.
23. Family dinners in 1969 and 1999: K. Bowman, "The Family Dinner, Alive and Well," *New York Times*, Aug. 25, 1999. Family dinners in 2014: J. Hook, "WSJ/NBC Poll Suggests Social Media Aren't Replacing Direct Interactions," *Wall Street Journal*, May 2, 2014. Gallup poll: L. Saad, "Most U.S. Families Still Routinely Dine Together at Home," *Gallup*, Dec. 26, 2013, http://www.gallup .com/poll/166628/families-routinely-dine-together-home.aspx?g_source=family%20and%20dinner &g_medium=search&g_campaign=tiles. Fischer 2011 comes to a similar conclusion.
24. Parents spend more time with their children: Sayer, Bianchi, & Robinson 2004; see also notes 25–27 below.
25. Parents and children: Caplow, Hicks, & Wattenberg 2001, pp. 88–89.
26. Mothers and children: Coontz 1992/2016, p. 24.
27. Increased child care, decreased leisure: Aguiar & Hurst 2007, pp. 980–82.
28. Electronic versus face-to-face contact: Susan Pinker 2014.
29. Pork and starch: N. Irwin, "What Was the Greatest Era for Innovation? A Brief Guided Tour," *New York Times*, May 13, 2016. See also D. Thompson, "America in 1915: Long Hours, Crowded Houses, Death by Trolley," *The Atlantic*, Feb. 11, 2016.
30. Grocery items, 1920s–1980s: N. Irwin, "What Was the Greatest Era for Innovation? A Brief Guided Tour," *New York Times*, May 13, 2016. Items in 2015: Food Marketing Institute 2017.
31. Loneliness and boredom: Bettmann 1974, pp. 62–63.
32. Newspapers and saloons: N. Irwin, "What Was the Greatest Era for Innovation? A Brief Guided Tour," *New York Times*, May 13, 2016.
33. Accuracy of *Wikipedia*: Giles 2005; Greenstein & Zhu 2014; Kräenbring et al. 2014.

CHAPTER 18: HAPPINESS

1. Transcribed and lightly edited from https://www.youtube.com/watch?v=q8LaT5Iiwo4 and other Internet clips.
2. Mueller 1999, p. 14.
3. Easterlin 1973.
4. Hedonic treadmill: Brickman & Campbell 1971.
5. Social comparison theory: See chapter 9, note 11; Kelley & Evans 2017.
6. G. Monbiot, "Neoliberalism Is Creating Loneliness. That's What's Wrenching Society Apart," *The Guardian*, Oct. 12, 2016.
7. Axial Age and origin of deepest questions: Goldstein 2013. Philosophy and history of happiness: Haidt 2006; Haybron 2013; McMahon 2006. Science of happiness: Gilbert 2006; Haidt 2006; Helliwell, Layard, & Sachs 2016; Layard 2005; Ortiz-Ospina & Roser 2017.
8. Human capabilities: Nussbaum 2000, 2008; Sen 1987, 1999.
9. Choosing what doesn't make you happy: Gilbert 2006.
10. Freedom makes people happy: Helliwell, Layard, & Sachs 2016; Inglehart et al. 2008.
11. Freedom makes life meaningful: Baumeister, Vohs, et al. 2013.

12. Validity of happiness reports: Gilbert 2006; Helliwell, Layard, & Sachs 2016; Layard 2005.
13. Experience versus evaluation of happiness: Baumeister, Vohs, et al. 2013; Helliwell, Layard, & Sachs 2016; Kahneman 2011; Veenhoven 2010.
14. Context-sensitivity of ratings happiness versus satisfaction versus good life: Deaton 2011; Helliwell, Layard, & Sachs 2016; Veenhoven 2010. Just average them: Helliwell, Layard, & Sachs 2016; Kelley & Evans 2017; Stevenson & Wolfers 2009.
15. Helliwell, Layard, & Sachs 2016, p. 4, table 2.1, pp. 16, 18.
16. *Eudaemonia* or meaningfulness: Baumeister, Vohs, et al. 2013; Haybron 2013; McMahon 2006; R. Baumeister, "The Meanings of Life," *Aeon*, Sept. 16, 2013.
17. Adaptive function of happiness: Pinker 1997/2009, chap. 6. Different adaptive functions of happiness and meaningfulness: R. Baumeister, "The Meanings of Life," *Aeon*, Sept. 16, 2013.
18. Percent happy: cited in Ipsos 2016; see also Veenhoven 2010. Average ladder placement: 5.4 on a 1–10 scale, Helliwell, Layard, & Sachs 2016, p. 3.
19. Happiness gap: Ipsos 2016.
20. Money does buy happiness: Deaton 2013; Helliwell, Layard, & Sachs 2016; Inglehart et al. 2008; Stevenson & Wolfers 2008a; Ortiz-Ospina & Roser 2017.
21. Independence of happiness and inequality: Kelley & Evans 2017.
22. Helliwell, Layard, & Sachs 2016, pp. 12–13.
23. Winning the lottery: Stephens-Davidowitz 2017, p. 229.
24. National happiness increases over time: Sacks, Stevenson, & Wolfers 2012; Stevenson & Wolfers 2008a; Stokes 2007; Veenhoven 2010; Ortiz-Ospina & Roser 2017.
25. World Values Survey shows increasing happiness: Inglehart et al. 2008.
26. Happiness, health, and freedom: Helliwell, Layard, & Sachs 2016; Inglehart et al. 2008; Veenhoven 2010.
27. Culture and happiness: Inglehart et al. 2008.
28. Non-monetary contributors to happiness: Helliwell, Layard, & Sachs 2016.
29. American happiness: Deaton 2011; Helliwell, Layard, & Sachs 2016; Inglehart et al. 2008; Sacks, Stevenson, & Wolfers 2012; Smith, Son, & Schapiro 2015.
30. *World Happiness Report 2016* rankings: 1. Denmark (7.5 rungs up from the worst possible life); 2. Switzerland; 3. Iceland; 4. Norway; 5. Finland; 6. Canada; 7. Netherlands; 8. New Zealand; 9. Australia; 10. Sweden; 11. Israel; 12. Austria; 13. United States; 14. Costa Rica; 15. Puerto Rico. The unhappiest countries are Benin, Afghanistan, Togo, Syria, and Burundi (157th place, 2.9 rungs up from the worst possible life).
31. American happiness: A fall and rise is seen in the World Database of Happiness (Veenhoven undated), which includes data from the World Values Survey; see the online appendix to Inglehart et al. 2008. A slight decline is seen in the General Social Survey (gss.norc.org); see Smith, Son, & Schapiro 2015 and figure 18-4 in this chapter, which plots the "Very happy" trend.
32. Restriction of range in American happiness: Deaton 2011.
33. Inequality as part of the explanation for the American happiness stagnation: Sacks, Stevenson, & Wolfers 2012.
34. America as a happiness trend outlier: Inglehart et al. 2008; Sacks, Stevenson, & Wolfers 2012.
35. African American happiness increase: Stevenson & Wolfers 2009; Twenge, Sherman, & Lyubomirsky 2016.
36. Declining female happiness: Stevenson & Wolfers 2009.
37. Distinguishing age, period, and cohort: Costa & McCrae 1982; Smith 2008.
38. Older people are happier overall: Deaton 2011; Smith, Son, & Schapiro 2015; Sutin et al. 2013.
39. Dips in middle age and in the final years: Bardo, Lynch, & Land, 2017; Fukuda 2013.
40. Great Recession trough: Bardo, Lynch, & Land 2017.
41. Each successive cohort happier through the Baby Boomers: Sutin et al. 2013.
42. Gen X and Millennials happier than Baby Boomers: Bardo, Lynch, & Land 2017; Fukuda 2013; Stevenson & Wolfers 2009; Twenge, Sherman, & Lyubomirsky 2016.
43. Loneliness, longevity, and health: Susan Pinker 2014.
44. Both quotes are from Fischer 2011, p. 110.
45. Fischer 2011, p. 114. See also Susan Pinker 2014, for a judicious analysis of the changes and constancies.
46. Fischer 2011, p. 114. Fischer cites "a few sources of social support" in full awareness of a highly publicized 2006 report which announced that from 1985 to 2004 Americans reported a third fewer people with whom they could discuss important matters, with a quarter of them saying they had no one at all. He concluded that the result was an artifact of the survey methods: Fischer 2009.
47. Fischer 2011, p. 112.
48. Hampton, Rainie, et al. 2015.
49. Connectedness of social media users: Hampton, Goulet, et al. 2011.
50. Stress in social media users: Hampton, Rainie, et al. 2015.
51. Changes and constancies in social interaction: Fischer 2005, 2011; Susan Pinker 2014.
52. Suicide rates depend on availability of methods: Miller, Azrael, & Barber 2012; Thomas & Gunnell 2010.

53. Risk factors for suicide: Ortiz-Ospina, Lee, & Roser 2016; World Health Organization 2016d.
54. Happiness-suicide paradox: Daly et al. 2010.
55. US suicides in 2014 (42,773, to be exact): Data from National Vital Statistics, Kochanek et al. 2016, table B. World suicides in 2012: Data from World Health Organization, Värnik 2012 and World Health Organization 2016d.
56. "20 graphs to celebrate women's progress around the world," *HumanProgress*, http://human progress.org/blog/20-graphs-to-celebrate-womens-progress-around-the-world.
57. Suicide by age and period for England: Thomas & Gunnell 2010. Suicide by age, cohort, and period for Switzerland: Ajdacic-Gross et al. 2006. For the United States: Phillips 2014.
58. Falling adolescent suicide rates: Costello, Erkanli, & Angold 2006; Twenge 2015.
59. Negative spin on suicide figures: M. Nock, "Five Myths About Suicide," *Washington Post*, May 6, 2016.
60. Eisenhower and Swedish suicide: http://fed.wiki.org/journal.hapgood.net/eisenhower-on-sweden.
61. Suicide rates for 1960 are from Ortiz-Ospina, Lee, & Roser 2016. Suicide rates for 2012 (age-adjusted) are from World Health Organization 2017b.
62. Medium suicide rates in Western Europe: Värnik 2012, p. 768. Decline of Swedish suicide: Ohlander 2010.
63. Generational increase in depression: Lewinsohn et al. 1993.
64. Triggers for PTSD: McNally 2016.
65. Expanding empire of psychopathology: Haslam 2016; Horwitz & Wakefield 2007; McNally 2016; PLOS Medicine Editors 2013.
66. R. Rosenberg, "Abnormal Is the New Normal," *Slate*, April 12, 2013, based on Kessler et al. 2005.
67. Expanding concepts of harm as moral progress: Haslam 2016.
68. Evidence-based psychological treatment: Barlow et al. 2013.
69. Global burden of depression: Murray et al. 2012. Adult risks: Kessler et al. 2003.
70. The paradox of mental health: PLOS Medicine Editors 2013.
71. Lack of gold standard: Twenge 2015.
72. No rise in depression over a century: Mattisson et al. 2005; Murphy et al. 2000.
73. Twenge et al. 2010.
74. Twenge & Nolen-Hoeksema 2002: Between 1980 and 1998, successive cohorts of Generation X and Millennial boys aged 8–16 became *less* depressed, with no change in the girls. Twenge 2015: Between the 1980s and the 2010s, teenagers had fewer suicidal thoughts; college students and adults were less likely to report they were depressed. Olfson, Druss, & Marcus 2015: Rates of mental illness in children and adolescents fell.
75. Costello, Erkanli, & Angold 2006.
76. Baxter et al. 2014.
77. Jacobs 2011.
78. Baxter et al. 2014; Twenge 2015; Twenge et al. 2010.
79. Stein's Law and anxiety: Sage 2010.
80. Terracciano 2010; Trzesniewski & Donnellan 2010.
81. Baxter et al. 2014.
82. For example, "Depression as a Disease of Modernity: Explanations for Increasing Prevalence," Hidaka 2012.
83. Stevenson & Wolfers 2009.
84. Excerpted from the book version: Allen 1987, pp. 131–33.
85. Johnston & Davey 1997; see also Jackson 2016; Otieno, Spada, & Renkl 2013; Unz, Schwab, & Winterhoff-Spurk 2008.
86. Statement: Cornwall Alliance for the Stewardship of Creation 2000. "So-called climate crisis": Cornwall Alliance, "Sin, Deception, and the Corruption of Science: A Look at the So-Called Climate Crisis," 2016, http://cornwallalliance.org/2016/07/sin-deception-and-the-corruption-of-science-a-look-at-the-so-called-climate-crisis/. See also Bean & Teles 2016; L. Vox, "Why Don't Christian Conservatives Worry About Climate Change? God," *Washington Post*, June 2, 2017.
87. Garbage barge: M. Winerip, "Retro Report: Voyage of the Mobro 4000," *New York Times*, May 6, 2013.
88. Environmental friendliness of landfills: J. Tierney, "The Reign of Recycling," *New York Times*, Oct. 3, 2015. The *New York Times* "Retro Report" series, including the story cited in the preceding note, is an exception to the lack of follow-ups on crisis reporting.
89. Boredom crisis: Nisbet 1980/2009, pp. 349–51. The two main alarmists were scientists: Dennis Gabor and Harlow Shapley.
90. See the references in notes 15 and 16 above.
91. Anxiety over the life cycle: Baxter et al. 2014.

CHAPTER 19: EXISTENTIAL THREATS

1. Mythical missile gap: Berry et al. 2010; Preble 2004.
2. Nuclear retaliation for cyberattacks: Sagan 2009c, p. 164. See also the comments from Keith Payne reproduced in P. Sonne, G. Lubold, & C. E. Lee, " 'No First Use' Nuclear Policy Proposal Assailed by U.S. Cabinet Officials, Allies," *Wall Street Journal*, Aug. 12, 2016.

3. K. Bird, "How to Keep an Atomic Bomb from Being Smuggled into New York City? Open Every Suitcase with a Screwdriver," *New York Times*, Aug. 5, 2016.
4. Randle & Eckersley 2015.
5. Quoted on the home page for Ocean Optimism, http://www.oceanoptimism.org/about/.
6. 2012 Ipsos poll: C. Michaud, "One in Seven Thinks End of World Is Coming: Poll," *Reuters*, May 1, 2012, http://www.reuters.com/article/us-mayancalendar-poll-idUSBRE8400XH20120501. The rate for the United States was 22 percent, and in a 2015 YouGov poll, 31 percent: http://cdn.yougov.com/cumulus_uploads/document/i7p20mektl/toplines_OPI_disaster_20150227.pdf.
7. Power-law distributions: Johnson et al. 2006; Newman 2005; see Pinker 2011, pp. 210–22, for a review. See the references in note 17 of chapter 11 for an explanation of the complexities in estimating the risks from the data.
8. Overestimating the probability of extreme risks: Pinker 2011, pp. 368–73.
9. End-of-the-world predictions: "Doomsday Forecasts," *The Economist*, Oct. 7, 2015, http://www.economist.com/blogs/graphicdetail/2015/10/predicting-end-world.
10. Apocalyptic movies: "List of Apocalyptic Films," *Wikipedia*, https://en.wikipedia.org/wiki/List_of_apocalyptic_films, retrieved Dec. 15, 2016.
11. Quoted in Ronald Bailey, "Everybody Loves a Good Apocalypse," *Reason*, Nov. 2015.
12. Y2K bug: M. Winerip, "Revisiting Y2K: Much Ado About Nothing?" *New York Times*, May 27, 2013.
13. G. Easterbrook, "We're All Gonna Die!" *Wired*, July 1, 2003.
14. P. Ball, "Gamma-Ray Burst Linked to Mass Extinction," *Nature*, Sept. 24, 2003.
15. Denkenberger & Pearce 2015.
16. Rosen 2016.
17. D. Cox, "NASA's Ambitious Plan to Save Earth from a Supervolcano," *BBC Future*, Aug. 17, 2017, http://www.bbc.com/future/story/20170817-nasas-ambitious-plan-to-save-earth-from-a-super volcano.
18. Deutsch 2011, p. 207.
19. "More dangerous than nukes": Tweeted in Aug. 2014, quoted in A. Elkus, "Don't Fear Artificial Intelligence," *Slate*, Oct. 31, 2014. "End of the human race": Quoted in R. Cellan-Jones, "Stephen Hawking Warns Artificial Intelligence Could End Mankind," *BBC News*, Dec. 2, 2014, http://www.bbc.com/news/technology-30290540.
20. In a 2014 poll of the hundred most-cited AI researchers, just 8 percent feared that high-level AI posed the threat of "an existential catastrophe": Müller & Bostrom 2014. AI experts who are publicly skeptical include Paul Allen (2011), Rodney Brooks (2015), Kevin Kelly (2017), Jaron Lanier (2014), Nathan Myhrvold (2014), Ramez Naam (2010), Peter Norvig (2015), Stuart Russell (2015), and Roger Schank (2015). Skeptical psychologists and biologists include Roy Baumeister (2015), Dylan Evans (2015), Gary Marcus (2015), Mark Pagel (2015), and John Tooby (2015). See also A. Elkus, "Don't Fear Artificial Intelligence," *Slate*, Oct. 31, 2014; M. Chorost, "Let Artificial Intelligence Evolve," *Slate*, April 18, 2016.
21. Modern scientific understanding of intelligence: Pinker 1997/2009, chap. 2; Kelly 2017.
22. Foom: Hanson & Yudkowsky 2008.
23. The technology expert Kevin Kelly (2017) recently made the same argument.
24. Intelligence as a contraption: Brooks 2015; Kelly 2017; Pinker 1997/2009, 2007a; Tooby 2015.
25. AI doesn't progress by Moore's Law: Allen 2011; Brooks 2015; Deutsch 2011; Kelly 2017; Lanier 2014; Naam 2010. Many of the commentators in Lanier 2014 and Brockman 2015 make this point as well.
26. AI researchers vs. AI hype: Brooks 2015; Davis & Marcus 2015; Kelly 2017; Lake et al. 2017; Lanier 2014; Marcus 2016; Naam 2010; Schank 2015. See also note 25 above.
27. Shallowness and brittleness of current AI: Brooks 2015; Davis & Marcus 2015; Lanier 2014; Marcus 2016; Schank 2015.
28. Naam 2010.
29. Robots turning us into paper clips and other Value Alignment Problems: Bostrom 2016; Hanson & Yudkowsky 2008; Omohundro 2008; Yudkowsky 2008; P. Torres, "Fear Our New Robot Overlords: This Is Why You Need to Take Artificial Intelligence Seriously," *Salon*, May 14, 2016.
30. Why we won't be turned into paper clips: B. Hibbard, "Reply to AI Risk," http://www.ssec.wisc.edu/~billh/g/AIRisk_Reply.html; R. Loosemore, "The Maverick Nanny with a Dopamine Drip: Debunking Fallacies in the Theory of AI Motivation," *Institute for Ethics and Emerging Technologies*, July 24, 2014, http://ieet.org/index.php/IEET/more/loosemore20140724; A. Elkus, "Don't Fear Artificial Intelligence," *Slate*, Oct. 31, 2014; R. Hanson, "I Still Don't Get Foom," *Humanity+*, July 29, 2014, http://hplusmagazine.com/2014/07/29/i-still-dont-get-foom/; Hanson & Yudkowsky 2008. See also Kelly 2017, and notes 26 and 27 above.
31. Quoted in J. Bohannon, "Fears of an AI Pioneer," *Science*, July 17, 2016.
32. Quoted in Brynjolfsson & McAfee 2015.
33. Self-driving cars not quite ready: Brooks 2016.
34. Robots and jobs: Brynjolfsson & McAfee 2016; see also chapter 9, notes 67 and 68.
35. The bet is registered on the "Long Bets" Web site, http://longbets.org/9/.
36. Improving computer security: Schneier 2008; B. Schneier, "Lessons from the Dyn DDoS Attack,"

Schneier on Security, Nov. 1, 2016, https://www.schneier.com/blog/archives/2016/11/lessons _from_th_5.html.

37. Strengthening bioweapon security: Bradford Project on Strengthening the Biological and Toxin Weapons Convention, http://www.bradford.ac.uk/acad/sbtwc/.

38. Protection against infectious disease protects against bioterrorism: Carlson 2010. Preparing for pandemics: Bill & Melinda Gates Foundation, "Preparing for Pandemics," http://nyti.ms /256CNNc; World Health Organization 2016b.

39. Standard antiterrorist measures: Mueller 2006, 2010a; Mueller & Stewart 2016a; Schneier 2008.

40. Kelly 2010, 2013.

41. Personal communication, May 21, 2017; see also Kelly 2013, 2016.

42. Easy to commit murder and mayhem: Branwen 2016.

43. Branwen 2016 lists several real-life examples of product sabotage with damage ranging from $150 million to $1.5 billion.

44. B. Schneier, "Where Are All the Terrorist Attacks?" *Schneier on Security*, https://www.schneier .com/essays/archives/2010/05/where_are_all_the_te.html. Similar points: Mueller 2004b; M. Abrahms, "A Few Bad Men: Why America Doesn't Really Have a Terrorism Problem," *Foreign Policy*, April 17, 2013.

45. Most terrorists are schlemiels: Mueller 2006; Mueller & Stewart 2016a, chap. 4; Branwen 2016; M. Abrahms, "Does Terrorism Work as a Political Strategy? The Evidence Says No," *Los Angeles Times*, April 1, 2016; J. Mueller & M. Stewart, "Hapless, Disorganized, and Irrational: What the Boston Bombers Had in Common with Most Would-Be Terrorists," *Slate*, April 22, 2013; D. Kenner, "Mr. Bean to Jihadi John," *Foreign Policy*, Sept. 12, 2014.

46. D. Adnan & T. Arango, "Suicide Bomb Trainer in Iraq Accidentally Blows Up His Class," *New York Times*, Feb. 10, 2014.

47. "Saudi Suicide Bomber Hid IED in His Anal Cavity," *Homeland Security News Wire*, Sept. 9, 2009, http://www.homelandsecuritynewswire.com/saudi-suicide-bomber-hid-ied-his-anal-cavity.

48. Terrorism is ineffective: Abrahms 2006, 2012; Brandwen 2016; Cronin 2009; Fortna 2015; Mueller 2006; Mueller & Stewart 2010; see also note 45 above. IQ is negatively correlated with criminality and psychopathy: Beaver, Schwartz, et al. 2013; Beaver, Vaughn, et al. 2012; de Ribera, Kavish, & Boutwell 2017.

49. Hazards of larger terrorist plots: Mueller 2006.

50. Serious cybercrime requires a state: B. Schneier, "Someone Is Learning How to Take Down the Internet," *Lawfare*, Sept. 13, 2016.

51. Skepticism about cyberwar: Lawson 2013; Mueller & Friedman 2014; Rid 2012; B. Schneier, "Threat of 'Cyberwar' Has Been Hugely Hyped," *CNN.com*, July 7, 2010, http://www.cnn.com /2010/OPINION/07/07/schneier.cyberwar.hyped/; E. Morozov, "Cyber-Scare: The Exaggerated Fears over Digital Warfare," *Boston Review*, July/Aug. 2009; E. Morozov, "Battling the Cyber Warmongers," *Wall Street Journal*, May 8, 2010; R. Singel, "Cyberwar Hype Intended to Destroy the Open Internet," *Wired*, March 1, 2010; R. Singel, "Richard Clarke's *Cyberwar*: File Under Fiction," *Wired*, April 22, 2010; P. W. Singer, "The Cyber Terror Bogeyman," *Brookings*, Nov. 1, 2012, https://www.brookings.edu/articles/the-cyber-terror-bogeyman/.

52. From Schneier's article cited in the preceding note.

53. Resilience: Lawson 2013; Quarantelli 2008.

54. Quarantelli 2008, p. 899.

55. Societies don't collapse under disasters: Lawson 2013; Quarantelli 2008.

56. Modern societies are resilient: Lawson 2013.

57. Biological warfare and terrorism: Ewald 2000; Mueller 2006.

58. Terrorism as theater: Abrahms 2006; Branwen 2016; Cronin 2009; Ewald 2000; Y. N. Harari, "The Theatre of Terror," *The Guardian*, Jan. 31, 2015.

59. Evolution of virulence and contagion: Ewald 2000; Walther & Ewald 2004.

60. Rarity of bioterrorism: Mueller 2006; Parachini 2003.

61. Difficulty of designing a pathogen even with gene-editing: Paul Ewald, personal communication, Dec. 27, 2016.

62. Comment in Kelly 2013, summarizing arguments in Carlson 2010.

63. New antibiotics: Meeske et al. 2016; Murphy, Zeng, & Herzon 2017; Seiple et al. 2016. Identifying potentially hazardous pathogens: Walther & Ewald 2004.

64. Ebola vaccine: Henao-Restrepo et al. 2017. False predictions of catastrophic pandemics: Norberg 2016; Ridley 2010; M. Ridley, "Apocalypse Not: Here's Why You Shouldn't Worry About End Times," *Wired*, Aug. 17, 2012; D. Bornstein & T. Rosenberg, "When Reportage Turns to Cynicism," *New York Times*, Nov. 14, 2016.

65. Bet on bioterror with Martin Rees: http://longbets.org/9/.

66. Reviews of nuclear weapons today: Evans, Ogilvie-White, & Thakur 2015; Federation of American Scientists (undated); Rhodes 2010; Scoblic 2010.

67. World's nuclear stockpile: Kristensen & Norris 2016a; see also note 113 below.

68. Nuclear winter: Robock & Toon 2012; A. Robock & O. B. Toon, "Let's End the Peril of a Nuclear Winter," *New York Times*, Feb. 11, 2016. History of nuclear winter/autumn controversy: Morton 2015.

69. Doomsday Clock: *Bulletin of the Atomic Scientists* 2017.
70. Eugene Rabinowitch, quoted in Mueller 2010a, p. 26.
71. Doomsday Clock: *Bulletin of the Atomic Scientists*, "A Timeline of Conflict, Culture, and Change," Nov. 13, 2013, http://thebulletin.org/multimedia/timeline-conflict-culture-and-change.
72. Quoted in Mueller 1989, p. 98.
73. Quoted in Mueller 1989, p. 271, note 2.
74. Snow 1961, p. 259.
75. Address to the incoming graduate students, Faculty of Arts and Sciences, Harvard University, September 1976.
76. Quoted in Mueller 1989, p. 271, note 2.
77. Close call lists: Future of Life Institute 2017; Schlosser 2013; Union of Concerned Scientists 2015a.
78. Union of Concerned Scientists, "To Russia with Love," http://www.ucsusa.org/nuclear-weapons/close-calls#.WGQC1lMrJEY.
79. Skepticism on close-call lists: Mueller 2010a; J. Mueller, "Fire, Fire (Review of E. Schlosser's 'Command and Control')," *Times Literary Supplement*, March 7, 2014.
80. The Google Ngram Viewer (https://books.google.com/ngrams) indicates that in 2008 (the most recent year displayed) mentions of *nuclear war* in published books were outnumbered by mentions of *racism, terrorism,* and *inequality* tenfold to twentyfold. The Corpus of Contemporary American English (http://corpus.byu.edu/coca/) indicates that in American newspapers in 2015, *nuclear war* appeared 0.65 times per million words of text, compared with 13.13 times for *inequality,* 19.5/million for *racism,* and 30.93/million for *terrorism.*
81. Quotes taken from Morton 2015, p. 324.
82. Letter dated 17 April 2003 to the Security Council, written when he was the US representative to the UN, quoted in Mueller 2012.
83. Collection of terror predictions: Mueller 2012.
84. Warren B. Rudman, Stephen E. Flynn, Leslie H. Gelb, and Gary Hart, Dec. 16, 2004, reproduced in Mueller 2012.
85. Quoted in Boyer 1985/2005, p. 72.
86. Scare tactics backfiring: Boyer 1986.
87. From a 1951 editorial in the *Bulletin of the Atomic Scientists*, quoted in Boyer 1986.
88. What motivates activism: Sandman & Valenti 1986. See chapter 10, note 55, for similar observations on climate change.
89. Quoted in Mueller 2016.
90. Quoted in Mueller 2016. The term *nuclear metaphysics* comes from the political scientist Robert Johnson.
91. Disarmament without treaties: Kristensen & Norris 2016a; Mueller 2010a.
92. Odds next to zero: Welch & Blight 1987–88, p. 27; see also Blight, Nye, & Welch 1987, p. 184; Frankel 2004; Mueller 2010a, pp. 38–40, p. 248, notes 31–33.
93. Nuclear safety features prevent accidents: Mueller 2010a, pp. 100–102; Evans, Ogilvie-White, & Thakur 2015, p. 56; J. Mueller, "Fire, Fire (Review of E. Schlosser's 'Command and Control')," *Times Literary Supplement*, March 7, 2014. Note that the common claim that the Soviet navy officer Vasili Arkhipov "saved the world" during the Cuban Missile Crisis by overruling an embattled submarine captain who was about to fire a nuclear-tipped torpedo at American ships is cast in doubt by Aleksandr Mozgovoi's 2002 book *Kubinskaya Samba Kvarteta Fokstrotov* (Cuban Samba of the Quartet of Foxtrots), in which Vadim Pavlovich Orlov, a communications officer who took part in the events, reports that the captain had spontaneously backed off from his impulse: Mozgovoi 2002. Note as well that a single tactical weapon detonated at sea would not necessarily have escalated into all-out war; see Mueller 2010a, pp. 100–102.
94. Union of Concerned Scientists 2015a.
95. The history of chemical weapons after they were banned following World War I suggests that accidental and one-time uses don't automatically lead to mutual escalation; see Pinker 2011, pp. 273–74.
96. Predictions of nuclear proliferation: Mueller 2010a, p. 90; T. Graham, "Avoiding the Tipping Point," *Arms Control Today*, 2004, https://www.armscontrol.org/act/2004_11/BookReview. Lack of proliferation: Bluth 2011; Sagan 2009b, 2010.
97. States that gave up nukes: Sagan 2009b, 2010, and personal communication, Dec. 30, 2016; see Pinker 2011, pp. 272–73.
98. G. Evans 2015.
99. Quoted in Pinker 2013a.
100. Poison gas from airplanes: Mueller 1989. Geophysical warfare: Morton 2015, p. 136.
101. The USSR, not Hiroshima, made Japan surrender: Berry et al. 2010; Hasegawa 2006; Mueller 2010a; Wilson 2007.
102. Nobel to the nukes: Suggested by Elspeth Rostow, quoted in Pinker 2011, p. 268. Nuclear weapons are poor deterrents: Pinker 2011, p. 269; Berry et al. 2010; Mueller 2010a; Ray 1989.
103. Nuclear taboo: Mueller 1989; Sechser & Fuhrmann 2017; Tannenwald 2005; Ray 1989, pp. 429–31; Pinker 2011, chap. 5, "Is the Long Peace a Nuclear Peace?" pp. 268–78.
104. Effectiveness of conventional deterrence: Mueller 1989, 2010a.

105. Nuclear states and armed burglars: Schelling 1960.
106. Berry et al. 2010, pp. 7–8.
107. George Shultz, William Perry, Henry Kissinger, & Sam Nunn, "A World Free of Nuclear Weapons," *Wall Street Journal*, Jan. 4, 2007; William Perry, George Shultz, Henry Kissinger, & Sam Nunn, "Toward a Nuclear-Free World," *Wall Street Journal*, Jan. 15, 2008.
108. "Remarks by President Barack Obama in Prague as Delivered," White House, April 5, 2009, https://obamawhitehouse.archives.gov/the-press-office/remarks-president-barack-obama-prague-delivered.
109. United Nations Office for Disarmament Affairs (undated).
110. Public opinion on Global Zero: Council on Foreign Relations 2009.
111. Getting to zero: Global Zero Commission 2010.
112. Global Zero skeptics: H. Brown & J. Deutch, "The Nuclear Disarmament Fantasy," *Wall Street Journal*, Nov. 19, 2007; Schelling 2009.
113. The Pentagon has reported that in 2015 the US nuclear stockpile contained 4,571 weapons (United States Department of Defense 2016). The Federation of American Scientists (Kristensen & Norris 2016b, updated in Kristensen 2016) estimates that about 1,700 of the warheads are deployed on ballistic missiles and at bomber bases, 180 consist of tactical bombs deployed in Europe, and the remaining 2,700 are kept in storage. (The term *stockpile* usually embraces both deployed and stored missiles, though sometimes it refers just to the stored ones.) In addition, approximately 2,340 warheads are retired and awaiting dismantlement.
114. A. E. Kramer, "Power for U.S. from Russia's Old Nuclear Weapons," *New York Times*, Nov. 9, 2009.
115. The Federation of American Scientists estimates the 2015 Russian stockpile at 4,500 warheads (Kristensen & Norris 2016b). New START: Woolf 2017.
116. Stockpile reductions will continue in tandem with modernization: Kristensen 2016.
117. Nuclear arsenals: Estimates from Kristensen 2016; they include warheads that are deployed or kept in storage and deployable; they exclude warheads that are retired, and bombs that cannot be deployed by the nation's delivery platforms.
118. No imminent new nuclear states: Sagan 2009b, 2010, and personal communication, Dec. 30, 2016; see also Pinker 2011, pp. 272–73. Fewer states with fissile materials: "Sam Nunn Discusses Today's Nuclear Risks," Foreign Policy Association blogs, http://foreignpolicyblogs.com/2016/04/06/sam-nunn-discusses-todays-nuclear-risks/.
119. Disarmament without treaties: Kristensen & Norris 2016a; Mueller 2010a.
120. GRIT: Osgood 1962.
121. Small arsenal, no nuclear winter: A. Robock & O. B. Toon, "Let's End the Peril of a Nuclear Winter," *New York Times*, Feb. 11, 2016. The authors recommend that the United States reduce its arsenal to 1,000 warheads, but they don't say whether this would rule out the possibility of nuclear winter. The number 200 comes from a presentation by Robock at MIT, April 2, 2016, "Climatic Consequences of Nuclear War," http://futureoflife.org/wp-content/uploads/2016/04/Alan_Robock_MIT_April2.pdf.
122. No hair trigger: Evans, Ogilvie-White, & Thakur 2015, p. 56.
123. Against launch on warning: Evans, Ogilvie-White, & Thakur 2015; J. E. Cartwright & V. Dvorkin, "How to Avert a Nuclear War," *New York Times*, April 19, 2015; B. Blair, "How Obama Could Revolutionize Nuclear Weapons Strategy Before He Goes," *Politico*, June 22, 2016; Long fuse: Brown & Lewis 2013.
124. Takes nukes off "hair trigger": Union of Concerned Scientists 2015b.
125. No First Use: Sagan 2009a; J. E. Cartwright & B. G. Blair, "End the First-Use Policy for Nuclear Weapons," *New York Times*, Aug. 14, 2016. Rebuttals of arguments against No First Use: Global Zero Commission 2016; B. Blair, "The Flimsy Case Against No-First-Use of Nuclear Weapons," *Politico*, Sept. 28, 2016.
126. Incremental pledges: J. G. Lewis & S. D. Sagan, "The Common-Sense Fix That American Nuclear Policy Needs," *Washington Post*, Aug. 24, 2016.
127. D. Sanger & W. J. Broad, "Obama Unlikely to Vow No First Use of Nuclear Weapons," *New York Times*, Sept. 5, 2016.

CHAPTER 20: THE FUTURE OF PROGRESS

1. The data in these paragraphs come from chapters 5–19.
2. All declines calculated as a proportion of their 20th-century peaks.
3. For evidence that war in particular is not cyclical, see Pinker 2011, p. 207.
4. From the *Review of Southey's Colloquies on Society*, quoted in Ridley 2010, chap. 1.
5. See the references at the end of chapters 8 and 16; pp. 124 and 130 of chapter 10; p. 228 of chapter 15; and the discussion of the Easterlin paradox in chapter 18.
6. Average of the years 1961 through 1973; World Bank 2016c.
7. Average of the years 1974 through 2015; World Bank 2016c. Rates for the United States for these two periods are 3.3 percent and 1.7 percent, respectively.
8. Estimates are of Total Factor Productivity, taken from Gordon 2014, fig. 1.
9. Secular stagnation: Summers 2014b, 2016. For analysis and commentaries, see Teulings & Baldwin 2014.
10. No one knows: M. Levinson, "Every US President Promises to Boost Economic Growth. The

Catch: No One Knows How," *Vox*, Dec. 22, 2016; G. Ip, "The Economy's Hidden Problem: We're Out of Big Ideas," *Wall Street Journal*, Dec. 20, 2016; Teulings & Baldwin 2014.

11. Gordon 2014, 2016.
12. American complacency: Cowen 2017; Glaeser 2014; F. Erixon & B. Weigel, "Risk, Regulation, and the Innovation Slowdown," *Cato Policy Report*, Sept./Oct. 2016; G. Ip, "The Economy's Hidden Problem: We're Out of Big Ideas," *Wall Street Journal*, Dec. 20, 2016.
13. World Bank 2016c. American GDP per capita has grown in all but eight of the past fifty-five years.
14. Sleeper effect in technological development: G. Ip, "The Economy's Hidden Problem: We're Out of Big Ideas," *Wall Street Journal*, Dec. 20, 2016; Eichengreen 2014.
15. Technologically driven age of abundance: Brand 2009; Bryce 2014; Brynjolfsson & McAfee 2016; Diamandis & Kotler 2012; Eichengreen 2014; Mokyr 2014; Naam 2013; Reese 2013.
16. Interview with Ezra Klein, "Bill Gates: The Energy Breakthrough That Will 'Save Our Planet' Is Less Than 15 Years Away," *Vox*, Feb. 24, 2016, http://www.vox.com/2016/2/24/11100702/billgates energy. Gates casually alluded to the "'peace breaks out' book that was written in 1940." I'm guessing he was referring to Norman Angell's *The Great Illusion*, commonly misremembered as having predicted that war was impossible on the eve of World War I. In fact the pamphlet, first published in 1909, argued that war was unprofitable, not that it was obsolete.
17. Diamandis & Kotler 2012, p. 11.
18. Fossil power, guilt-free: Service 2017.
19. Jane Langdale, "Radical Ag: C4 Rice and Beyond," Seminars About Long-Term Thinking, Long Now Foundation, March 14, 2016.
20. Second Machine Age: Brynjolfsson & McAfee 2016. See also Diamandis & Kotler 2012.
21. Mokyr 2014, p. 88; see also Feldstein 2017; T. Aeppel, "Silicon Valley Doesn't Believe U.S. Productivity Is Down," *Wall Street Journal*, July 16, 2015; K. Kelly, "The Post-Productive Economy," *The Technium*, Jan. 1, 2013.
22. Demonetization: Diamandis & Kotler 2012.
23. G. Ip, "The Economy's Hidden Problem: We're Out of Big Ideas," *Wall Street Journal*, Dec. 20, 2016.
24. Authoritarian populism: Inglehart & Norris 2016; Norris & Inglehart 2016; see also chapter 23 in this book.
25. Norris & Inglehart 2016.
26. History of Trump through his election: J. Fallows, "The Daily Trump: Filling a Time Capsule," *The Atlantic*, Nov. 20, 2016, http://www.theatlantic.com/notes/2016/11/on-the-future-of-the-time -capsules/508268/. History of Trump in his first half-year as president: E. Levitz, "All the Terrifying Things That Donald Trump Did Lately," *New York*, June 9, 2017.
27. "Donald Trump's File," *PolitiFact*, http://www.politifact.com/personalities/donald-trump/. See also D. Dale, "Donald Trump: The Unauthorized Database of False Things," *The Star*, Nov. 4, 2016, which lists 560 false claims he made in a span of two months, about twenty per day; M. Yglesias, "The Bullshitter-in-Chief," *Vox*, May 30, 2017; and D. Leonhardt & S. A. Thompson, "Trump's Lies," *New York Times*, June 23, 2017.
28. Adapted from the science-fiction writer Philip K. Dick: "Reality is that which, when you stop believing in it, doesn't go away."
29. S. Kinzer, "The Enlightenment Had a Good Run," *Boston Globe*, Dec. 23, 2016.
30. Obama approval: J. McCarthy, "President Obama Leaves White House with 58% Favorable Rating," *Gallup*, Jan. 16, 2017, http://www.gallup.com/poll/202349/president-obama-leaves-white -house-favorable-rating.aspx. Farewell address: Obama referred to the "essential spirit of innovation and practical problem-solving that guided our Founders" that was "born of the Enlightenment" and which he defined as "a faith in reason, and enterprise, and the primacy of right over might" ("President Obama's Farewell Address, Jan. 10, 2017," *The White House*, https://www .cnn.com/2017/01/10/politics/president-obama-farewell-speech/index.html).
31. Trump ratings: J. McCarthy, "Trump's Pre-Inauguration Favorables Remain Historically Low," *Gallup*, Jan. 16, 2017; "How Unpopular Is Donald Trump?" *FiveThirtyEight*, https://projects .fivethirtyeight.com/trump-approval-ratings/; "Presidential Approval Ratings—Donald Trump," *Gallup*, Aug. 25, 2017.
32. G. Aisch, A. Pearce, & B. Rousseau, "How Far Is Europe Swinging to the Right?" *New York Times*, Dec. 5, 2016. Of the twenty countries whose parliamentary elections were tracked, nine had an increase in the representation of right-wing parties since the preceding election, nine had a decrease, and two (Spain and Portugal) had no representation at all.
33. A. Chrisafis, "Emmanuel Macron Vows Unity After Winning French Presidential Election," *The Guardian*, May 8, 2017.
34. US election exit poll data, *New York Times* 2016. N. Carnes & N. Lupu, "It's Time to Bust the Myth: Most Trump Voters Were Not Working Class," *Washington Post*, June 5, 2017. See also the references in notes 35 and 36 below.
35. N. Silver, "Education, Not Income, Predicted Who Would Vote for Trump," *FiveThirtyEight*, Nov. 22, 2016, http://fivethirtyeight.com/features/education-not-income-predicted-who-would-vote -for-trump/; N. Silver, "The Mythology of Trump's 'Working Class' Support: His Voters Are Better Off Economically Compared with Most Americans," *FiveThirtyEight*, May 3, 2016, https://

fivethirtyeight.com/features/the-mythology-of-trumps-working-class-support/. Confirmation from Gallup polls: J. Rothwell, "Economic Hardship and Favorable Views of Trump," *Gallup*, July 22, 2016, http://www.gallup.com/opinion/polling-matters/193898/economic-hardship-favorable-views-trump.aspx.

36. N. Silver, "Strongest correlate I've found for Trump support is Google searches for the n-word. Others have reported this too," *Twitter*, https://twitter.com/natesilver538/status/703975062500732932?lang=en; N. Cohn, "Donald Trump's Strongest Supporters: A Certain Kind of Democrat," *New York Times*, Dec. 31, 2015; Stephens-Davidowitz 2017. See also G. Lopez, "Polls Show Many—Even Most—Trump Supporters Really Are Deeply Hostile to Muslims and Nonwhites," *Vox*, Sept. 12, 2016.

37. Exit poll data: *New York Times* 2016.

38. European populism: Inglehart & Norris 2016.

39. Inglehart & Norris 2016; based on their Model C, the one with the combination of best fit and fewest predictors, endorsed by the authors.

40. A. B. Guardia, "How Brexit Vote Broke Down," *Politico*, June 24, 2016.

41. Inglehart & Norris 2016, p. 4.

42. Quoted in I. Lapowsky, "Don't Let Trump's Win Fool You—America's Getting More Liberal," *Wired*, Dec. 19, 2016.

43. Populist party representation in different countries: Inglehart & Norris 2016; G. Aisch, A. Pearce, & B. Rousseau, "How Far Is Europe Swinging to the Right?" *New York Times*, Dec. 5, 2016.

44. Tininess of the alt-right movement: Alexander 2016. Seth Stephens-Davidowitz notes that Google searches for "Stormfront," the most prominent white nationalist Internet forum, have been in steady decline since 2008 (other than a few news-related blips).

45. Young liberal, old conservative meme: G. O'Toole, "If You Are Not a Liberal at 25, You Have No Heart. If You Are Not a Conservative at 35 You Have No Brain," *Quote Investigator*, Feb. 24, 2014, http://quoteinvestigator.com/2014/02/24/heart-head/; B. Popik, "If You're Not a Liberal at 20 You Have No Heart, If Not a Conservative at 40 You Have No Brain," *BarryPopik.com*, http://www.barrypopik.com/index.php/new_york_city/entry/if_youre_not_a_liberal_at_20_you_have_no_heart_if_not_a_conservative_at_40.

46. Ghitza & Gelman 2014; see also Kohut et al. 2011; Taylor 2016a, 2016b.

47. Based loosely on a quotation from the physicist Max Planck.

48. Voter turnout: H. Enten, "Registered Voters Who Stayed Home Probably Cost Clinton the Election," *FiveThirtyEight*, Jan. 5, 2017, https://fivethirtyeight.com/features/registered-voters-who-stayed-home-probably-cost-clinton-the-election/. A. Payne, "Brits Who Didn't Vote in the EU Referendum Now Wish They Voted Against Brexit," *Business Insider*, Sept. 23, 2016. A. Rhodes, "Young People—If You're So Upset by the Outcome of the EU Referendum, Then Why Didn't You Get Out and Vote?" *The Independent*, June 27, 2016.

49. Publius Decius Mus 2016. In 2017, the author of the pseudonymous piece, Michael Anton, joined the Trump administration as a national security official.

50. C. R. Ketcham, "Anarchists for Donald Trump—Let the Empire Burn," *Daily Beast*, June 9, 2016, http://www.thedailybeast.com/articles/2016/06/09/anarchists-for-donald-trump-let-the-empire-burn.html.

51. A similar argument was made by D. Bornstein & T. Rosenberg, "When Reportage Turns to Cynicism," *New York Times*, Nov. 15, 2016, quoted in chapter 4.

52. Berlin 1988/2013, p. 15.

53. Excerpt from a talk, shared in a personal communication; adapted from Kelly 2016, pp. 13–14.

54. "Pessimistic hopefulness" is from the journalist Yuval Levin (2017). "Radical incrementalism" is originally from the political scientist Aaron Wildavsky, recently revived by Halpern & Mason 2015.

55. The term *possibilism* had previously been coined by the economist Albert Hirschman (1971). Rosling was quoted in "Making Data Dance," *The Economist*, Dec. 9, 2010.

CHAPTER 21: REASON

1. Recent examples (not from psychologists): J. Gray, "The Child-Like Faith in Reason," *BBC News Magazine*, July 18, 2014; C. Bradatan, "Our Delight in Destruction," *New York Times*, March 27, 2017.

2. Nagel 1997, pp. 14–15. "One can't criticize something with nothing": p. 20.

3. Transcendental arguments: Bardon (undated).

4. Nagel 1997, p. 35, attributes the phrase "One thought too many" to the philosopher Bernard Williams, who used it to make a different point. For more on why "believing in reason" is one thought too many, and why explicit deduction has to stop somewhere, see Pinker 1997/2009, pp. 98–99.

5. See the references in chapter 2, notes 22–25.

6. See the references in chapter 1, notes 4 and 9. Kant's metaphor refers to the "unsocial sociability" of humans, who differ from trees in a crowded forest that grow straight to stay out of each other's shadows. It has been interpreted as applying to reason insofar as humans have difficulty seeing the advantages of cooperation. (Thanks to Anthony Pagden for pointing this out to me.)

7. Selection for rationality: Pinker 1997/2009, chaps. 2 and 5; Pinker 2010; Tooby & DeVore 1987; Norman 2016.
8. Personal communication, Jan. 5, 2017; for supporting detail, see Liebenberg 1990, 2014.
9. Liebenberg 2014, pp. 191–92.
10. Shtulman 2006; see also Rice, Olson, & Colbert 2011.
11. Evolution as a litmus for religiosity: Roos 2014.
12. Kahan 2015.
13. Climate literacy: Kahan 2015; Kahan, Wittlin, et al. 2011. Ozone hole, toxic waste dumps, and climate change: Bostrom et al. 1994.
14. Pew Research Center 2015b; see Jones, Cox, & Navarro-Rivera 2014, for similar data.
15. Kahan: Braman et al. 2007; Eastop 2015; Kahan 2015; Kahan, Jenkins-Smith, & Braman 2011; Kahan, Jenkins-Smith, et al. 2012; Kahan, Wittlin, et al. 2011.
16. Kahan, Wittlin, et al. 2011, p. 15.
17. Tragedy of the Belief Commons: Kahan 2012; Kahan, Wittlin, et al. 2011. Kahan calls it the Tragedy of the Risk-Perception Commons.
18. A. Marcotte, "It's Science, Stupid: Why Do Trump Supporters Believe So Many Things That Are Crazy and Wrong?" *Salon*, Sept. 30, 2016.
19. Blue lies: J. A. Smith, "How the Science of 'Blue Lies' May Explain Trump's Support," *Scientific American*, March 24, 2017.
20. Tooby 2017.
21. Motivated reasoning: Kunda 1990. My-Side bias: Baron 1993. Biased evaluation: Lord, Ross, & Lepper 1979; Taber & Lodge 2006. See also Mercier & Sperber 2011, for a review.
22. Hastorf & Cantril 1954.
23. Testosterone and elections: Stanton et al. 2009.
24. Polarizing effect of evidence: Lord, Ross, & Lepper 1979. For updates, see Taber & Lodge 2006 and Mercier & Sperber 2011.
25. Political engagement as sports fandom: Somin 2016.
26. Kahan, Peters, et al. 2012; Kahan, Wittlin, et al. 2011.
27. Kahan, Braman, et al. 2009.
28. M. Kaplan, "The Most Depressing Discovery About the Brain, Ever," *Alternet*, Sept. 16, 2013, http://www.alternet.org/media/most-depressing-discovery-about-brain-ever. Study itself: Kahan, Peters, et al. 2013.
29. E. Klein, "How Politics Makes Us Stupid," *Vox*, April 6, 2014; C. Mooney, "Science Confirms: Politics Wrecks Your Ability to Do Math," *Grist*, Sept. 8, 2013.
30. Bias bias (actually called the "bias blind spot"): Pronin, Lin, & Ross 2002.
31. Verhulst, Eaves, & Hatemi 2016.
32. Rigged studies on prejudice: Duarte et al. 2015.
33. Economic illiteracy among leftists: Buturovic & Klein 2010; see also Caplan 2007.
34. Economic illiteracy follow-up and retraction: Klein & Buturovic 2011.
35. D. Klein, "I Was Wrong, and So Are You," *The Atlantic*, Dec. 2011.
36. See Pinker 2011, chaps. 3–5.
37. Deaths from communism: Courtois et al. 1999; Rummel 1997; White 2011; see also Pinker 2011, chaps. 4–5.
38. Marxists among social scientists: Gross & Simmons 2014.
39. According to the 2016 *Index of Economic Freedom* compiled by the *Wall Street Journal* and the Heritage Foundation (http://www.heritage.org/index/ranking), New Zealand, Canada, Ireland, the United Kingdom, and Denmark equal or exceed the United States in economic freedom. All but Canada exceed the United States in the proportion of GDP devoted to social spending (OECD 2014).
40. The trouble with right-wing libertarianism: Friedman 1997; J. Taylor, "Is There a Future for Libertarianism?" *RealClearPolicy*, Feb. 23, 2016, http://www.realclearpolicy.com/blog/2016/02/23/is_there_a_future_for_libertarianism_1563.html; M. Lind, "The Question Libertarians Just Can't Answer," *Salon*, June 4, 2013; B. Lindsey, "Liberaltarians," *New Republic*, Dec. 4, 2006; W. Wilkinson, "Libertarian Principles, Niskanen, and Welfare Policy," Niskanen blog, March 29, 2016, https://niskanencenter.org/blog/libertarian-principles-niskanen-and-welfare-policy/.
41. The road to totalitarianism: Payne 2005.
42. Though the United States has the world's highest GDP, it falls in 13th place in happiness (Helliwell, Layard, & Sachs 2016), 8th in the UN's Human Development Index (Roser 2016h), and 19th in the Social Progress Index (Porter, Stern, & Green 2016). Recall that social transfers boost the Human Development Index up to around 25–30 percent of GDP (Prados de la Escosura 2015); the United States allocates around 19 percent.
43. Visions of the left and right: Pinker 2002/2016; Sowell 1987, chap. 16.
44. The problems with predictions: Gardner 2010; Mellers et al. 2014; Silver 2015; Tetlock & Gardner 2015; Tetlock, Mellers, & Scoblic 2017.
45. N. Silver, "Why FiveThirtyEight Gave Trump a Better Chance Than Almost Anyone Else," *FiveThirtyEight*, Nov. 11, 2016, http://fivethirtyeight.com/features/why-fivethirtyeight-gave-trump-a-better-chance-than-almost-anyone-else/.

46. Tetlock & Gardner 2015, p. 68.
47. Tetlock & Gardner 2015, p. 69.
48. Active open-mindedness: Baron 1993.
49. Tetlock 2015.
50. Increasing political polarization: Pew Research Center 2014.
51. Data from the General Social Survey, http://gss.norc.org, compiled in Abrams 2016.
52. Abrams 2016.
53. Political orientations of college faculty: Eagan et al. 2014; Gross & Simmons 2014; E. Schwitz-gebel, "Political Affiliations of American Philosophers, Political Scientists, and Other Academics," *Splintered Mind*, http://schwitzsplinters.blogspot.hk/2008/06/political-affiliations-of-american.html. See also N. Kristof, "A Confession of Liberal Intolerance," *New York Times*, May 7, 2016.
54. Liberal tilt of journalism: In 2013, the ratio of Democrats to Republicans among American journalists was four to one, though a majority were Independent (50.2 percent) or Other (14.6 percent); Willnat & Weaver 2014, p. 11. A recent content analysis suggests that newspapers slant a bit to the left, but so do their readers; Gentzkow & Shapiro 2010.
55. Social forces congenial to liberals versus conservatives: Sowell 1987.
56. Intellectual liberals at the forefront: Grayling 2007; Hunt 2007.
57. We are all liberals: Courtwright 2010; Nash 2009; Welzel 2013.
58. Political bias in science: Jussim et al. 2017. Political bias in medicine: Satel 2000.
59. Duarte et al. 2015.
60. "Look different but think alike": from the civil liberties lawyer Harvey Silverglate.
61. Duarte et al. 2015 includes thirty-three commentaries, many critical but all respectful, and the authors' response. *The Blank Slate* won prizes from two divisions of the American Psychological Association.
62. N. Kristof, "A Confession of Liberal Intolerance," *New York Times*, May 7, 2016; N. Kristof, "The Liberal Blind Spot," *New York Times*, May 28, 2016.
63. J. McWhorter, "Antiracism, Our Flawed New Religion," *Daily Beast*, July 27, 2015.
64. Illiberalism on campus and social justice warriors: Lukianoff 2012, 2014; G. Lukianoff & J. Haidt, "The Coddling of the American Mind," *The Atlantic*, Sept. 2015; L. Jussim, "Mostly Leftist Threats to Mostly Campus Speech," *Psychology Today* blog, Nov. 23, 2015, https://www.psychologytoday.com/us/blog/rabble-rouser/201511/mostly-leftist-threats-mostly-campus-speech.
65. Public shaming: D. Lat, "The Harvard Email Controversy: How It All Began," *Above the Law*, May 3, 2010, http://abovethelaw.com/2010/05/the-harvard-email-controversy-how-it-all-began/.
66. Stalinesque investigations: Dreger 2015; A. Reese & C. Maltby, "In Her Own Words: L. Kipnis' 'Title IX Inquisition' at Northwestern," *TheFire.org*, https://www.thefire.org/in-her-own-words-laura-kipnis-title-ix-inquisition-at-northwestern-video/; see also note 64 above.
67. Unintended comedy: G. Lukianoff & J. Haidt, "The Coddling of the American Mind," *The Atlantic*, Sept. 2015; C. Friedersdorf, "The New Intolerance of Student Activism," *The Atlantic*, Nov. 9, 2015; J. W. Moyer, "University Yoga Class Canceled Because of 'Oppression, Cultural Genocide,'" *Washington Post*, Nov. 23, 2015.
68. Comedians are not amused: G. Lukianoff & J. Haidt, "The Coddling of the American Mind," *The Atlantic*, Sept. 2015; T. Kingkade, "Chris Rock Stopped Playing Colleges Because They're 'Too Conservative,'" *Huffington Post*, Dec. 2, 2014. See also the 2015 documentary, *Can We Take a Joke?*
69. Diversity of opinions within academia: Shields & Dunn 2016.
70. The earliest version was expressed by Samuel Johnson; see G. O'Toole, "Academic Politics Are So Vicious Because the Stakes Are So Small," *Quote Investigator*, Aug. 18, 2013, http://quoteinvestigator.com/2013/08/18/acad-politics/.
71. Extremist, antidemocratic Republicans: Mann & Ornstein 2012/2016.
72. Cynicism about democracy: Foa & Mounk 2016; Inglehart 2016.
73. Right-wing anti-intellectualism has been deplored by conservatives themselves in books like Charlie Sykes's *How the Right Lost Its Mind* (2017) and Matt Lewis's *Too Dumb to Fail* (2016).
74. Centrality of reason: Nagel 1997; Norman 2016.
75. Extraordinary popular delusions: Mackay 1841/1995; see also K. Malik, "All the Fake News That Was Fit to Print," *New York Times*, Dec. 4, 2016.
76. A. D. Holan, "All Politicians Lie. Some Lie More Than Others," *New York Times*, Dec. 11, 2015.
77. In analyzing history's deadliest conflicts, Matthew White comments, "I'm amazed at how often the immediate cause of a conflict is a mistake, unfounded suspicion, or rumor." In addition to the first two listed here he includes the First World War, Sino-Japanese War, Seven Years' War, Second French War of Religion, An Lushan Rebellion in China, Indonesian Purge, and Russia's Time of Troubles; White 2011, p. 537.
78. Opinion of Judge Leon M. Bazile, Jan. 22, 1965, *Encyclopedia Virginia*, http://www.encyclopediavirginia.org/Opinion_of_Judge_Leon_M_Bazile_January_22_1965.
79. S. Sontag, "Some Thoughts on the Right Way (for Us) to Love the Cuban Revolution," *Ramparts*, April 1969, pp. 6–19. Sontag went on to claim that the homosexuals "have long since been sent home," but gays continued to be sent to forced labor camps in Cuba throughout the 1960s and 1970s. See "Concentration Camps in Cuba: The UMAP," *Totalitarian Images*, Feb. 6, 2010, http://

totalitarianimages.blogspot.com/2010/02/concentration-camps-in-cuba-umap.html, and J. Halatyn, "From Persecution to Acceptance? The History of LGBT Rights in Cuba," *Cutting Edge*, Oct. 24, 2012, http://www.thecuttingedgenews.com/index.php?article=76818.

80. Affective tipping point: Redlawsk, Civettini, & Emmerson 2010.
81. Naked emperors and common knowledge: Pinker 2007a; Thomas et al. 2014; Thomas, DeScioli, & Pinker 2018.
82. For an excellent summary of common fallacies, see the Web site and poster "Thou shalt not commit logical fallacies," https://yourlogicalfallacyis.com/. Critical thinking curricula: Willingham 2007.
83. Debiasing: Bond 2009; Gigerenzer 1991; Gigerenzer & Hoffrage 1995; Lilienfeld, Ammirati, & Landfield 2009; Mellers et al. 2014; Morewedge et al. 2015.
84. The trouble with critical-thinking curricula: Willingham 2007.
85. Effective debiasing: Bond 2009; Gigerenzer 1991; Gigerenzer & Hoffrage 1995; Lilienfeld, Ammirati, & Landfield 2009; Mellers et al. 2014; Mercier & Sperber 2011; Morewedge et al. 2015; Tetlock & Gardner 2015; Willingham 2007.
86. Giving debiasing away: Lilienfeld, Ammirati, & Landfield 2009.
87. Anonymous, quoted in P. Voosen, "Striving for a Climate Change," *Chronicle Review of Higher Education*, Nov. 3, 2014.
88. Improving argument: Kuhn 1991; Mercier & Sperber 2011, 2017; Sloman & Fernbach 2017.
89. Truth wins: Mercier & Sperber 2011.
90. Adversarial collaboration: Mellers, Hertwig, & Kahneman 2001.
91. The Illusion of Explanatory Depth: Rozenblit & Keil 2002. Using the illusion to debias: Sloman & Fernbach 2017.
92. Mercier & Sperber 2011, p. 72; Mercier & Sperber 2017.
93. More rational journalism: Silver 2015; A. D. Holan, "All Politicians Lie. Some Lie More Than Others," *New York Times*, Dec. 11, 2015.
94. More rational intelligence-gathering: Tetlock & Gardner 2015; Tetlock, Mellers, & Scoblic 2017.
95. More rational medicine: Topol 2012.
96. More rational psychotherapy: T. Rousmaniere, "What Your Therapist Doesn't Know," *The Atlantic*, April 2017.
97. More rational crimefighting: Abt & Winship 2016; Latzer 2016.
98. More rational international development: Banerjee & Duflo 2011.
99. More rational altruism: MacAskill 2015.
100. More rational sports: Lewis 2016.
101. "What Exactly Is the 'Rationality Community'?" *LessWrong*, http://lesswrong.com/lw/ov2/what_exactly_is_the_rationality_community/.
102. More rational governance: Behavioral Insights Team 2015; Haskins & Margolis 2014; Schuck 2015; Sunstein 2013; D. Leonhardt, "The Quiet Movement to Make Government Fail Less Often," *New York Times*, July 15, 2014.
103. Democracy versus rationality: Achen & Bartels 2016; Brennan 2016; Caplan 2007; Mueller 1999; Somin 2016.
104. Plato and democracy: Goldstein 2013.
105. Kahan, Wittlin, et al. 2011, p. 16.
106. HPV versus hep B: E. Klein, "How Politics Makes Us Stupid," *Vox*, April 6, 2014.
107. Party over policy: Cohen 2003.
108. Evidence that same-side spokespeople can change minds: Nyhan 2013.
109. Kahan, Jenkins-Smith, et al. 2012.
110. Depoliticized Florida compact: Kahan 2015.
111. Chicago Way: Sean Connery's Jim Malone in *The Untouchables* (1987). GRIT: Osgood 1962.

CHAPTER 22: SCIENCE

1. The example is from Murray 2003.
2. Carroll 2016, p. 426.
3. Naming species: Costello, May, & Stork 2013. The estimate refers to eukaryotic species (those with a nucleus, excluding viruses and bacteria).
4. The party of stupid: See chapter 21, notes 71 and 73.
5. Mooney 2005; see also Pinker 2008b.
6. Lamar Smith and the House Science Committee: J. D. Trout, "The House Science Committee Hates Science and Should Be Disbanded," *Salon*, May 17, 2016.
7. J. Mervis, "Updated: U.S. House Passes Controversial Bill on NSF Research," *Science*, Feb. 11, 2016.
8. From *Note-book of Anton Chekhov*. The quote continues, "What is national is no longer science."
9. J. Lears, "Same Old New Atheism: On Sam Harris," *The Nation*, April 27, 2011.
10. L. Kass, "Keeping Life Human: Science, Religion, and the Soul," Wriston Lecture, Manhattan Institute, Oct. 18, 2007, https://www.manhattan-institute.org/html/2007-wriston-lecture-keeping-life-human-science-religion-and-soul-8894.html. See also L. Kass, "Science, Religion, and the Human Future," *Commentary*, April 2007, pp. 36–48.

11. On the numbering of the Two Cultures, see chapter 3, note 12.
12. D. Linker, "Review of Christopher Hitchens's 'And Yet . . .' and Roger Scruton's 'Fools, Frauds and Firebrands,'" *New York Times Book Review*, Jan. 8, 2016.
13. Snow introduced the term "Third Culture" in a postscript to *The Two Cultures* called "A Second Look." He was vague about who he had in mind, referring to them as "social historians," by which he seems to have meant social scientists; Snow 1959/1998, pp. 70, 80.
14. Revival of "Third Culture": Brockman 1991. Consilience: Wilson 1998.
15. L. Wieseltier, "Crimes Against Humanities," *New Republic*, Sept. 3, 2013.
16. Hume as cognitive psychologist: See the references in Pinker 2007a, chap. 4. Kant as cognitive psychologist: Kitcher 1990.
17. The definition is from the *Stanford Encyclopedia of Philosophy*, Papineau 2015, which adds, "The great majority of contemporary philosophers would accept naturalism as just characterized." In a survey of 931 philosophy professors (mainly analytic/Anglo-American), 50 percent endorsed "naturalism," 26 percent endorsed "non-naturalism," and 24 percent indicated "other," including "The question is too unclear to answer" (10 percent), "Insufficiently familiar with the issue" (7 percent), and "Agnostic/undecided" (3 percent); Bourget & Chalmers 2014.
18. No "scientific method": Popper 1983.
19. Falsificationism versus Bayesian inference: Howson & Urbach 1989/2006; Popper 1983.
20. In 2012–13, the *New Republic* published four denunciations of scientism, and others appeared in *Bookforum*, the *Claremont Review*, the *Huffington Post*, *The Nation*, *National Review Online*, the *New Atlantis*, the *New York Times*, and *Standpoint*.
21. According to the Open Syllabus Project (http://opensyllabusproject.org/), which has analyzed more than a million university syllabuses, *Structure* is the twentieth most assigned book overall, well ahead of *The Origin of Species*. A classic book with a more realistic take on how science works, Karl Popper's *The Logic of Scientific Discovery*, is not in the top 200.
22. Kuhn controversy: Bird 2011.
23. Wootton 2015, p. 16, note ii.
24. The quotes come from J. De Vos, "The Iconographic Brain. A Critical Philosophical Inquiry into (the Resistance of) the Image," *Frontiers in Human Neuroscience*, May 15, 2014. This was not the researcher I heard (a transcript of his talk is not available), but the content was essentially the same.
25. Carey et al. 2016. Similar examples may be found in the Twitter stream *New Real PeerReview*, @RealPeerReview.
26. From the first page of Horkheimer & Adorno 1947/2007.
27. Foucault 1999; see Menschenfreund 2010; Merquior 1985.
28. Bauman 1989, p. 91. See Menschenfreund 2010, for analysis.
29. Ubiquity of premodern genocide and autocracy, and their decline after 1945: See the references in chapters 11 and 14, and in Pinker 2011, chaps. 4–6. On Foucault's neglect of totalitarianism before the Enlightenment, see Merquior 1985.
30. Ubiquity of slavery: Patterson 1985; Payne 2004; see also Pinker 2011, chap. 4. Religious justifications for slavery: Price 2006.
31. Greeks and Arabs on Africans: Lewis 1990/1992. Cicero on Britons: B. Delong, "Cicero: The Britons Are Too Stupid to Make Good Slaves," http://www.bradford-delong.com/2009/06/cicero-the-britons-are-too-stupid-to-make-good-slaves.html.
32. Gobineau, Wagner, Chamberlain, and Hitler: Herman 1997, chap. 2; see also Hellier 2011; Richards 2013. Many misconceptions about the link between "racial science" and Darwinism were spread by the biologist Stephen Jay Gould in his tendentious 1981 bestseller *The Mismeasure of Man;* see Blinkhorn 1982; Davis 1983; Lewis et al. 2011.
33. Darwinian versus traditional, religious, and Romantic theories of race: Hellier 2011; Johnson 2009; Price 2006.
34. Hitler was not a Darwinian: Richards 2013; see also Hellier 2011; Price 2006.
35. Evolution as Rorschach test: Montgomery & Chirot 2015. Social Darwinism: Degler 1991; Leonard 2009; Richards 2013.
36. The misapplication of the term *social Darwinism* to a variety of right-wing movements was begun by the historian Richard Hofstadter in his 1944 book *Social Darwinism in American Thought;* see Johnson 2010; Leonard 2009; Price 2006.
37. An example is an article on evolutionary psychology in *Scientific American* by John Horgan entitled "The New Social Darwinists" (October 1995).
38. Glover 1998, 1999; Proctor 1988.
39. As in the title of another *Scientific American* article by John Horgan, "Eugenics Revisited: Trends in Behavioral Genetics" (June 1993).
40. Degler 1991; Kevles 1985; Montgomery & Chirot 2015; Ridley 2000.
41. Tuskegee reexamined: Benedek & Erlen 1999; Reverby 2000; Shweder 2004; Lancet Infectious Diseases Editors 2005.
42. Review boards abridge free speech: American Association of University Professors 2006; Schneider 2015; C. Shea, "Don't Talk to the Humans: The Crackdown on Social Science Research," *Lingua Franca*, Sept. 2000, http://linguafranca.mirror.theinfo.org/print/0009/humans

.html. Review boards as ideological weapons: Dreger 2008. Review boards bog down research without protecting subjects: Atran 2007; Gunsalus et al. 2006; Hyman 2007; Klitzman 2015; Schneider 2015; Schrag 2010.
43. Moss 2005.
44. Protecting suicide bombers: Atran 2007.
45. Philosophers against bioethics: Glover 1998; Savulescu 2015. For other critiques of contemporary bioethics, see Pinker 2008b; Satel 2010; S. Pinker, "The Case Against Bioethocrats and CRISPR Germline Ban," *The Niche*, Aug. 10, 2015, https://ipscell.com/2015/08/stevenpinker/8/; S. Pinker, "The Moral Imperative for Bioethics," *Boston Globe*, Aug. 1, 2015; H. Miller, "When 'Bioethics' Is Not Ethical," *Forbes*, Nov. 9, 2016, reprinted in http://dailycaller.com/2018/06/04/when-bioeth ics-not-ethical. See also the references in note 42 above.
46. See the references in chapter 21, notes 93-102.
47. Dawes, Faust, & Meehl 1954/2013; Meehl 1954/2013. Recent replications: Mental health, Ægisdóttir et al. 2006; Lilienfeld et al. 2013; selection and admission decisions, Kuncel et al. 2013; violence, Singh, Grann, & Fazel 2011.
48. Blessed are the peacekeepers: Fortna 2008, p. 173. See also Hultman, Kathman, & Shannon 2013, and Goldstein 2011, who credits peacekeeping forces with much of the post-1945 decline of war.
49. Ethnic neighbors rarely fight: Fearon & Laitin 1996, 2003; Mueller 2004a.
50. Chenoweth 2016; Chenoweth & Stephan 2011.
51. Revolutionary leaders are educated: Chirot 1996. Suicide terrorists are educated: Atran 2003.
52. Trouble in the humanities: American Academy of Arts and Sciences 2015; Armitage et al. 2013. For earlier lamentations, see Pinker 2002/2016, opening to chap. 20.
53. Why democracy needs the humanities: Nussbaum 2016.
54. Cultural pessimism in the humanities: Herman 1997; Lilla 2001, 2016; Nisbet 1980/2009; Wolin 2004.
55. The framers and human nature: McGinnis 1996, 1997. Politics and human nature: Pinker 2002/2016, chap. 16; Pinker 2011, chaps. 8 and 9; Haidt 2012; Sowell 1987.
56. Art and science: Dutton 2009; Livingstone 2014.
57. Music and science: Bregman 1990; Lerdahl & Jackendoff 1983; Patel 2008; see also Pinker 1997/2009, chap. 8.
58. Literature and science: Boyd, Carroll, & Gottschall 2010; Connor 2016; Gottschall 2012; Gottschall & Wilson 2005; Lodge 2002; Pinker 2007b; Slingerland 2008; see also Pinker 1997/2009, chap. 8, and William Benzon's blog *New Savanna*, http://new-savanna.blogspot.com.
59. Digital humanities: Michel et al. 2010; see the e-journal *Digital Humanities Now* (http://digital humanitiesnow.org/), the Stanford Humanities Center (http://shc.stanford.edu/digital -humanities), and the journal *Digital Humanities Quarterly* (http://www.digitalhumanities.org /dhq/).
60. Gottschall 2012; A. Gopnik, "Can Science Explain Why We Tell Stories?" *New Yorker*, May 18, 2012.
61. Wieseltier 2013, "Crimes Against Humanities," which was a reply to my essay "Science Is Not Your Enemy" (Pinker 2013b); see also "Science vs. the Humanities, Round III" (Pinker & Wieseltier 2013).
62. Pre-Darwinian, pre-Copernican: L. Wieseltier, "Among the Disrupted," *New York Times*, Jan. 7, 2015.
63. In "A Letter Addressed to the Abbe Raynal," Paine 1782/2016, quoted in Shermer 2015.

CHAPTER 23: HUMANISM

1. "Good without God": From the 19th century, revived by the Harvard Humanist chaplain Greg Epstein (Epstein 2009). Other recent explanations of humanism: Grayling 2013; Law 2011. History of American Humanism: Jacoby 2005. Major Humanist organizations include the American Humanist Association (https://americanhumanist.org/) and the other members of the Secular Coalition of America (https://www.secular.org/member_orgs), the British Humanist Association (https://humanism.org.uk/), the International Humanist and Ethical Union (http://iheu .org/), and the Freedom from Religion Foundation (www.ffrf.org).
2. *Humanist Manifesto III*: American Humanist Association 2003. Predecessors: *Humanist Manifesto I* (mainly by Raymond B. Bragg, 1933), American Humanist Association 1933/1973. *Humanist Manifesto II* (mainly by Paul Kurtz and Edwin H. Wilson, 1973), American Humanist Association 1973. Other Humanist manifestoes include Paul Kurtz's *Secular Humanist Declaration*, Council for Secular Humanism 1980, and *Humanist Manifesto 2000*, Council for Secular Humanism 2000, and the Amsterdam Declarations of 1952 and 2002, International Humanist and Ethical Union 2002.
3. R. Goldstein, "Speaking Prose All Our Lives," *The Humanist*, Dec. 21, 2012, https://thehumanist .com/magazine/january-february-2013/features/speaking-prose-all-our-lives.
4. The rights declarations of 1688, 1776, 1789, and 1948: Hunt 2007.
5. Morality as impartiality: de Lazari-Radek & Singer 2012; Goldstein 2006; Greene 2013; Nagel 1970; Railton 1986; Singer 1981/2011, Smart & Williams 1973. The "impartiality" umbrella was articulated most explicitly by the philosopher Henry Sidgwick (1838-1900).
6. For an exhaustive (if eccentric) list of Golden, Silver, and Platinum rules across cultures and history, see Terry 2008.

7. Evolution explains the existence of mind despite entropy: Tooby, Cosmides, & Barrett 2003. Natural selection is the only explanation of nonrandom design: Dawkins 1983.
8. Curiosity and sociality as concomitants of the evolution of intelligence: Pinker 2010; Tooby & DeVore 1987.
9. Evolutionary conflicts of interest within and among people: Pinker 1997/2009, chaps. 6 and 7; Pinker 2002/2016, chap. 14; Pinker 2011, chaps. 8 and 9. Many of these ideas originated with the biologist Robert Trivers (2002).
10. The Pacifist's Dilemma and the historical decline of violence: Pinker 2011, chap. 10.
11. DeScioli 2016.
12. Evolution of sympathy: Dawkins 1976/1989; McCullough 2008; Pinker 1997/2009; Trivers 2002; Pinker 2011, chap. 9.
13. Expanding circle of sympathy: Pinker 2011; Singer 1981/2011.
14. For example, T. Nagel, "The Facts Fetish (Review of Sam Harris's *The Moral Landscape*)," *New Republic*, Oct. 20, 2010.
15. Utilitarianism, for and against: Rachels & Rachels 2010; Smart & Williams 1973.
16. Compatibility of deontological and consequential meta-ethics: Parfit 2011.
17. Track record of utilitarianism: Pinker 2011, chaps. 4 and 6; Greene 2013.
18. From *Notes on the State of Virginia*, Jefferson 1785/1955, p. 159.
19. Unintuitiveness of classical liberalism: Fiske & Rai 2015; Haidt 2012; Pinker 2011, chap. 9.
20. Greene 2013.
21. The importance of philosophical thinness: Berlin 1988/2013; Gregg 2003; Hammond 2017.
22. Hammond 2017.
23. Maritain 1949. Original typescript available at the UNESCO Web site, http://unesdoc.unesco .org/images/0015/001550/155042eb.pdf.
24. Universal Declaration of Human Rights: United Nations 1948. History of the Declaration: Glendon 1999, 2001; Hunt 2007.
25. Quoted in Glendon 1999.
26. Human rights not particularly Western: Glendon 1998; Hunt 2007; Sikkink 2017.
27. R. Cohen, "The Death of Liberalism," *New York Times*, April 14, 2016.
28. S. Kinzer, "The Enlightenment Had a Good Run," *Boston Globe*, Dec. 23, 2016.
29. ISIS more appealing than Enlightenment: R. Douthat, "The Islamic Dilemma," *New York Times*, Dec. 13, 2015; R. Douthat, "Among the Post-Liberals," *New York Times*, Oct. 8, 2016; M. Khan, "This Is What Happens When Modernity Fails All of Us," *New York Times*, Dec. 6, 2015; P. Mishra, "The Western Model Is Broken," *The Guardian*, Oct. 14, 2014.
30. Universality of proscriptions of murder, rape, and violence: Brown 2000.
31. God as enforcer: Atran 2002; Norenzayan 2015.
32. Fatal flaws in arguments for the existence of God: Goldstein 2010; see also Dawkins 2006 and Coyne 2015.
33. Coyne draws in part on arguments from the astronomer Carl Sagan and the philosophers Yonatan Fishman and Maarten Boudry. For a review, see S. Pinker, "The Untenability of Faitheism," *Current Biology*, Aug. 23, 2015, pp. R638–640.
34. Debunking the soul: Blackmore 1991; Braithwaite 2008; Musolino 2015; Shermer 2002; Stein 1996. See also the magazines *Skeptical Inquirer* (http://www.csicop.org/si) and *The Skeptic* (http:// www.skeptic.com/) for regular updates.
35. Stenger 2011.
36. The multiverse: Carroll 2016; Tegmark 2003; B. Greene, "Welcome to the Multiverse," *Newsweek*, May 21, 2012.
37. A universe from nothing: Krauss 2012.
38. B. Greene, "Welcome to the Multiverse," *Newsweek*, May 21, 2012.
39. Easy and hard problems of consciousness: Block 1995; Chalmers 1996; McGinn 1993; Nagel 1974; see also Pinker 1997/2009, chaps. 2 and 8, and S. Pinker, "The Mystery of Consciousness," *Time*, Jan. 29, 2007.
40. Adaptive nature of consciousness: Pinker, 1997/2009, chap. 2.
41. Dehaene 2009; Dehaene & Changeux 2011; Gaillard et al. 2009.
42. For an extended defense of this distinction, see Goldstein 1976.
43. Nagel 1974, p. 441. Nearly four decades later, Nagel changed his mind (see Nagel 2012), but like most philosophers and scientists, I think he got it right the first time. See, for example, S. Carroll, Review of *Mind and Cosmos*, http://www.preposterousuniverse.com/blog/2013/08/22/mind -and-cosmos/; E. Sober, "Remarkable Facts: Ending Science as We Know It," *Boston Review*, Nov. 7, 2012; B. Leiter & M. Weisberg, "Do You Only Have a Brain?" *The Nation*, Oct. 3, 2012.
44. McGinn 1993.
45. Moral realism: Sayre-McCord 1988, 2015. Moral realists: Boyd 1988; Brink 1989; de Lazari-Radek & Singer 2012; Goldstein 2006, 2010; Nagel 1970; Parfit 2011; Railton 1986; Singer 1981/2011.
46. Examples are the European wars of religion (Pinker 2011, pp. 234, 676–77) and even the American Civil War (Montgomery & Chirot 2015, p. 350).
47. White 2011, pp. 107–11.
48. S. Bannon, remarks to a conference at the Vatican, 2014, transcribed in J. L. Feder, "This Is How

Steve Bannon Sees the Entire World," *BuzzFeed*, Nov. 16, 2016, https://www.buzzfeed.com/lester feder/this-is-how-steve-bannon-sees-the-entire-world.

49. Nazis sympathetic to Christianity and vice versa: Ericksen & Heschel 1999; Hellier 2011; Heschel 2008; Steigmann-Gall 2003; White 2011. Hitler was not an atheist: Hellier 2011; Murphy 1999; Richards 2013; see also "Hitler Was a Christian," http://www.evilbible.com/evil-bible-home -page/hitler-was-a-christian/.

50. The final sentences of *Mein Kampf*, vol. I chapter 2. For similar quotations, see the references in the preceding note.

51. Sam Harris, *The End of Faith* (2004); Richard Dawkins, *The God Delusion* (2006); Daniel Dennett, *Breaking the Spell* (2006); Christopher Hitchens, *God Is Not Great* (2007).

52. Randall Munroe, "Atheists," https://xkcd.com/774/.

53. The claim that people treat scripture allegorically (for example, Wieseltier 2013) is untrue: A 2005 Rasmussen poll found that 63 percent of Americans believed that the Bible is literally true (http://legacy.rasmussenreports.com/2005/Bible.htm); a 2014 Gallup poll found that 28 percent of Americans believed that "the Bible is the actual word of God and is to be taken literally, word for word," and another 47 percent believed it was "the inspired word of God" (L. Saad, "Three in Four in U.S. Still See the Bible as Word of God," *Gallup*, June 4, 2014, http://www.gallup.com/poll /170834/three-four-bible-word-god.aspx).

54. Psychology of religion: Pinker 1997/2009, chap. 8; Atran 2002; Bloom 2012; Boyer 2001; Dawkins 2006; Dennett 2006; Goldstein 2010.

55. Why there is no "God module": Pinker 1997/2009, chap. 8; Bloom 2012; Pinker 2005.

56. Community participation, not religious belief, explains the benefits of religious belonging: Putnam & Campbell 2010; see Bloom 2012 and Susan Pinker 2014 for reviews. For a recent study finding the same pattern for mortality, see Kim, Smith, & Kang 2015.

57. Regressive religious policies: Coyne 2015.

58. God and climate: Bean & Teles 2016; see also chapter 18, note 86.

59. Trump support from Evangelicals: See *New York Times* 2016 and chapter 20, note 34.

60. A. Wilkinson, "Trump Wants to 'Totally Destroy' a Ban on Churches Endorsing Political Candidates," *Vox*, Feb. 7, 2017.

61. "*The Oprah Winfrey Show* Finale," *oprah.com*, http://www.oprah.com/oprahshow/the-oprah -winfrey-show-finale_1/all.

62. Excerpted and lightly edited from "The Universe—Uncensored," *Inside Amy Schumer*, https:// www.youtube.com/watch?v=6eqCaiwmr_M.

63. Hostility to atheists: G. Paul & P. Zuckerman, "Don't Dump On Us Atheists," *Washington Post*, April 30, 2011; Gervais & Najle 2018.

64. From the *World Christian Encyclopedia* (2001), cited in Paul & Zuckerman 2007.

65. Global Index of Religiosity and Atheism: WIN-Gallup International 2012. The Index's sample of countries in 2005 was smaller (thirty-nine countries) and more religious (68 percent still identifying themselves as religious in 2005, as opposed to 59 percent in the full 2012 sample). In the longitudinal subset, the percentage of atheists grew from 4 to 7 percent, a 75 percent increase in seven years. It would be dubious to generalize this multiplier to larger samples, because of the nonlinearity of the percentage scale at the low end, so in estimating the increase in atheism in the fifty-seven-country sample over this period, I assumed a more conservative 30 percent increase.

66. Secularization Thesis: Inglehart & Welzel 2005; Voas & Chaves 2016.

67. Correlation of irreligion with income and education: Barber 2011; Lynn, Harvey, & Nyborg 2009; WIN-Gallup International 2012.

68. WIN-Gallup International 2012. Other minority-religious countries in the sample are Austria and the Czech Republic, and those in which the percentage just squeaks past 50 percent include Finland, Germany, Spain, and Switzerland. Other secular Western countries such as Denmark, New Zealand, Norway, and the United Kingdom were not surveyed. According to a different set of surveys from around 2004 (Zuckerman 2007, reproduced in Lynn, Harvey, & Nyborg 2009), more than a quarter of respondents in fifteen developed countries say they don't believe in God, together with more than half of Czechs, Japanese, and Swedes.

69. Pew Research Center 2012a.

70. The Methodology Appendix to Pew Research Center 2012a, particularly note 85, indicates that their fertility estimates are current snapshots, and are not adjusted for anticipated changes. Muslim fertility decline: Eberstadt & Shah 2011.

71. Religious change in the Anglosphere: Voas & Chaves 2016.

72. American religious exceptionalism: Paul 2014; Voas & Chaves 2016. These numbers are from WIN-Gallup International 2012.

73. Lynn, Harvey, & Nyborg 2009; Zuckerman 2007.

74. American secularization: Hout & Fischer 2014; Jones et al. 2016b; Pew Research Center 2015a; Voas & Chaves 2016.

75. The preceding figures are from Jones et al. 2016b. Another sign of the underreported decline in religion in the United States is that the proportion of white Evangelicals in the PRRI surveys fell from 20 percent in 2012 to 16 percent in 2016.

76. Younger irreligious more likely to stay irreligious: Hout & Fischer 2014; Jones et al. 2016b; Voas & Chaves 2016.

77. Blatant nonbelievers: D. Leonhardt, "The Rise of Young Americans Who Don't Believe in God," *New York Times,* May 12, 2015, based on data from Pew Research Center 2015a. Little nonbelief in the 1950s: Voas & Chaves 2016, based on data from the General Social Survey.

78. Gervais & Najle 2018.

79. Jones et al. 2016b, p. 18.

80. Explanations for secularization: Hout & Fischer 2014; Inglehart & Welzel 2005; Jones et al. 2016b; Paul & Zuckerman 2007; Voas & Chaves 2016.

81. Secularization and declining trust in institutions: Twenge, Campbell, & Carter 2014. Trust in institutions peaked in the 1960s: Mueller 1999, pp. 167–68.

82. Secularization and emancipative values: Hout & Fischer 2014; Inglehart & Welzel 2005; Welzel 2013.

83. Secularization and existential security: Inglehart & Welzel 2005; Welzel 2013. Secularization and the social safety net: Barber 2011; Paul 2014; Paul & Zuckerman 2007.

84. Main reason Americans leave religion: Jones et al. 2016b. Note also that belief in the literal truth of the Bible among respondents in the Gallup poll described in note 53 above has decreased over time, from 40 percent in 1981 to 28 percent in 2014, while belief that it is a book of "fables, legends, history, and moral precepts recorded by man" rose from 10 percent to 21 percent.

85. Secularization and rising IQ: Kanazawa 2010; Lynn, Harvey, & Nyborg 2009.

86. "Total eclipse": From a quote by Friedrich Nietzsche.

87. Happiness: See chapter 18, and Helliwell, Layard, & Sachs 2016. Indicators of social well-being: See Porter, Stern, & Green 2016; chapter 21, note 42; and note 90 below. In a regression analysis of 116 countries, Keehup Yong and I found that the correlation between the Social Progress Index and the percentage of the population not believing in God (taken from Lynn, Harvey, & Nyborg 2009) was .63, which was statistically significant ($p < .0001$) holding constant GDP per capita.

88. Unfortunate American exceptionalism: See chapter 21, note 42; Paul 2009, 2014.

89. Religious state, dysfunctional state: Delamontagne 2010.

90. Though more than a quarter of the world's 195 countries are Muslim-majority, none are found among the thirty-eight ranked as "Very High" and "High" on the Social Progress Index (Porter, Stern, & Green 2016, pp. 19–20) or among the twenty-five happiest (Helliwell, Layard, & Sachs 2016). None is a "full democracy," just three are "flawed democracies," and more than forty are "authoritarian" or "hybrid" regimes: *The Economist* Intelligence Unit, https://infographics.economist .com/2017/DemocracyIndex/. For similar assessments, see Marshall & Gurr 2014; Marshall, Gurr, & Jaggers 2016; Pryor 2007.

91. Wars in 2016: See chapter 11, note 9; and Gleditsch & Rudolfsen 2016. Terrorism: Institute for Economics and Peace 2016, using data from the National Consortium for the Study of Terrorism and Responses to Terrorism, http://www.start.umd.edu/.

92. Precocious scientific revolution: Al-Khalili 2010; Huff 1993. Tolerance in the Arab and Ottoman Empires: Lewis 2002; Pelham 2016.

93. Regressive passages in the Quran, Hadith, and Sunna: Rizvi 2017, chap. 2; Hirsi Ali 2015a, 2015b; S. Harris, "Verses from the Koran," *Truthdig,* http://www.truthdig.com/images/diguploads /verses.html; *The Skeptic's Annotated Quran,* http://skepticsannotatedbible.com/quran/int/long .html. Recent discussion by journalists include R. Callimachi, "ISIS Enshrines a Theology of Rape," *New York Times,* Aug. 13, 2015; G. Wood, "What ISIS Really Wants," *The Atlantic,* March 2015; and Wood 2017. Recent scholarly discussions include Cook 2014 and Bowering 2015.

94. Alexander & Welzel 2011, pp. 256–58.

95. Alexander and Welzel cite the Bertelsmann Foundation's *Religious Monitor.* See also Pew Research Center 2012c; WIN-Gallup International 2012, for comparable figures (though with regional variation).

96. Quotes from Pew Research Center 2013, pp. 24 and 15, and Pew Research Center 2012c, pp. 11 and 12. The countries asked about interpreting the Quran word for word were the United States and fifteen countries in sub-Saharan Africa, which probably bracket the range. The exceptions to wanting sharia as national law include Turkey, Lebanon, and formerly communist regions.

97. Welzel 2013; see also Alexander & Welzel 2011 and Inglehart 2017.

98. Alexander & Welzel 2011. See also Pew Research Center 2013, which found higher support for sharia law among devout Muslims.

99. Religious stranglehold: Huff 1993; Kuran 2010; Lewis 2002; United Nations Development Programme 2003; Montgomery & Chirot 2015, chap. 7; see also Rizvi 2016 and Hirsi Ali 2015a for first-person accounts.

100. Reactionary Islam: Montgomery & Chirot 2015, chap. 7; Lilla 2016; Hathaway & Shapiro 2017.

101. Western intellectuals apologizing for repression in the Islamic world: Berman 2010; J. Palmer, "The Shame and Disgrace of the Pro-Islamist Left," *Quillette,* Dec. 6, 2015; J. Tayler, "The Left Has Islam All Wrong," *Salon,* May 10, 2015; J. Tayler, "On Betrayal by the Left—Talking with Ex-Muslim Sarah Haider," *Quillette,* March 16, 2017.

102. Quoted in J. Tayler, "On Betrayal by the Left—Talking with Ex-Muslim Sarah Haider," *Quillette*, March 16, 2017.

103. Al-Khalili 2010; Huff 1993.

104. Sen 2000, 2005, 2009; see also Pelham 2016, for examples in the Ottoman Empire.

105. Esposito & Mogahed 2007; Inglehart 2017; Welzel 2013.

106. Islamic modernization: Mahbubani & Summers 2016. Cohort replacement: See chapter 15, especially figure 15-7; Inglehart 2017; Welzel 2013. Inglehart notes, however, that while thirteen of the Muslim-majority countries in the World Values Survey show a generational shift toward gender equality, fourteen do not; the reasons for the split are unclear.

107. J. Burke, "Osama bin Laden's bookshelf: Noam Chomsky, Bob Woodward, and Jihad," *The Guardian*, May 20, 2015.

108. Extramural drivers of moral progress: Appiah 2010; Hunt 2007.

109. Nietzsche's famous works, many of whose titles have become highbrow memes, include *The Birth of Tragedy, Beyond Good and Evil, Thus Spake Zarathustra, The Genealogy of Morals, Twilight of the Idols, Ecce Homo,* and *The Will to Power*. For critical discussion, see Anderson 2017; Glover 1999; Herman 1997; Russell 1945/1972; Wolin 2004.

110. The first three quotations are taken from Russell 1945/1972, pp. 762–66, the last two from Wolin 2004, pp. 53, 57.

111. *Relativismo e Fascismo,* quoted in Wolin 2004, p. 27.

112. Rosenthal 2002.

113. Nietzsche's influence on Rand and her cover-up: Burns 2009.

114. From *The Genealogy of Morals* and *The Will to Power,* quoted in Wolin 2004, pp. 32–33.

115. Tyrannophilia: Lilla 2001. The syndrome was first identified in *The Treason of the Intellectuals* by the French philosopher Julien Benda (Benda 1927/2006). More recent histories include Berman 2010; Herman 1997; Hollander 1981/2014; Sesardić 2016; Sowell 2010; Wolin 2004. See also Humphrys (undated).

116. Scholars and Writers for America, "Statement of Unity," Oct. 30, 2016, https://scholarsandwriters foramerica.org/.

117. J. Baskin, "The Academic Home of Trumpism," *Chronicle of Higher Education,* March 17, 2017.

118. Nietzsche influenced not only Mussolini but the Fascist theoretician Julius Evola, discussed below. He also influenced the philosopher Leo Strauss, a major influence on the Claremont school and reactionary theoconservatism; see J. Baskin, "The Academic Home of Trumpism," *Chronicle of Higher Education,* March 17, 2017; Lampert 1996.

119. Nationalism and counter-Enlightenment Romanticism: Berlin 1979; Garrard 2006; Herman 1997; Howard 2001; McMahon 2001; Sternhell 2010; Wolin 2004.

120. Rediscovery of early Fascists: J. Horowitz, "Steve Bannon Cited Italian Thinker Who Inspired Fascists," *New York Times,* Feb. 10, 2017; P. Levy, "Stephen Bannon Is a Fan of a French Philosopher . . . Who Was an Anti-Semite and a Nazi Supporter," *Mother Jones,* March 16, 2017; M. Crowley, "The Man Who Wants to Unmake the West," *Politico,* March/April 2017. Alt-right: A. Bokhari & M. Yiannopoulos, "An Establishment Conservative's Guide to the Alt-Right," *Breitbart.com,* March 29, 2016, http://www.breitbart.com/tech/2016/03/29/an-establishment -conservatives-guide-to-the-alt-right/. Nietzsche's influence on the alt-right: G. Wood, "His Kampf," *The Atlantic,* June 2017; S. Illing, "The Alt-Right Is Drunk on Bad Readings of Nietzsche. The Nazis Were Too," *Vox,* Aug. 17, 2017, https://www.vox.com/2017/8/17/16140846 /nietzsche-richard-spencer-alt-right-nazism.

121. Naïve evolutionary-psychological explanation of nationalism, and its problems: Pinker 2012.

122. Theoconservatism: Lilla 2016; Linker 2007; Pinker 2008b.

123. Written under the pseudonym Publius Decius Mus; see Publius Decius Mus 2016. See also M. Warren, "The Anonymous Pro-Trump 'Decius' Now Works Inside the White House," *Weekly Standard,* Feb. 2, 2017.

124. The reactionary mindset: Lilla 2016. For more on reactionary Islam, see Montgomery & Chirot 2015 and Hathaway & Shapiro 2017.

125. A. Restuccia & J. Dawsey, "How Bannon and Pruitt Boxed In Trump on Climate Pact," *Politico,* May 31, 2017.

126. Cognitive flexibility of "tribe": Kurzban, Tooby, & Cosmides 2001; Sidanius & Pratto 1999; see also Center for Evolutionary Psychology, UCSB, Erasing Race FAQ, http://www.cep.ucsb.edu /erasingrace.htm.

127. Manipulating group intuitions: Pinker 2012.

128. Tribalism and cosmopolitanism: Appiah 2006.

129. Diamond 1997; Sowell 1994, 1996, 1998.

130. Glaeser 2011; Sowell 1996.

REFERENCES

Abrahms, M. 2006. Why terrorism does not work. *International Security, 31*, 42–78.
Abrahms, M. 2012. The political effectiveness of terrorism revisited. *Comparative Political Studies, 45*, 366–93.
Abrams, S. 2016. Professors moved left since 1990s, rest of country did not. *Heterodox Academy.* http://heterodoxacademy.org/2016/01/09/professors-moved-left-but-country-did-not/.
Abt, T., & Winship, C. 2016. *What works in reducing community violence: A meta-review and field study for the Northern Triangle.* Washington: US Agency for International Development.
Acemoglu, D., & Robinson, J. A. 2012. *Why nations fail: The origins of power, prosperity, and poverty.* New York: Crown.
Achen, C. H., & Bartels, L. M. 2016. *Democracy for realists: Why elections do not produce responsive government.* Princeton, NJ: Princeton University Press.
Adriaans, P. 2013. Information. In E. N. Zalta, ed., *Stanford Encyclopedia of Philosophy.* http://plato.stanford.edu/archives/fall2013/entries/information/.
Ægisdóttir, S., White, M. J., Spengler, P. M., Maugherman, A. S., Anderson, L. A., et al. 2006. The Meta-Analysis of Clinical Judgment Project: Fifty-six years of accumulated research on clinical versus statistical prediction. *The Counseling Psychologist, 34*, 341–82.
Aguiar, M., & Hurst, E. 2007. Measuring trends in leisure: The allocation of time over five decades. *Quarterly Journal of Economics, 122*, 969–1006.
Ajdacic-Gross, V., Bopp, M., Gostynski, M., Lauber, C., Gutzwiller, F., & Rössler, W. 2006. Age–period–cohort analysis of Swiss suicide data, 1881–2000. *European Archives of Psychiatry and Clinical Neuroscience, 256*, 207–14.
Al-Khalili, J. 2010. *Pathfinders: The golden age of Arabic science.* New York: Penguin.
Alesina, A., Glaeser, E. L., & Sacerdote, B. 2001. Why doesn't the United States have a European-style welfare state? *Brookings Papers on Economic Activity, 2*, 187–277.
Alexander, A. C., & Welzel, C. 2011. Islam and patriarchy: How robust is Muslim support for patriarchal values? *International Review of Sociology, 21*, 249–75.
Alexander, S. 2016. You are still crying wolf. *Slate Star Codex*, Nov. 16. http://slatestarcodex.com/2016/11/16/you-are-still-crying-wolf/.
Alferov, Z. I., Altman, S., & 108 other Nobel Laureates. 2016. Laureates letter supporting precision agriculture (GMOs). http://supportprecisionagriculture.org/nobel-laureate-gmo-letter_rjr.html.
Allen, P. G. 2011. The singularity isn't near. *Technology Review*, Oct. 12.
Allen, W. 1987. *Hannah and her sisters.* New York: Random House.
Alrich, M. 2001. History of workplace safety in the United States, 1880–1970. In R. Whaples, ed., *EH.net Encyclopedia.* http://eh.net/encyclopedia/history-of-workplace-safety-in-the-united-states-1880-1970/.
Amabile, T. M. 1983. Brilliant but cruel: Perceptions of negative evaluators. *Journal of Experimental Social Psychology, 19*, 146–56.
American Academy of Arts and Sciences. 2015. *The heart of the matter: The humanities and social sciences for a vibrant, competitive, and secure nation.* Cambridge, MA: American Academy of Arts and Sciences.
American Association of University Professors. 2006. *Research on human subjects: Academic freedom and the institutional review board.* https://www.aaup.org/report/research-human-subjects-academic-freedom-and-institutional-review-board.
American Humanist Association. 1933/1973. *Humanist Manifesto I.* https://americanhumanist.org/what-is-humanism/manifesto1/.
American Humanist Association. 1973. *Humanist Manifesto II.* https://americanhumanist.org/what-is-humanism/manifesto2/.
American Humanist Association. 2003. *Humanism and its aspirations: Humanist Manifesto III.* http://americanhumanist.org/humanism/humanist_manifesto_iii.

Anderson, J. R. 2007. *How can the human mind occur in the physical universe?* New York: Oxford University Press.

Anderson, R. L. 2017. Friedrich Nietzsche. In E. N. Zalta, ed., *Stanford Encyclopedia of Philosophy.* https://plato.stanford.edu/entries/nietzsche/.

Appiah, K. A. 2006. *Cosmopolitanism: Ethics in a world of strangers.* New York: Norton.

Appiah, K. A. 2010. *The honor code: How moral revolutions happen.* New York: Norton.

Ariely, D. 2010. *Predictably irrational: The hidden forces that shape our decisions* (rev. ed.). New York: HarperCollins.

Armitage, D., Bhabha, H., Dench, E., Hamburger, J., Hamilton, J., et al. 2013. *The teaching of the arts and humanities at Harvard College: Mapping the future.* https://harvardmagazine.com/sites/default/files/Mapping%20the%20Future%20of%20the%20Humanities.pdf.

Arrow, K., Jorgenson, D., Krugman, P., Nordhaus, W., & Solow, R. 1997. The economists' statement on climate change. *Redefining Progress.* http://rprogress.org/publications/1997/econstatement.htm.

Asafu-Adjaye, J., Blomqvist, L., Brand, S., DeFries, R., Ellis, E., et al. 2015. *An Ecomodernist Manifesto.* http://www.ecomodernism.org/manifesto-english/.

Asal, V., & Pate, A. 2005. The decline of ethnic political discrimination, 1950–2003. In M. G. Marshall & T. R. Gurr, eds., *Peace and conflict 2005: A global survey of armed conflicts, self-determination movements, and democracy.* College Park: Center for International Development and Conflict Management, University of Maryland.

Atkins, P. 2007. *Four laws that drive the universe.* New York: Oxford University Press.

Atkinson, A. B., Hasell, J., Morelli, S., & Roser, M. 2017. *The chartbook of economic inequality.* https://www.chartbookofeconomicinequality.com/.

Atran, S. 2002. *In gods we trust: The evolutionary landscape of religion.* New York: Oxford University Press.

Atran, S. 2003. Genesis of suicide terrorism. *Science, 299,* 1534–39.

Atran, S. 2007. Research police—how a university IRB thwarts understanding of terrorism. *Institutional Review Blog.* http://www.institutionalreviewblog.com/2007/05/scott-atran-research-police-how.html.

Ausubel, J. H. 1996. The liberation of the environment. *Daedalus, 125,* 1–18.

Ausubel, J. H. 2007. Renewable and nuclear heresies. *International Journal of Nuclear Governance, Economy, and Ecology, 1,* 229–43.

Ausubel, J. H. 2015. *Nature rebounds.* San Francisco: Long Now Foundation. https://phe.rockefeller.edu/docs/Nature_Rebounds.pdf.

Ausubel, J. H., & Grübler, A. 1995. Working less and living longer: Long-term trends in working time and time budgets. *Technological Forecasting and Social Change, 50,* 195–213.

Ausubel, J. H., & Marchetti, C. 1998. Wood's H:C ratio. https://phe.rockefeller.edu/docs/WoodsHtoCratio.pdf.

Ausubel, J. H., Wernick, I. K., & Waggoner, P. E. 2012. Peak farmland and the prospect for land sparing. *Population and Development Review, 38,* 221–242.

Autor, D. H. 2014. Skills, education, and the rise of earnings inequality among the "other 99 percent." *Science, 344,* 843–51.

Aviation Safety Network. 2017. Fatal airliner (14+ passengers) hull-loss accidents. https://aviation-safety.net/statistics/period/stats.php?cat=A1.

Bailey, R. 2015. *The end of doom: Environmental renewal in the 21st century.* New York: St. Martin's Press.

Balmford, A. 2012. *Wild hope: On the front lines of conservation success.* Chicago: University of Chicago Press.

Balmford, A., & Knowlton, N. 2017. Why Earth Optimism? *Science, 356,* 225.

Banerjee, A. V., & Duflo, E. 2011. *Poor economics: A radical rethinking of the way to fight global poverty.* New York: PublicAffairs.

Barber, N. 2011. A cross-national test of the uncertainty hypothesis of religious belief. *Cross-Cultural Research, 45,* 318–33.

Bardo, A. R., Lynch, S. M., & Land, K. C. 2017. The importance of the Baby Boom cohort and the Great Recession in understanding age, period, and cohort patterns in happiness. *Social Psychological and Personality Science, 8,* 341–50.

Bardon, A. (Undated.) Transcendental arguments. *Internet Encyclopedia of Philosophy.* http://www.iep.utm.edu/trans-ar/.

Barlow, D. H., Bullis, J. R., Comer, J. S., & Ametaj, A. A. 2013. Evidence-based psychological treatments: An update and a way forward. *Annual Review of Clinical Psychology, 9,* 1–27.

Baron, J. 1993. Why teach thinking? *Applied Psychology, 42,* 191–237.

Basu, K. 1999. Child labor: Cause, consequence, and cure, with remarks on international labor standards. *Journal of Economic Literature, 37,* 1083–1119.

Bauman, Z. 1989. *Modernity and the Holocaust.* Cambridge, UK: Polity.

Baumard, N., Hyafil, A., Morris, I., & Boyer, P. 2015. Increased affluence explains the emergence of ascetic wisdoms and moralizing religions. *Current Biology, 25,* 10–15.

Baumeister, R. 2015. Machines think but don't want, and hence aren't dangerous. *Edge.* https://www.edge.org/response-detail/26282.

Baumeister, R., Bratslavsky, E., Finkenauer, C., & Vohs, K. D. 2001. Bad is stronger than good. *Review of General Psychology, 5*, 323–70.

Baumeister, R., Vohs, K. D., Aaker, J. L., & Garbinsky, E. N. 2013. Some key differences between a happy life and a meaningful life. *Journal of Positive Psychology, 8*, 505–16.

Baxter, A. J., Scott, K. M., Ferrari, A. J., Norman, R. E., Vos, T., et al. 2014. Challenging the myth of an "epidemic" of common mental disorders: Trends in the global prevalence of anxiety and depression between 1990 and 2010. *Depression and Anxiety, 31*, 506–16.

Bean, L., & Teles, S. 2016. God and climate. *Democracy: A Journal of Ideas, 40*.

Beaver, K. M., Schwartz, J. A., Nedelec, J. L., Connolly, E. J., Boutwell, B. B., et al. 2013. Intelligence is associated with criminal justice processing: Arrest through incarceration. *Intelligence, 41*, 277–88.

Beaver, K. M., Vaughn, M. G., DeLisi, M., Barnes, J. C., & Boutwell, B. B. 2012. The neuropsychological underpinnings to psychopathic personality traits in a nationally representative and longitudinal sample. *Psychiatric Quarterly, 83*, 145–59.

Behavioral Insights Team. 2015. *EAST: Four simple ways to apply behavioral insights.* London: Behavioral Insights.

Benda, J. 1927/2006. *The treason of the intellectuals.* New Brunswick, NJ: Transaction.

Benedek, T. G., & Erlen, J. 1999. The scientific environment of the Tuskegee Study of Syphilis, 1920–1960. *Perspectives in Biology and Medicine, 43*, 1–30.

Berlin, I. 1979. The Counter-Enlightenment. In I. Berlin, ed., *Against the current: Essays in the history of ideas.* Princeton, NJ: Princeton University Press.

Berlin, I. 1988/2013. The pursuit of the ideal. In I. Berlin, ed., *The crooked timber of humanity.* Princeton, NJ: Princeton University Press.

Berman, P. 2010. *The flight of the intellectuals.* New York: Melville House.

Bernanke, B. S. 2016. How do people really feel about the economy? *Brookings Blog.* https://www.brookings.edu/blog/ben-bernanke/2016/06/30/how-do-people-really-feel-about-the-economy/.

Berry, K., Lewis, P., Pelopidas, B., Sokov, N., & Wilson, W. 2010. *Delegitimizing nuclear weapons: Examining the validity of nuclear deterrence.* Monterey, CA: Monterey Institute of International Studies.

Besley, T. & Kudamatsu, M. 2006. Health and democracy. *American Economic Review, 96*, 313–18.

Bettmann, O. L. 1974. *The good old days—they were terrible!* New York: Random House.

Betzig, L. 1986. *Despotism and differential reproduction.* Hawthorne, NY: Aldine de Gruyter.

Bird, A. 2011. Thomas Kuhn. In E. N. Zalta, ed., *Stanford Encyclopedia of Philosophy.* https://plato.stanford.edu/entries/thomas-kuhn/.

Blackmore, S. 1991. Near-death experiences: In or out of the body? *Skeptical Inquirer, 16*, 34–45.

Blair, J. P., & Schweit, K. W. 2014. *A study of active shooter incidents, 2000–2013.* Washington: Federal Bureau of Investigation.

Blees, T. 2008. *Prescription for the planet: The painless remedy for our energy and environmental crises.* North Charleston, SC: Booksurge.

Blight, J. G., Nye, J. S., & Welch, D. A. 1987. The Cuban Missile Crisis revisited. *Foreign Affairs, 66*, 170–88.

Blinkhorn, S. 1982. Review of S. J. Gould's "The mismeasure of man." *Nature, 296*, 506.

Block, N. 1986. Advertisement for a semantics for psychology. In P. A. French, T. E. Uehling, & H. K. Wettstein, eds., *Midwest studies in philosophy: Studies in the philosophy of mind* (vol. 10). Minneapolis: University of Minnesota Press.

Block, N. 1995. On a confusion about a function of consciousness. *Behavioral and Brain Sciences, 18*, 227–87.

Bloom, P. 2012. Religion, morality, evolution. *Annual Review of Psychology, 63*, 179–99.

Bloomberg, M., & Pope, C. 2017. *Climate of hope: How cities, businesses, and citizens can save the planet.* New York: St. Martin's Press.

Bluth, C. 2011. *The myth of nuclear proliferation.* School of Politics and International Studies, University of Leeds.

Bohle, R. H. 1986. Negativism as news selection predictor. *Journalism Quarterly, 63*, 789–96.

Bond, M. 2009. Risk school. *Nature, 461*, 1189–1192.

Bostrom, A., Morgan, M. G., Fischhoff, B., & Read, D. 1994. What do people know about global climate change? 1. Mental models. *Risk Analysis, 14*, 959–71.

Bostrom, N. 2016. *Superintelligence: Paths, dangers, strategies.* New York: Oxford University Press.

Botello, M. A. 2016. Mexico, tasa de homicidios por 100 mil habitantes desde 1931 a 2015. *Mexico-Maxico.* http://www.mexicomaxico.org/Voto/Homicidios100M.htm.

Bourget, D., & Chalmers, D. J. 2014. What do philosophers believe? *Philosophical Studies, 170*, 465–500.

Bourguignon, F., & Morrisson, C. 2002. Inequality among world citizens, 1820–1992. *American Economic Review, 92*, 727–44.

Bowering, G. 2015. *Islamic political thought: An introduction.* Princeton, NJ: Princeton University Press.

Boyd, B., Carroll, J., & Gottschall, J., eds. 2010. *Evolution, literature, and film: A reader.* New York: Columbia University Press.

Boyd, R. 1988. How to be a moral realist. In G. Sayre-McCord, ed., *Essays on moral realism.* Ithaca, NY: Cornell University Press.

Boyer, Pascal. 2001. *Religion explained: The evolutionary origins of religious thought*. New York: Basic Books.
Boyer, Paul. 1985/2005. *By the bomb's early light: American thought and culture at the dawn of the Atomic Age*. Chapel Hill: University of North Carolina Press.
Boyer, Paul. 1986. A historical view of scare tactics. *Bulletin of the Atomic Scientists*, 17–19.
Braithwaite, J. 2008. Near death experiences: The dying brain. *Skeptic, 21* (2). http://www.critical -thinking.org.uk/paranormal/near-death-experiences/the-dying-brain.php.
Braman, D., Kahan, D. M., Slovic, P., Gastil, J., & Cohen, G. L. 2007. The Second National Risk and Culture Study: Making sense of—and making progress in—the American culture war of fact. *GW Law Faculty Publications and Other Works, 211*. http://scholarship.law.gwu.edu/faculty _publications/211.
Branch, T. 1988. *Parting the waters: America in the King years, 1954–63*. New York: Simon & Schuster.
Brand, S. 2009. *Whole Earth discipline: Why dense cities, nuclear power, transgenic crops, restored wildlands, and geoengineering are necessary*. New York: Penguin.
Branwen, G. 2016. Terrorism is not effective. *Gwern.net*. https://www.gwern.net/Terrorism-is-not -Effective.
Braudel, F. 2002. *Civilization and capitalism, 15th–18th century* (vol. 1: *The structures of everyday life*). London: Phoenix Press.
Bregman, A. S. 1990. *Auditory scene analysis: The perceptual organization of sound*. Cambridge, MA: MIT Press.
Bregman, R. 2016. *Utopia for realists: The case for a universal basic income, open borders, and a 15-hour workweek*. Boston: Little, Brown.
Brennan, J. 2016. Against democracy. *National Interest*, Sept. 7.
Brickman, P., & Campbell, D. T. 1971. Hedonic relativism and planning the good society. In M. H. Appley, ed., *Adaptation-level theory: A symposium*. New York: Academic Press.
Briggs, J. C. 2015. Re: Accelerated modern human-induced species losses: Entering the sixth mass extinction. *Science*. http://advances.sciencemag.org/content/1/5/e1400253.e-letters.
Briggs, J. C. 2016. Global biodiversity loss: Exaggerated versus realistic estimates. *Environmental Skeptics and Critics, 5*, 20–27.
Brink, D. O. 1989. *Moral realism and the foundations of ethics*. New York: Cambridge University Press.
British Petroleum. 2016. *BP Statistical Review of World Energy 2016*, June.
Brockman, J. 1991. The third culture. *Edge*. https://www.edge.org/conversation/john_brockman-the -third-culture.
Brockman, J., ed. 2003. *The new humanists: Science at the edge*. New York: Sterling.
Brockman, J., ed. 2015. *What to think about machines that think? Today's leading thinkers on the age of machine intelligence*. New York: HarperPerennial.
Brooks, R. 2015. Mistaking performance for competence misleads estimates of AI's 21st century promise and danger. *Edge*. https://www.edge.org/response-detail/26057.
Brooks, R. 2016. Artificial intelligence. *Edge*. https://www.edge.org/response-detail/26678.
Brown, A., & Lewis, J. 2013. Reframing the nuclear de-alerting debate: Towards maximizing presidential decision time. *Nuclear Threat Initiative*. http://nti.org/3521A.
Brown, D. E. 1991. *Human universals*. New York: McGraw-Hill.
Brown, D. E. 2000. Human universals and their implications. In N. Roughley, ed., *Being humans: Anthropological universality and particularity in transdisciplinary perspectives*. New York: Walter de Gruyter.
Brunnschweiler, C. N., & Lujala, P. 2015. Economic backwardness and social tension. University of East Anglia. https://ideas.repec.org/p/uea/aepppr/2012_72.html.
Bryce, R. 2014. *Smaller faster lighter denser cheaper: How innovation keeps proving the catastrophists wrong*. New York: Perseus.
Brynjolfsson, E., & McAfee, A. 2015. Will humans go the way of horses? *Foreign Affairs*, July/Aug.
Brynjolfsson, E., & McAfee, A. 2016. *The Second Machine Age: Work, progress, and prosperity in a time of brilliant technologies*. New York: Norton.
Bulletin of the Atomic Scientists. 2017. Doomsday Clock timeline. http://thebulletin.org/timeline.
Bunce, V. 2017. The prospects for a color revolution in Russia. *Daedalus, 146*, 19–29.
Bureau of Labor Statistics. 2016a. Census of fatal occupational injuries. https://www.bls.gov/iif/oshcfoi1 .htm.
Bureau of Labor Statistics. 2016b. Charts from the American Time Use Survey. https://www.bls.gov /tus/charts/.
Bureau of Labor Statistics. 2016c. Time spent in primary activities and percent of the civilian population engaging in each activity, averages per day by sex, 2015. https://www.bls.gov/news .release/atus.t01.htm.
Bureau of Labor Statistics. 2017. College enrollment and work activity of 2016 high school graduates. https://www.bls.gov/news.release/hsgec.nr0.htm.
Buringh, E., & van Zanden, J. 2009. Charting the "rise of the West": Manuscripts and printed books in Europe, a long-term perspective from the sixth through eighteenth centuries. *Journal of Economic History, 69*, 409–45.
Burney, D. A., & Flannery, T. F. 2005. Fifty millennia of catastrophic extinctions after human contact. *Trends in Ecology and Evolution, 20*, 395–401.

Burns, J. 2009. *Goddess of the market: Ayn Rand and the American right.* New York: Oxford University Press.

Burtless, G. 2014. Income growth and income inequality: The facts may surprise you. *Brookings Blog.* https://www.brookings.edu/opinions/income-growth-and-income-inequality-the-facts-may-surprise-you/.

Buturovic, Z., & Klein, D. B. 2010. Economic enlightenment in relation to college-going, ideology, and other variables: A Zogby survey of Americans. *Economic Journal Watch, 7,* 174–96.

Caplan, B. 2007. *The myth of the rational voter: Why democracies choose bad policies.* Princeton, NJ: Princeton University Press.

Caplow, T., Hicks, L., & Wattenberg, B. 2001. *The first measured century: An illustrated guide to trends in America, 1900–2000.* Washington: AEI Press.

CarbonBrief. 2016. Explainer: 10 ways "negative emissions" could slow climate change. https://www.carbonbrief.org/explainer-10-ways-negative-emissions-could-slow-climate-change.

Carey, J. 1993. *The intellectuals and the masses: Pride and prejudice among the literary intelligentsia, 1880–1939.* New York: St. Martin's Press.

Carey, M., Jackson, M., Antonello, A., & Rushing, J. 2016. Glaciers, gender, and science. *Progress in Human Geography, 40,* 770–93.

Carey, S. 2009. *The origin of concepts.* Cambridge, MA: MIT Press.

Carlson, R. H. 2010. *Biology is technology: The promise, peril, and new business of engineering life.* Cambridge, MA: Harvard University Press.

Carroll, S. M. 2016. *The big picture: On the origins of life, meaning, and the universe itself.* New York: Dutton.

Carter, R. 1966. *Breakthrough: The saga of Jonas Salk.* Trident Press.

Carter, S. B., Gartner, S. S., Haines, M. R., Olmstead, A. L., Sutch, R., et al., eds. 2006. *Historical statistics of the United States: Earliest times to the present* (vol. 1, part A: Population). New York: Cambridge University Press.

Case, A., & Deaton, A. 2015. Rising morbidity and mortality in midlife among white non-Hispanic Americans in the 21st century. *Proceedings of the National Academy of Sciences, 112,* 15078–83.

Center for Systemic Peace. 2015. Integrated network for societal conflict research data page. http://www.systemicpeace.org/inscrdata.html.

Centers for Disease Control. 1999. Improvements in workplace safety—United States, 1900–1999. *CDC Morbidity and Mortality Weekly Report, 48,* 461–69.

Centers for Disease Control. 2015. Injury prevention and control: Data and statistics (WISQARS). https://www.cdc.gov/injury/wisqars/.

Central Intelligence Agency. 2016. The world factbook. https://www.cia.gov/library/publications/the-world-factbook/.

Chalk, F., & Jonassohn, K. 1990. *The history and sociology of genocide: Analyses and case studies.* New Haven: Yale University Press.

Chalmers, D. J. 1996. *The conscious mind: In search of a fundamental theory.* New York: Oxford University Press.

Chang, L. T. 2009. *Factory girls: From village to city in a changing China.* New York: Spiegel & Grau.

Chen, D. H. C., & Dahlman, C. J. 2006. *The knowledge economy, the KAM methodology and World Bank operations.* Washington: World Bank. http://documents.worldbank.org/curated/en/69521 1468153873436/The-knowledge-economy-the-KAM-methodology-and-World-Bank-operations.

Chenoweth, E. 2016. Why is nonviolent resistance on the rise? *Diplomatic Courier.* http://www.diplomaticourier.com/2016/06/28/nonviolent-resistance-rise/.

Chenoweth, E., & Stephan, M. J. 2011. *Why civil resistance works: The strategic logic of nonviolent conflict.* New York: Columbia University Press.

Chernew, M., Cutler, D. M., Ghosh, K., & Landrum, M. B. 2016. *Understanding the improvement in disability free life expectancy in the U.S. elderly population.* Cambridge, MA: National Bureau of Economic Research.

Chirot, D. 1996. *Modern tyrants.* Princeton, NJ: Princeton University Press.

Cipolla, C. 1994. *Before the Industrial Revolution: European society and economy, 1000–1700* (3rd ed.). New York: Norton.

Clark, A. M., & Sikkink, K. 2013. Information effects and human rights data: Is the good news about increased human rights information bad news for human rights measures? *Human Rights Quarterly, 35,* 539–68.

Clark, D. M. T., Loxton, N. J., & Tobin, S. J. 2015. Declining loneliness over time: Evidence from American colleges and high schools. *Personality and Social Psychology Bulletin, 41,* 78–89.

Clark, G. 2007. *A farewell to alms: A brief economic history of the world.* Princeton, NJ: Princeton University Press.

Cohen, G. L. 2003. Party over policy: The dominating impact of group influence on political beliefs. *Journal of Personality and Social Psychology, 85,* 808–22.

Collier, P. 2007. *The bottom billion: Why the poorest countries are failing and what can be done about it.* New York: Oxford University Press.

Collier, P., & Rohner, D. 2008. Democracy, development and conflict. *Journal of the European Economic Association, 6,* 531–40.

Collini, S. 1998. Introduction. In C. P. Snow, *The two cultures*. New York: Cambridge University Press.

Collini, S. 2013. Introduction. In F. R. Leavis, *Two cultures? The significance of C. P. Snow*. New York: Cambridge University Press.

Combs, B., & Slovic, P. 1979. Newspaper coverage of causes of death. *Journalism & Mass Communication Quarterly, 56*, 837–43.

Connor, S. 2014. *The horror of number: Can humans learn to count?* Paper presented at the Alexander Lecture. http://stevenconnor.com/horror.html.

Connor, S. 2016. *Living by numbers: In defence of quantity*. London: Reaktion Books.

Conrad, S. 2012. Enlightenment in global history: A historiographical critique. *American Historical Review, 117*, 999–1027.

Cook, M. 2014. *Ancient religions, modern politics: The Islamic case in comparative perspective*. Princeton, NJ: Princeton University Press.

Coontz, S. 1992/2016. *The way we never were: American families and the nostalgia trap* (rev. ed.). New York: Basic Books.

Corlett, A. 2016. *Examining an elephant: Globalisation and the lower middle class of the rich world*. London: Resolution Foundation.

Cornwall Alliance for the Stewardship of Creation. 2000. The Cornwall Declaration on Environmental Stewardship. http://cornwallalliance.org/landmark-documents/the-cornwall-declaration-on -environmental-stewardship/.

Cosmides, L., & Tooby, J. 1992. Cognitive adaptations for social exchange. In J. H. Barkow, L. Cosmides, & J. Tooby, eds., *The adapted mind: Evolutionary psychology and the generation of culture*. New York: Oxford University Press.

Costa, D. L. 1998. *The evolution of retirement: An American economic history, 1880–1990*. Chicago: University of Chicago Press.

Costa, P. T., & McCrae, R. R. 1982. An approach to the attribution of aging, period, and cohort effects. *Psychological Bulletin, 92*, 238–50.

Costello, E. J., Erkanli, A., & Angold, A. 2006. Is there an epidemic of child or adolescent depression? *Journal of Child Psychology and Psychiatry, 47*, 1263–71.

Costello, M. J., May, R. M., & Stork, N. E. 2013. Can we name Earth's species before they go extinct? *Science, 339*, 413–16.

Council for Secular Humanism. 1980. *A Secular Humanist Declaration*. https://www.secularhumanism .org/index.php/11.

Council for Secular Humanism. 2000. *Humanist Manifesto 2000*. https://www.secularhumanism.org /index.php/1169.

Council on Foreign Relations. 2009. World opinion on proliferation of weapons of mass destruction. https://www.cfr.org/backgrounder/world-opinion-proliferation-weapons-mass-de struction.

Council on Foreign Relations. 2011. World opinion on human rights. *Public Opinion on Global Issues*. https://www.cfr.org/backgrounder/world-opinion-human-rights.

Courtois, S., Werth, N., Panné, J.-L., Paczkowski, A., Bartosek, K., et al. 1999. *The Black Book of Communism: Crimes, terror, repression*. Cambridge, MA: Harvard University Press.

Courtwright, D. 2010. *No right turn: Conservative politics in a liberal America*. Cambridge, MA: Harvard University Press.

Cowen, T. 2017. *The complacent class: The self-defeating quest for the American dream*. New York: St. Martin's Press.

Coyne, J. A. 2015. *Faith versus fact: Why science and religion are incompatible*. New York: Penguin.

Cravens, G. 2007. *Power to save the world: The truth about nuclear energy*. New York: Knopf.

Cronin, A. K. 2009. *How terrorism ends: Understanding the decline and demise of terrorist campaigns*. Princeton, NJ: Princeton University Press.

Cronon, W. 1995. The trouble with wilderness; or, getting back to the wrong nature. In W. Cronon, ed., *Uncommon ground: Rethinking the human place in nature*. New York: Norton.

Cunningham, H. 1996. Combating child labour: The British experience. In H. Cunningham & P. P Viazzo, eds., *Child labour in historical perspective, 1800–1985: Case studies from Europe, Japan and Colombia*. Florence: UNICEF.

Cunningham, T. J., Croft, J. B., Liu, Y., Lu, H., Eke, P. I., et al. 2017. Vital signs: Racial disparities in age-specific mortality among Blacks or African Americans—United States, 1999–2015. *Morbidity and Mortality Weekly Report, 66*, 444–56.

Daly, M. C., Oswald, A. J., Wilson, D., & Wu, S. 2010. The happiness-suicide paradox. *Federal Reserve Bank of San Francisco Working Papers, 2010*.

Davis, B. D. 1983. Neo-Lysenkoism, IQ, and the press. *Public Interest, 73*, 41–59.

Davis, E., & Marcus, G. F. 2015. Commonsense reasoning and commonsense knowledge in artificial intelligence. *Communications of the ACM, 58*, 92–103.

Dawes, R. M., Faust, D., & Meehl, P. E. 1989. Clinical versus actuarial judgment. *Science, 243*, 1668–74.

Dawkins, R. 1976/1989. *The selfish gene* (new ed.). New York: Oxford University Press.

Dawkins, R. 1983. Universal Darwinism. In D. S. Bendall, ed., *Evolution from molecules to men*. New York: Cambridge University Press.

Dawkins, R. 1986. *The blind watchmaker: Why the evidence of evolution reveals a universe without design.* New York: Norton.

Dawkins, R. 2006. *The God delusion.* New York: Houghton Mifflin.

de Lazari-Radek, K., & Singer, P. 2012. The objectivity of ethics and the unity of practical reason. *Ethics, 123,* 9–31.

de Ribera, O. S., Kavish, N., & Boutwell, B. B. 2017. On the relationship between psychopathy and general intelligence: A meta-analytic review. *bioRχiv,* doi: https://doi.org/10.1101/100693.

Deary, I. J. 2001. *Intelligence: A very short introduction.* New York: Oxford University Press.

Death Penalty Information Center. 2017. Facts about the death penalty. http://www.deathpenaltyinfo .org/documents/FactSheet.pdf.

Deaton, A. 2011. The financial crisis and the well-being of Americans. *Oxford Economic Papers,* 1–26.

Deaton, A. 2013. *The Great Escape: Health, wealth, and the origins of inequality.* Princeton, NJ: Princeton University Press.

Deaton, A. 2017. Thinking about inequality. *Cato's Letter, 15,* 1–5.

Deep Decarbonization Pathways Project 2015. *Pathways to deep decarbonization.* Paris: Institute for Sustainable Development and International Relations.

DeFries, R. 2014. *The big ratchet: How humanity thrives in the face of natural crisis.* New York: Basic Books.

Degler, C. N. 1991. *In search of human nature: The decline and revival of Darwinism in American social thought.* New York: Oxford University Press.

Dehaene, S. 2009. Signatures of consciousness. *Edge.* http://www.edge.org/3rd_culture/dehaene09/ dehaene09_index.html.

Dehaene, S., & Changeux, J.-P. 2011. Experimental and theoretical approaches to conscious processing. *Neuron, 70,* 200–227.

Delamontagne, R. G. 2010. High religiosity and societal dysfunction in the United States during the first decade of the twenty-first century. *Evolutionary Psychology, 8,* 617–57.

Denkenberger, D., & Pearce, J. 2015. *Feeding everyone no matter what: Managing food security after global catastrophe.* New York: Academic Press.

Dennett, D. C. 2006. *Breaking the spell: Religion as a natural phenomenon.* New York: Penguin Books.

DeScioli, P. 2016. The side-taking hypothesis for moral judgment. *Current Opinion in Psychology, 7,* 23–27.

DeScioli, P., & Kurzban, R. 2009. Mysteries of morality. *Cognition, 112,* 281–99.

Desvousges, W. H., Johnson, F. R., Dunford, R. W., Boyle, K. J., Hudson, S. P., et al. 1992. *Measuring nonuse damages using contingent valuation: An experimental evaluation of accuracy.* Research Triangle Park, NC: RTI International.

Deutsch, D. 2011. *The beginning of infinity: Explanations that transform the world.* New York: Viking.

Devereux, S. 2000. *Famine in the twentieth century.* Sussex, UK: Institute of Development Studies. http://www.ids.ac.uk/publication/famine-in-the-twentieth-century.

Diamandis, P., & Kotler, S. 2012. *Abundance: The future is better than you think.* New York: Free Press.

Diamond, J. M. 1997. *Guns, germs, and steel: The fates of human societies.* New York: Norton.

Dinda, S. 2004. Environmental Kuznets curve hypothesis: A survey. *Ecological Economics, 49,* 431–55.

Dobbs, R., Madgavkar, A., Manyika, J., Woetzel, J., Bughin, J., et al. 2016. *Poorer than their parents? Flat or falling incomes in advanced economies.* McKinsey Global Institute.

Dreger, A. 2008. The controversy surrounding "The man who would be queen": A case history of the politics of science, identity, and sex in the Internet age. *Archives of Sexual Behavior, 37,* 366–421.

Dreger, A. 2015. *Galileo's middle finger: Heretics, activists, and the search for justice in science.* New York: Penguin.

Dretske, F. I. 1981. *Knowledge and the flow of information.* Cambridge, MA: MIT Press.

Duarte, J. L., Crawford, J. T., Stern, C., Haidt, J., Jussim, L., & Tetlock, P. E. 2015. Political diversity will improve social psychological science. *Behavioral and Brain Sciences, 38,* 1–13.

Dunlap, R. E., Gallup, G. H., & Gallup, A. M. 1993. Of global concern. *Environment: Science and Policy for Sustainable Development, 35,* 7–39.

Duntley, J. D., & Buss, D. M. 2011. Homicide adaptations. *Aggression and Violent Behavior, 16,* 399–410.

Dutton, D. 2009. *The art instinct: Beauty, pleasure, and human evolution.* New York: Bloomsbury Press.

Eagan, K., Stolzenberg, E. B., Lozano, J. B., Aragon, M. C., Suchard, M. R., et al. 2014. *Undergraduate teaching faculty: The 2013–2014 HERI faculty survey.* Los Angeles: Higher Education Research Institute at UCLA.

Easterbrook, G. 2003. *The progress paradox: How life gets better while people feel worse.* New York: Random House.

Easterlin, R. A. 1973. Does money buy happiness? *Public Interest, 30,* 3–10.

Easterlin, R. A. 1981. Why isn't the whole world developed? *Journal of Economic History, 41,* 1–19.

Easterly, W. 2006. *The white man's burden: Why the West's efforts to aid the rest have done so much ill and so little good.* New York: Penguin.

Eastop, E.-R. 2015. *Subcultural cognition: Armchair oncology in the age of misinformation.* Master's thesis, University of Oxford.

Eberstadt, N., & Shah, A. 2011. *Fertility decline in the Muslim world: A veritable sea-change, still curiously unnoticed.* Washington: American Enterprise Institute.

Eddington, A. S. 1928/2015. *The nature of the physical world.* Andesite Press.

Eibach, R. P., & Libby, L. K. 2009. Ideology of the good old days: Exaggerated perceptions of moral decline and conservative politics. In J. T. Jost, A. Kay, & H. Thorisdottir, eds., *Social and psychological bases of ideology and system justification.* New York: Oxford University Press.

Eichengreen, B. 2014. Secular stagnation: A review of the issues. In C. Teulings & R. Baldwin, eds., *Secular stagnation: Facts, causes and cures.* London: Centre for Economic Policy Research.

Eisner, M. 2001. Modernization, self-control and lethal violence: The long-term dynamics of European homicide rates in theoretical perspective. *British Journal of Criminology, 41,* 618–38.

Eisner, M. 2003. Long-term historical trends in violent crime. *Crime and Justice, 30,* 83–142.

Eisner, M. 2014a. From swords to words: Does macro-level change in self-control predict long-term variation in levels of homicide? *Crime and Justice, 43,* 65–134.

Eisner, M. 2014b. *Reducing homicide by 50% in 30 years: Universal mechanisms and evidence-based public policy.* In M. Krisch, M. Eisner, C. Mikton, & A. Butchart, eds., *Global strategies to reduce violence by 50% in 30 years: Findings from the WHO and University of Cambridge Global Violence Reduction Conference 2014.* Cambridge, UK: Institute of Criminology, University of Cambridge.

Eisner, M. 2015. *How to reduce homicide by 50% in the next 30 years.* Rio de Janeiro: Igarapé Institute.

Elias, N. 1939/2000. *The Civilizing Process: Sociogenetic and psychogenetic investigations* (rev. ed.). Cambridge, MA: Blackwell.

England, J. L. 2015. Dissipative adaptation in driven self-assembly. *Nature Nanotechnology, 10,* 919–23.

Epstein, A. 2014. *The moral case for fossil fuels.* New York: Penguin.

Epstein, G. 2009. *Good without God: What a billion nonreligious people do believe.* New York: William Morrow.

Ericksen, R. P., & Heschel, S. 1999. *Betrayal: German churches and the Holocaust.* Minneapolis: Fortress Press.

Erwin, D. 2015. *Extinction: How life on Earth nearly ended 250 million years ago* (updated ed.). Princeton, NJ: Princeton University Press.

Esposito, J. L., & Mogahed, D. 2007. *Who speaks for Islam? What a billion Muslims really think.* New York: Gallup Press.

Evans, D. 2015. The great AI swindle. *Edge.* https://www.edge.org/response-detail/26073.

Evans, G. 2015. Challenges for the *Bulletin of the Atomic Scientists* at 70: Restoring reason to the nuclear debate. Paper presented at the Annual Clock Symposium, *Bulletin of the Atomic Scientists.*

Evans, G., Ogilvie-White, T., & Thakur, R. 2015. *Nuclear weapons: The state of play 2015.* Canberra: Centre for Nuclear Non-Proliferation and Disarmament, Australian National University.

Everett, D. 2008. *Don't sleep, there are snakes: Life and language in the Amazonian jungle.* New York: Vintage.

Ewald, P. 2000. *Plague time: The new germ theory of disease.* New York: Anchor.

Faderman, L. 2015. *The Gay Revolution: The story of the struggle.* New York: Simon & Schuster.

Fariss, C. J. 2014. Respect for human rights has improved over time: Modeling the changing standard of accountability. *American Political Science Review, 108,* 297–318.

Fawcett, A. A., Iyer, G. C., Clarke, L. E., Edmonds, J. A., Hultman, N. E., et al. 2015. Can Paris pledges avert severe climate change? *Science, 350,* 1168–69.

Fearon, J. D., & Laitin, D. D. 1996. Explaining interethnic cooperation. *American Political Science Review, 90,* 715–35.

Fearon, J. D., & Laitin, D. D. 2003. Ethnicity, insurgency, and civil war. *American Political Science Review, 97,* 75–90.

Federal Bureau of Investigation. 2016a. Crime in the United States by volume and rate, 1996–2015. https://ucr.fbi.gov/crime-in-the-u.s/2015/crime-in-the-u.s.-2015/tables/table-1.

Federal Bureau of Investigation. 2016b. Hate crime. *FBI Uniform Crime Reports.* https://ucr.fbi.gov/hate-crime.

Federal Highway Administration. 2003. *A review of pedestrian safety research in the United States and abroad: Final report.* Washington: US Department of Transportation. https://www.fhwa.dot.gov/publications/research/safety/pedbike/03042/part2.cfm.

Federation of American Scientists. (Undated.) Nuclear weapons. https://fas.org/issues/nuclear-weapons/.

Feinberg, M., & Willer, R. 2011. Apocalypse soon? Dire messages reduce belief in global warming by contradicting just-world beliefs. *Psychological Science, 22,* 34–38.

Feldstein, M. 2017. Underestimating the real growth of GDP, personal income, and productivity. *Journal of Economic Perspectives, 31,* 145–64.

Ferreira, F., Jolliffe, D. M., & Prydz, E. B. 2015. The international poverty line has just been raised to $1.90 a day, but global poverty is basically unchanged. How is that even possible? http://blogs.worldbank.org/developmenttalk/international-poverty-line-has-just-been-raised-190-day-global-poverty-basically-unchanged-how-even.

Finkelhor, D. 2014. Trends in child welfare. Paper presented at the Carsey Institute Policy Series, Department of Sociology, University of New Hampshire.

Finkelhor, D., Shattuck, A., Turner, H. A., & Hamby, S. L. 2014. Trends in children's exposure to violence, 2003–2011. *JAMA Pediatrics, 168,* 540–46.

Fischer, C. S. 2005. Bowling alone: What's the score? *Social Networks, 27,* 155–67.

Fischer, C. S. 2009. The 2004 GSS finding of shrunken social networks: An artifact? *American Sociological Review, 74,* 657–69.

Fischer, C. S. 2011. *Still connected: Family and friends in America since 1970.* New York: Russell Sage Foundation.

Fiske, A. P., & Rai, T. 2015. *Virtuous violence: Hurting and killing to create, sustain, end, and honor social relationships.* New York: Cambridge University Press.

Fletcher, J. 1997. *Violence and civilization: An introduction to the work of Norbert Elias.* Cambridge, UK: Polity.

Flynn, J. R. 2007. *What is intelligence?* New York: Cambridge University Press.

Flynn, J. R. 2012. *Are we getting smarter? Rising IQ in the twenty-first century.* New York: Cambridge University Press.

Foa, R. S., & Mounk, Y. 2016. The danger of deconsolidation: The democratic disconnect. *Journal of Democracy, 27,* 5–17.

Fodor, J. A. 1987. *Psychosemantics: The problem of meaning in the philosophy of mind.* Cambridge, MA: MIT Press.

Fodor, J. A. 1994. *The elm and the expert: Mentalese and its semantics.* Cambridge, MA: MIT Press.

Fogel, R. W. 2004. *The escape from hunger and premature death, 1700–2100.* New York: Cambridge University Press.

Food Marketing Institute. 2017. Supermarket facts. https://www.fmi.org/our-research/supermarket-facts.

Foreman, C. 2013. On justice movements: Why they fail the environment and the poor. *The Breakthrough,* http://thebreakthrough.org/index.php/journal/past-issues/issue-3/on-justice-movements.

Fortna, V. P. 2008. *Does peacekeeping work? Shaping belligerents' choices after civil war.* Princeton, NJ: Princeton University Press.

Fortna, V. P. 2015. Do terrorists win? Rebels' use of terrorism and civil war outcomes. *International Organization, 69,* 519–56.

Foucault, M. 1999. *The history of sexuality.* New York: Vintage.

Fouquet, R., & Pearson, P. J. G. 2012. The long run demand for lighting: Elasticities and rebound effects in different phases of economic development. *Economics of Energy and Environmental Policy, 1,* 83–100.

Francis. 2015. *Laudato Si': Encyclical letter of the Holy Father Francis on care for our common home.* Vatican City: The Vatican. http://w2.vatican.va/content/francesco/en/encyclicals/documents/papa-francesco_20150524_enciclica-laudato-si.html.

Frankel, M. 2004. *High noon in the Cold War: Kennedy, Khrushchev, and the Cuban Missile Crisis.* New York: Ballantine Books.

Frankfurt, H. G. 2015. *On inequality.* Princeton, NJ: Princeton University Press.

Freed, J. 2014. *Back to the future: Advanced nuclear energy and the battle against climate change.* Washington: Brookings Institution.

Freilich, J. D., Chermak, S. M., Belli, R., Gruenewald, J., & Parkin, W. S. 2014. Introducing the United States Extremist Crime Database (ECDB). *Terrorism and Political Violence, 26,* 372–84.

Friedman, J. 1997. What's wrong with libertarianism. *Critical Review, 11,* 407–67.

Fryer, R. G. 2016. An empirical analysis of racial differences in police use of force. *National Bureau of Economic Research Working Papers,* 1–63.

Fukuda, K. 2013. A happiness study using age-period-cohort framework. *Journal of Happiness Studies, 14,* 135–53.

Fukuyama, F. 1989. The end of history? *National Interest,* Summer.

Furman, J. 2005. Wal-Mart: A progressive success story. https://www.mackinac.org/archives/2006/walmart.pdf.

Furman, J. 2014. Poverty and the tax code. *Democracy: A Journal of Ideas, 32,* 8–22.

Future of Life Institute. 2017. Accidental nuclear war: A timeline of close calls. https://futureoflife.org/background/nuclear-close-calls-a-timeline/.

Fyfe, J. J. 1988. Police use of deadly force: Research and reform. *Justice Quarterly, 5,* 165–205.

Gaillard, R., Dehaene, S., Adam, C., Clémenceau, S., Hasboun, D., et al. 2009. Converging intracranial markers of conscious access. *PLOS Biology, 7,* 472–92.

Gallup. 2002. Acceptance of homosexuality: A youth movement. http://www.gallup.com/poll/5341/Acceptance-Homosexuality-Youth-Movement.aspx.

Gallup. 2010. Americans' acceptance of gay relations crosses 50% threshold. http://www.gallup.com/poll/135764/Americans-Acceptance-Gay-Relations-Crosses-Threshold.aspx.

Gallup. 2016. Death penalty. http://www.gallup.com/poll/1606/death-penalty.aspx.

Galor, O., & Moav, O. 2007. The neolithic origins of contemporary variations in life expectancy. http://dx.doi.org/10.2139/ssrn.1012650.

Galtung, J., & Ruge, M. H. 1965. The structure of foreign news. *Journal of Peace Research, 2,* 64–91.

Gardner, D. 2008. *Risk: The science and politics of fear.* London: Virgin Books.

Gardner, D. 2010. *Future babble: Why expert predictions fail—and why we believe them anyway.* New York: Dutton.

Garrard, G. 2006. *Counter-enlightenments: From the eighteenth century to the present.* New York: Routledge.

Gash, T. 2016. *Criminal: The truth about why people do bad things.* London: Allen Lane.

Gat, A. 2015. Proving communal warfare among hunter-gatherers: The quasi-Rousseauan error. *Evolutionary Anthropology, 24,* 111–26.

Gauchat, G. 2012. Politicization of science in the public sphere: A study of public trust in the United States, 1974 to 2010. *American Sociological Review, 77,* 167–87.

Gell-Mann, M. 1994. *The quark and the jaguar: Adventures in the simple and the complex.* New York: W. H. Freeman.

Gentzkow, M., & Shapiro, J. M. 2010. What drives media slant? Evidence from U.S. daily newspapers. *Econometrica, 78,* 35–71.

Gervais, W. M., & Najle, M. B. 2017. How many atheists are there? *Social Psychological and Personality Science,* 9, 3–10.

Ghitza, Y., & Gelman, A. 2014. The Great Society, Reagan's revolution, and generations of presidential voting. http://www.stat.columbia.edu/~gelman/research/unpublished/cohort_voting_2014 0605.pdf.

Gigerenzer, G. 1991. How to make cognitive illusions disappear: Beyond "heuristics and biases." *European Review of Social Psychology, 2,* 83–115.

Gigerenzer, G. 2015. *Simply rational: Decision making in the real world.* New York: Oxford University Press.

Gigerenzer, G. 2016. Fear of dread risks. *Edge.* https://www.edge.org/response-detail/26645.

Gigerenzer, G., & Hoffrage, U. 1995. How to improve Bayesian reasoning without instruction: Frequency formats. *Psychological Review, 102,* 684–704.

Gilbert, D. T. 2006. *Stumbling on happiness.* New York: Knopf.

Giles, J. 2005. Internet encyclopaedias go head to head. *Nature, 438,* 900–901.

Glaeser, E. L. 2011. *Triumph of the city: How our greatest invention makes us richer, smarter, greener, healthier, and happier.* New York: Penguin.

Glaeser, E. L. 2014. *Secular joblessness.* London: Centre for Economic Policy Research.

Glaeser, E. L., Ponzetto, G. A. M., & Shleifer, A. 2007. Why does democracy need education? *Journal of Economic Growth, 12,* 77–99.

Glaeser, E. L., La Porta, R., Lopez-de-Silanes, F., & Shleifer, A. 2004. Do institutions cause growth? *Journal of Economic Growth, 9,* 271–303.

Gleditsch, N. P. 2008. The liberal moment fifteen years on. *International Studies Quarterly, 52,* 691–712.

Gleditsch, N. P., & Rudolfsen, I. 2016. Are Muslim countries more prone to violence? Paper presented at the 57th Annual Convention of the International Studies Association, Atlanta.

Gleditsch, N. P., Wallensteen, P., Eriksson, M., Sollenberg, M., & Strand, H. 2002. Armed conflict, 1946–2001: A new dataset. *Journal of Peace Research, 39,* 615–37.

Gleick, J. 2011. *The information: A history, a theory, a flood.* New York: Pantheon.

Glendon, M. A. 1998. Knowing the Universal Declaration of Human Rights. *Notre Dame Law Review, 73,* 1153–90.

Glendon, M. A. 1999. Foundations of human rights: The unfinished business. *American Journal of Jurisprudence, 44,* 1–14.

Glendon, M. A. 2001. *A world made new: Eleanor Roosevelt and the Universal Declaration of Human Rights.* New York: Random House.

Global Zero Commission. 2010. Global Zero action plan. http://globalzero.org/files/GZAP _6.0.pdf.

Global Zero Commission. 2016. US adoption of no-first-use and its effects on nuclear proliferation by allies. http://www.globalzero.org/files/gzap_6.0.pdf.

Glover, J. 1998. Eugenics: Some lessons from the Nazi experience. In J. R. Harris & S. Holm, eds., *The future of human reproduction: Ethics, choice, and regulation.* New York: Oxford University Press.

Glover, J. 1999. *Humanity: A moral history of the twentieth century.* London: Jonathan Cape.

Goertz, G., Diehl, P. F., & Balas, A. 2016. *The puzzle of peace: The evolution of peace in the international system.* New York: Oxford University Press.

Goklany, I. M. 2007. *The improving state of the world: Why we're living longer, healthier, more comfortable lives on a cleaner planet.* Washington: Cato Institute.

Goldin, C., & Katz, L. F. 2010. *The race between education and technology.* Cambridge, MA: Harvard University Press.

Goldstein, J. S. 2011. *Winning the war on war: The decline of armed conflict worldwide.* New York: Penguin.

Goldstein, J. S. 2015. Is the current refugee crisis the worst since World War II? (Unpublished manuscript.) http://www.joshuagoldstein.com/.

Goldstein, R. N. 1976. *Reduction, realism, and the mind.* Ph.D. dissertation, Princeton University.

Goldstein, R. N. 2006. *Betraying Spinoza: The renegade Jew who gave us modernity.* New York: Nextbook /Schocken.

Goldstein, R. N. 2010. *Thirty-six arguments for the existence of God: A work of fiction.* New York: Pantheon.

Goldstein, R. N. 2013. *Plato at the Googleplex: Why philosophy won't go away.* New York: Pantheon.

Gómez, J. M., Verdú, M., González-Megías, A., & Méndez, M. 2016. The phylogenetic roots of human lethal violence. *Nature, 538,* 233–37.

Gordon, R. J. 2014. The turtle's progress: Secular stagnation meets the headwinds. In C. Teulings & R. Baldwin, eds., *Secular stagnation: Facts, causes and cures.* London: Centre for Economic Policy Research.

Gordon, R. J. 2016. *The rise and fall of American growth.* Princeton, NJ: Princeton University Press.

Gottfredson, L. S. 1997. Why *g* matters: The complexity of everyday life. *Intelligence, 24,* 79–132.

Gottlieb, A. 2016. *The dream of enlightenment: The rise of modern philosophy.* New York: Norton.

Gottschall, J. 2012. *The storytelling animal: How stories make us human.* Boston: Houghton Mifflin Harcourt.

Gottschall, J., & Wilson, D. S., eds. 2005. *The literary animal: Evolution and the nature of narrative.* Evanston, IL: Northwestern University Press.

Graham, P. 2016. The refragmentation. *Paul Graham Blog.* http://www.paulgraham.com/re.html.

Grayling, A. C. 2007. *Toward the light of liberty: The struggles for freedom and rights that made the modern Western world.* New York: Walker.

Grayling, A. C. 2013. *The God argument: The case against religion and for humanism.* London: Bloomsbury.

Greene, J. 2013. *Moral tribes: Emotion, reason, and the gap between us and them.* New York: Penguin.

Greenstein, S., & Zhu, F. 2014. Do experts or collective intelligence write with more bias? Evidence from *Encyclopædia Britannica* and Wikipedia. *Harvard Business School Working Paper, 15-023.*

Greenwood, J., Seshadri, A., & Yorukoglu, M. 2005. Engines of liberation. *Review of Economic Studies, 72,* 109–33.

Gregg, B. 2003. *Thick moralities, thin politics: Social integration across communities of belief.* Durham, NC: Duke University Press.

Gross, N., & Simmons, S. 2014. The social and political views of American college and university professors. In N. Gross & S. Simmons, eds., *Professors and their politics.* Baltimore: Johns Hopkins University Press.

Guerrero Velasco, R. 2015. An antidote to murder. *Scientific American, 313,* 46–50.

Gunsalus, C. K., Bruner, E. M., Burbules, N., Dash, L. D., Finkin, M., et al. 2006. *Improving the system for protecting human subjects: Counteracting IRB mission creep* (No. LE06-016). University of Illinois, Urbana. https://papers.ssrn.com/sol3/papers2.cfm?abstract_id=902995.

Gurr, T. R. 1981. Historical trends in violent crime: A critical review of the evidence. In N. Morris & M. Tonry, eds., *Crime and Justice* (vol. 3). Chicago: University of Chicago Press.

Gyldensted, C. 2015. *From mirrors to movers: Five elements of positive psychology in constructive journalism.* GGroup Publishers.

Hafer, R. W. 2017. New estimates on the relationship between IQ, economic growth and welfare. *Intelligence, 61,* 92–101.

Hahn, R., Bilukha, O., Crosby, A., Fullilove, M. T., Liberman, A., et al. 2005. Firearms laws and the reduction of violence: A systematic review. *American Journal of Preventive Medicine, 28,* 40–71.

Haidt, J. 2006. *The happiness hypothesis: Finding modern truth in ancient wisdom.* New York: Basic Books.

Haidt, J. 2012. *The righteous mind: Why good people are divided by politics and religion.* New York: Pantheon.

Halpern, D., & Mason, D. 2015. Radical incrementalism. *Evaluation, 21,* 143–49.

Hammel, A. 2010. *Ending the death penalty: The European experience in global perspective.* Basingstoke: Palgrave Macmillan.

Hammond, S. 2017. The future of liberalism and the politicization of everything. *Niskanen Center Blog.* https://niskanencenter.org/blog/future-liberalism-politicization-everything/.

Hampton, K., Goulet, L. S., Rainie, L., & Purcell, K. 2011. *Social networking sites and our lives.* Washington: Pew Research Center.

Hampton, K., Rainie, L., Lu, W., Shin, I., & Purcell, K. 2015. *Social media and the cost of caring.* Washington: Pew Research Center.

Hanson, R., & Yudkowsky, E. 2008. *The Hanson-Yudkowsky AI-foom debate ebook.* Machine Intelligence Research Institute, Berkeley.

Harff, B. 2003. No lessons learned from the Holocaust? Assessing risks of genocide and political mass murder since 1955. *American Political Science Review, 97,* 57–73.

Harff, B. 2005. Assessing risks of genocide and politicide. In M. G. Marshall & T. R. Gurr, eds., *Peace and conflict 2005: A global survey of armed conflicts, self-determination movements, and democracy.* College Park, MD: Center for International Development and Conflict Management, University of Maryland.

Hargraves, R. 2012. *Thorium: Energy cheaper than coal.* North Charleston, SC: CreateSpace.

Hasegawa, T. 2006. *Racing the enemy: Stalin, Truman, and the surrender of Japan.* Cambridge, MA: Harvard University Press.

Hasell, J., & Roser, M. 2017. Famines. *Our World in Data.* https://ourworldindata.org/famines/.

Haskins, R., & Margolis, G. 2014. *Show me the evidence: Obama's fight for rigor and results in social policy.* Washington: Brookings Institution.

Haslam, N. 2016. Concept creep: Psychology's expanding concepts of harm and pathology. *Psychological Inquiry, 27,* 1–17.

Hassett, K. A., & Mathur, A. 2012. *A new measure of consumption inequality.* Washington: American Enterprise Institute.

Hastorf, A. H., & Cantril, H. 1954. They saw a game; a case study. *Journal of Abnormal and Social Psychology, 49,* 129–34.

Hathaway, O., & Shapiro, S. 2017. *The internationalists: How a radical plan to outlaw war remade the world.* New York: Simon & Schuster.

Haybron, D. M. 2013. *Happiness: A very short introduction.* New York: Oxford University Press.

Hayek, F. A. 1945. The use of knowledge in society. *American Economic Review, 35,* 519–30.

Hayek, F. A. 1960/2011. *The constitution of liberty: The definitive edition.* Chicago: University of Chicago Press.

Hayflick, L. 2000. The future of aging. *Nature, 408,* 267–69.

Hedegaard, H., Chen, L.-H., & Warner, M. 2015. Drug-poisoning deaths involving heroin: United States, 2000–2013. *NCHS Data Brief, 190.*

Hegre, H. 2014. Democracy and armed conflict. *Journal of Peace Research, 51,* 159–72.

Hegre, H., Karlsen, J., Nygård, H. M., Strand, H., & Urdal, H. 2013. Predicting armed conflict, 2010–2050. *International Studies Quarterly, 57,* 250–70.

Hellier, C. 2011. Nazi racial ideology was religious, creationist and opposed to Darwinism. *Coelsblog: Defending scientism.* https://coelsblog.wordpress.com/2011/11/08/nazi-racial-ideology-was-religious-creationist-and-opposed-to-darwinism/#sec4.

Helliwell, J. F., Layard, R., & Sachs, J., eds. 2016. *World Happiness Report 2016.* New York: Sustainable Development Solutions Network.

Henao-Restrepo, A. M., Camacho, A., Longini, I. M., Watson, C. H., Edmunds, W. J., et al. 2017. Efficacy and effectiveness of an rVSV-vectored vaccine in preventing Ebola virus disease: Final results from the Guinea ring vaccination, open-label, cluster-randomised trial. *The Lancet, 389,* 505–18.

Henry, M., Shivji, A., de Sousa, T., & Cohen, R. 2015. *The 2015 annual homeless assessment report to Congress.* Washington: US Department of Housing and Urban Development.

Herman, A. 1997. *The idea of decline in Western history.* New York: Free Press.

Heschel, S. 2008. *The Aryan Jesus: Christian theologians and the Bible in Nazi Germany.* Princeton, NJ: Princeton University Press.

Hidaka, B. H. 2012. Depression as a disease of modernity: Explanations for increasing prevalence. *Journal of Affective Disorders, 140,* 205–14.

Hidalgo, C. A. 2015. *Why information grows: The evolution of order, from atoms to economies.* New York: Basic Books.

Hirschl, T. A., & Rank, M. R. 2015. The life course dynamics of affluence. *PLOS ONE, 10 (1):* e0116370/.

Hirschman, A. O. 1971. *A bias for hope: Essays on development and Latin America.* New Haven: Yale University Press.

Hirschman, A. O. 1991. *The rhetoric of reaction: Perversity, futility, jeopardy.* Cambridge, MA: Harvard University Press.

Hirsi Ali, A. 2015a. *Heretic: Why Islam needs a reformation now.* New York: HarperCollins.

Hirsi Ali, A. 2015b. Islam is a religion of violence. *Foreign Policy,* Nov. 9.

Hoffmann, M., Hilton-Taylor, C., Angulo, A., Böhm, M., Brooks, T. M., et al. 2010. The impact of conservation on the status of the world's vertebrates. *Science, 330,* 1503–9.

Hollander, P. 1981/2014. *Political pilgrims: Western intellectuals in search of the good society.* New Brunswick, NJ: Transaction.

Horkheimer, M., & Adorno, T. W. 1947/2007. *Dialectic of Enlightenment.* Stanford: Stanford University Press.

Horwitz, A. V., & Wakefield, J. C. 2007. *The loss of sadness: How psychiatry transformed normal sorrow into depressive disorder.* New York: Oxford University Press.

Horwitz, S. 2015. Inequality, mobility, and being poor in America. *Social Philosophy and Policy, 31,* 70–91.

Housel, M. 2013. Everything is amazing and nobody is happy. *The Motley Fool.* http://www.fool.com/investing/general/2013/11/29/everything-is-great-and-nobody-is-happy.aspx.

Hout, M., & Fischer, C. S. 2014. Explaining why more Americans have no religious preference: Political backlash and generational succession, 1987–2012. *Sociological Science, 1,* 423–47.

Howard, M. 2001. *The invention of peace and the reinvention of war.* London: Profile Books.

Howson, C., & Urbach, P. 1989/2006. *Scientific reasoning: The Bayesian approach* (3rd ed.). Chicago: Open Court Publishing.

Hu, G., & Baker, S. P. 2012. An explanation for the recent increase in the fall death rate among older Americans: A subgroup analysis. *Public Health Reports, 127,* 275–81.

Hu, G., & Mamady, K. 2014. Impact of changes in specificity of data recording on cause-specific injury mortality in the United States, 1999–2010. *BMC Public Health, 14,* 1010.

Huberman, M., & Minns, C. 2007. The times they are not changin': Days and hours of work in old and new worlds, 1870–2000. *Explorations in Economic History, 44,* 538–67.

Huff, T. E. 1993. *The rise of early modern science: Islam, China, and the West.* New York: Cambridge University Press.

Hultman, L., Kathman, J., & Shannon, M. 2013. United Nations peacekeeping and civilian protection in civil war. *American Journal of Political Science, 57,* 875–91.

Human Security Centre. 2005. *Human Security Report 2005: War and peace in the 21st century.* New York: Oxford University Press.

Human Security Report Project. 2007. *Human Security Brief 2007*. Vancouver, BC: Human Security Report Project.

Human Security Report Project. 2009. *Human Security Report 2009: The shrinking costs of war*. New York: Oxford University Press.

Human Security Report Project. 2011. *Human Security Report 2009/2010: The causes of peace and the shrinking costs of war*. New York: Oxford University Press.

Humphrys, M. (Undated.) The left's historical support for tyranny and terrorism. http://markhumphrys.com/left.tyranny.html.

Hunt, L. 2007. *Inventing human rights: A history*. New York: Norton.

Huntington, S. P. 1991. *The third wave: Democratization in the late twentieth century*. Norman: University of Oklahoma Press.

Hyman, D. A. 2007. The pathologies of institutional review boards. *Regulation, 30*, 42–49.

Inglehart, R. 1997. *Modernization and postmodernization: Cultural, economic, and political change in 43 societies*. Princeton, NJ: Princeton University Press.

Inglehart, R. 2016. How much should we worry? *Journal of Democracy, 27*, 18–23.

Inglehart, R. 2017. Changing values in the Islamic world and the West. In M. Moaddel & M. J. Gelfand, eds., *Values, political action, and change in the Middle East and the Arab Spring*. New York: Oxford University Press.

Inglehart, R., Foa, R., Peterson, C., & Welzel, C. 2008. Development, freedom, and rising happiness: A global perspective (1981–2007). *Perspectives on Psychological Science, 3*, 264–85.

Inglehart, R., & Norris, P. 2016. *Trump, Brexit, and the rise of populism: Economic have-nots and cultural backlash*. Paper presented at the Annual Meeting of the American Political Science Association, Philadelphia.

Inglehart, R., & Welzel, C. 2005. *Modernization, cultural change, and democracy*. New York: Cambridge University Press.

Institute for Economics and Peace. 2016. *Global Terrorism Index 2016*. New York: Institute for Economics and Peace.

Instituto Nacional de Estadística y Geografía. 2016. Registros administrativos: Mortalidad. http://www.inegi.org.mx/est/contenidos/proyectos/registros/vitales/mortalidad/default.aspx.

Insurance Institute for Highway Safety. 2016. General statistics. http://www.iihs.org/iihs/topics/t/general-statistics/fatalityfacts/overview-of-fatality-facts.

Intergovernmental Panel on Climate Change. 2014. *Climate change 2014: Synthesis report. Contribution of working groups I, II and III to the fifth assessment report of the Intergovernmental Panel on Climate Change*. Geneva: IPCC.

International Humanist and Ethical Union. 2002. The Amsterdam Declaration. http://iheu.org/humanism/the-amsterdam-declaration/.

International Labour Organization. 2013. *Marking progress against child labour: Global estimates and trends 2000–2012*. Geneva: International Labour Organization.

Ipsos. 2016. The perils of perception 2016. https://perils.ipsos.com/.

Irwin, D. A. 2016. The truth about trade. *Foreign Affairs*, June 13.

Israel, J. I. 2001. *Radical enlightenment: Philosophy and the making of modernity 1650–1750*. New York: Oxford University Press.

Jackson, J. 2016. Publishing the positive: Exploring the motivations for and the consequences of reading solutions-focused journalism. https://www.constructivejournalism.org/wp-content/uploads/2016/11/Publishing-the-Positive_MA-thesis-research-2016_Jodie-Jackson.pdf.

Jacobs, A. 2011. Introduction. In W. H. Auden, *The age of anxiety: A Baroque eclogue*. Princeton, NJ: Princeton University Press.

Jacobson, M. Z., & Delucchi, M. A. 2011. Providing all global energy with wind, water, and solar power. *Energy Policy, 39*, 1154–69.

Jacoby, S. 2005. *Freethinkers: A history of American secularism*. New York: Henry Holt.

Jamison, D. T., Summers, L. H., Alleyne, G., Arrow, K. J., Berkley, S., et al. 2013. Global health 2035: A world converging within a generation. *The Lancet, 382*, 1898–1955.

Jefferson, T. 1785/1955. *Notes on the state of Virginia*. Chapel Hill: University of North Carolina Press.

Jensen, R. 2007. The digital provide: Information (technology), market performance, and welfare in the South Indian fisheries sector. *Quarterly Journal of Economics, 122*, 879–924.

Jervis, R. 2011. Force in our times. *International Relations, 25*, 403–25.

Johnson, D. D. P. 2004. *Overconfidence and war: The havoc and glory of positive illusions*. Cambridge, MA: Harvard University Press.

Johnson, E. M. 2010. Deconstructing social Darwinism: Parts I–IV. *The Primate Diaries*. http://scienceblogs.com/primatediaries/2010/01/05/deconstructing-social-darwinis/.

Johnson, E. M. 2009. Darwin's connection to Nazi eugenics exposed. *The Primate Diaries*. http://scienceblogs.com/primatediaries/2009/07/14/darwins-connection-to-nazi-eug/.

Johnson, N. F., Spagat, M., Restrepo, J. A., Becerra, O., Bohorquez, J. C., et al. 2006. Universal patterns underlying ongoing wars and terrorism. *arXiv.org*. http://arxiv.org/abs/physics/0605035.

Johnston, W. M., & Davey, G. C. L. 1997. The psychological impact of negative TV news bulletins: The catastrophizing of personal worries. *British Journal of Psychology, 88*, 85–91.

Jones, R. P., Cox, D., Cooper, B., & Lienesch, R. 2016a. *The divide over America's future: 1950 or 2050? Findings from the 2016 American Values Survey.* Washington: Public Religion Research Institute.

Jones, R. P., Cox, D., Cooper, B., & Lienesch, R. 2016b. *Exodus: Why Americans are leaving religion—and why they're unlikely to come back.* Washington: Public Religion Research Institute.

Jones, R. P., Cox, D., & Navarro-Rivera, J. 2014. *Believers, sympathizers, and skeptics: Why Americans are conflicted about climate change, environmental policy, and science.* Washington: Public Religion Research Institute.

Jussim, L., Krosnick, J., Vazire, S., Stevens, S., Anglin, S., et al. 2017. Political bias. *Best Practices in Science.* https://bps.stanford.edu/?page_id=3371.

Kahan, D. M. 2012. Cognitive bias and the constitution of the liberal republic of science. Yale Law School, Public Law Working Paper 270. https://papers.ssrn.com/sol3/papers.cfm?abstract_id=2174032.

Kahan, D. M. 2015. Climate-science communication and the measurement problem. *Political Psychology, 36,* 1–43.

Kahan, D. M., Braman, D., Slovic, P., Gastil, J., & Cohen, G. 2009. Cultural cognition of the risks and benefits of nanotechnology. *Nature Nanotechnology, 4,* 87–90.

Kahan, D. M., Jenkins-Smith, H., & Braman, D. 2011. Cultural cognition of scientific consensus. *Journal of Risk Research, 14,* 147–74.

Kahan, D. M., Jenkins-Smith, H., Tarantola, T., Silva, C. L., & Braman, D. 2012. Geoengineering and climate change polarization: Testing a two-channel model of science communication. *Annals of the American Academy of Political and Social Science, 658,* 193–222.

Kahan, D. M., Peters, E., Dawson, E. C., & Slovic, P. 2013. Motivated numeracy and enlightened self-government. https://papers.ssrn.com/sol3/papers.cfm?abstract_id=2319992.

Kahan, D. M., Peters, E., Wittlin, M., Slovic, P., Ouellette, L. L., et al. 2012. The polarizing impact of science literacy and numeracy on perceived climate change risks. *Nature Climate Change, 2,* 732–35.

Kahan, D. M., Wittlin, M., Peters, E., Slovic, P., Ouellette, L. L., et al. 2011. The tragedy of the risk-perception commons: Culture conflict, rationality conflict, and climate change. Cultural Cognition Project Working Paper 89. https://papers.ssrn.com/sol3/papers.cfm?abstract_id=1871503.

Kahneman, D. 2011. *Thinking, fast and slow.* New York: Farrar, Straus & Giroux.

Kahneman, D., Krueger, A., Schkade, D., Schwarz, N., & Stone, A. 2004. A survey method for characterizing daily life experience: The day reconstruction method. *Science, 3,* 1776–80.

Kanazawa, S. 2010. Why liberals and atheists are more intelligent. *Social Psychology Quarterly, 73,* 33–57.

Kane, T. 2016. Piketty's crumbs. *Commentary,* April 14.

Kant, I. 1784/1991. *An answer to the question: What is enlightenment?* London: Penguin.

Kant, I. 1795/1983. Perpetual peace: A philosophical sketch. In I. Kant, *Perpetual peace and other essays.* Indianapolis: Hackett. http://www.mtholyoke.edu/acad/intrel/kant/kant1.htm.

Kasturiratne, A., Wickremasinghe, A. R., de Silva, N., Gunawardena, N. K., Pathmeswaran, A., et al. 2008. The global burden of snakebite: A literature analysis and modelling based on regional estimates of envenoming and deaths. *PLOS Medicine, 5,* e218.

Keith, D. 2013. *A case for climate engineering.* Cambridge, MA: MIT Press.

Keith, D. 2015. Patient geoengineering. Paper presented at the Seminars About Long-Term Thinking, San Francisco. http://longnow.org/seminars/02015/feb/17/patient-geoengineering/.

Keith, D., Weisenstein, D., Dykema, J., & Keutsch, F. 2016. Stratospheric solar geoengineering without ozone loss. *Proceedings of the National Academy of Sciences, 113,* 14910–14.

Kelley, J., & Evans, M. D. R. 2017. Societal inequality and individual subjective well-being: Results from 68 societies and over 200,000 individuals, 1981–2008. *Social Science Research, 62,* 1–23.

Kelly, K. 2010. *What technology wants.* New York: Penguin.

Kelly, K. 2013. Myth of the lone villain. *The Technium.* http://kk.org/thetechnium/myth-of-the-lon/.

Kelly, K. 2016. *The inevitable: Understanding the 12 technological forces that will shape our future.* New York: Viking.

Kelly, K. 2017. The AI cargo cult: The myth of a superhuman AI. *Wired.* https://www.wired.com/2017/04/the-myth-of-a-superhuman-ai/.

Kennedy, D. 2011. *Don't shoot: One man, a street fellowship, and the end of violence in inner-city America.* New York: Bloomsbury.

Kenny, C. 2011. *Getting better: Why global development is succeeding—and how we can improve the world even more.* New York: Basic Books.

Kessler, R. C., Berglund, P., Demler, O., Jin, R., Koretz, D., et al. 2003. The epidemiology of major depressive disorder: Results from the National Comorbidity Survey Replication (NCS-R). *Journal of the American Medical Association, 289,* 3095–3105.

Kessler, R. C., Berglund, P., Demler, O., Jin, R., Merikangas, K. R., et al. 2005. Lifetime prevalence and age-of-onset distributions of DSM-IV disorders in the National Comorbidity Survey Replication. *Archives of General Psychiatry, 62,* 593–602.

Kevles, D. J. 1985. *In the name of eugenics: Genetics and the uses of human heredity.* Cambridge, MA: Harvard University Press.

Kharecha, P. A., & Hansen, J. E. 2013. Prevented mortality and greenhouse gas emissions from historical and projected nuclear power. *Environmental Science & Technology, 47,* 4889–95.

Kharrazi, R. J., Nash, D., & Mielenz, T. J. 2015. Increasing trend of fatal falls in older adults in the United States, 1992 to 2005; Coding practice or reporting quality? *Journal of the American Geriatrics Society, 63*, 1913–17.

Kim, J., Smith, T. W., & Kang, J.-H. 2015. Religious affiliation, religious service attendance, and mortality. *Journal of Religion and Health, 54*, 2052–72.

King, D., Schrag, D., Dadi, Z., Ye, Q., & Ghosh, A. 2015. *Climate change: A risk assessment*. Cambridge, UK: University of Cambridge Centre for Science and Policy.

Kitcher, P. 1990. *Kant's transcendental psychology*. New York: Oxford University Press.

Klein, D. B., & Buturovic, Z. 2011. Economic enlightenment revisited: New results again find little relationship between education and economic enlightenment but vitiate prior evidence of the left being worse. *Economic Journal Watch, 8*, 157–73.

Klitzman, R. L. 2015. *The ethics police? The struggle to make human research safe*. New York: Oxford University Press.

Kochanek, K. D., Murphy, S. L., Xu, J., & Tejada-Vera, B. 2016. Deaths: Final data for 2014. *National Vital Statistics Reports, 65* (4). http://www.cdc.gov/nchs/data/nvsr/nvsr65/nvsr65_04.pdf.

Kohut, A., Taylor, P. J., Keeter, S., Doherty, C., Dimock, M., et al. 2011. *The generation gap and the 2012 election*. Washington: Pew Research Center. http://www.people-press.org/files/legacy-pdf/11-3-11%20Generations%20Release.pdf.

Kolosh, K. 2014. Injury facts statistical highlights. http://www.nsc.org/SafeCommunitiesDocuments/Conference-2014/Injury-Facts-Statistical-Analysis-Kolosh.pdf.

Koningstein, R., & Fork, D. 2014. What it would really take to reverse climate change. *IEEE Spectrum*. http://spectrum.ieee.org/energy/renewables/what-it-would-really-take-to-reverse-climate-change.

Kräenbring, J., Monzon Penza, T., Gutmann, J., Muehlich, S., Zolk, O., et al. 2014. Accuracy and completeness of drug information in Wikipedia: A comparison with standard textbooks of pharmacology. *PLOS ONE, 9*, e106930.

Krauss, L. M. 2012. *A universe from nothing: Why there is something rather than nothing*. New York: Free Press.

Krisch, M., Eisner, M., Mikton, C., & Butchart, A., eds. 2015. *Global strategies to reduce violence by 50% in 30 years: Findings from the WHO and University of Cambridge Global Violence Reduction Conference 2014*. Cambridge, UK: Institute of Criminology, University of Cambridge.

Kristensen, H. M. 2016. U.S. nuclear stockpile numbers published enroute to Hiroshima. *Federation of American Scientists Strategic Security Blog*. https://fas.org/blogs/security/2016/05/hiroshima-stockpile/.

Kristensen, H. M., & Norris, R. S. 2016a. Status of world nuclear forces. *Federation of American Scientists*. https://fas.org/issues/nuclear-weapons/status-world-nuclear-forces/.

Kristensen, H. M., & Norris, R. S. 2016b. United States nuclear forces, 2016. *Bulletin of the Atomic Scientists, 72*, 63–73.

Krug, E. G., Dahlberg, L. L., Mercy, J. A., Zwi, A. B., & Lozano, R., eds. 2002. *World report on violence and health*. Geneva: World Health Organization.

Kuhn, D. 1991. *The skills of argument*. New York: Cambridge University Press.

Kuncel, N. R., Klieger, D. M., Connelly, B. S., & Ones, D. S. 2013. Mechanical versus clinical data combination in selection and admissions decisions: A meta-analysis. *Journal of Applied Psychology, 98*, 1060–72.

Kunda, Z. 1990. The case for motivated reasoning. *Psychological Bulletin, 108*, 480–98.

Kuran, T. 2004. Why the Middle East is economically underdeveloped: Historical mechanisms of institutional stagnation. *Journal of Economic Perspectives, 18*, 71–90.

Kurlansky, M. 2006. *Nonviolence: Twenty-five lessons from the history of a dangerous idea*. New York: Modern Library.

Kurzban, R., Tooby, J., & Cosmides, L. 2001. Can race be erased? Coalitional computation and social categorization. *Proceedings of the National Academy of Sciences, 98*, 15387–92.

Kuznets, S. 1955. Economic growth and income inequality. *American Economic Review, 45*, 1–28.

Lacina, B. 2006. Explaining the severity of civil wars. *Journal of Conflict Resolution, 50*, 276–89.

Lacina, B., & Gleditsch, N. P. 2005. Monitoring trends in global combat: A new dataset in battle deaths. *European Journal of Population, 21*, 145–66.

Lake, B. M., Ullman, T. D., Tenenbaum, J. B., & Gershman, S. J. 2017. Building machines that learn and think like people. *Behavioral and Brain Sciences, 39*, 1–101.

Lakner, C., & Milanović, B. 2016. Global income distribution: From the fall of the Berlin Wall to the Great Recession. *World Bank Economic Review, 30*, 203–232.

Lampert, L. 1996. *Leo Strauss and Nietzsche*. Chicago: University of Chicago Press.

Lancet Infectious Diseases Editors. 2005. Clearing the myths of time: Tuskegee revisited. *The Lancet Infectious Diseases, 5*, 127.

Land, K. C., Michalos, A. C., & Sirgy, J., eds. 2012. *Handbook of social indicators and quality of life research*. New York: Springer.

Lane, N. 2015. *The vital question: Energy, evolution, and the origins of complex life*. New York: Norton.

Lanier, J. 2014. The myth of AI. *Edge*. https://www.edge.org/conversation/jaron_lanier-the-myth-of-ai.

Lankford, A. 2013. *The myth of martyrdom*. New York: Palgrave Macmillan.

Lankford, A., & Madfis, E. 2018. Don't name them, don't show them, but report everything else: A pragmatic proposal for denying mass killers the attention they seek and deterring future offenders. *American Behavioral Scientist, 62*, 260–279.

Latzer, B. 2016. *The rise and fall of violent crime in America*. New York: Encounter Books.

Laudan, R. 2016. Was the agricultural revolution a terrible mistake? Not if you take food processing into account. http://www.rachellaudan.com/2016/01/was-the-agricultural-revolution-a-terrible-mistake.html.

Law, S. 2011. *Humanism: A very short introduction*. New York: Oxford University Press.

Lawson, S. 2013. Beyond cyber-doom: Assessing the limits of hypothetical scenarios in the framing of cyber-threats. *Journal of Information Technology & Politics, 10*, 86–103.

Layard, R. 2005. *Happiness: Lessons from a new science*. New York: Penguin.

Le Quéré, C., Andrew, R. M., Canadell, J. G., Sitch, S., Korsbakken, J. I., et al. 2016. Global carbon budget 2016. *Earth System Science Data, 8*, 605–49.

Leavis, F. R. 1962/2013. *Two cultures? The significance of C. P. Snow*. New York: Cambridge University Press.

Lee, J.-W., & Lee, H. 2016. Human capital in the long run. *Journal of Development Economics, 122*, 147–69.

Leetaru, K. 2011. Culturomics 2.0: Forecasting large-scale human behavior using global news media tone in time and space. *First Monday, 16* (9). http://firstmonday.org/article/view/3663/3040.

Leon, C. B. 2016. The life of American workers in 1915. *Monthly Labor Review*. http://www.bls.gov/opub/mlr/2016/article/the-life-of-american-workers-in-1915.htm.

Leonard, T. C. 2009. Origins of the myth of social Darwinism: The ambiguous legacy of Richard Hofstadter's "Social Darwinism in American thought." *Journal of Economic Behavior & Organization, 71*, 37–51.

Lerdahl, F., & Jackendoff, R. 1983. *A generative theory of tonal music*. Cambridge, MA: MIT Press.

Levin, Y. 2017. Conservatism in an age of alienation. *Modern Age*, Spring. https://eppc.org/publications/conservatism-in-an-age-of-alienation/.

Levinson, A. 2008. Environmental Kuznets curve. In S. N. Durlauf & L. E. Blume, eds., *The New Palgrave Dictionary of Economics* (2nd ed.). New York: Palgrave Macmillan.

Levitsky, S., & Way, L. 2015. The myth of democratic recession. *Journal of Democracy, 26*, 45–58.

Levitt, S. D. 2004. Understanding why crime fell in the 1990s: Four factors that explain the decline and six that do not. *Journal of Economic Perspectives, 18*, 163–90.

Levy, J. S. 1983. *War in the modern great power system 1495–1975*. Lexington: University Press of Kentucky.

Levy, J. S., & Thompson, W. R. 2011. *The arc of war: Origins, escalation, and transformation*. Chicago: University of Chicago Press.

Lewinsohn, P. M., Rohde, P., Seeley, J. R., & Fischer, S. A. 1993. Age-cohort changes in the lifetime occurrence of depression and other mental disorders. *Journal of Abnormal Psychology, 102*, 110–20.

Lewis, B. 1990/1992. *Race and slavery in the Middle East: An historical enquiry*. New York: Oxford University Press.

Lewis, B. 2002. *What went wrong? The clash between Islam and modernity in the Middle East*. New York: HarperPerennial.

Lewis, J. E., DeGusta, D., Meyer, M. R., Monge, J. M., Mann, A. E., et al. 2011. The mismeasure of science: Stephen Jay Gould versus Samuel George Morton on skulls and bias. *PLOS Biology, 9*, e1001071.

Lewis, M. 2016. *The undoing project: A friendship that changed our minds*. New York: Norton.

Liebenberg, L. 1990. *The art of tracking: The origin of science*. Cape Town: David Philip.

Liebenberg, L. 2014. *The origin of science: On the evolutionary roots of science and its implications for self-education and citizen science*. Cape Town: CyberTracker. http://www.cybertracker.org/science-the-origin-of-science.

Lilienfeld, S. O., Ammirati, R., & Landfield, K. 2009. Giving debiasing away. *Perspectives on Psychological Science, 4*, 390–98.

Lilienfeld, S. O., Ritschel, L. A., Lynn, S. J., Cautin, R. L., & Latzman, R. D. 2013. Why many clinical psychologists are resistant to evidence-based practice: Root causes and constructive remedies. *Clinical Psychology Review, 33*, 883–900.

Lilla, M. 2001. *The reckless mind: Intellectuals in politics*. New York: New York Review of Books.

Lilla, M. 2016. *The shipwrecked mind: On political reaction*. New York: New York Review of Books.

Lindert, P. 2004. *Growing public: Social spending and economic growth since the eighteenth century* (vol. 1: *The story*). New York: Cambridge University Press.

Linker, D. 2007. *The theocons: Secular America under siege*. New York: Random House.

Liu, L., Oza, S., Hogan, D., Perin, J., Rudan, I., et al. 2014. Global, regional, and national causes of child mortality in 2000–13, with projections to inform post-2015 priorities: An updated systematic analysis. *The Lancet, 385*, 430–40.

Livingstone, M. S. 2014. *Vision and art: The biology of seeing* (updated ed.). New York: Harry Abrams.

Lloyd, S. 2006. *Programming the universe: A quantum computer scientist takes on the cosmos*. New York: Vintage.

Lodge, D. 2002. *Consciousness and the novel*. Cambridge, MA: Harvard University Press.

López, R. E., & Holle, R. L. 1998. Changes in the number of lightning deaths in the United States during the twentieth century. *Journal of Climate, 11*, 2070–77.

Lord, C. G., Ross, L., & Lepper, M. R. 1979. Biased assimilation and attitude polarization: The effects of prior theories on subsequently considered evidence. *Journal of Personality and Social Psychology, 37,* 2098–2109.

Lovering, J., Trembath, A., Swain, M., & Lavin, L. 2015. Renewables and nuclear at a glance. *The Breakthrough.* http://thebreakthrough.org/index.php/issues/energy/renewables-and-nuclear-at-a-glance.

Luard, E. 1986. *War in international society.* New Haven: Yale University Press.

Lucas, R. E. 1988. On the mechanics of economic development. *Journal of Monetary Economics, 22,* 3–42.

Lukianoff, G. 2012. *Unlearning liberty: Campus censorship and the end of American debate.* New York: Encounter Books.

Lukianoff, G. 2014. *Freedom from speech.* New York: Encounter Books.

Luria, A. R. 1976. *Cognitive development: Its cultural and social foundations.* Cambridge, MA: Harvard University Press.

Lutz, W., Butz, W. P., & Samir, K. C., eds. 2014. *World population and human capital in the twenty-first century.* New York: Oxford University Press.

Lutz, W., Cuaresma, J. C., & Abbasi-Shavazi, M. J. 2010. Demography, education, and democracy: Global trends and the case of Iran. *Population and Development Review, 36,* 253–81.

Lynn, R., Harvey, J., & Nyborg, H. 2009. Average intelligence predicts atheism rates across 137 nations. *Intelligence, 37,* 11–15.

MacAskill, W. 2015. *Doing good better: How effective altruism can help you make a difference.* New York: Penguin.

Macnamara, J. 1999. *Through the rearview mirror: Historical reflections on psychology.* Cambridge, MA: MIT Press.

Maddison Project. 2014. Maddison Project. http://www.ggdc.net/maddison/maddison-project/home.htm.

Mahbubani, K. 2013. *The great convergence: Asia, the West, and the logic of one world.* New York: Public-Affairs.

Mahbubani, K., & Summers, L. H. 2016. The fusion of civilizations. *Foreign Affairs,* May/June.

Makari, G. 2015. *Soul machine: The invention of the modern mind.* New York: Norton.

Makel, M. C., Kell, H. J., Lubinski, D., Putallaz, M., & Benbow, C. P. 2016. When lightning strikes twice: Profoundly gifted, profoundly accomplished. *Psychological Science, 27,* 1004–18.

Mankiw, G. 2013. Defending the one percent. *Journal of Economic Perspectives, 27,* 21–34.

Mann, T. E., & Ornstein, N. J. 2012/2016. *It's even worse than it looks: How the American constitutional system collided with the new politics of extremism* (new ed.). New York: Basic Books.

Marcus, G. 2015. Machines won't be thinking anytime soon. *Edge.* https://www.edge.org/response-detail/26175.

Marcus, G. 2016. Is big data taking us closer to the deeper questions in artificial intelligence? *Edge.* https://www.edge.org/conversation/gary_marcus-is-big-data-taking-us-closer-to-the-deeper-questions-in-artificial.

Maritain, J. 1949. Introduction. In UNESCO, *Human rights: Comments and interpretations.* New York: Columbia University Press.

Marlowe, F. 2010. *The Hadza: Hunter-gatherers of Tanzania.* Berkeley: University of California Press.

Marshall, M. G. 2016. Major episodes of political violence, 1946–2015. Vienna, VA: Center for Systemic Peace. http://www.systemicpeace.org/warlist/warlist.htm.

Marshall, M. G., & Gurr, T. R. 2014. Polity IV individual country regime trends, 1946–2013. Vienna, VA: Center for Systemic Peace. http://www.systemicpeace.org/polity/polity4x.htm.

Marshall, M. G., Gurr, T. R., & Harff, B. 2009. *PITF State Failure Problem Set: Internal wars and failures of governance, 1955–2008. Dataset and coding guidelines.* Vienna, VA: Center for Systemic Peace. http://www.systemicpeace.org/inscr/PITFProbSetCodebook2014.pdf.

Marshall, M. G., Gurr, T. R., & Jaggers, K. 2016. *Polity IV project: Political regime characteristics and transitions, 1800–2015, dataset users' manual.* Vienna, VA: Center for Systemic Peace. http://systemic peace.org/inscrdata.html.

Mathers, C. D., Sadana, R., Salomon, J. A., Murray, C. J. L., & Lopez, A. D. 2001. Healthy life expectancy in 191 countries, 1999. *The Lancet, 357,* 1685–91.

Mattisson, C., Bogren, M., Nettelbladt, P., Munk-Jörgensen, P., & Bhugra, D. 2005. First incidence depression in the Lundby study: A comparison of the two time periods 1947–1972 and 1972–1997. *Journal of Affective Disorders, 87,* 151–60.

McCloskey, D. N. 1994. Bourgeois virtue. *American Scholar, 63,* 177–91.

McCloskey, D. N. 1998. Bourgeois virtue and the history of P and S. *Journal of Economic History, 58,* 297–317.

McCloskey, D. N. 2014. Measured, unmeasured, mismeasured, and unjustified pessimism: A review essay of Thomas Piketty's "Capital in the twenty-first century." *Erasmus Journal for Philosophy and Economics, 7,* 73–115.

McCullough, M. E. 2008. *Beyond revenge: The evolution of the forgiveness instinct.* San Francisco: Jossey-Bass.

McEvedy, C., & Jones, R. 1978. *Atlas of world population history.* London: Allen Lane.

McGinn, C. 1993. *Problems in philosophy: The limits of inquiry.* Cambridge, MA: Blackwell.

McGinnis, J. O. 1996. The original constitution and our origins. *Harvard Journal of Law and Public Policy, 19,* 251–61.

McGinnis, J. O. 1997. The human constitution and constitutive law: A prolegomenon. *Journal of Contemporary Legal Issues, 8,* 211–39.

McKay, C. 1841/1995. *Extraordinary popular delusions and the madness of crowds.* New York: Wiley.

McMahon, D. M. 2001. *Enemies of the Enlightenment: The French counter-Enlightenment and the making of modernity.* New York: Oxford University Press.

McMahon, D. M. 2006. *Happiness: A history.* New York: Grove/Atlantic.

McNally, R. J. 2016. The expanding empire of psychopathology: The case of PTSD. *Psychological Inquiry, 27,* 46–49.

McNaughton-Cassill, M. E. 2001. The news media and psychological distress. *Anxiety, Stress, and Coping, 14,* 193–211.

McNaughton-Cassill, M. E., & Smith, T. 2002. My world is OK, but yours is not: Television news, the optimism gap, and stress. *Stress and Health, 18,* 27–33.

Meehl, P. E. 1954/2013. *Clinical versus statistical prediction: A theoretical analysis and a review of the evidence.* Brattleboro, VT: Echo Point Books.

Meeske, A. J., Riley, E. P., Robins, W. P., Uehara, T., Mekalanos, J. J., et al. 2016. SEDS proteins are a widespread family of bacterial cell wall polymerases. *Nature, 537,* 634–38.

Melander, E., Pettersson, T., & Themnér, L. 2016. Organized violence, 1989–2015. *Journal of Peace Research, 53,* 727–42.

Mellers, B. A., Hertwig, R., & Kahneman, D. 2001. Do frequency representations eliminate conjunction effects? An exercise in adversarial collaboration. *Psychological Science, 12,* 269–75.

Mellers, B. A., Ungar, L., Baron, J., Ramos, J., Gurcay, B., et al. 2014. Psychological strategies for winning a geopolitical forecasting tournament. *Psychological Science, 25,* 1106–1115.

Menschenfreund, Y. 2010. The Holocaust and the trial of modernity. *Azure, 39,* 58–83. http://azure.org.il/include/print.php?id=526.

Mercier, H., & Sperber, D. 2011. Why do humans reason? Arguments for an argumentative theory. *Behavioral and Brain Sciences, 34,* 57–111.

Mercier, H., & Sperber, D. 2017. *The enigma of reason.* Cambridge, MA: Harvard University Press.

Merquior, J. G. 1985. *Foucault.* Berkeley: University of California Press.

Merton, R. K. 1942/1973. The normative structure of science. In R. K. Merton, ed., *The sociology of science: Theoretical and empirical investigations.* Chicago: University of Chicago Press.

Meyer, B. D., & Sullivan, J. X. 2011. *The material well-being of the poor and middle class since 1980.* Washington: American Enterprise Institute.

Meyer, B. D., & Sullivan, J. X. 2012. Winning the war: Poverty from the Great Society to the Great Recession. *Brookings Papers on Economic Activity,* 133–200.

Meyer, B. D., & Sullivan, J. X. 2017a. Consumption and income inequality in the U.S. since the 1960s. NBER Working Paper 23655. https://www3.nd.edu/~jsulliv4/jrs_papers/Inequality6.5.pdf.

Meyer, B. D., & Sullivan, J. X. 2017b. Annual report on U.S. consumption poverty: 2016. http://www.aei.org/publication/annual-report-on-us-consumption-poverty-2016/.

Michel, J.-B., Shen, Y. K., Aiden, A. P., Veres, A., Gray, M. K., The Google Books Team, Pickett, J. P., Hoiberg, D., Clancy, D., Norvig, P., Orwant, J., Pinker, S., Nowak, M., & Lieberman-Aiden, E. 2011. Quantitative analysis of culture using millions of digitized books. *Science, 331,* 176–82.

Milanović, B. 2012. *Global income inequality by the numbers: In history and now—an overview.* Washington: World Bank Development Research Group.

Milanović, B. 2016. *Global inequality: A new approach for the age of globalization.* Cambridge, MA: Harvard University Press.

Miller, M., Azrael, D., & Barber, C. 2012. Suicide mortality in the United States: The importance of attending to method in understanding population-level disparities in the burden of suicide. *Annual Review of Public Health, 33,* 393–408.

Miller, R. A., & Albert, K. 2015. If it leads, it bleeds (and if it bleeds, it leads): Media coverage and fatalities in militarized interstate disputes. *Political Communication, 32,* 61–82.

Miller, T. R., Lawrence, B. A., Carlson, N. N., Hendrie, D., Randall, S., et al. 2016. Perils of police action: A cautionary tale from US data sets. *Injury Prevention.*

Moatsos, M., Baten, J., Foldvari, P., van Leeuwen, B., & van Zanden, J. L. 2014. Income inequality since 1820. In J. van Zanden, J. Baten, M. M. d'Ercole, A. Rijpma, C. Smith, & M. Timmer, eds., *How was life? Global well-being since 1820.* Paris: OECD Publishing.

Mokyr, J. 2012. *The enlightened economy: An economic history of Britain, 1700–1850.* New Haven: Yale University Press.

Mokyr, J. 2014. Secular stagnation? Not in your life. In C. Teulings & R. Baldwin, eds., *Secular stagnation: Facts, causes and cures.* London: Centre for Economic Policy Research.

Montgomery, S. L., & Chirot, D. 2015. *The shape of the new: Four big ideas and how they made the modern world.* Princeton, NJ: Princeton University Press.

Mooney, C. 2005. *The Republican war on science.* New York: Basic Books.

Morewedge, C. K., Yoon, H., Scopelliti, I., Symborski, C. W., Korris, J. H., et al. 2015. Debiasing decisions: Improved decision making with a single training intervention. *Policy Insights from the Behavioral and Brain Sciences, 2,* 129–40.

Morton, O. 2015. *The planet remade: How geoengineering could change the world.* Princeton, NJ: Princeton University Press.

Moss, J. 2005. Could Morton do it today? *University of Chicago Record, 40,* 27–28.

Mozgovoi, A. 2002. Recollections of Vadim Orlov (USSR submarine B-59). *The Cuban Samba of the Quartet of Foxtrots: Soviet submarines in the Caribbean crisis of 1962.* http://nsarchive.gwu.edu /nsa/cuba_mis_cri/020000%20Recollections%20of%20Vadim%20Orlov.pdf.

Mueller, J. 1989. *Retreat from doomsday: The obsolescence of major war.* New York: Basic Books.

Mueller, J. 1999. *Capitalism, democracy, and Ralph's Pretty Good Grocery.* Princeton, NJ: Princeton University Press.

Mueller, J. 2004a. *The remnants of war.* Ithaca, NY: Cornell University Press.

Mueller, J. 2004b. Why isn't there more violence? *Security Studies, 13,* 191–203.

Mueller, J. 2006. *Overblown: How politicians and the terrorism industry inflate national security threats, and why we believe them.* New York: Free Press.

Mueller, J. 2009. War has almost ceased to exist: An assessment. *Political Science Quarterly, 124,* 297–321.

Mueller, J. 2010a. *Atomic obsession: Nuclear alarmism from Hiroshima to Al-Qaeda.* New York: Oxford University Press.

Mueller, J. 2010b. Capitalism, peace, and the historical movement of ideas. *International Interactions, 36,* 169–84.

Mueller, J. 2012. Terror predictions. https://politicalscience.osu.edu/faculty/jmueller/PREDICT.pdf.

Mueller, J. 2014. Did history end? Assessing the Fukuyama thesis. *Political Science Quarterly, 129,* 35–54.

Mueller, J. 2016. Embracing threatlessness: US military spending, Newt Gingrich, and the Costa Rica option. https://politicalscience.osu.edu/faculty/jmueller/CNArestraintCato16.pdf.

Mueller, J., & Friedman, B. 2014. The cyberskeptics. https://www.cato.org/research/cyberskeptics.

Mueller, J., & Stewart, M. G. 2010. Hardly existential: Thinking rationally about terrorism. *Foreign Affairs,* April 2.

Mueller, J., & Stewart, M. G. 2016a. *Chasing ghosts: The policing of terrorism.* New York: Oxford University Press.

Mueller, J., & Stewart, M. G. 2016b. Conflating terrorism and insurgency. *Lawfare.* https://www.lawfare blog.com/conflating-terrorism-and-insurgency.

Muggah, R. 2015. Fixing fragile cities. *Foreign Affairs,* Jan. 15.

Muggah, R. 2016. Terrorism is on the rise—but there's a bigger threat we're not talking about. *World Economic Forum Global Agenda.* https://www.weforum.org/agenda/2016/04/terrorism-is-on -the-rise-but-there-s-a-bigger-threat-we-re-not-talking-about/.

Muggah, R., & Szabo de Carvalho, I. 2016. The end of homicide. *Foreign Affairs,* Sept. 7.

Müller, J.-W. 2016. *What is populism?* Philadelphia: University of Pennsylvania Press.

Müller, V. C., & Bostrom, N. 2014. Future progress in artificial intelligence: A survey of expert opinion. In V. C. Müller, ed., *Fundamental issues of artificial intelligence.* New York: Springer.

Mulligan, C. B., Gil, R., & Sala-i-Martin, X. 2004. Do democracies have different public policies than nondemocracies? *Journal of Economic Perspectives, 18,* 51–74.

Munck, G. L., & Verkuilen, J. 2002. Conceptualizing and measuring democracy: Evaluating alternative indices. *Comparative Political Studies, 35,* 5–34.

Murphy, J. M., Laird, N. M., Monson, R. R., Sobol, A. M., & Leighton, A. H. 2000. A 40-year perspective on the prevalence of depression: The Stirling County study. *Archives of General Psychiatry, 57,* 209–215.

Murphy, J. P. M. 1999. Hitler was *not* an atheist. *Free Inquiry, 19*(2).

Murphy, S. K., Zeng, M., & Herzon, S. B. 2017. A modular and enantioselective synthesis of the pleuromutilin antibiotics. *Science, 356,* 956–59.

Murray, C. 2003. *Human accomplishment: The pursuit of excellence in the arts and sciences, 800 B.C. to 1950.* New York: HarperPerennial.

Murray, C. J. L., et al. (487 coauthors). 2012. Disability-adjusted life years (DALYs) for 291 diseases and injuries in 21 regions, 1990–2010: A systematic analysis for the Global Burden of Disease study 2010. *The Lancet, 380,* 2197–2223.

Musolino, J. 2015. *The soul fallacy: What science shows we gain from letting go of our soul beliefs.* Amherst, NY: Prometheus Books.

Myhrvold, N. 2014. Commentary on Jaron Lanier's "The myth of AI." *Edge.* https://www.edge.org /conversation/jaron_lanier-the-myth-of-ai#25983.

Naam, R. 2010. Top five reasons "the singularity" is a misnomer. *Humanity+.* http://hplusmagazine.com /2010/11/11/top-five-reasons-singularity-misnomer/.

Naam, R. 2013. *The infinite resource: The power of ideas on a finite planet.* Lebanon, NH: University Press of New England.

Nadelmann, E. A. 1990. Global prohibition regimes: The evolution of norms in international society. *International Organization, 44,* 479–526.

Nagdy, M., & Roser, M. 2016a. Military spending. *Our World in Data.* https://ourworldindata.org /military-spending/.

Nagdy, M., & Roser, M. 2016b. Optimism and pessimism. *Our World in Data.* https://ourworldindata .org/optimism-pessimism/.

Nagdy, M., & Roser, M. 2016c. Projections of future education. *Our World in Data*. https://ourworldin data.org/projections-of-future-education/.

Nagel, T. 1970. *The possibility of altruism*. Princeton, NJ: Princeton University Press.

Nagel, T. 1974. What is it like to be a bat? *Philosophical Review, 83,* 435–50.

Nagel, T. 1997. *The last word*. New York: Oxford University Press.

Nagel, T. 2012. *Mind and cosmos: Why the materialist neo-Darwinian conception of nature is almost certainly false*. New York: Oxford University Press.

Nash, G. H. 2009. *Reappraising the right: The past and future of American conservatism*. Wilmington, DE: Intercollegiate Studies Institute.

National Assessment of Adult Literacy. (Undated.) Literacy from 1870 to 1979. https://nces.ed.gov /naal/lit_history.asp.

National Center for Health Statistics. 2014. *Health, United States, 2013*. Hyattsville, MD: National Center for Health Statistics.

National Center for Statistics and Analysis. 1995. *Traffic safety facts 1995—pedestrians*. Washington: National Highway Traffic Safety Administration. https://crashstats.nhtsa.dot.gov/Api/Public /ViewPublication/95F9.

National Center for Statistics and Analysis. 2006. *Pedestrians: 2005 data*. Washington: National Highway Traffic Safety Administration. https://crashstats.nhtsa.dot.gov/Api/Public/ViewPublication /810624.

National Center for Statistics and Analysis. 2016. *Pedestrians: 2014 data*. Washington: National Highway Traffic Safety Administration. https://crashstats.nhtsa.dot.gov/Api/Public/ViewPublication /812270.

National Center for Statistics and Analysis. 2017. *Pedestrians: 2015 data*. Washington: National Highway Traffic Safety Administration. https://crashstats.nhtsa.dot.gov/Api/Public/Publication /812375.

National Consortium for the Study of Terrorism and Responses to Terrorism. 2016. *Global Terrorism Database*. https://www.start.umd.edu/gtd/.

National Institute on Drug Abuse. 2016. DrugFacts: High school and youth trends. https://www .drugabuse.gov/publications/drugfacts/high-school-youth-trends.

National Safety Council. 2011. *Injury facts, 2011 edition*. Itasca, IL: National Safety Council.

National Safety Council. 2016. *Injury facts, 2016 edition*. Itasca, IL: National Safety Council.

Nemirow, J., Krasnow, M., Howard, R., & Pinker, S. 2016. Ineffective charitable altruism suggests adaptations for partner choice. Presented at the Annual Meeting of the Human Behavior and Evolution Society, Vancouver.

New York Times. 2016. Election 2016: Exit polls. https://www.nytimes.com/interactive/2016/11/08/us /politics/election-exit-polls.html?_r=0.

Newman, M. E. J. 2005. Power laws, Pareto distributions and Zipf's law. *Contemporary Physics, 46,* 323–51.

Nietzsche, F. 1887/2014. *On the genealogy of morals*. New York: Penguin.

Nisbet, R. 1980/2009. *History of the idea of progress*. New Brunswick, NJ: Transaction.

Norberg, J. 2016. *Progress: Ten reasons to look forward to the future*. London: Oneworld.

Nordhaus, T. 2016. Back from the energy future: What decades of failed forecasts say about clean energy and climate change. *Foreign Affairs*, Oct. 18.

Nordhaus, T., & Lovering, J. 2016. Does climate policy matter? Evaluating the efficacy of emissions caps and targets around the world. *The Breakthrough*. http://thebreakthrough.org/issues/Climate -Policy/does-climate-policy-matter.

Nordhaus, T., & Shellenberger, M. 2007. *Break through: From the death of environmentalism to the politics of possibility*. Boston: Houghton Mifflin.

Nordhaus, T., & Shellenberger, M. 2011. The long death of environmentalism. *The Breakthrough*. http://thebreakthrough.org/archive/the_long_death_of_environmenta.

Nordhaus, T., & Shellenberger, M. 2013. How the left came to reject cheap energy for the poor: The great progressive reversal, part two. *The Breakthrough*. http://thebreakthrough.org/index .php/voices/michael-shellenberger-and-ted-nordhaus/the-great-progressive-reversal.

Nordhaus, W. 1974. Resources as a constraint on growth. *American Economic Review, 64,* 22–26.

Nordhaus, W. 1996. Do real-output and real-wage measures capture reality? The history of lighting suggests not. In T. F. Bresnahan & R. J. Gordon, eds., *The economics of new goods*. Chicago: University of Chicago Press.

Nordhaus, W. 2013. *The climate casino: Risk, uncertainty, and economics for a warming world*. New Haven: Yale University Press.

Norenzayan, A. 2015. *Big gods: How religion transformed cooperation and conflict*. Princeton, NJ: Princeton University Press.

Norman, A. 2016. Why we reason: Intention-alignment and the genesis of human rationality. *Biology and Philosophy, 31,* 685–704.

Norris, P., & Inglehart, R. 2016. Populist-authoritarianism. https://www.electoralintegrityproject.com /populistauthoritarianism/.

North, D. C., Wallis, J. J., & Weingast, B. R. 2009. *Violence and social orders: A conceptual framework for interpreting recorded human history*. New York: Cambridge University Press.

Norvig, P. 2015. Ask not can machines think, ask how machines fit into the mechanisms we design. *Edge.* https://www.edge.org/response-detail/26055.

Nozick, R. 1974. *Anarchy, state, and utopia.* New York: Basic Books.

Nussbaum, M. 2000. *Women and human development: The capabilities approach.* New York: Cambridge University Press.

Nussbaum, M. 2008. Who is the happy warrior? Philosophy poses questions to psychology. *Journal of Legal Studies, 37,* 81–113.

Nussbaum, M. 2016. *Not for profit: Why democracy needs the humanities* (updated ed.). Princeton, NJ: Princeton University Press.

Nyhan, B. 2013. Building a better correction. *Columbia Journalism Review,* http://archives.cjr.org /united_states_project/building_a_better_correction_nyhan_new_misperception_research .php.

Ó Gráda, C. 2009. *Famine: A short history.* Princeton, NJ: Princeton University Press.

O'Neill, S., & Nicholson-Cole, S. 2009. "Fear won't do it": Promoting positive engagement with climate change through visual and iconic representations. *Science Communication, 30,* 355–79.

O'Neill, W. L. 1989. *American high: The years of confidence, 1945–1960.* New York: Simon & Schuster.

OECD. 1985. *Social expenditure 1960–1990: Problems of growth and control.* Paris: OECD Publishing.

OECD. 2014. Social expenditure update—social spending is falling in some countries, but in many others it remains at historically high levels. https://www.oecd.org/els/soc/OECD2014-Soci alExpenditure_Update19Nov_Rev.pdf

OECD. 2015a. *Education at a glance 2015: OECD indicators.* Paris: OECD Publishing.

OECD. 2015b. Suicide rates. https://data.oecd.org/healthstat/suicide-rates.htm.

OECD. 2016. Income distribution and poverty. http://stats.oecd.org/Index.aspx?DataSetCode=IDD.

OECD. 2017. Social expenditure: Aggregated data. http://stats.oecd.org/Index.aspx?datasetcode=SOCX _AGG.

Oeppen, J., & Vaupel, J. W. 2002. Broken limits to life expectancy. *Science, 296,* 1029–31.

Oesterdiekhoff, G. W. 2015. The nature of "premodern" mind: Tylor, Frazer, Lévy-Bruhl, Evans-Pritchard, Piaget, and beyond. *Anthropos, 110,* 15–25.

Office for National Statistics. 2016. UK environmental accounts: How much material is the UK consuming? https://www.ons.gov.uk/economy/environmentalaccounts/articles/ukenvironmental accountshowmuchmaterialistheukconsuming/ukenvironmentalaccountshowmuchmaterialis theukconsuming.

Office for National Statistics. 2017. Homicide. https://www.ons.gov.uk/peoplepopulationandcommunity /crimeandjustice/compendium/focusonviolentcrimeandsexualoffences/yearendingmarch2016 /homicide.

Ohlander, J. 2010. *The decline of suicide in Sweden, 1950–2000.* Ph.D. dissertation, Pennsylvania State University.

Olfson, M., Druss, B. G., & Marcus, S. C. 2015. Trends in mental health care among children and adolescents. *New England Journal of Medicine, 372,* 2029–38.

Omohundro, S. M. 2008. The basic AI drives. In P. Wang, B. Goertzel, & S. Franklin, eds., *Artificial general intelligence 2008: Proceedings of the first AGI conference.* Amsterdam: IOS Press.

Oreskes, N., & Conway, E. 2010. *Merchants of doubt: How a handful of scientists obscured the truth on issues from tobacco smoke to global warming.* New York: Bloomsbury Press.

Ortiz-Ospina, E., Lee, L., & Roser, M. 2016. Suicide. *Our World in Data.* https://ourworldindata.org /suicide/.

Ortiz-Ospina, E., & Roser, M. 2016a. Child labor. *Our World in Data.* https://ourworldindata.org/child -labor/.

Ortiz-Ospina, E., & Roser, M. 2016b. Public spending. *Our World in Data.* https://ourworldindata.org /public-spending/.

Ortiz-Ospina, E., & Roser, M. 2016c. Trust. *Our World in Data.* https://ourworldindata.org/trust/.

Ortiz-Ospina, E., & Roser, M. 2016d. World population growth. *Our World in Data.* https://ourworld indata.org/world-population-growth/.

Ortiz-Ospina, E., & Roser, M. 2017. Happiness and life satisfaction. *Our World in Data.* https:// ourworldindata.org/happiness-and-life-satisfaction/.

Osgood, C. E. 1962. *An alternative to war or surrender.* Urbana: University of Illinois Press.

Otieno, C., Spada, H., & Renkl, A. 2013. Effects of news frames on perceived risk, emotions, and learning. *PLOS ONE, 8,* 1–12.

Otterbein, K. F. 2004. *How war began.* College Station: Texas A&M University Press.

Ottosson, D. 2006. *LGBT world legal wrap up survey.* Brussels: International Lesbian and Gay Association.

Ottosson, D. 2009. *State-sponsored homophobia.* Brussels: International Lesbian, Gay, Bisexual, Trans, and Intersex Association.

Pacala, S., & Socolow, R. 2004. Stabilization wedges: Solving the climate problem for the next 50 years with current technologies. *Science, 305,* 968–72.

Pagden, A. 2013. *The Enlightenment: And why it still matters.* New York: Random House.

Pagel, M. 2015. Machines that can think will do more good than harm. *Edge.* https://www.edge.org /response-detail/26038.

Paine, T. 1782/2016. *Thomas Paine ultimate collection: Political works, philosophical writings, speeches, letters and biography*. Prague: e-artnow.

Papineau, D. 2015. Naturalism. In E. N. Zalta, ed., *Stanford Encyclopedia of Philosophy*. https://plato.stanford.edu/entries/naturalism/.

Parachini, J. 2003. Putting WMD terrorism into perspective. *Washington Quarterly, 26*, 37–50.

Parfit, D. 1997. Equality and priority. *Ratio, 10*, 202–21.

Parfit, D. 2011. *On what matters*. New York: Oxford University Press.

Patel, A. 2008. *Music, language, and the brain*. New York: Oxford University Press.

Patterson, O. 1985. *Slavery and social death*. Cambridge, MA: Harvard University Press.

Paul, G. S. 2009. The chronic dependence of popular religiosity upon dysfunctional psychosociological conditions. *Evolutionary Psychology, 7*, 398–441.

Paul, G. S. 2014. The health of nations. *Skeptic, 19*, 10–16.

Paul, G. S., & Zuckerman, P. 2007. Why the gods are not winning. *Edge*. https://www.edge.org/conversation/gregory_paul-phil_zuckerman-why-the-gods-are-not-winning.

Payne, J. L. 2004. *A history of force: Exploring the worldwide movement against habits of coercion, bloodshed, and mayhem*. Sandpoint, ID: Lytton Publishing.

Payne, J. L. 2005. The prospects for democracy in high-violence societies. *Independent Review, 9*, 563–72.

PBL Netherlands Environmental Assessment Agency. (Undated.) *History database of the global environment: Population*. http://themasites.pbl.nl/tridion/en/themasites/hyde/basicdrivingfactors/population/index-2.html.

Pegula, S., & Janocha, J. 2013. Death on the job: Fatal work injuries in 2011. *Beyond the Numbers, 2* (22). http://www.bls.gov/opub/btn/volume-2/death-on-the-job-fatal-work-injuries-in-2011.htm.

Pelham, N. 2016. *Holy lands: Reviving pluralism in the Middle East*. New York: Columbia Global Reports.

Pentland, A. 2007. The human nervous system has come alive. *Edge*. https://www.edge.org/response-detail/11497.

Perlman, J. E. 1976. *The myth of marginality: Urban poverty and politics in Rio de Janeiro*. Berkeley: University of California Press.

Peterson, M. B. 2015. Evolutionary political psychology: On the origin and structure of heuristics and biases in politics. *Advances in Political Psychology, 36*, 45–78.

Pettersson, T., & Wallensteen, P. 2015. Armed conflicts, 1946–2014. *Journal of Peace Research, 52*, 536–50.

Pew Research Center. 2010. *Gender equality universally embraced, but inequalities acknowledged*. Washington: Pew Research Center.

Pew Research Center. 2012a. *The global religious landscape*. Washington: Pew Research Center.

Pew Research Center. 2012b. *Trends in American values, 1987–2012*. Washington: Pew Research Center.

Pew Research Center. 2012c. *The world's Muslims: Unity and diversity*. Washington: Pew Research Center.

Pew Research Center. 2013. *The world's Muslims: Religion, politics, and society*. Washington: Pew Research Center.

Pew Research Center. 2014. *Political polarization in the American public*. Washington: Pew Research Center.

Pew Research Center. 2015a. *America's changing religious landscape*. Washington: Pew Research Center.

Pew Research Center. 2015b. *Views about climate change, by education and science knowledge*. Washington: Pew Research Center.

Phelps, E. S. 2013. *Mass flourishing: How grassroots innovation created jobs, challenge, and change*. Princeton, NJ: Princeton University Press.

Phillips, J. A. 2014. A changing epidemiology of suicide? The influence of birth cohorts on suicide rates in the United States. *Social Science and Medicine, 114*, 151–60.

Pietschnig, J., & Voracek, M. 2015. One century of global IQ gains: A formal meta-analysis of the Flynn effect (1909–2013). *Perspectives on Psychological Science, 10*, 282–306.

Piketty, T. 2013. *Capital in the twenty-first century*. Cambridge, MA: Harvard University Press.

Pinker, S. 1994/2007. *The language instinct*. New York: HarperCollins.

Pinker, S. 1997/2009. *How the mind works*. New York: Norton.

Pinker, S. 1999/2011. *Words and rules: The ingredients of language*. New York: HarperCollins.

Pinker, S. 2002/2016. *The blank slate: The modern denial of human nature*. New York: Penguin.

Pinker, S. 2005. The evolutionary psychology of religion. *Freethought Today*. https://ffrf.org/about/getting-acquainted/item/13184-the-evolutionary-psychology-of-religion.

Pinker, S. 2006. Preface to "Dangerous Ideas." *Edge*. https://www.edge.org/conversation/steven_pinker-preface-to-dangerous-ideas.

Pinker, S. 2007a. *The stuff of thought: Language as a window into human nature*. New York: Penguin.

Pinker, S. 2007b. Toward a consilient study of literature: Review of J. Gottschall & D. S. Wilson's "The literary animal: Evolution and the nature of narrative." *Philosophy and Literature, 31*, 162–78.

Pinker, S. 2008a. The moral instinct. *New York Times Magazine*, January 13.

Pinker, S. 2008b. The stupidity of dignity. *New Republic*, May 28.

Pinker, S. 2010. The cognitive niche: Coevolution of intelligence, sociality, and language. *Proceedings of the National Academy of Sciences, 107*, 8993–99.

Pinker, S. 2011. *The better angels of our nature: Why violence has declined*. New York: Penguin.

Pinker, S. 2012. The false allure of group selection. *Edge*. http://edge.org/conversation/steven_pinker -the-false-allure-of-group-selection.

Pinker, S. 2013a. George A. Miller (1920–2012). *American Psychologist, 68*, 467–68.

Pinker, S. 2013b. Science is not your enemy. *New Republic*, Aug. 6.

Pinker, S., & Wieseltier, L. 2013. Science vs. the humanities, round III. *New Republic*, Sept. 26.

Pinker, Susan. 2014. *The village effect: How face-to-face contact can make us healthier, happier, and smarter*. New York: Spiegel & Grau.

Plomin, R., & Deary, I. J. 2015. Genetics and intelligence differences: Five special findings. *Molecular Psychiatry, 20*, 98–108.

PLOS Medicine Editors. 2013. The paradox of mental health: Over-treatment and under-recognition. *PLOS Medicine, 10*, e1001456.

Plumer, B. 2015. Global warming, explained. *Vox*. http://www.vox.com/cards/global-warming/what -is-global-warming.

Popper, K. 1945/2013. *The open society and its enemies*. Princeton, NJ: Princeton University Press.

Popper, K. 1983. *Realism and the aim of science*. London: Routledge.

Porter, M. E., Stern, S., & Green, M. 2016. *Social Progress Index 2016*. Washington: Social Progress Imperative.

Porter, R. 2000. *The creation of the modern world: The untold story of the British Enlightenment*. New York: Norton.

Potts, M., & Hayden, T. 2008. *Sex and war: How biology explains warfare and terrorism and offers a path to a safer world*. Dallas, TX: Benbella Books.

Powell, J. L. 2015. Climate scientists virtually unanimous: Anthropogenic global warming is true. *Bulletin of Science, Technology & Society, 35*, 121–24.

Prados de la Escosura, L. 2015. World human development, 1870–2007. *Review of Income and Wealth, 61*, 220–47.

Pratto, F., Sidanius, J., & Levin, S. 2006. Social dominance theory and the dynamics of intergroup relations: Taking stock and looking forward. *European Review of Social Psychology, 17*, 271–320.

Preble, C. 2004. *John F. Kennedy and the missile gap*. DeKalb: Northern Illinois University Press.

Price, R. G. 2006. The mis-portrayal of Darwin as a racist. *RationalRevolution.net*. http://www.rational revolution.net/articles/darwin_nazism.htm.

Proctor, B. D., Semega, J. L., & Kollar, M. A. 2016. *Income and poverty in the United States: 2015*. Washington: United States Census Bureau. http://www.census.gov/content/dam/Census/library /publications/2016/demo/p60-256.pdf.

Proctor, R. N. 1988. *Racial hygiene: Medicine under the Nazis*. Cambridge, MA: Harvard University Press.

Pronin, E., Lin, D. Y., & Ross, L. 2002. The bias blind spot: Perceptions of bias in self versus others. *Personality and Social Psychology Bulletin, 28*, 369–81.

Pryor, F. L. 2007. Are Muslim countries less democratic? *Middle East Quarterly, 14*, 53–58.

Publius Decius Mus (Michael Anton). 2016. The flight 93 election. *Claremont Review of Books Digital*. http://www.claremont.org/crb/basicpage/the-flight-93-election/.

Putnam, R. D., & Campbell, D. E. 2010. *American grace: How religion divides and unites us*. New York: Simon & Schuster.

Quarantelli, E. L. 2008. Conventional beliefs and counterintuitive realities. *Social Research, 75*, 873–904.

Rachels, J., & Rachels, S. 2010. *The elements of moral philosophy*. Columbus, OH: McGraw-Hill.

Radelet, S. 2015. *The great surge: The ascent of the developing world*. New York: Simon & Schuster.

Railton, P. 1986. Moral realism. *Philosophical Review, 95*, 163–207.

Randle, M., & Eckersley, R. 2015. Public perceptions of future threats to humanity and different societal responses: A cross-national study. *Futures, 72*, 4–16.

Rawcliffe, C. 1998. *Medicine and society in later medieval England*. Stroud, UK: Sutton.

Rawls, J. 1976. *A theory of justice*. Cambridge, MA: Harvard University Press.

Ray, J. L. 1989. The abolition of slavery and the end of international war. *International Organization, 43*, 405–39.

Redlawsk, D. P., Civettini, A. J. W., & Emmerson, K. M. 2010. The affective tipping point: Do motivated reasoners ever "get it"? *Political Psychology, 31*, 563–93.

Reese, B. 2013. *Infinite progress: How the internet and technology will end ignorance, disease, poverty, hunger, and war*. Austin, TX: Greenleaf Book Group Press.

Reverby, S. M., ed. 2000. *Tuskegee's truths: Rethinking the Tuskegee syphilis study*. Chapel Hill: University of North Carolina Press.

Rhodes, R. 2010. *The twilight of the bombs*. New York: Knopf.

Rice, J. W., Olson, J. K., & Colbert, J. T. 2011. University evolution education: The effect of evolution instruction on biology majors' content knowledge, attitude toward evolution, and theistic position. *Evolution: Education and Outreach, 4*, 137–44.

Richards, R. J. 2013. *Was Hitler a Darwinian? Disputed questions in the history of evolutionary theory*. Chicago: University of Chicago Press.

Rid, T. 2012. Cyber war will not take place. *Journal of Strategic Studies, 35*, 5–32.

Ridley, M. 2000. *Genome: The autobiography of a species in 23 chapters*. New York: HarperCollins.

Ridley, M. 2010. *The rational optimist: How prosperity evolves*. New York: HarperCollins.

Ridout, T. N., Grosse, A. C., & Appleton, A. M. 2008. News media use and Americans' perceptions of global threat. *British Journal of Political Science, 38*, 575–93.

Rijpma, A. 2014. A composite view of well-being since 1820. In J. van Zanden, J. Baten, M. M. d'Ercole, A. Rijpma, C. Smith, & M. Timmer, eds., *How was life? Global well-being since 1820*. Paris: OECD Publishing.

Riley, J. C. 2005. Estimates of regional and global life expectancy, 1800–2001. *Population and Development Review, 31*, 537–43.

Rindermann, H. 2008. Relevance of education and intelligence for the political development of nations: Democracy, rule of law and political liberty. *Intelligence, 36*, 306–22.

Rindermann, H. 2012. Intellectual classes, technological progress and economic development: The rise of cognitive capitalism. *Personality and Individual Differences, 53*, 108–13.

Risso, M. I. 2014. Intentional homicides in São Paulo city: A new perspective. *Stability: International Journal of Security & Development, 3*, art. 19.

Ritchie, H., & Roser, M. 2017. CO_2 and other greenhouse gas emissions. *Our World in Data*. https://ourworldindata.org/co2-and-other-greenhouse-gas-emissions/.

Ritchie, S. 2015. *Intelligence: All that matters*. London: Hodder & Stoughton.

Ritchie, S., Bates, T. C., & Deary, I. J. 2015. Is education associated with improvements in general cognitive ability, or in specific skills? *Developmental Psychology, 51*, 573–82.

Rizvi, A. A. 2016. *The atheist Muslim: A journey from religion to reason*. New York: St. Martin's Press.

Robinson, F. S. 2009. *The case for rational optimism*. New Brunswick, NJ: Transaction.

Robinson, J. 2013. Americans less rushed but no happier: 1965–2010 trends in subjective time and happiness. *Social Indicators Research, 113*, 1091–1104.

Robock, A., & Toon, O. B. 2012. Self-assured destruction: The climate impacts of nuclear war. *Bulletin of the Atomic Scientists, 68*, 66–74.

Romer, P. 2016. Conditional optimism about progress and climate. *Paul Romer.net*. https://paulromer.net/conditional-optimism-about-progress-and-climate/.

Romer, P., & Nelson, R. R. 1996. Science, economic growth, and public policy. In B. L. R. Smith & C. E. Barfield, eds., *Technology, R&D, and the economy*. Washington: Brookings Institution.

Roos, J. M. 2014. Measuring science or religion? A measurement analysis of the National Science Foundation sponsored Science Literacy Scale, 2006–2010. *Public Understanding of Science, 23*, 797–813.

Ropeik, D., & Gray, G. 2002. *Risk: A practical guide for deciding what's really safe and what's really dangerous in the world around you*. Boston: Houghton Mifflin.

Rose, S. J. 2016. *The growing size and incomes of the upper middle class*. Washington: Urban Institute.

Rosen, J. 2016. Here's how the world could end—and what we can do about it. *Science*. http://www.sciencemag.org/news/2016/07/here-s-how-world-could-end-and-what-we-can-do-about-it.

Rosenberg, N., & Birdzell, L. E., Jr. 1986. *How the West grew rich: The economic transformation of the industrial world*. New York: Basic Books.

Rosenthal, B. G. 2002. *New myth, new world: From Nietzsche to Stalinism*. University Park: Penn State University Press.

Roser, M. 2016a. Child mortality. *Our World in Data*. https://ourworldindata.org/child-mortality/.

Roser, M. 2016b. Democracy. *Our World in Data*. https://ourworldindata.org/democracy/.

Roser, M. 2016c. Economic growth. *Our World in Data*. https://ourworldindata.org/economic-growth/.

Roser, M. 2016d. Food per person. *Our World in Data*. https://ourworldindata.org/food-per-person/.

Roser, M. 2016e. Food prices. *Our World in Data*. https://ourworldindata.org/food-prices/.

Roser, M. 2016f. Forests. *Our World in Data*. https://ourworldindata.org/forests/.

Roser, M. 2016g. Global economic inequality. *Our World In Data*. https://ourworldindata.org/global-economic-inequality/.

Roser, M. 2016h. Human Development Index (HDI). *Our World in Data*. https://ourworldindata.org/human-development-index/.

Roser, M. 2016i. Human rights. *Our World in Data*. https://ourworldindata.org/human-rights/.

Roser, M. 2016j. Hunger and undernourishment. *Our World in Data*. https://ourworldindata.org/hunger-and-undernourishment/.

Roser, M. 2016k. Income inequality. *Our World in Data*. https://ourworldindata.org/income-inequality/.

Roser, M. 2016l. Indoor air pollution. *Our World in Data*. https://ourworldindata.org/indoor-air-pollution/.

Roser, M. 2016m. Land use in agriculture. *Our World in Data*. https://ourworldindata.org/yields-and-land-use-in-agriculture/.

Roser, M. 2016n. Life expectancy. *Our World in Data*. https://ourworldindata.org/life-expectancy/.

Roser, M. 2016o. Light. *Our World in Data*. https://ourworldindata.org/light/.

Roser, M. 2016p. Maternal mortality. *Our World in Data*. https://ourworldindata.org/maternal-mortality/.

Roser, M. 2016q. Natural catastrophes. *Our World in Data*. https://ourworldindata.org/natural-catastrophes/.

Roser, M. 2016r. Oil spills. *Our World in Data*. https://ourworldindata.org/oil-spills/.

Roser, M. 2016s. Treatment of minorities. *Our World in Data*. https://ourworldindata.org/treatment-of-minorities/.

Roser, M. 2016t. Working hours. *Our World in Data.* https://ourworldindata.org/working-hours/.

Roser, M. 2016u. Yields. *Our World in Data.* https://ourworldindata.org/yields-and-land-use-in-agri culture/.

Roser, M., & Ortiz-Ospina, E. 2016a. Global rise of education. *Our World in Data.* https://ourworld indata.org/global-rise-of-education/.

Roser, M., & Ortiz-Ospina, E. 2016b. Literacy. *Our World in Data.* https://ourworldindata.org/literacy/.

Roser, M., & Ortiz-Ospina, E. 2017. Global extreme poverty. *Our World in Data.* https://ourworldindata .org/extreme-poverty/.

Roser, M. & Ortiz-Ospina, E. 2018. Primary and secondary education. *Our World in Data.* https:// ourworldindata.org/primary-and-secondary-education.

Roth, R. 2009. *American homicide.* Cambridge, MA: Harvard University Press.

Rozenblit, L., & Keil, F. C. 2002. The misunderstood limits of folk science: An illusion of explanatory depth. *Cognitive Science, 26,* 521–62.

Rozin, P., & Royzman, E. B. 2001. Negativity bias, negativity dominance, and contagion. *Personality and Social Psychology Review, 5,* 296–320.

Ruddiman, W. F., Fuller, D. Q., Kutzbach, J. E., Tzedakis, P. C., Kaplan, J. O., et al. 2016. Late Holocene climate: Natural or anthropogenic? *Reviews of Geophysics, 54,* 93–118.

Rummel, R. J. 1994. *Death by government.* New Brunswick, NJ: Transaction.

Rummel, R. J. 1997. *Statistics of democide.* New Brunswick, NJ: Transaction.

Russell, B. 1945/1972. *A history of Western philosophy.* New York: Simon & Schuster.

Russell, S. 2015. Will they make us better people? *Edge.* https://www.edge.org/response-detail/26157.

Russett, B. 2010. Capitalism or democracy? Not so fast. *International Interactions, 36,* 198–205.

Russett, B., & Oneal, J. 2001. *Triangulating peace: Democracy, interdependence, and international organiza- tions.* New York: Norton.

Sacerdote, B. 2017. *Fifty years of growth in American consumption, income, and wages.* Cambridge, MA: National Bureau of Economic Research. http://www.nber.org/papers/w23292.

Sacks, D. W., Stevenson, B., & Wolfers, J. 2012. *The new stylized facts about income and subjective well- being.* Bonn: IZA Institute for the Study of Labor.

Sagan, S. D. 2009a. The case for No First Use. *Survival, 51,* 163–82.

Sagan, S. D. 2009b. The global nuclear future. *Bulletin of the American Academy of Arts and Sciences, 62,* 21–23.

Sagan, S. D. 2009c. Shared responsibilities for nuclear disarmament. *Daedalus, 138,* 157–68.

Sagan, S. D. 2010. Nuclear programs with sources. Center for International Security and Cooperation, Stanford University.

Sage, J. C. 2010. *Birth cohort changes in anxiety from 1993–2006: A cross-temporal meta-analysis.* Master's thesis, San Diego State University, San Diego.

Sanchez, D. L., Nelson, J. H., Johnston, J. C., Mileva, A., & Kammen, D. M. 2015. Biomass enables the transition to a carbon-negative power system across western North America. *Nature Climate Change, 5,* 230–34.

Sandman, P. M., & Valenti, J. M. 1986. Scared stiff—or scared into action. *Bulletin of the Atomic Scien- tists, 42,* 12–16.

Satel, S. L. 2000. *PC, M.D.: How political correctness is corrupting medicine.* New York: Basic Books.

Satel, S. L. 2010. The limits of bioethics. *Policy Review,* Feb. & March.

Satel, S. L. 2017. Taking on the scourge of opioids. *National Affairs,* Summer, 1–19.

Saunders, P. 2010. *Beware false prophets: Equality, the good society and the spirit level.* London: Policy Exchange.

Savulescu, J. 2015. Bioethics: Why philosophy is essential for progress. *Journal of Medical Ethics, 41,* 28–33.

Sayer, L. C., Bianchi, S. M., & Robinson, J. P. 2004. Are parents investing less in children? Trends in mothers' and fathers' time with children. *American Journal of Sociology, 110,* 1–43.

Sayre-McCord, G. 1988. *Essays on moral realism.* Ithaca, NY: Cornell University Press.

Sayre-McCord, G. 2015. Moral realism. In E. N. Zalta, ed., *Stanford Encyclopedia of Philosophy.* https:// plato.stanford.edu/entries/moral-realism/.

Schank, R. C. 2015. Machines that think are in the movies. *Edge.* https://www.edge.org/response-detail /26037.

Scheidel, W. 2017. *The great leveler: Violence and the history of inequality from the Stone Age to the twenty- first century.* Princeton, NJ: Princeton University Press.

Schelling, T. C. 1960. *The strategy of conflict.* Cambridge, MA: Harvard University Press.

Schelling, T. C. 2009. A world without nuclear weapons? *Daedalus, 138,* 124–29.

Schlosser, E. 2013. *Command and control: Nuclear weapons, the Damascus accident, and the illusion of safety.* New York: Penguin.

Schneider, C. E. 2015. *The censor's hand: The misregulation of human-subject research.* Cambridge, MA: MIT Press.

Schneider, G., & Gleditsch, N. P. 2010. The capitalist peace: The origins and prospects of a liberal idea. *International Interactions, 36,* 107–14.

Schneier, B. 2008. *Schneier on security.* New York: Wiley.

Schrag, D. 2009. Coal as a low-carbon fuel? *Nature Geoscience, 2,* 818–20.

Schrag, Z. M. 2010. *Ethical imperialism: Institutional review boards and the social sciences, 1965–2009.* Baltimore: Johns Hopkins University Press.

Schrauf, R. W., & Sanchez, J. 2004. The preponderance of negative emotion words in the emotion lexicon: A cross-generational and cross-linguistic study. *Journal of Multilingual and Multicultural Development, 25,* 266–84.

Schuck, P. H. 2015. *Why government fails so often: And how it can do better.* Princeton, NJ: Princeton University Press.

Scoblic, J. P. 2010. What are nukes good for? *New Republic,* April 7.

Scott, J. C. 1998. *Seeing like a state: How certain schemes to improve the human condition failed.* New Haven: Yale University Press.

Scott, R. A. 2010. *Miracle cures: Saints, pilgrimage, and the healing powers of belief.* Berkeley: University of California Press.

Sechser, T. S., & Fuhrmann, M. 2017. *Nuclear weapons and coercive diplomacy.* New York: Cambridge University Press.

Sehu, Y., Chen, L.-H., & Hedegaard, H. 2015. Death rates from unintentional falls among adults aged ≥ 65 years, by sex—United States, 2000–2013. *CDC Morbidity and Mortality Weekly Report, 64,* 450.

Seiple, I. B., Zhang, Z., Jakubec, P., Langlois-Mercier, A., Wright, P. M., et al. 2016. A platform for the discovery of new macrolide antibiotics. *Nature, 533,* 338–45.

Semega, J. L., Fontenot, K. R., & Kollar, M. A. 2017. Income and poverty in the United States: 2016. Washington: United States Census Bureau. https://www.census.gov/library/publications /2017/demo/p60-259.html.

Sen, A. 1984. *Poverty and famines: An essay on entitlement and deprivation.* New York: Oxford University Press.

Sen, A. 1987. *On ethics and economics.* Oxford: Blackwell.

Sen, A. 1999. *Development as freedom.* New York: Knopf.

Sen, A. 2000. East and West: The reach of reason. *New York Review of Books,* July 20.

Sen, A. 2005. *The argumentative Indian: Writings on Indian history, culture and identity.* New York: Farrar, Straus & Giroux.

Sen, A. 2009. *The idea of justice.* Cambridge, MA: Harvard University Press.

Service, R. F. 2017. Fossil power, guilt free. *Science, 356,* 796–99.

Sesardić, N. 2016. *When reason goes on holiday: Philosophers in politics.* New York: Encounter.

Sheehan, J. J. 2008. *Where have all the soldiers gone? The transformation of modern Europe.* Boston: Houghton Mifflin.

Shellenberger, M. 2017. Nuclear technology, innovation and economics. *Environmental Progress.* http://www.environmentalprogress.org/nuclear-technology-innovation-economics/.

Shellenberger, M., & Nordhaus, T. 2013. Has there been a great progressive reversal? How the left abandoned cheap electricity. AlterNet. https://www.alternet.org/environment/how-progressives -abandoned-cheap-electricity.

Shermer, M., ed. 2002. *The Skeptic Encyclopedia of Pseudoscience* (vols. 1 and 2). Denver: ABC-CLIO.

Shermer, M. 2015. *The moral arc: How science and reason lead humanity toward truth, justice, and freedom.* New York: Henry Holt.

Shermer, M. 2018. *Heavens on earth: The scientific search for the afterlife, immortality, and utopia.* New York: Henry Holt.

Shields, J. A., & Dunn, J. M. 2016. *Passing on the right: Conservative professors in the progressive university.* New York: Oxford University Press.

Shtulman, A. 2006. Qualitative differences between naive and scientific theories of evolution. *Cognitive Psychology, 52,* 170–94.

Shweder, R. A. 2004. Tuskegee re-examined. *Spiked.* http://www.spiked-online.com/newsite/article /14972#.WUdPYOvysYM.

Sidanius, J., & Pratto, F. 1999. *Social dominance.* New York: Cambridge University Press.

Siebens, J. 2013. *Extended measures of well-being: Living conditions in the United States, 2011.* Washington: US Census Bureau. https://www.census.gov/prod/2013pubs/p70-136.pdf.

Siegel, R., Naishadham, D., & Jemal, A. 2012. Cancer statistics, 2012. *CA: A Cancer Journal for Clinicians, 62,* 10–29.

Sikkink, K. 2017. *Evidence for hope: Making human rights work in the 21st century.* Princeton, NJ: Princeton University Press.

Silver, N. 2015. *The signal and the noise: Why so many predictions fail—but some don't.* New York: Penguin.

Simon, J. 1981. *The ultimate resource.* Princeton, NJ: Princeton University Press.

Singer, P. 1981/2011. *The expanding circle: Ethics and sociobiology.* Princeton, NJ: Princeton University Press.

Singer, P. 2010. *The life you can save: How to do your part to end world poverty.* New York: Random House.

Singh, J. P., Grann, M., & Fazel, S. 2011. A comparative study of violence risk assessment tools: A systematic review and metaregression analysis of 68 studies involving 25,980 participants. *Clinical Psychology Review, 31,* 499–513.

Slingerland, E. 2008. *What science offers the humanities: Integrating body and culture.* New York: Cambridge University Press.

Sloman, S., & Fernbach, P. 2017. *The knowledge illusion: Why we never think alone.* New York: Penguin.

Slovic, P. 1987. Perception of risk. *Science, 236,* 280–85.

Slovic, P., Fischhoff, B., & Lichtenstein, S. 1982. Facts versus fears: Understanding perceived risk. In D. Kahneman, P. Slovic, & A. Tversky, eds., *Judgment under uncertainty: Heuristics and biases.* New York: Cambridge University Press.

Smart, J. J. C., & Williams, B. 1973. *Utilitarianism: For and against.* New York: Cambridge University Press.

Smith, A. 1776/2009. *The wealth of nations.* New York: Classic House Books.

Smith, E. A., Hill, K., Marlowe, F., Nolin, D., Wiessner, P., et al. 2010. Wealth transmission and inequality among hunter-gatherers. *Current Anthropology, 51,* 19–34.

Smith, H. L. 2008. Advances in age-period-cohort analysis. *Sociological Methods and Research, 36,* 287–96.

Smith, T. W., Son, J., & Schapiro, B. 2015. *General Social Survey final report: Trends in psychological well-being, 1972–2014.* Chicago: National Opinion Research Center at the University of Chicago.

Snow, C. P. 1959/1998. *The two cultures.* New York: Cambridge University Press.

Snow, C. P. 1961. The moral un-neutrality of science. *Science, 133,* 256–59.

Snowdon, C. 2010. *The spirit level delusion: Fact-checking the left's new theory of everything.* Ripon, UK: Little Dice.

Snowdon, C. 2016. *The Spirit Level Delusion* (blog). http://spiritleveldelusion.blogspot.co.uk/.

Snyder, T. D., ed. 1993. *120 years of American education: A statistical portrait.* Washington: National Center for Education Statistics.

Somin, I. 2016. *Democracy and political ignorance: Why smaller government is smarter* (2nd ed.). Stanford, CA: Stanford University Press.

Sowell, T. 1980. *Knowledge and decisions.* New York: Basic Books.

Sowell, T. 1987. *A conflict of visions: Ideological origins of political struggles.* New York: Quill.

Sowell, T. 1994. *Race and culture: A world view.* New York: Basic Books.

Sowell, T. 1995. *The vision of the anointed: Self-congratulation as a basis for social policy.* New York: Basic Books.

Sowell, T. 1996. *Migrations and cultures: A world view.* New York: Basic Books.

Sowell, T. 1998. *Conquests and cultures: An international history.* New York: Basic Books.

Sowell, T. 2010. *Intellectuals and society.* New York: Basic Books.

Sowell, T. 2015. *Wealth, poverty, and politics: An international perspective.* New York: Basic Books.

Spagat, M. 2015. Is the risk of war declining? *Sense About Science USA.* http://www.senseaboutscience usa.org/is-the-risk-of-war-declining/.

Spagat, M. 2017. Pinker versus Taleb: A non-deadly quarrel over the decline of violence. http://personal .rhul.ac.uk/uhte/014/York%20talk%20Spagat.pdf.

Stansell, C. 2010. *The feminist promise: 1792 to the present.* New York: Modern Library.

Stanton, S. J., Beehner, J. C., Saini, E. K., Kuhn, C. M., & LaBar, K. S. 2009. Dominance, politics, and physiology: Voters' testosterone changes on the night of the 2008 United States presidential election. *PLOS ONE, 4,* e7543.

Starmans, C., Sheskin, M., & Bloom, P. 2017. Why people prefer unequal societies. *Nature Human Behavior, 1,* 1–7.

Statistics Times. 2015. List of European countries by population (2015). http://statisticstimes.com /population/european-countries-by-population.php.

Steigmann-Gall, R. 2003. *The Holy Reich: Nazi conceptions of Christianity, 1919–1945.* New York: Cambridge University Press.

Stein, G., ed. 1996. *The Encyclopedia of the Paranormal.* Amherst, NY: Prometheus Books.

Stenger, V. J. 2011. *The fallacy of fine-tuning: Why the universe is not designed for us.* Amherst, NY: Prometheus Books.

Stephens-Davidowitz, S. 2014. The cost of racial animus on a black candidate: Evidence using Google search data. *Journal of Public Economics, 118,* 26–40.

Stephens-Davidowitz, S. 2017. *Everybody lies: Big data, new data, and what the internet reveals about who we really are.* New York: HarperCollins.

Stern, D. 2014. The environmental Kuznets curve: A primer. Centre for Climate Economics and Policy, Crawford School of Public Policy, Australian National University.

Sternhell, Z. 2010. *The anti-Enlightenment tradition.* New Haven: Yale University Press.

Stevens, J. A., & Rudd, R. A. 2014. Circumstances and contributing causes of fall deaths among persons aged 65 and older: United States, 2010. *Journal of the American Geriatrics Society, 62,* 470–75.

Stevenson, B., & Wolfers, J. 2008a. Economic growth and subjective well-being: Reassessing the Easterlin paradox. *Brookings Papers on Economic Activity,* Spring, 1–87.

Stevenson, B., & Wolfers, J. 2008b. Happiness inequality in the United States. *Journal of Legal Studies, 37,* S33–S79.

Stevenson, B., & Wolfers, J. 2009. The paradox of declining female happiness. *American Economic Journal: Economic Policy, 1,* 190–225.

Stevenson, L., & Haberman, D. L. 1998. *Ten theories of human nature.* New York: Oxford University Press.

Stokes, B. 2007. *Happiness is increasing in many countries—but why?* Washington: Pew Reseach Center. http://www.pewglobal.org/2007/07/24/happiness-is-increasing-in-many-countries -but-why/#rich-and-happy.

Stork, N. E. 2010. Re-assessing current extinction rates. *Biodiversity and Conservation, 19,* 357–71.

Stuermer, M., & Schwerhoff, G. 2016. Non-renewable resources, extraction technology, and endogenous growth. National Bureau of Economic Research. https://paulromer.net/wp-content/uploads/2016/07/Stuermer-Schwerhoff-160716.pdf.

Suckling, K., Mehrhof, L. A., Beam, R., & Hartl, B. 2016. *A wild success: A systematic review of bird recovery under the Endangered Species Act.* Tucson, AZ: Center for Biological Diversity. http://www.esasuccess.org/pdfs/WildSuccess.pdf.

Summers, L. H. 2014a. The inequality puzzle. *Democracy: A Journal of Ideas, 33.*

Summers, L. H. 2014b. Reflections on the "new secular stagnation hypothesis." In C. Teulings & R. Baldwin, eds., *Secular stagnation: Facts, causes and cures.* London: Centre for Economic Policy Research.

Summers, L. H. 2016. The age of secular stagnation. *Foreign Affairs,* Feb. 15.

Summers, L. H., & Balls, E. 2015. *Report of the Commission on Inclusive Prosperity.* Washington: Center for American Progress.

Sunstein, C. R. 2013. *Simpler: The future of government.* New York: Simon & Schuster.

Sutherland, R. 2016. The dematerialization of consumption. *Edge.* https://www.edge.org/response-detail/26750.

Sutherland, S. 1992. *Irrationality: The enemy within.* London: Penguin.

Sutin, A. R., Terracciano, A., Milaneschi, Y., An, Y., Ferrucci, L., et al. 2013. The effect of birth cohort on well-being: The legacy of economic hard times. *Psychological Science, 24,* 379–85.

Taber, C. S., & Lodge, M. 2006. Motivated skepticism in the evaluation of political beliefs. *American Journal of Political Science, 50,* 755–69.

Tannenwald, N. 2005. Stigmatizing the bomb: Origins of the nuclear taboo. *International Security, 29,* 5–49.

Taylor, P. 2016a. *The next America: Boomers, millennials, and the looming generational showdown.* Washington: PublicAffairs.

Taylor, P. 2016b. *The demographic trends shaping American politics in 2016 and beyond.* Washington: Pew Research Center.

Tebeau, M. 2016. Accidents. *Encyclopedia of Children and Childhood in History and Society.* http://www.faqs.org/childhood/A-Ar/Accidents.html.

Tegmark, M. 2003. Parallel universes. *Scientific American, 288,* 41–51.

Teixeira, R., Halpin, J., Barreto, M., & Pantoja, A. 2013. *Building an all-in nation: A view from the American public.* Washington: Center for American Progress.

Terracciano, A. 2010. Secular trends and personality: Perspectives from longitudinal and cross-cultural studies—commentary on Trzesniewski & Donnellan (2010). *Perspectives on Psychological Science, 5,* 93–96.

Terry, Q. C. 2008. *Golden Rules and Silver Rules of humanity: Universal wisdom of civilization.* Berkeley: AuthorHouse.

Tetlock, P. E. 2002. Social functionalist frameworks for judgment and choice: Intuitive politicians, theologians, and prosecutors. *Psychological Review, 109,* 451–71.

Tetlock, P. E. 2015. All it takes to improve forecasting is keep score. Paper presented at the Seminars About Long-Term Thinking, San Francisco. http://longnow.org/seminars/02015/nov/23/super forecasting/.

Tetlock, P. E., & Gardner, D. 2015. *Superforecasting: The art and science of prediction.* New York: Crown.

Tetlock, P. E., Mellers, B. A., & Scoblic, J. P. 2017. Bringing probability judgments into policy debates via forecasting tournaments. *Science, 355,* 481–83.

Teulings, C., & Baldwin, R., eds. 2014. *Secular stagnation: Facts, causes and cures.* London: Centre for Economic Policy Research.

Thomas, C. D. 2017. *Inheritors of the Earth: How nature is thriving in an age of extinction.* New York: PublicAffairs.

Thomas, K. A., DeScioli, P., Haque, O. S., & Pinker, S. 2014. The psychology of coordination and common knowledge. *Journal of Personality and Social Psychology, 107,* 657–76.

Thomas, K. A., DeScioli, P., & Pinker, S. 2018. Common knowledge, coordination, and the logic of self-conscious emotions. Department of Psychology, Harvard University.

Thomas, K. H., & Gunnell, D. 2010. Suicide in England and Wales 1861–2007: A time trends analysis. *International Journal of Epidemiology, 39,* 1464–75.

Thompson, D. 2013. How airline ticket prices fell 50% in 30 years (and why nobody noticed). *The Atlantic,* Feb. 28.

Thyne, C. L. 2006. ABC's, 123's, and the Golden Rule: The pacifying effect of education on civil war, 1980–1999. *International Studies Quarterly, 50,* 733–54.

Toniolo, G., & Vecchi, G. 2007. Italian children at work, 1881–1961. *Giornale degli Economisti e Annali di Economia, 66,* 401–27.

Tooby, J. 2015. The iron law of intelligence. *Edge.* https://www.edge.org/response-detail/26197.

Tooby, J. 2017. Coalitional instincts. *Edge.* https://www.edge.org/response-detail/27168.

Tooby, J., Cosmides, L., & Barrett, H. C. 2003. The second law of thermodynamics is the first law of psychology: Evolutionary developmental psychology and the theory of tandem, coordinated inheritances. *Psychological Bulletin, 129,* 858–65.

Tooby, J., & DeVore, I. 1987. The reconstruction of hominid behavioral evolution through strategic modeling. In W. G. Kinzey, ed., *The evolution of human behavior: Primate models.* Albany, NY: SUNY Press.

Topol, E. 2012. *The creative destruction of medicine: How the digital revolution will create better health care.* New York: Basic Books.

Trivers, R. L. 2002. *Natural selection and social theory: Selected papers of Robert Trivers.* New York: Oxford University Press.

Trzesniewski, K. H., & Donnellan, M. B. 2010. Rethinking "generation me": A study of cohort effects from 1976–2006. *Perspectives on Psychological Science, 5,* 58–75.

Tupy, M. L. 2016. We work less, have more leisure time and earn more money. *HumanProgress.* http://humanprogress.org/blog/we-work-less-have-more-leisure-time-and-earn-more-money.

Tversky, A., & Kahneman, D. 1973. Availability: A heuristic for judging frequency and probability. *Cognitive Psychology, 4,* 207–32.

Twenge, J. M. 2000. The age of anxiety? The birth cohort change in anxiety and neuroticism, 1952–1993. *Journal of Personality and Social Psychology 79,* 1007–21.

Twenge, J. M. 2015. Time period and birth cohort differences in depressive symptoms in the U.S., 1982–2013. *Social Indicators Research, 121,* 437–54.

Twenge, J. M., Campbell, W. K., & Carter, N. T. 2015. Declines in trust in others and confidence in institutions among American adults and late adolescents, 1972–2012. *Psychological Science, 25,* 1914–23.

Twenge, J. M., Gentile, B., DeWall, C. N., Ma, D., Lacefield, K., et al. 2010. Birth cohort increases in psychopathology among young Americans, 1938–2007: A cross-temporal meta-analysis of the MMPI. *Clinical Psychology Review, 30,* 145–54.

Twenge, J. M., & Nolen-Hoeksema, S. 2002. Age, gender, race, socioeconomic status, and birth cohort differences on the children's depression inventory: A meta-analysis. *Journal of Abnormal Psychology, 111,* 578–88.

Twenge, J. M., Sherman, R. A., & Lyubomirsky, S. 2016. More happiness for young people and less for mature adults: Time period differences in subjective well-being in the United States, 1972–2014. *Social Psychological and Personality Science, 7,* 131–41.

ul Haq, M. 1996. *Reflections on human development.* New York: Oxford University Press.

UNAIDS: Joint United Nations Program on HIV/AIDS. 2016. *Fast-track: Ending the AIDS epidemic by 2030.* Geneva: UNAIDS.

Union of Concerned Scientists. 2015a. Close calls with nuclear weapons. http://www.ucsusa.org/sites/default/files/attach/2015/04/Close%20Calls%20with%20Nuclear%20Weapons.pdf.

Union of Concerned Scientists. 2015b. Leaders urge taking weapons off hair-trigger alert. http://www.ucsusa.org/nuclear-weapons/hair-trigger-alert/leaders#.WUXs6evysYN.

United Nations. 1948. Universal Declaration of Human Rights. http://www.un.org/en/universal-declaration-human-rights/index.html.

United Nations. 2015a. *The Millennium Development Goals Report 2015.* New York: United Nations.

United Nations. 2015b. Millennium Development Goals, goal 3: Promote gender equality and empower women. http://www.un.org/millenniumgoals/gender.shtml.

United Nations Children's Fund. 2014. *Female genital mutilation/cutting: What might the future hold?* New York: UNICEF.

United Nations Development Programme. 2003. *Arab Human Development Report 2002: Creating opportunities for future generations.* New York: Oxford University Press.

United Nations Development Programme. 2011. *Human Development Report 2011.* New York: United Nations.

United Nations Development Programme. 2016. Human Development Index (HDI). http://hdr.undp.org/en/content/human-development-index-hdi.

United Nations Economic and Social Council. 2014. World crime trends and emerging issues and responses in the field of crime prevention and criminal justice. https://www.unodc.org/documents/data-and-analysis/statistics/crime/ECN.1520145_EN.pdf.

United Nations Food and Agriculture Organization. 2012. *State of the world's forests 2012.* Rome: FAO.

United Nations Food and Agriculture Organization. 2014. *The state of food insecurity in the world.* Rome: FAO.

United Nations Office for Disarmament Affairs. (Undated.) Treaty on the non-proliferation of nuclear weapons (NPT). https://www.un.org/disarmament/wmd/nuclear/npt/text.

United Nations Office of the High Commissioner for Human Rights. 1966. International covenant on economic, social and cultural rights. http://www.ohchr.org/EN/ProfessionalInterest/Pages/CESCR.aspx.

United Nations Office on Drugs and Crime. 2013. Global study on homicide. https://www.unodc.org/gsh/en/data.html.

United Nations Office on Drugs and Crime. 2014. *Global study on homicide 2013.* Vienna: United Nations.

United States Census Bureau. 2016. Educational attainment in the United States, 2015. https://www.census.gov/content/dam/Census/library/publications/2016/demo/p20-578.pdf.

United States Census Bureau. 2017. Population and housing unit estimates. https://www.census.gov /programs-surveys/popest/data.html.

United States Department of Defense. 2016. Stockpile numbers, end of fiscal years 1962–2015. http:// open.defense.gov/Portals/23/Documents/frddwg/2015_Tables_UNCLASS.pdf.

United States Department of Labor. 2016. Women in the labor force. https://www.dol.gov/wb/stats /NEWSTATS/facts.htm.

United States Environmental Protection Agency. 2016. Air quality—national summary. https://www .epa.gov/air-trends/air-quality-national-summary.

Unz, D., Schwab, F., & Winterhoff-Spurk, P. 2008. TV news—the daily horror? Emotional effects of violent television news. *Journal of Media Psychology, 20,* 141–55.

Uppsala Conflict Data Program. 2017. UCDP datasets. http://ucdp.uu.se/downloads/.

van Bavel, B., & Rijpma, A. 2016. How important were formalized charity and social spending before the rise of the welfare state? A long-run analysis of selected Western European cases, 1400–1850. *Economic History Review, 69,* 159–87.

van Leeuwen, B., & van Leeuwen-Li, J. 2014. Education since 1820. In J. van Zanden, J. Baten, M. M. d'Ercole, A. Rijpma, C. Smith, & M. Timmer, eds., *How was life? Global well-being since 1820.* Paris: OECD Publishing.

van Zanden, J., Baten, J., d'Ercole, M. M., Rijpma, A., Smith, C., & Timmer, M., eds. 2014. *How was life? Global well-being since 1820.* Paris: OECD Publishing.

Värnik, P. 2012. Suicide in the world. *International Journal of Environmental Research and Public Health, 9,* 760–71.

Veenhoven, R. 2010. Life is getting better: Societal evolution and fit with human nature. *Social Indicators Research 97,* 105–22.

Veenhoven, R. (Undated.) World Database of Happiness. http://worlddatabaseofhappiness.eur.nl/.

Verhulst, B., Eaves, L., & Hatemi, P. K. 2016. Erratum to "Correlation not causation: The relationship between personality traits and political ideologies." *American Journal of Political Science, 60,* E3–E4.

Voas, D., & Chaves, M. 2016. Is the United States a counterexample to the secularization thesis? *American Journal of Sociology, 121,* 1517–56.

Walther, B. A., & Ewald, P. W. 2004. Pathogen survival in the external environment and the evolution of virulence. *Biological Review, 79,* 849–69.

Watson, W. 2015. *The inequality trap: Fighting capitalism instead of poverty.* Toronto: University of Toronto Press.

Weaver, C. L. 1987. Support of the elderly before the Depression: Individual and collective arrangements. *Cato Journal, 7,* 503–25.

Welch, D. A., & Blight, J. G. 1987–88. The eleventh hour of the Cuban Missile Crisis: An introduction to the ExComm transcripts. *International Security, 12,* 5–29.

Welzel, C. 2013. *Freedom rising: Human empowerment and the quest for emancipation.* New York: Cambridge University Press.

Whaples, R. 2005. Child labor in the United States. In R. Whaples, ed., *EH.net Encyclopedia.* http://eh.net /encyclopedia/child-labor-in-the-united-states/.

White, M. 2011. *Atrocities: The 100 deadliest episodes in human history.* New York: Norton.

Whitman, D. 1998. *The optimism gap: The I'm OK—They're Not syndrome and the myth of American decline.* New York: Bloomsbury USA.

Wieseltier, L. 2013. Crimes against humanities. *New Republic,* Sept. 3.

Wilkinson, R., & Pickett, K. 2009. *The spirit level: Why more equal societies almost always do better.* London: Allen Lane.

Wilkinson, W. 2016a. Revitalizing liberalism in the age of Brexit and Trump. *Niskanen Center Blog.* https://niskanencenter.org/blog/revitalizing-liberalism-age-brexit-trump/.

Wilkinson, W. 2016b. What if we can't make government smaller? *Niskanen Center Blog.* https://niskanen center.org/blog/cant-make-government-smaller/.

Williams, J. H., Haley, B., Kahrl, F., Moore, J., Jones, A. D., et al. 2014. *Pathways to deep decarbonization in the United States* (rev. ed.). San Francisco: Institute for Sustainable Development and International Relations.

Willingham, D. T. 2007. Critical thinking: Why is it so hard to teach? *American Educator,* Summer, 8–19.

Willnat, L., & Weaver, D. H. 2014. *The American journalist in the digital age.* Bloomington: Indiana University School of Journalism.

Wilson, E. O. 1998. *Consilience: The unity of knowledge.* New York: Knopf.

Wilson, M., & Daly, M. 1992. The man who mistook his wife for a chattel. In J. H. Barkow, L. Cosmides, & J. Tooby, eds., *The adapted mind: Evolutionary psychology and the generation of culture.* New York: Oxford University Press.

Wilson, W. 2007. The winning weapon? Rethinking nuclear weapons in light of Hiroshima. *International Security, 31,* 162–79.

WIN-Gallup International. 2012. Global Index of Religiosity and Atheism. https://sidmennt.is /wp-content/uploads/Gallup-International-um-tr%C3%BA-og-tr%C3%BAaleysi-2012.pdf

Winship, S. 2013. Overstating the costs of inequality. *National Affairs*, Spring.

Wolf, M. 2007. *Proust and the squid: The story and science of the reading brain*. New York: HarperCollins.

Wolin, R. 2004. *The seduction of unreason: The intellectual romance with fascism from Nietzsche to post-modernism*. Princeton, NJ: Princeton University Press.

Wood, G. 2017. *The way of the strangers: Encounters with the Islamic State*. New York: Random House.

Woodley, M. A., te Nijenhuis, J., & Murphy, R. 2013. Were the Victorians cleverer than us? The decline in general intelligence estimated from a meta-analysis of the slowing of simple reaction time. *Intelligence, 41*, 843–50.

Woodward, B., Shurkin, J., & Gordon, D. 2009. *Scientists greater than Einstein: The biggest lifesavers of the twentieth century*. Fresno, CA: Quill Driver.

Woolf, A. F. 2017. *The New START treaty: Central limits and key provisions*. Washington: Congressional Research Service. https://fas.org/sgp/crs/nuke/R41219.pdf.

Wootton, D. 2015. *The invention of science: A new history of the Scientific Revolution*. New York: Harper-Collins.

World Bank. 2012a. *Turn down the heat: Why a 4°C warmer world must be avoided*. Washington: World Bank.

World Bank. 2012b. *World Development Report 2013: Jobs*. Washington: World Bank.

World Bank. 2016a. Adult literacy rate, population 15+ years, both sexes (%). http://data.worldbank.org/indicator/SE.ADT.LITR.ZS.

World Bank. 2016b. Air transport, passengers carried. http://data.worldbank.org/indicator/IS.AIR.PSGR.

World Bank. 2016c. GDP per capita growth (annual %). http://data.worldbank.org/indicator/NY.GDP.PCAP.KD.ZG.

World Bank. 2016d. Gini index (World Bank estimate). http://data.worldbank.org/indicator/SI.POV.GINI?locations=US.

World Bank. 2016e. International tourism, number of arrivals. http://data.worldbank.org/indicator/ST.INT.ARVL.

World Bank. 2016f. Literacy rate, youth (ages 15–24), gender parity index (GPI). http://data.worldbank.org/indicator/SE.ADT.1524.LT.FM.ZS.

World Bank. 2016g. PovcalNet: An online analysis tool for global poverty monitoring. http://iresearch.worldbank.org/PovcalNet/home.aspx.

World Bank. 2016h. Terrestrial protected areas (% of total land area). http://data.worldbank.org/indicator/ER.LND.PTLD.ZS.

World Bank. 2016i. Youth literacy rate, population 15–24 years, both sexes (%). http://data.worldbank.org/indicator/SE.ADT.1524.LT.ZS.

World Bank. 2017. World development indicators: Deforestation and biodiversity. http://wdi.worldbank.org/table/3.4.

World Health Organization. 2014. *Injuries and violence: The facts 2014*. Geneva: World Health Organization. http://www.who.int/violence_injury_prevention/media/news/2015/Injury_violence_facts_2014/en/.

World Health Organization. 2015a. European Health for All database (HFA-DB). https://gateway.euro.who.int/en/datasets/european-health-for-all-database/.

World Health Organization. 2015b. *Global technical strategy for malaria, 2016–2030*. Geneva: World Health Organization. http://apps.who.int/iris/bitstream/10665/176712/1/9789241564991_eng.pdf?ua=1&ua=1.

World Health Organization. 2015c. *Trends in maternal mortality, 1990 to 2015*. Geneva: World Health Organization. http://apps.who.int/iris/bitstream/10665/194254/1/9789241565141_eng.pdf?ua=1.

World Health Organization. 2016a. Global Health Observatory (GHO) data. http://www.who.int/gho/mortality_burden_disease/life_tables/situation_trends/en/.

World Health Organization. 2016b. A research and development blueprint for action to prevent epidemics. http://www.who.int/blueprint/en/.

World Health Organization. 2016c. Road safety: Estimated number of road traffic deaths, 2013. http://gamapserver.who.int/gho/interactive_charts/road_safety/road_traffic_deaths/atlas.html.

World Health Organization. 2016d. Suicide. http://www.who.int/mediacentre/factsheets/fs398/en/.

World Health Organization. 2017a. European health information gateway: Deaths (#), all causes. https://gateway.euro.who.int/en/indicators/hfamdb-indicators/hfamdb_98-deaths-all-causes/.

World Health Organization. 2017b. Suicide rates, crude: Data by country. http://apps.who.int/gho/data/node.main.MHSUICIDE?lang=en.

World Health Organization. 2017c. The top 10 causes of death. http://www.who.int/mediacentre/factsheets/fs310/en/.

Wrangham, R. W. 2009. *Catching fire: How cooking made us human*. New York: Basic Books.

Wrangham, R. W., & Glowacki, L. 2012. Intergroup aggression in chimpanzees and war in nomadic hunter-gatherers. *Human Nature, 23*, 5–29.

Young, O. R. 2011. Effectiveness of international environmental regimes: Existing knowledge, cutting-edge themes, and research strategies. *Proceedings of the National Academy of Sciences, 108*, 19853–60.

Yudkowsky, E. 2008. Artificial intelligence as a positive and negative factor in global risk. In N. Bos-
    trom & M. Ćirković, eds., *Global catastrophic risks*. New York: Oxford University Press.
Zelizer, V. A. 1985. *Pricing the priceless child: The changing social value of children*. New York: Basic Books.
Zimring, F. E. 2007. *The Great American Crime Decline*. New York: Oxford University Press.
Zuckerman, P. 2007. Atheism: Contemporary numbers and patterns. In M. Martin, ed., *The Cambridge
    Companion to Atheism*. New York: Cambridge University Press.

INDEX

NOTE: Page numbers in *italics* indicate a graph or table.

fascist movements inspired by, 445, 448
intellectuals and artists as fans, 445, 446–7, 452
and nationalism, 165–6, 445, 447, 448, 449–51
rejection of Enlightenment, 33, 444–5
and relativism, 445, 446
sexism of, 444
and totalitarian dictators, 445, 446–7, 491n118
Trumpism as influenced by, 448–50, 491n118
and will to power, 33, 296, 444, 445
See also Nietzsche, Friedrich
Romanticism, 11, 30, 33
counter-Enlightenment, 30, 351
environmental movement and, 32, 121, 122
and factory work, 92
heroic struggle as the greatest good, 30, 33, 448
progress, version of, 11. *See also* progress: vs.
dialectics and other mystical forces, arcs,
and struggles
race theory of, 398, 400
violence as glorified in, 30, 33
See also environmental movement (traditional);
romantic heroism
romantic militarism, 165–6, 445
romantic nationalism, 165, 447, 448, 449–51
Romantic poetry, 433
Roma people, 399
Rome, ancient. *See* classical Greece and Rome
Romer, Paul, 154–5
Roosevelt, Eleanor, 419
Roosevelt, Franklin D., 63
Roosevelt, Theodore, 400
Rose, Stephen (economist), 114
Rose, Steven (neuroscientist), 447
Rosenberg, Nathan, 79
Rosenberg, Robin, 282
Rosenberg, Tina, 50
Rosenthal, Bernice Glatzer, 445
Roser, Max, 52, 53, 88–9
Rosling, Hans, 52, 53, 74, 251–2, 345
Rosling, Ola, 86
Ross, Lee, 359–60
Rotblat, Joseph, 308
Roth, Philip, 284
Roth, Randolph, 174
Rothschild, Nathan Mayer, 62
Rousseau, Jean-Jacques, 10, 30, 230
Rowling, J. K., 99, 100, 118
Ruddiman, William, 123
rule of law
democracy as dependent on, 335–6
establishment of, in early modern Europe, 43
integrity of, and emancipative values, 228
violent crime reductions and, 43, 168–70, 174
Rushdie, Salman, 443
Ruskin, John, 165
Russell, Bertrand, 421, 445
Russell, Stuart, 300, 477n20
Russia
as autocracy, 201, 203, 205, 335
civil war, 78
conflict with Georgia, 335
conflict with Ukraine, 158, 159, 335
Crimea annexation (2014), 164, 335
cyberattacks by, 335
democracy, undermining of, 335
famine in, 72
homicide rates in, 172, 174

homophobia in, 223
legitimacy of government, and crime wave, 174
nationalism of, 159
nuclear power and, 147, 150
nuclear weapons, 308, 315, 316–17, *318*, 320–21
revolution, 78
secularization and, 436
Time of Troubles, 199, 484n77
Trump administration's collusion with, 335
See also Cold War; nuclear war; Soviet Union
Rwanda, 69, *85*, 86, 161–2

safety, 167–90, 323, 480n2
auto safety, 177–8, 190
fall prevention, 181–2
fire safety, 183
flood control, 188
gas and vapor, 183
government regulations, 177–8, 186, *187*
natural disasters and, 187–9
opioid addiction, 184
Trump and, 335
in the workplace, 185–7, *187*
See also accidental deaths; motor vehicles
Sagan, Carl, 308, 310
Sahel, 73
Said, Edward, 39–40
Saint-Pierre, Abbé de, 13
Sale, Kirkpatrick, 456n1
Salk, Jonas, 63–4, 65
Sanders, Bernie, 97
Sanger, Margaret, 400
sanitation, 63, 67, 331
San Pedro Sula, Honduras, 172
San people, 249, 353–4
São Paulo, Brazil, homicide rate in, 172
Sartre, Jean-Paul, 39–40, 446, 447
Saturday Night Live, 266
Satyarthi, Kailash, 232
Saudi Arabia, 209–210, 336, 419
Savulescu, Julian, 402
Scalia, Antonin, 336
Scandinavia. *See* Nordic countries
Schank, Roger, 477n20
Scheidel, Walter, 106–7
Schell, Jonathan, 309–310, 456n1
Schelling, Friedrich, 30
Schelling, Thomas, 480nn105,112
Schmitt, Carl, 447
Schneier, Bruce, 303, 304
Schopenhauer, Arthur, 39–40, 165
Schrag, Daniel, 151
Schumer, Amy, 434
Schwartz, Richard, 274
science
application to wealth creation, 82–3, 94–5
beauty and, 34, 260, 386, 407–8, 433–4
climate change, consensus on, 137–8, 464n45
collaboration in, 64, 409
cosmopolitan virtues of, 409
definition of, 9, 391–3
depth of achievements of, 385–7
doubt as first principle of, 390
and errors and prejudices, discrediting own,
391
heroes of, 63–4
ideals of, 27, 387–8, 390, 392–3, 409

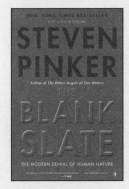